Studies in Systems, Decision and Control

Volume 573

The series "Studies in Systems, Decision and Control" (SSDC) covers both new developments and advances, as well as the state of the art, in the various areas of broadly perceived systems, decision making and control–quickly, up to date and with a high quality. The intent is to cover the theory, applications, and perspectives on the state of the art and future developments relevant to systems, decision making, control, complex processes and related areas, as embedded in the fields of engineering, computer science, physics, economics, social and life sciences, as well as the paradigms and methodologies behind them. The series contains monographs, textbooks, lecture notes and edited volumes in systems, decision making and control spanning the areas of Cyber-Physical Systems, Autonomous Systems, Sensor Networks, Control Systems, Energy Systems, Automotive Systems, Biological Systems, Vehicular Networking and Connected Vehicles, Aerospace Systems, Automation, Manufacturing, Smart Grids, Nonlinear Systems, Power Systems, Robotics, Social Systems, Economic Systems and other. Of particular value to both the contributors and the readership are the short publication timeframe and the world-wide distribution and exposure which enable both a wide and rapid dissemination of research output.

Indexed by SCOPUS, DBLP, WTI Frankfurt eG, zbMATH, SCImago.

All books published in the series are submitted for consideration in Web of Science.

Vitalii Babak · Sergii Babak · Artur Zaporozhets

Statistical Diagnostics of Electric Power Equipment

Springer

Vitalii Babak
General Energy Institute
National Academy of Sciences of Ukraine
Kyiv, Ukraine

Sergii Babak
Verkhovna Rada of Ukraine
Kyiv, Ukraine

Artur Zaporozhets
General Energy Institute
National Academy of Sciences of Ukraine
Kyiv, Ukraine

Green Technology Research Center
Yuan Ze University
Taoyuan, Taiwan

Center for Information-analytical
and Technical Support of Nuclear Power
Facilities Monitoring of the National
Academy of Sciences of Ukraine
Kyiv, Ukraine

National Technical University of Ukraine
"Igor Sikorsky Kyiv Polytechnic Institute"
Kyiv, Ukraine

ISSN 2198-4182 ISSN 2198-4190 (electronic)
Studies in Systems, Decision and Control
ISBN 978-3-031-76252-9 ISBN 978-3-031-76253-6 (eBook)
https://doi.org/10.1007/978-3-031-76253-6

This Springer imprint is published by the registered company Springer Nature Switzerland AG
The registered company address is: Gewerbestrasse 11, 6330 Cham, Switzerland

Preface

Today, according to various estimates, from 70 to 90% of the main and auxiliary equipment in the electric power complex of Ukraine has reached the end of its service life. In addition, the domestic energy sector suffers from the Russian invasion and constant shelling of energy facilities. Ukraine's energy infrastructure suffered losses in the amount of $11 billion due to military actions. More than 50% of the energy infrastructure was damaged, causing blackouts across the country.

Under such conditions, further operation of energy-intensive and in some cases extremely dangerous (e.g., NPP) equipment requires the creation of special, scientifically based methods and tools that allow such operation, while ensuring the necessary level of reliability and safety. Recently, thanks to new information technologies and the Internet, a sufficient number of such methods and means have appeared. Among them one of the most effective is methods of non-destructive control, monitoring and diagnostics of electrical equipment (EE) units. In all these methods the carrier of information about the technical condition of the investigated object is a diagnostic signal. It is through a comprehensive study of the diagnostic signal (taking into account its measurement, transformation, processing and subsequent analysis) that the researcher obtains the necessary information about the object of interest.

To confidently solve the above tasks, the researcher, firstly, needs a mathematical model of the diagnostic signal, which, taking into account the physical features of its formation in the object being diagnosed, would make it possible to obtain objective diagnostic information about this object. Secondly, it should have methods and algorithms of diagnostic signal processing built on their basis, which can be realised on the basis of modern electronic devices and information technologies. In this case, in chronological terms, the model is primary, and the choice of methods and technical means of measurement and processing of diagnostic signals is secondary.

The construction of the mathematical model of the diagnostic signal is conditioned by the specific physical process selected by the researcher as a carrier of diagnostic information, as well as by the EE node which is the object of diagnostics. Besides, the character of the model of the diagnostic signal depends on the chosen type of diagnostics—test or functional.

It is known that each node of the diagnosed EE has its own specific features of diagnostic signals formation. In this connection, considerable attention in the proposed work is paid to the issues of development and analysis of mathematical models of these signals, which are based on the corresponding physical processes occurring in the nodes of EE. Two main approaches can be used in constructing these models—deterministic and statistical.

In the first approach, a deterministic function of time is chosen as the initial mathematical model of the signal coming from primary sensors. In this case, amplitude-frequency, phase-frequency parameters of the diagnostic signal characterising the operation of the studied EE unit are most often informative. Diagnostics by deterministic methods is mainly reduced to theoretical determination of possible diagnostic features and their comparison with the results of experimental data analysis. If the latter differ significantly from the results obtained theoretically, a conclusion is made about the presence of a defect in the investigated node. In essence, these methods are applicable in the case when the identical results of all observations under the realisation of the same initial conditions are identical, and also if the same parameter (attribute) is measured under the same conditions. Such a situation is either strongly idealised or is observed at very low accuracy of the measuring instruments used, when random influences on the measurements are not perceived by these instruments. Consequently, the deterministic approach eliminates the need for multiple measurements, as they are all the same and a conclusion about the technical condition of a node can be drawn from a single measurement. However, it is always possible, by improving the accuracy of each individual measurement, to detect non-repeatability of the measurement results in the situation described above. Then the question arises, which of the obtained series of numbers is to be considered as the result of the measurement. Therefore the application of deterministic methods cannot be considered satisfactory and argumented, as many physical processes (diagnostic signals) occurring in different units of EE are random in nature, i.e., their realisations change from observation to observation. Thus, it is impossible to obtain a practically reliable answer about the technical condition of the diagnosed units based on a single observation.

The statistical approach makes it possible to recommend an algorithm based on a series of results of a certain measurement experiment, which calculates the best approximation of the measured parameter to its true value in the probabilistic sense. In addition, when using statistical methods of diagnostics, the measure of possible wrong decisions about the technical condition of the EE diagnosed units is taken into account, and it is also possible to estimate the average number of possible wrong conclusions, their scatter, etc.

In the majority of works, no proper attention is paid to the construction of mathematical and physical models of reference diagnostic signals, and without this, in our opinion, a comparative analysis of various statistical methods of control and diagnostics is impossible. In reasoning about possibilities of statistical approach for control and diagnostics, the most initial moment of diagnostics—carrying out of measurements on objects of diagnostics—is often absent. These questions also found their reflection in the present work.

Summarising the results of a brief comparison of deterministic and statistical approaches in the construction of models, methods and systems of diagnostics, the authors in this paper use the statistical approach, since the vast majority of physical processes occurring in the studied units of EE are random in nature. A number of reasons related to the presence of strong electromagnetic, thermal, acoustic, etc., fields in the operating EE, which act as interference in the measurement, transformation, processing and analysis of the information diagnostic signal, require the urgent application of statistical methods in diagnostics of EE units.

An essential point in building a mathematical model of a diagnostic signal and then a diagnostic information-measuring system (IMS), which is based on this model, is the type of diagnostic signal used in the study of various objects. In this work, various types of diagnostic signals—vibration (vibration acceleration), acoustic emission, as well as such a specific diagnostic signal as deviation of the angular velocity of the crankshaft of the diesel-electric generator—were used in the diagnosis of EE units. The latter signal was used in the diagnostics of the diesel cylinder-piston group. In addition, the results of thermal and radiation fields measured in the areas where the electric power facilities are located were used as information in the application of the unmanned aerial complex.

Working on the development of IMS of diagnostics of EE units, the authors of this monograph considered in complex the issues of constructing mathematical probabilistic models of diagnostic signals, the development of statistical methods of their analysis in order to make diagnostic decisions and, finally, the technical implementation of the proposed diagnostic methods. Following the concept of primacy of the mathematical model of the diagnostic signal, the authors considered it expedient to consider first of all the questions connected with the theory of random processes possessing infinitely divisible distribution laws, linear and linear periodic random processes. Considerable attention is paid to the issues of simulation modelling of diagnostic signals and their statistical evaluation. Modern element base and new information technologies allowed to develop, build and practically test a number of experimental samples of information-measuring systems of statistical diagnostics of electric power engineering objects. Among these IMS the systems realised by means of unmanned diagnostic complexes and also IMS of vibrodiagnostics of moving units of electric machines represent an important role. A large amount of experimental research has shown the operability and efficiency of the built IMS samples.

Particular attention is paid to the selection of diagnostic spaces, formation of training sets, construction of solving rules for diagnosis and classification of EE defects.

The authors do not pretend to a comprehensive consideration of the issues of EE diagnostics using statistical methods and IMS realised on their basis. At the same time, the results of researches, stated in this monograph, were a natural continuation of the subject of application of statistical methods in the field of control, monitoring and diagnostics for objects of electric power industry.

Kyiv, Ukraine
June 2024

Vitalii Babak
Sergii Babak
Artur Zaporozhets

Contents

1 Tasks and Main Methods of Statistical Diagnostics of Electric Power Equipment 1
1.1 Main Types of Electrical Equipment 2
1.2 Physical Processes Characterizing the Technical Condition of Electrical Equipment 13
1.2.1 Vibrations of Electrical Machines 13
1.2.2 Acoustic Emission 16
1.2.3 Electrical Insulation Breakdown 17
1.2.4 Non-uniformity of Angular Rotation Speed of the Shaft 18
1.3 From the History of Development of Means of Measuring Characteristics of Physical Processes 20
1.4 Key Concepts and Definitions in Measurement Technology and Diagnostics 22
1.5 Regulation of EE Diagnostics 24
1.5.1 EE's Technical Maintenance 25
1.5.2 General Requirements for Technical Diagnostics Systems 28
1.5.3 Structure and Features of Modern IMSs for EE Diagnostics 31
1.6 Features of Applying Deterministic and Statistical Methods in Diagnostics 32
1.7 Methods of Forecasting Failures of EE 38
1.7.1 Relationship Between Diagnostic and Forecasting Models in Reliability Assurance Tasks 38
1.7.2 Using Statistical Spline Function Methods for Failure Forecasting of EE 41
References 43

2 Linear Random Functions as Models of Diagnostic Signals 51
2.1 Infinitely Divisible Distribution Laws and Random Processes 52
2.2 White Noise, Random Processes and Fields 55
2.3 Linear Random Processes and Fields 60
References 68

3 Stochastic Models of Diagnostic Signals Arising During the Operation of Electrical Equipment 75
3.1 Representation of Vibrations of EM Components Using LRP Model 77
3.2 Mechanical Model of a Rolling Bearing Operating on the Shaft of an EM, and Its Electrical Analogy 81
3.3 Vibration Models of Laminated Magnetic Circuits of EM 83
3.4 Mathematical Model of Linear RLC-Noise in Describing Vibrations of Rolling Bearings and Laminated Magnetic Circuits of Electrical Machines 89
3.5 Linear Model of AE Signals Accompanying the Operation of Electrical Equipment Units 94
3.6 Diagnostic Signs Based on a Statistical Approach to the Diagnosis of Electrical Equipment 103
3.6.1 Rationale of Diagnostic Signs Based on the Probabilistic Characteristics of Vibrations of EM Rolling Bearings 103
3.6.2 Selection of Diagnostic Features Based on the Probabilistic Characteristics of AE Signals 108
References 115

4 Linear Periodic Random Processes in Constructing Models Characterizing the Operation of Electrical Equipment 123
4.1 Models of Cyclic Processes in the Description of Diagnostic Signals 124
4.2 Vibrations of Rotating Components of Electrical Machines 127
4.3 Dynamics of Changes in Cylinder Power on the Shaft of a Diesel-Electric Generator 131
4.4 Construction of Estimates of LPRP Describing the Dynamics of Cylinder Power on the Shaft of DEG 135
References 140

5 Statistical Assessment of Diagnostic Parameters 145
5.1 Numerical Parameters of Diagnostic Signals 146
5.2 Spectral Correlation Analysis 149
5.3 Empirical Densities and Probability Distribution Functions 154
5.4 Statistical Criteria for Assessing the Stationarity of Diagnostic Signals 162

5.5 Identification and Classification of Defects in Electrical Equipment 167
5.5.1 Testing of Statistical Hypotheses 167
5.5.2 Diagnostic Decisions Based on a Set of Diagnostic Parameters 170
5.5.3 Diagnostic Decisions in the One-Dimensional Gaussian Case 172
References 174

6 Simulation of Diagnostic Signals of Electric Equipment 181
6.1 Formation of Random Sequences with Specified Parameters 182
6.2 Modeling Realizations of Linear Random Fields and Processes 187
6.3 Simulation of Diagnostic Signals in Electrical Equipment 189
6.3.1 Acoustic Emission Signals 189
6.3.2 Vibrations in Massive Assemblies of EE During Shock Diagnostics 193
6.3.3 Dynamics of Cylinder Powers on the Shaft of the Diesel-Electric Generator 202
References 209

7 Information and Measuring Systems for Statistical Diagnostics of Electrical Equipment 215
7.1 Peculiarities of Construction of Statistical Diagnostic Systems 215
7.2 IMS of Vibration Diagnostics Rolling Bearings of Electric Machines 218
7.2.1 IMS of Shock Diagnostics of Charge Cores 222
7.2.2 IMS of Vibration Diagnostics of Moving Parts of Electric Machines 225
7.2.3 IMS of Acoustic Emission Diagnostics 230
7.2.4 IMS of Diagnostics of Cylinder-Piston Group DEGs 235
7.3 IMS Software of Statistical Diagnostics of EE 240
7.3.1 Software for Controlling the Operation of the IMS 241
7.3.2 Software Statistical Processing of Diagnostic Signals 244
7.3.3 Software for Making Diagnostic Decisions 250
7.4 Statistical Diagnostics IMS Errors 251
7.4.1 Instrumental Errors of the IMS, Due to the Presence of Analog and Digital Components 253
7.4.2 Features of Errors of Analog-to-Digital Conversion of Realizations of Random Processes 259
7.4.3 Estimation of Errors of IMS of Statistical Diagnostics 264
7.5 Training of OMS of Statistical Diagnostic 268
References 270

8 Experimental Studies of Statistical Diagnostics Information and Measuring Systems ... 279
8.1 Installation for Studying Vibrations of Rolling Bearings ... 280
8.2 Research of Diagnostic Signs Condition of EM Bearing Units ... 291
8.2.1 Peculiarities of EM Bearing Units Vibrations ... 292
8.2.2 Correlation and Spectral Diagnostic Features ... 292
8.2.3 Diagnostic Features Based on Estimates of Vibration Moments ... 299
8.3 Analysis of Shock Vibration Signals, Excited in the Charging Coil ... 301
8.4 Analysis of AE Diagnostic Signals in EE's Assemblies ... 304
8.5 Analysis of Diagnostic Signs Condition of the DEG's CPG ... 308
8.6 Construction of Solving Rules for EE Diagnostics ... 313
8.6.1 Decision Rules for Vibration Diagnostics and Classification of Defects in EM Rolling Bearings ... 316
8.6.2 Diagnostics of the Press Condition Packed Magnetic Core with the Help of IMS of Shock Diagnostics ... 323
8.6.3 Determination of the Degree of Mechanical Loading by Analysis of AE Signal ... 327
8.6.4 Diagnostics of DEG's CPG ... 332
8.7 Prediction of EE Failures ... 335
References ... 339

9 Tasks for Creating the Environmental Monitoring Systems for Energy Objects ... 345
9.1 Scientific and Technical Problems and Tasks of Environmental Monitoring of Power Engineering Facilities ... 346
9.2 Information Support of Environmental Monitoring System ... 351
9.3 Realization of Information Support at the Life Cycle Stages of the Monitoring System ... 354
9.4 Generalized Models of Information Fields of Environmental Air ... 357
9.5 Stochastic Model of Formation of Meteorological Fields ... 359
9.6 Models for Assessing the Consequences of Atmospheric Releases NPPs Under Conditions of Different Types of Radionuclide Fields in the Air ... 365
9.7 Mathematical Model of the Atmospheric Transport of Radionuclides for the Measuring Their Volumetric Activity by a Monitoring System on the Basis of an Unmanned Aerial System ... 372
References ... 381

10 Unmanned Aerial Measurement Systems for Monitoring the Environmental Area of NPP and TPP 387
10.1 Development of Information and Measurement Systems Based on UAV Complexes 387
10.2 History and Current State of Unmanned Aviation 392
10.3 Experience in the Use of UAVs at the Fukushima-1 NPP 400
10.4 UASs in the Tasks of Control of Power Engineering Facilities 403
10.5 Examples of Construction of Measuring Channels Systems of UAV-Based NPP and TPP Environmental Characteristics Monitoring 415
10.6 Experimental Results of Radiation Monitoring at the Border of the Chornobyl Exclusion Zone 421
References 423

Abbreviations

AC	Alternating Current
ADC	Analog-to-Digital Converter
ADS	Automatic Dependent Surveillance
AE	Acoustic Emission
AFB	Analog Filtering Block
AFC	Amplitude-Frequency Characteristic
AFCS	Automatic Flight Control System
AT	Autotransformer
CADS	Computer-Aided Design System
CF	Characteristic Function
CHPP	Combined Heat and Power Plant
CISU	Control and Information Support Unit
CPG	Cylinder-Piston Group
CS	Channel Selector
DC	Direct Current
DCID	Data Conversion and Input Device
DCM	Duty Cycle Modulation
DDR	Diagnostic Decision Rules
DEG	Diesel-Electric Generator
DIS	Diagnostic Information Source
DO	Diagnostic Object
DR	Diagnostic Result
EDR	Exposure Dose Rate
EDS	Evaluation of Diagnostic Signs
EE	Electrical Equipment
EM	Electric Machine
FA	Fuel Assembly
FB	Filter Block
FED	Filter Ejection Device
FFT	Fast Fourier Transform
GCE	Ground Control Equipment

GIS	Geographic Information System
GPS	Global Positioning System
HAP	Histogram Analysis Program
HPCTR	High-Power Channel-Type Reactor
HPF	High-Pass Filter
HPP	Hydroelectric Power Plant
HV	High Voltage
ICE	Internal Combustion Engine
IMS	Information-Measuring System
INS	Inertial Navigation System
IRS	Information and Reference Support
IH	Impact Hammer
LPF	Low-Pass Filter
LPRP	Linear Periodic Random Process
LRP	Linear Random Process
LV	Low Voltage
MP	Microprocessor
MS	Mode Switch
NPP	Nuclear Power Plant
OA	Output Amplifier
OCE	Onboard Control Equipment
OS	Observation Space
OCC	Onboard Computing Complex
PA	Preliminary Amplifier
PAP	Package of Application Programs
PC	Personal Computer
PET	Piezoelectric Transducer
PFC	Phase-Frequency Characteristic
PU	Preamplifier Unit
RC	Remote Control
RF	Refrigeration Filter
RM	Rudder Machine
RO	Research Object
SCAP	Spectral-Correlation Analysis Program
SCP	Stationarity Check Program
SCS	Safety-Critical System
SD	Starting Device
SINS	Strapdown Inertial Navigation System
SP	Synchronous Pulses
SPP	Statistical Preliminary Processing
SPU	Secondary Processing Unit
SRN	Satellite Radio Navigation
TDP	Top Dead Point
TPP	Thermal Power Plant
TS	Training Sets

UAMS	Unmanned Aviation Measurement System
UAS	Unmanned Aerial Complex
UAV	Unmanned Aerial Vehicle
UES	Unified Energy System
USB	Universal Serial Bus
VA	Voltage Amplifier
WWER	Water-Water Energetic Reactor
D	Dispersion operator
$F(x)$	Probability distribution function of a random variable
$F_{\xi}(x; t)$	One-dimensional probability distribution function of a random process $\xi(t)$
$f(u)$	Characteristic function of a random variable
$f_{\xi}(u, t)$	One-dimensional characteristic function of a random process $\xi(t)$
H_n	Hypothesis n
M	Expectation operator
P{**A**}	Probability of a random event **A**
$\mathrm{P}_{\Theta}\{H_0\}$	Probability of accepting the hypothesis H_0, provided that the true value of the distribution parameter is equal to Θ
$p(x)$	One-dimensional probability density function
$R(t_1, t_2)$	Correlation function of a random process
$R(s)$	Correlation function of a stationary random process
$\hat{R}(s)$	Normalized correlation function of a stationary random process
$S(\omega)$	Power spectral density of a stationary random process
$U(t)$	Heaviside unit function
u_{α}	Quantile of the standard normal distribution level α
α	Probability of type I error
β	Probability of type II error
$\delta(t)$	Dirac delta function
δ_{τ}	Discrete delta function
$\phi(\tau, t)$	Kernel of a linear random process with continuous time
$\eta(t)$	Random process with independent (uncorrelated) increments
κ_k	k-th order cumulant
μ_k	k-th order central moment
ν_k	k-th order initial moment
$\pi_1(t)$	Generalized Poisson process with independent increments
Θ	Distribution parameter regarding which the hypothesis is formulated
Ξ_m	Observation vector
$\xi(t)$	Random process with continuous time
ξ_t	Random process with discrete time
$\zeta(t)$	White noise (random process with independent or uncorrelated values)
ζ_j	Element of a sequence of independent random variables

Chapter 1
Tasks and Main Methods of Statistical Diagnostics of Electric Power Equipment

Abstract This chapter delves into the intricate field of statistical diagnostics for electric power equipment, emphasizing the integration of modern computational and statistical methods rooted in pattern recognition theory. It provides an extensive overview of the Ukrainian power industry, highlighting the distinct roles of thermal, hydro, and nuclear power plants and their associated reliability challenges. Key diagnostic methods—non-destructive testing, vibration, and acoustic emission signals—are discussed, focusing on their critical application in identifying faults in electric machines, particularly winding and bearing damages which contribute to the majority of failures. The chapter also examines the diagnostic processes, touching upon the historical evolution of measurement technologies, the development of diagnostic systems, and compliance with industry standards. Furthermore, it explores mathematical modeling and statistical spline functions for forecasting failures, demonstrating how these methods can enhance the prediction and reliability of industrial equipment by monitoring gradual changes in diagnostic parameters. The comprehensive analysis presented integrates theoretical foundations with practical applications, offering substantial insights into the methodologies for ensuring the optimal functioning and longevity of electrical machinery.

Keywords Statistical diagnostics · Electric power equipment · Pattern recognition theory · Non-destructive testing · Vibration analysis · Acoustic emission signals · Winding damage · Bearing damage · Reliability indicators · Technical diagnostics · Diagnostic systems

Technical diagnostics, as a separate branch of knowledge, began to form in the 1960s-1970s due to the increasing complexity of technical objects' design and the growing requirements for their reliability. Traditional methods of checking the technical condition required the object to be taken out of operation for inspection, leading to significant material losses. Disassembly and assembly operations accounted for the majority of the time spent on checking the object's condition. Therefore, research on creating methods and corresponding means of non-destructive

V. Babak et al., *Statistical Diagnostics of Electric Power Equipment*, Studies in Systems, Decision and Control 573, https://doi.org/10.1007/978-3-031-76253-6_1

diagnostics, which allow obtaining information about the actual technical condition of equipment without taking it out of operation, is of great practical importance.

At the current stage of technological development, ensuring the reliability and durability of complex equipment is impossible without constant or episodic assessment of its actual technical condition. This is the main task of technical diagnostics. Today, this scientific field covers a wide range of issues related to obtaining and processing diagnostic information. Electronic data processing machines (computers) and methods of information processing using them play a significant role in this, leading to the development of cutting-edge diagnostic directions. To define the type of technical diagnostics based on the application of modern information technologies for obtaining, storing, and processing diagnostic information using computers, V.A. Gulyaev introduced the term "computational diagnostics" in his monograph.

In recent decades, statistical diagnostics [1–5] has undergone significant development, based on methods of making diagnostic decisions and classifying defects using pattern recognition theory. The term "statistical diagnostics" was apparently first used in the work [6].

Let's consider the basic concepts used in technical diagnostics, its main tasks, methods, and means of their solution.

1.1 Main Types of Electrical Equipment

Main *electrical equipment* (EE) is defined as equipment intended for generation (electric power, steam, hot water), conversion (chemical energy of burned fuel into thermal energy of steam or hot water), transportation, or transmission of mechanical energy of the energy carrier (water, gas, steam, compressed air, oxygen, nitrogen, etc.).

Main *electrical equipment* is conventionally subdivided into:

Thermo-mechanical: steam and hot water boilers; waste heat boilers (cooling boilers); steam and gas turbines; auxiliary equipment for boiler installations; air separation units; refrigeration units; equipment for gas distribution stations; centrifugal and reciprocating compressors; blowers (air blowers, gas blowers, and exhausters), coke blowers; smoke exhausters; pumps; vessels operating under pressure (power); water pipelines (drinking, hot, technical, circulation, sludge, water lowering), gas (natural, blast furnace, coke oven, etc.), steam, district heating, air, oxygen, nitrogen, hydrogen, and other media; storm, technological, and wastewater channels; masts and supports, power transmission lines; fittings (shut-off, regulating), platforms for servicing pipeline fittings located at height [7–12]; *and energy*: generators; motors; transformers; synchronous compensators; switching equipment; power transmission lines and other network equipment; means of control, protection, and automation; computer equipment [13–19].

The technological process of production, distribution, and consumption of electrical and thermal energy in the form of a generalized scheme is shown in Fig. 1.1.

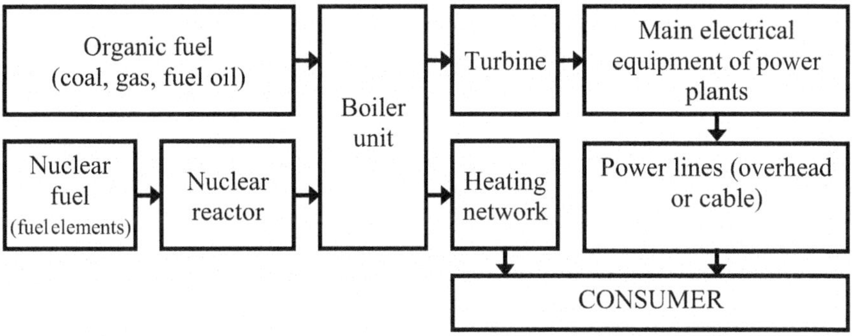

Fig. 1.1 Generalized scheme of the technological process of production, distribution, and consumption of electric and thermal energy

Following the presented scheme, this work will focus on the issues of diagnosing typical types of main EE. In addition, we will consider the issues of using special diagnostic tools installed on unmanned aerial vehicles (UAVs) and used to monitor the condition of environmental characteristics in the areas where energy facilities are located [20–26].

Below, we will briefly describe the current state of Ukraine's power industry.

The power industry is the most important sector of Ukraine's economy. Electricity production is based on the combustion of coal, oil, natural gas, the use of nuclear energy, water, wind, and solar energy [27–34].

The most powerful *thermal* power plants (TPP) are located in the Donbas region (Uglehirska, Starobeshivska, Myronivska, Kurakhivska), in the Dnipro region (Prydniprovska, Kryvyorizka), in the Kharkiv region (Zmiivska), Kyiv region (Trypilska), Ivano-Frankivsk region (Burshtynska), Lviv region (Dobrotvirska), in Zaporizhzhia, Odesa, and others.

In the development of *hydro energy*, a significant role in the 20th and early twenty-first centuries was played by the cascade of hydroelectric power plants (HPP) built on the Dnipro River: Dniproges, Kakhovska, Serednodniprovska, Kremenchutska, Kyivska, and Kanivska HPP. HPPs have also been built on the Dniester River (Dniester), on the Ros River, in the Zakarpattia region.

In recent decades, *nuclear* energy has been developing rapidly in terms of the number of energy reactors, which currently 15, Ukraine ranks tenth in the world and fifth in Europe. In total, Ukraine operates 4 nuclear power plants (NPPs): South Ukrainian, Rivne, Khmelnytskyi, and Zaporizhzhia. The latter, containing 6 power units with a total installed capacity of 6000 MW, is the largest in Europe. But in 2022 it was occupied by Russian troops [35–39].

Ukraine inherited twelve power units from the USSR, and three more were commissioned after its dissolution, in 1995 and 2004. All the power units with RBMK reactors operating in Ukraine were part of the Chornobyl NPP. As a result of the Chornobyl accident, power unit 4 was destroyed, and the rest were gradually shut down from 1991 to 2000. Thus, all remaining reactors in the country belong to the VVER type: 2—VVER 440 and 13—VVER 1000.

Until 2011, all nuclear fuel for Ukrainian NPPs was supplied from Russia by fuel assembly (FA) company. According to the contract concluded in 2008, the American company Westinghouse Electric Company was supposed to supply at least 630 fuel assemblies (FAs) during 2011–2015 for the phased replacement of Russian fuel in at least 3 power units with VVER-1000 reactors. In 2011, Westinghouse began supplying its FAs to Ukraine.

In April 2012, defects were found in American FAs on the 3rd power unit of the South Ukrainian NPP, which operated in an experimental-industrial mode using Westinghouse fuel, and their use was suspended [40–42]. According to the investigation by the Ukrainian commission, it was concluded that there were design flaws in the American FAs. The losses of "Energoatom" amounted to at least 175 million dollars. The problem was solved by replacing the defective FAs with Russian ones and continuing to operate undamaged FAs-W.

In 2012, the contribution of nuclear energy amounted to 46% of the total electricity production in the country, with the total capacity of nuclear power plants being 13,107 MW. Recently, due to a shortage of organic fuel sources (coal, gas, fuel oil), there has been a noticeable increase in energy production at nuclear power plants in Ukraine (during the summer-autumn period of 2014, output exceeded 50% of the total production). Apparently, this trend will continue into the near future, as in 2006 the Ukrainian government planned to construct 11 new energy units by 2030.

Increasing the share of electricity production at nuclear power plants requires the development of modern methods and means to ensure safe operation, as well as monitoring, control, and diagnostics of nuclear power plant equipment.

Decades of experience (about a thousand reactor-years) in operating nuclear power plants with pressurized water reactors (VVER) have demonstrated a high level of reliability and safety of VVER-type reactor installations. However, well-known disasters, as well as the continued wariness towards nuclear energy in the global community, underscore the importance of further enhancing NPP safety. The main avenues for improving the safety of VVER-type NPP include:

- enhancing the safety barriers of deep-tier protection;
- developing and improving the monitoring of equipment conditions and safety–critical system elements;
- modernizing and reconstructing safety–critical systems to enhance their reliability and expand capabilities for ensuring safe operation;
- qualifying nuclear power plant equipment under design operation regimes.

From the perspective of safety enhancement, the development of control systems at VVER-type NPPs should focus on expanding and improving operational and periodic diagnostics of the condition and reliability of safety–critical systems. This primarily concerns reactor plant control systems, including monitoring the condition of reactor vessel metal and structural characteristics, detecting leaks, monitoring vibration conditions, and detecting foreign objects in the reactor coolant system.

There's a need for in-depth scientific and technical substantiation in the development and implementation of promising reactor control systems such as acoustic and thermoacoustic diagnostic systems.

In addressing these issues, the task of developing modern information technologies to monitor the environment around nuclear power plants also arises. The ultimate goal of monitoring is to detect air pollution in the monitoring environment. One effective approach to solving this task is through the use of UAVs [20–26].

The backbone of Ukraine's electricity system is the Unified Power System (UPS), which centrally supplies electricity to internal consumers. Additionally, the UPS interacts with the energy systems of neighboring countries, facilitating the import and export of electricity [32, 43–48]. It consists of eight regional power systems interconnected by power transmission lines with voltages up to 750 kV.

As of 2021, the technical condition of the power system was unsatisfactory [49–53]. Modernization and the implementation of new energy-saving technologies are necessary, along with the development of alternative sources of electricity (such as solar, wind, biogas, and geothermal power stations). This fact necessitates work aimed at ensuring the reliability of the UPS. One solution to this challenge is the development of new methods and means for monitoring and diagnosing the specified equipment.

As noted above, the main elements of power objects include generators, various types of electrical machines (EM) used as motors for own needs, synchronous compensators, transformers (power, measuring, etc.), switching equipment (switches, disconnectors, etc.), and network equipment.

Average reliability indicators of the main power equipment [54–57], shown in Table 1.1, indicate that generators, electric motors, transformers, and their components can be considered the most unreliable power objects.

This is confirmed by the data provided in the table: synchronous generator (failure rate $\lambda = 1$, mean time to repair $T_R = 100$), asynchronous motor (failure rate $\lambda = 0.1$, mean time to repair for LV motors $T_R = 50$; and for HV motors $T_R = 160$). Therefore, this work will focus primarily on the diagnosis of these types of equipment. The methods and tools considered in this work are aimed at this equipment but can also be applied to other units and devices operating on similar principles of electrical energy conversion.

Various diagnostic methods for power objects are known and practically used [58–65]. These methods are determined by the measured physical process used as a diagnostic signal to obtain information about the technical condition of the unit under investigation.

Technical diagnostics of power system objects should be carried out using non-destructive testing methods, after which these objects can be used for their intended purpose. Non-destructive testing, depending on the physical phenomena underlying it, is classified into types (DSTU 2865-94): magnetic, electric, eddy current, radio wave, thermal, optical, radiographic, acoustic, and penetrant [66–71].

In this work, the main sources of diagnostic information considered are vibrational and acoustic emission diagnostic signals generated in the nodes of the diagnosed power objects. These signals are measured directly on operating equipment

Table 1.1 Average reliability indicators of some electrical equipment

The type of electrical equipment	λ, $1/Y$	T_R, h
Disconnector	0.01	2
Circuitbreake	0.02	10
Isolator	0.03	10
Automatic low-voltage (LV) circuit breaker	0.05	4
High-voltage fuse (HV)	0.1	2
Assembled busbars for voltages up to 10 kV	0.03	2
High-voltage cable line laid in: (HV)		
Trench	0.03	44
Ducts	0.005	18
LV cable line laid in a trench (per 1 km)	0.1	24
Overhead LV line (per 1 km)	0.02	5
Synchronous generator	1	100
Asynchronous electric motor:		
LV	0.1	50
HV	0.1	160

(functional diagnostics) as well as on equipment in non-operational condition. In the latter case, diagnostics are carried out using special impact actions (test diagnostics).

As operational experience shows, the main cause of failures and repairs of EM is *insulation breakdown*, both between turns and between the winding and the frame. The second most common cause of failures after winding is the bearing assembly. In the overwhelming majority of cases, generator and electric motor failures occur due to winding damage (85…95%). From 2 to 5% of EMs fail due to bearing damage. The remaining components (brush-collector assembly, ventilation system, etc.) account for 1…2% of failures. This work focuses on diagnosing rolling bearings and laminated magnetic cores of EMs. Issues related to the development of systems and methods for diagnosing rotating EM components will be discussed. It's worth noting the peculiarities of building diagnostic methods and systems for diesel-electric generators. In this case, the research object was chosen as the cylinder-piston group of the diesel engine, and the deviation of the crankshaft angular velocity was chosen as the diagnostic signal.

Let's briefly focus on the examination of EM as diagnostic objects. Firstly, it's necessary to identify specific components subject to diagnostics and determine the most informative physical processes that allow for reliable diagnosis of the EM's technical condition.

For EMs, especially low-power machines, there is a wide variety of types and designs. Each type has its weak points and requires a particular approach in developing diagnostic signal models and subsequently constructing diagnostic expert systems. In most cases, EM designs do not adhere to the principle of equal reliability of individual components. For example, according to [72–74], the primary failure source in asynchronous motors is the stator winding. In direct current machines, the weak spot is the brush-collector assembly. In low-power EMs, especially high-speed ones, a significant number of failures occur in bearings.

The operating conditions and modes of operation significantly influence the diagnostic results of EM components. For aviation machines, the most crucial requirement is faultlessness, i.e., a low probability of random failures; for power machines, it's substantial resource and durability.

Among the main components of a typical EM subject to diagnostics are [75–78]: rotor and stator windings; laminated magnetic cores of the rotor and stator; bearing assemblies with rolling bearings; frame and its attachment points to the foundation; brush-collector assembly; cooling system elements (fan impeller, air ducts, or channels for circulating cooling fluid, etc.).

To obtain information about the technical condition of electrical machine (EM) components, various processes occurring during their operation are utilized, specifically their numerical characteristics (parameters). For diagnosing specific EM components, the following parameters are typically used:

- *windings*: inter-turn insulation electrical resistance; surface temperature of windings; magnitude of magnetic induction;
- *laminated* magnetic cores: vibrations (displacement, velocity, acceleration) of core ends; magnitude of magnetic induction; surface temperature of the magnetic core;
- *bearing assembly*: vibrations (displacement, velocity, acceleration) of bearing shield; temperature of bearing shield;
- *frame* and its attachment points to the foundation: acoustic emission; frame vibration;
- *brush-collector assembly*: transitional resistance (conductivity) of sliding contact; vibrations (displacement, velocity, acceleration) of brush holder; temperature of brush holder;
- *cooling system elements of EM*: aerodynamic noise of fan impeller; vibration of surfaces of circulation channels for coolant movement of EM.

Let's briefly consider some issues related to the primary investigation (measurement) of the processes mentioned above, which serve as diagnostic signals in EM diagnostics.

EM windings: The reliability of EMs largely depends on the reliability of their windings, which, in turn, depends on the condition of their insulation. The insulation operates in conditions of strong electromagnetic, thermal, and vibrational fields. Additionally, external climatic conditions significantly influence the insulation condition during EM operation. Over time, these conditions collectively lead to progressive deterioration of insulation properties. It's worth noting that a considerable amount of research is dedicated to insulation control and diagnostics.

The main characteristic of insulation determining EM operation reliability is its electrical strength. However, this crucial insulation property can only be maintained during operation if many other qualities are present, a decrease in which leads to a reduction in electrical strength. For instance, insulation must maintain high thermal conductivity; otherwise, the inevitable occurrence of local overheating will lead to accelerated deterioration. Insulation must possess sufficient mechanical strength and elasticity, which would prevent the formation of residual deformation, cracks, or delamination under mechanical stresses. Insulation must maintain a stable chemical composition since changes lead to decreased electrical strength.

In high-voltage machines, the aging of insulation under the influence of the electric field is of significant importance. During operation, EM insulation remains under working voltage for prolonged periods and periodically experiences the effects of elevated voltages during preventive tests and various wave phenomena, the sources of which can be both external and internal to the machine.

Operating data and experimental research [75–78] show that noticeable effects of the electric field on insulation service life start to appear in machines with a nominal voltage of no less than 6 kV. In machines with lower voltages, electrical aging phenomena are not observed.

The slot insulation of windings may undergo compression due to electromagnetic forces, and if there are gaps in the slot, it is also subject to impacts and wear against the slot walls. If there is no freedom of movement in the slot, in addition to stretching and compression, insulation may also experience shear deformations. During the bending of winding end turns, the highest stresses occur at the points where the bars or coils exit the slots, subjecting insulation to compression and tension stresses. Furthermore, it is compressed against insulation barriers and at points of contact with retaining rings.

In most cases, the aforementioned stresses have a cyclic, alternating character, with vibration at 100 Hz being the most typical. During transient processes (starting and reversing of motors, short circuits, and asynchronous generator engagements), vibration amplitudes increase tens of times due to the increased currents in the windings and the quadratic dependence of electromagnetic forces on current. Particularly significant stresses can occur in windings of large machines, turbo- and hydro-generators.

The mechanical characteristics of insulation depend on its temperature. As the insulation heats up, its strength limit quickly decreases, while the insulation simultaneously becomes more elastic. This especially applies to insulation constructions based on thermoplastic compounds. For instance, the tensile strength limit of mica-compounded insulation is 3340 N/cm^2 (340 kgf/cm^2) at 20 °C and only 344 N/cm^2 (35 kgf/cm^2) at 100 °C. These values vary greatly depending on technological features, usage, and other factors; they also differ for different parts of the winding. As the temperature decreases, such insulation becomes brittle. Mechanical characteristics of insulation based on thermosetting binders are more stable.

According to [75–78], insulation breakdown occurs gradually, with the initiating role belonging to thermal aging processes. Even at relatively low temperatures, when thermo-oxidative degradation is insignificant, heat causes insulation to dry out, volatile components to evaporate from binders, elasticity of insulation to decrease, and its brittleness to increase. The latter contributes to the development of mechanical aging processes. Cracks and other defects appear in insulation, it delaminates and becomes loose, creating conditions for the onset of ionization phenomena. Insulation breakdown occurs non-uniformly and ends with breakdown at the weakest point. Moisture and aggressive environments accelerate and activate aging processes.

The main *diagnostic sign* for assessing the electrical strength of insulation in the windings of electric machines is usually the insulation resistance.

Currently, there are many instruments and systems on the market for determining this parameter.

Measurement of the insulation resistance between the windings of the rotor and stator windings of electric machines is carried out in accordance with GOST 17403-79. These measurements are performed on non-operating electric machines during routine or major repairs. For this purpose, megohmmeters of type M4100/4, F4102/2, as well as bridge R333, are used. Partial automation of this process is provided by the testing setup of type AID-70, where the measurement of the insulation resistance between the windings is carried out using special contact probes.

There are also methods and devices for monitoring and diagnosing inter-turn faults in the rotors of synchronous electric machines. They are based on the use of magnetic induction and are implemented using induction sensors located near the surface of the rotating rotor. When a direct current is applied to the rotor winding, the sensor generates a signal, which is then analyzed by a spectrum analyzer or computer.

Some modern methods, instruments and systems for diagnosing the insulation of asynchronous motors are discussed in [79–84].

Various types of thermocouples, such as type J (iron-constantan or iron-copper-nickel alloys), are used to estimate the temperature of the winding as an indirect indicator of the condition of the inter-turn insulation. These thermocouples are installed in the slots of laminated magnetic cores to locally monitor the temperature of the winding. An increase in temperature at these points may indicate a possible inter-turn short circuit in the windings. The use of thermocouples in electrical engineering is regulated by relevant standards such as GOST 6616-94, GOST R 8.585-2001, and IEC 60584.

Laminated magnetic core. The technical condition of the stator windings of EM is directly related to the condition of the laminated magnetic core. Currently, for laminated magnetic cores of powerful EMs, some of the most common faults are defects associated with the loosening of the press fit of the outer packets of the stator end parts (so-called "puffing"), as well as mechanical breakage of the tie rods or nuts securing the outer retaining flange. The results of recent studies [85–87] show that these defects are often associated with the manifestation of "abnormal vibromechanical phenomena" in the stator cores of powerful EMs. These phenomena become particularly noticeable when the diameter of the stator core approaches the frequency of the radial magnetic force in the air gap.

To effectively detect such defects, statistical spectral analysis is applied to process vibrational *diagnostic signals* (vibration displacement, vibration velocity, and vibration acceleration) measured using corresponding vibration sensors. Measurement, recording, and subsequent processing of these *vibrations* can be carried out using automated vibration diagnostics information and measuring systems (IMS). Descriptions of some of them are provided, for example, in [88]. In conducting these experiments, if the specified IMS for vibration diagnostics are used, primary sensors are located in the end parts, on the retaining flanges of the stator (or rotor barrel) of the EM. This approach also allows determining the degree of compression of the laminated magnetic core of the stator (and for some types of EM, the rotor).

Additional information about the technical condition of the EM core can be obtained from *temperature* data measured using various thermocouples located directly in the slots of the magnetic core of the EM.

Currently, there are enough methods that involve using various physical processes as diagnostic signals. For example, it is proposed to jointly use methods for controlling the inter-sheet insulation of individual sheets of the stator core and the temperature heating of the tooth zone for diagnosing the condition of the stator core. Some methods for controlling and diagnosing laminated magnetic cores of EMs are regulated by various instructions.

Indirectly, the technical condition of the magnetic core can be judged by measuring the magnitude of the electromagnetic field in the air gap. Its measurement is carried out using induction sensors, and further research is conducted using a special diagnostic IMS.

Bearing unit. The most informative physical process for diagnosing virtually all types of defects in rolling bearings is vibration (vibration displacement, vibration velocity, vibration acceleration) [89–93]. Following the requirements of GOST 12379-75, corresponding sensors are placed on the bearing units of EMs in three mutually orthogonal directions (X, Y, Z axes). Vibrations are measured radially (with respect to the rotating shaft of the EM) along the X and Y axes, and axially along the Z axis. Synchronous measurement of vibrations along these three directions allows investigating the vibration field using appropriate signal processing, significantly expanding diagnostic capabilities.

The high informativeness of the vibration diagnostic signal measured on the bearing unit of EM is because rolling bearings are the only contact link between the stationary and rotating parts of the EM, and vibrations originating from various parts of the machines are concentrated and superimposed on them.

Additional information for bearing diagnostics can be obtained by using other physical processes as diagnostic signals, for example, temperature measured using thermocouples, which can be located directly on the bearing shields of the EM.

Frame and its attachment points to the foundation. Some general information about the technical condition of EMs can be obtained from diagnosing the frame of the EM and its attachment points to the foundation.

Among the processes that allow judging the overall condition and mechanical load on massive mounting units of EMs are acoustic emission (AE) processes [18, 75, 94–96]. To measure the characteristics of AE processes, corresponding sensors should be located directly at the diagnostic points. A distinctive feature of AE processes is their relatively wide frequency range, the upper frequency limit of which reaches several megahertz for some materials.

In addition to AE processes, additional information about the technical condition of EMs during operation can be obtained by measuring vibration processes. In this case, according to DSTU ISO 5348:2009, vibration sensors are placed on bearing supports (for powerful EMs) as well as directly at the attachment points of the EM to the foundation.

Brush-collector unit. During the operation of most powerful EMs, which have an excitation system based on transferring current to a rotating excitation winding through sliding contacts, special attention is required for the operation of the brush-collector unit. However, as noted in [75–77, 97, 98], until now, the development of technical diagnostic tools both in Ukraine and abroad has been restrained primarily by the lack of theoretically substantiated and experimentally verified diagnostic signs and criteria that would allow conducting diagnostics of the brush-collector unit with the specified reliability.

The most successful attempt at theoretical substantiation and experimental verification of diagnostic features ensuring the determination of the technical condition of the brush-collector apparatus was made in the works [75–77]. In this work, for the control and diagnosis of this apparatus, it is recommended to use the following 5 physical processes:

- brush sparking;
- radio emission from the brush and its magnitude;
- voltage drop ΔU across the brush-ring junction;
- ambient air temperature near the brush apparatus;
- non-uniform distribution of current across individual brushes in the brush-collector apparatus of EM.

One of the main parameters used to monitor the condition of the brush-collector assembly is the transitional resistance (conductivity) of the sliding contact. This parameter is measured using special sensors installed directly on the brush holders. The theoretical justification and practical application of this method are described in [75–77].

The aforementioned works extensively discuss the theoretical foundation and practical application of each of the diagnostic features mentioned above for determining the technical condition of the brush-collector apparatus of large EMs, analyzing their advantages and disadvantages. An important aspect is the description of the possibility of using these features to build an automated diagnostic system for the brush-collector apparatus of EMs. The construction of diagnostic algorithms for this EM component is elaborated sufficiently in the works [75–77].

The work [75] is dedicated to the issues of monitoring the technical condition of the brush-collector assembly using brush sparking as a diagnostic feature. However, the technical implementation of the device for its implementation proved to be quite complex. Moreover, the accuracy of control results using such a device is conditional, as this result is based on video monitoring data of brush sparking, for which, according to GOST 183-78, a visual scale of spark rating was used.

Vibration measurement of brush holders for diagnosing sliding contact can be implemented on medium and large EMs. This is due to the requirement of GOST 17375-75, according to which the mass of the sensor installed on the brush holder should not exceed 10% of the mass of the brush holder on which it is installed. In this case, it is assumed that the sensor does not distort the vibrational "picture" of the brush holder itself.

It should be noted that this method has not found widespread practical application due to the complexity of its technical implementation.

Cooling system elements of electrical machines (EMs). Diagnosing the ventilation system of EMs can be implemented by measuring the acoustic noise and vibrations of its individual components (air duct surface, fan impeller, fan shaft rolling bearings) [99–101].

A wide range of acoustic measuring equipment can be used to measure the acoustic noise of the ventilation system of EMs, for example, devices like those produced by "Bruel and Kjaer", such as types 8609, 8637, 8752, and others. Practical measurements of EM acoustic noise are carried out in accordance with the requirements of standards such as GOST 17377-73, 17975-75, 17995-75, and others.

Vibration measurement of EMs is realized using primary sensors, which can be installed on both the air duct surface and the fan bearing units. It is necessary to ensure compliance with standards such as GOST 17375-75, 17377-73, and others.

Current state of electric power engineering in Ukraine. The energy industry in Ukraine and the world exists in a dynamic environment and changes in response to internal and external challenges. In recent years, Ukraine has focused on diversifying generation sources, developing renewable energy sources, reducing dependence on imported energy resources, especially gas, and improving energy efficiency. At the same time, the conflict with Russia posed serious challenges to Ukraine, because Russia's attempts to put pressure on the government of Ukraine through the destruction of the energy infrastructure were seen as one of the main tools in achieving the aggressor's goals. The military invasion resulted in damage to energy infrastructure and instability in energy supply in 2022–2024 [102].

According to World Bank estimates, by the end of 2023, Ukraine's energy infrastructure suffered losses in the amount of $11 billion due to military actions. More than 50% of the energy infrastructure was damaged, causing blackouts across the country.

Ukraine's energy system continues to operate in emergency mode with a limited safety margin, facing losses. In total, 42 of 94 (45%) vital high-voltage transformers in government-controlled areas have been damaged or destroyed by missile or drone strikes since the start of the current phase of the war.

Total energy generation capacity has been reduced to almost 50% of the level by 2022. Of the nearly 37 GW of installed capacity, more than 19 GW have been destroyed, damaged or occupied since February 2022. The situation is complicated by a significant reduction in maneuverability, in particular, the loss of more than 67% of thermal generation capacity.

1.2 Physical Processes Characterizing the Technical Condition of Electrical Equipment

Let's consider the main physical processes and the construction of their mathematical models (diagnostic signals) used in the diagnosis of electrical equipment (EE).

First of all, it should be noted that in operational EE, there are both movable and immovable units. This affects the methods of measuring diagnostic signals and the methods of their processing (filtering, taking into account the cyclic nature determined by the frequency of the rotating electromagnetic field, etc.). This circumstance must be taken into account when choosing a mathematical model to describe the studied physical processes in the units of EE. These issues will be considered in detail in the 2nd, 3rd, and 4th sections of this monograph. Here, we will focus on the main features of various diagnostic signals.

Many models are based on the fact that at the measurement point at any fixed moment in time, the studied process can be considered as a result of the superposition of a large number of disturbing factors characterizing the energy coming from various random sources located in space at some distance from this point. Such models can be used to describe noise and vibration processes and fields, processes of acoustic emission occurring in a solid body under mechanical loading, or, for example, processes of occurrence of electrical discharges in insulation under high voltage. Random processes in electrical equipment also include processes with some stochastic cyclicity. These can include uneven rotation of diesel-electric unit shafts, rolling bearings, and others.

As shown in [75–77], mechanical displacements or vibrations are one of the most informative processes that can be used in EE diagnostics. Based on this, let's consider the occurrence of disturbing forces in various EE units that cause their vibrations.

1.2.1 Vibrations of Electrical Machines

The disturbing influences causing vibrations in EMs include:

- electromagnetic forces;
- forces caused by the operation of rolling bearings;
- aerodynamic forces;
- forces caused by mechanical rotor imbalance;
- forces caused by the operation of the brush-collector assembly.

The intensity of vibrations due to ***electromagnetic forces*** is primarily determined by the fundamental frequency of the rotating magnetic field. The frequencies of their vibrations are concentrated in the area equal to the doubled frequency of the power supply network. The disturbing electromagnetic forces acting in the air gap of EMs can be represented in the form of sinusoidal force waves, each of which has its own frequency and amplitude; thus, the system of disturbing forces has a polyharmonic nature and is deterministic. When such a disturbing force acts on the casing and various units of EMs, the effect of so-called selective resonance is usually observed, i.e., the nature of the process observed at the sensor placement point significantly differs from the nature of the disturbing forces.

As noted above, ***vibrations of rolling bearings*** have a significant impact on the vibration characteristics of the entire EM. When accelerometers are installed on diagnosed EM units (housing, bearing shield, brush-collector assembly, etc.), realizations of a random process, which is an additive mixture of vibrations generated by various disturbing forces in EMs, are measured. Therefore, an important aspect in diagnosing a particular EE unit is the choice of the accelerometer location. For example, when diagnosing rolling bearings, accelerometers are usually placed on the bearing shields, and when diagnosing the pressing of a laminated core of the stator, they are placed on the pressure plate of the stator.

Let's consider the main points related to the occurrence of vibrations in rolling bearings.

According to [75–77], vibrations of a working rolling bearing are caused by four main factors:

- resonances of bearing elements and their mounts;
- acoustic emission;
- external vibrations.

The operation of the main bearing elements (inner and outer rings, rolling elements, and separator) causes the excitation of vibrational energy due to cyclic stresses or impacts on defects. Each of the listed elements has its characteristic frequency at which such energy excitation occurs. Analytical expressions [75–77] exist to determine these frequency bands.

The resonances of bearing elements are due to the oscillatory properties of each bearing element. These resonances occur at the main frequencies under the influence of load and in the presence of defects.

During the operation of rolling bearings, significant stresses occur at the contact points of the balls or rollers with the outer and inner rings. This is especially evident in the case of a bearing installed on a horizontally positioned EM shaft. In this case, significant stresses are created at the contact points of the balls located in the extreme lower position due to the action of the EM rotor's mass. This, in turn, leads to material displacement on an atomic scale and the formation of elastic stress waves, causing acoustic emission and bearing vibration.

External bearing vibration is due to the fact that it is the only natural link between the rotating and stationary parts of the EM. Therefore, the bearing assembly can be considered the main place for transmitting vibrations from moving to stationary parts.

Based on the results of studies [103–107], the elements of a rolling bearing during its rotation generate a pulse stream, the characteristics of which depend on the technical condition of the bearing tracks and rolling elements, the presence and condition of lubrication, the quality of bearing installation, etc. To justify the physical model of vibrations generated by the bearing assembly, it is necessary to have a reasoned basis to consider the process generating these vibrations as a pulse stream, namely, a pulse white noise [108–110], which will be discussed in detail in Chap. 2. Since such noise is an idealization, it is necessary to show that the real process in the frequency range of interest and pulse interval practically does not differ much from ideal pulse white noise. To do this, let's estimate the duration of pulses caused by irregularities on the surfaces of bearing tracks and rolling elements for an ideal bearing. Depending on the design and operational requirements of the bearing, its tracks and rolling elements (balls) are processed with a certain surface roughness, which is characterized by the height of roughness peaks R_z. For different types of bearings R_z, this value varies within the range from $6.3 \cdot 10^{-6}$ to $0.2 \cdot 10^{-7}$ m. The height of roughness peaks is of the same order as the distance between surface peaks. This means that for a bearing with $R_z = 6.3 \cdot 10^{-6}$ m, the average distance between two adjacent surface peaks is $d_a \approx 6 \cdot 10^{-6}$ m. If the bearing has a circumference diameter of the raceway of the inner ring $D_i = 0.1$ m and it rotates at a speed of $V = 3000$ rpm with a stationary outer ring, then the circumferential speed of the inner ring raceway points is $V_i = \pi D_i V / 60 \approx = 16$ m/s. The duration of an elementary pulse caused by a ball hitting an irregularity on the inner ring raceway (local peak of the surface) is $\Delta t \approx d_a / V_i = 3.75 \cdot 10^{-7}$ s. The frequency band of this pulse is approximately $\Delta f = 1 / \Delta t \approx 2.7$ MHz.

In most cases, rolling bearing vibrations are studied in the frequency range of 0.001–50 kHz [75–77]. Therefore, it is evident that a delta function can be chosen as the model of such a pulse. The intensity of individual pulses depends on the size of irregularities or the degree of defect and is random.

Thus, by its physical nature, the process causing bearing vibrations represents a stream of random pulses in time and intensity.

Aerodynamic noise is directly proportional to the increase in the amount of cooling air supplied to the EM to dissipate heat. The occurrence of aerodynamic noise is conditioned by [111–113]:

- the fan, where the air stream is cut by the edges of the fan blades and the fan disc;
- the rotation of the rotor, resulting in vortex shedding from its surface due to the cutting of the air stream by the coil ends of the rotor;
- the air flow caused by the shedding of vortices from stationary obstacles in the ventilation paths;
- periodic pressure oscillations at certain sections of the aerodynamic circuit, as well as pulsations of the airflow exiting the radial ventilation channels of the rotor and entering the radial ventilation channels of the stator.

There is a significant difference both in the level and in the spectral components of the aerodynamic noise of similar EMs. This is due to the various deviations encountered in the manufacturing of individual EM components, as well as the assembly of these components. From this, it can be inferred that the aerodynamic noise of EMs exhibits a random nature.

Research on aerodynamic noise in EMs is of considerable interest to designers in terms of reducing the vibroactivity of the machines being designed. In solving diagnostic problems, special attention should be paid to determining the frequency areas of vibroacoustic processes caused by aerodynamic noise, with the aim of their subsequent filtration.

Vibrations in EMs caused by rotor imbalance are primarily deterministic in nature and manifest at frequencies that are multiples of the rotation frequency. The main causes of rotor imbalance include poor balancing, thermal overheating, internal stresses, and detachment of the rotor attachment on the shaft. Thermal imbalance can be caused by coil short-circuits in the rotor winding, uneven insulation thickness of the winding, or non-uniform cooling.

As already noted, vibrations caused by rotor imbalance are primarily deterministic in nature. Their frequency is determined by the rotation frequency of the EM shaft. The main frequency components generated by imbalance are in the low-frequency range of the overall spectrum of frequencies excited in EMs by various sources (assuming that the vast majority of low- and medium-power EMs have a shaft rotation frequency of up to 3000 rpm).

The noise and vibrations of the brush-collector apparatus are mainly caused by technological and design factors such as the technical condition of the collector, unevenness of the collector, poor condition of the brushes and brush holders, physical processes occurring in sliding contact, and the operating mode of the EM.

The results of the research presented in [75–77] indicate that the frequency range of vibroacoustic processes caused by the operation of the brush-collector apparatus is in the range of 1000–8000 Hz, and the spectral composition of these processes is minimally dependent on the rotation frequency.

Based on this data, it can be inferred that the vibroacoustic processes occurring in the brush-collector apparatus are primarily random in nature. This is due to mechanical collisions between the surfaces of the brushes and collector plates, the generation of electrical sparks at the contact points between the brushes and the collector, surface roughness of the collector, and so on.

1.2.2 Acoustic Emission

Acoustic vibrations occurring in a solid body when it is loaded can be used as carriers of information about the condition of the diagnosed object. The analysis of acoustic emission processes is associated with the construction of their mathematical models. Naturally, these models are built using the apparatus of probability theory and mathematical statistics since the processes of acoustic emission are inherently random.

It is shown that the most commonly used mathematical models of the electrical signal of acoustic emission are Poissonian and Gaussian white noises [114–117].

When constructing mathematical models of acoustic emission processes, the following physical assumptions were used in [118]:

1. In a deformable volume of material under the influence of an external force, a large number of microdefects of the structure arise independently of each other. Moreover, the spatial position and the moment of occurrence of each defect are random and independent.
2. The formation of a defect is accompanied by the emission of an acoustic wave by some elementary source in the form of an impulse.

From these assumptions, it follows that a quantum approach should be used to construct mathematical models of processes, with the appropriate choice of the class of random processes that most adequately account for the physical nature of the emergence of acoustic emission processes.

The processes of acoustic emission during the operation of electric machines occur at the points of attachment of this equipment to the foundation, during the twisting of the shafts of the electric machines under load, during the pressing of the iron packs of the rotor and stator, and so on.

1.2.3 Electrical Insulation Breakdown

As an example of such processes, one can mention the development of avalanche breakdown in an air-insulated gap.

According to [75–77, 119], the occurrence of an avalanche is conditioned by an initial electron, which can be formed due to one of the following reasons:

- cosmic radiation;
- decay of a negative ion;
- emission from the cathode.

Such processes depend on time and the intensity of the electric field, have a purely random nature, and can be described by random processes with a Poisson distribution law.

The development process of the avalanche itself also has a random nature. Specifically, it can take on an avalanche-like character as a result of the formation of a large number of new electrons generated by the mutual collisions of initial electrons with gas molecules (ionization), or an electron can be lost (adhesion), and the process will not continue. In the first case, there is a breakdown of the air gap.

For the mathematical description of the avalanche formation process with discrete changes in the number of electrons, coefficients of ionization and adhesion are introduced, characterizing the intensity of electron formation and death. Expressions are obtained for determining the probability of reaching the critical number of electrons in the avalanche depending on the field intensity and the number of initial electrons. The probability models of insulation breakdown considered there also take into account many external factors affecting the breakdown voltage (presence of dust, humidity, etc.).

In the work [75–77], in addition to the processes of electrical breakdown of the air gap described above, issues of the formation of breakdown processes in liquids and solids are also described. It is shown that these processes have a quantum random nature.

1.2.4 Non-uniformity of Angular Rotation Speed of the Shaft

Special attention is required to solve the diagnostic task of a diesel-electric unit, as this object consists of interconnected internal combustion engines (ICE) and electric generators. Today, there is a vast number of methods and their implementing technical means for diagnosing both ICE and EM [120–123].

The joint investigation of diagnostic signals measured on a common shaft connecting the ICE and the electric generator opens up promising possibilities for diagnosing this object. Typically, the implementation of these diagnostic signals can be obtained by measuring them at the connection points of the crankshaft tail of the ICE and the electric generator shaft. As known [75–77], various physical processes (vibrations, temperature, gas-analytical composition of exhaust gases, non-uniformity of the angular rotation speed of the diesel-electric generator shaft, etc.) can be used as carriers of information about the technical condition of diesel-electric units.

Below, let's briefly describe such a specific physical process as the non-uniformity of the angular rotation speed of the diesel-electric generator shaft.

During the working cycle, a certain power develops on the shaft of the diesel-electric generator, which for multi-cylinder diesels can be considered as the cumulative result of the powers developed during the operation of each cylinder. Naturally, the discreteness of the working cycles in each cylinder, in turn, leads to the non-uniformity of rotation of the crankshaft of the diesel. It should be noted that the main elements of the theory of diagnosing rotating units of diesel-electric units are considered in the works [75–77].

During the operation of the cylinder-piston group, the crankshaft of the diesel generator is subject to the influence of many forces [124–126], and its resulting motion is observed.

The main of these forces are.

- inertia forces of the shaft;
- inertia forces of the parts attached to the shaft (pistons, connecting rods, flywheel, etc.);
- moments of forces caused by the operation of cylinders;
- moments of forces caused by the load of the generator;
- moments of forces caused by friction in the bearings.

Under the influence of these forces, each point of the crankshaft participates in rotary motion around a certain axis (further, the shaft is considered as a rigid mechanical system). In this case, the shaft performs both rotary motion around the axis and executes oscillatory movements, which are accompanied by its deformations (torsional vibrations, longitudinal bending).

To describe the mathematical functional relationship, some authors, for example [127–130], use a system of differential equations, the order of which is determined by the number of masses in the computational model of the crankshaft. By solving the system of equations, distributions of amplitudes and phases of torques on the cylinders of the unit are obtained. Deviation of these amplitudes from the mean value is an estimate of the degree of identity of successive working cycles. When creating such a model, authors used a deterministic approach that does not take into account the random nature of the non-uniformity of shaft rotation. This, in turn, leads to significant errors in measurements in practice, as disturbances that always occur in a working diesel generator are not taken into account. Amplification of the sensor signal does not lead to success in this case either, because with the amplification of the useful signal, the interference is amplified to the same extent, which, in turn, does not allow increasing the reliability of measuring the non-uniformity of the crankshaft rotation of the diesel generator.

The developed power on the diesel-generator shaft, and consequently the torque, fluctuates within certain limits around the mean value. In this case, the main frequency of the pulsations superimposed on the mean torque equals $f_p = n_r h / 60$, where n_r is the rotation frequency in revolutions per minute, and h is the number of impulses of the disturbing force per one revolution of the shaft, which depends on the number of diesel cylinders, the number of fuel injection cycles, the order of fuel ignition, etc. Based on this, for the process of non-uniform rotation of the diesel crankshaft, and therefore for the variation of cylinder power increments, stochastic periodicity is characteristic. This means that certain probabilistic characteristics of the signals behave periodically, although their realizations themselves are not obligated to be periodic. Approaching the construction of a mathematical model for such a process, it should be emphasized that despite the huge variety of signals, the number of mathematical stochastic models that could describe them is very limited. In mechanics, for example, stationary ergodic models are widely used, and due to the central limit theorem, Gaussian models. However, stationary models are not acceptable for describing the investigated periodic signals, since their initial and central moments are constant. To account for the properties of signals mentioned and subsequent quantitative assessments of their periodically changing probabilistic characteristics, qualitatively new models and corresponding methods of their statistical processing are necessary.

Having considered some physical processes arising in operating electrical equipment, let's proceed to discuss the general principles of building a diagnostic IMS, which are designed to measure the realizations of these processes.

1.3 From the History of Development of Means of Measuring Characteristics of Physical Processes

Measurement technology is one of the most important factors accelerating scientific and technological progress in practically all sectors of the economy. Various physical quantities are used to describe phenomena, processes, and properties of material objects, ranging from several thousand electrical, magnetic, spatial, and temporal to mechanical, acoustic, optical, chemical, biological, and others. These quantities differ not only qualitatively but also quantitatively and are evaluated by different numerical values. Establishing the numerical value of a physical quantity is achieved through measurement. The result of measurement is a quantitative characteristic in the form of a named number with a simultaneous assessment of the degree of approximation of the obtained value of the measured quantity to the true value of the physical quantity. It should be noted that finding the numerical value of the measured quantity is possible only experimentally, i.e., through physical experimentation. Technical means that perceive, transform, and represent numerical values of physical quantities are necessary for the realization of any measurement process. In practice, electrical and non-electrical methods (pneumatic, mechanical, chemical, etc.) are used for measuring physical quantities.

Electrical measurement methods have gained the widest acceptance because they allow for the relatively simple conversion, transmission, processing, storage, representation, and input of measurement information into computers. Technical means and various measurement methods form the basis of measurement technology. Any production process is characterized by a large number of parameters that vary widely. To maintain the required operating mode of a technological installation, it is necessary to measure these parameters. The more reliable the measurement of technological parameters is, the better the quality of the target output product. Modern enterprises, such as those in the petrochemical industry with continuous production, use measurement of various physical parameters to maintain the quality of the produced products. These parameters include temperature, volumetric and mass flow rates of substances, pressure, level and quantity of substances, time, substance composition, density, humidity, content of mechanical impurities, as well as voltage, current, speed, and others. The number of parameters required for measurement reaches several thousand. In nuclear energy, the number of parameters required for measurement reaches tens of thousands.

The acquisition and processing of measurement information are intended not only to achieve the required product quality but also for organizing production, accounting, and balancing the quantity of substances and energy. Currently, an important application area of measurement technology is the automation of scientific and technical experiments. For improving the economy of designed objects, mechanisms, and machines, experimental studies conducted on their physical models are of great importance. The task of acquiring and processing measurement information becomes so complex that its effective solution becomes possible only based on the application of specialized measurement and computing means.

Dmytro Mendeleev emphasized the role of measurement technology, stating that "science begins when measurement begins". Measurement technology began its development in the 1740s and is characterized by a sequential transition from indicating (mid-19th to late nineteenth century), analog self-recording (late 19th to early twentieth century), automatic, and digital instruments (1950s) to information-measuring systems (1960s to the present).

The end of the nineteenth century was characterized by the first successes of radio communication and radio electronics. Its development led to the need for the creation of a new type of measurement technology, designed for small input signals, high frequencies, and high-impedance inputs. Radio-electronic components such as rectifiers, amplifiers, modulators, and generators were used in these new measurement devices, which at various stages of electronic technology development could be vacuum tube-based, transistor-based, or based on integrated circuits, etc. Thus, the expansion of the range and qualitative indicators of measurement technology is closely linked to the achievements of radio electronics.

One of the modern directions of development of measurement technology is digital instruments with a discrete form of information representation, which has proven convenient for conversion, transmission, processing, and storage of information.

The development of discrete measurement technology has led to the creation of digital direct current voltmeters, with measurement errors below 0.0001, and the speed of voltage-to-code converters reaching several billion measurements per second. The upper limit of measurement of modern digital frequency meters has reached gigahertz, digital time interval meters have a lower measurement limit down to fractions of a picosecond, electric currents are measured in the range from 10^{-16} to 10^{-5} A, and lengths are measured in the range from 10^{-12} to $3.086 \cdot 10^{-16}$ m (the atom size).

Measurement technology has seen broad opportunities with the emergence of microprocessors (MP) and microcomputers. Thanks to this, the areas of application of measurement technology have significantly expanded, their technical characteristics have improved, reliability and speed have increased, and ways to address tasks that were previously impossible have been opened. In terms of breadth and effectiveness of application, measurement technology occupies one of the leading positions with the increasing use of microprocessors in control systems. The importance of microprocessors and microcomputers in creating automated measurement tools for management, research, control, and testing of complex objects cannot be overestimated. The development of science and technology requires constant improvement of measurement technology, whose role is steadily increasing.

1.4 Key Concepts and Definitions in Measurement Technology and Diagnostics

The concepts and definitions used in measurement technology are regulated in Ukraine by DSTU 2681-94 "Metrology. Terms and Definitions", which was developed to replace GOST 16263-70. *Measurement* is an informational process of obtaining, by experimental means, a numerical relationship between a given physical quantity and some value of it taken as a unit of comparison. *The measurement result* is a named number found by measuring the physical quantity. The measurement result can be accepted as the actual value of the measured quantity.

One of the main tasks of measurement is *to assess the degree of approximation* or difference between the true and actual values of the measured physical quantity—measurement error. Measurement error is the deviation of the measurement result from the true value of the measured quantity. *Measurement error* is a direct characteristic of measurement accuracy. *Accuracy of measurement* is the degree of closeness of the measurement result to the true value of the measured physical quantity. Measurement reduces the initial uncertainty of the value of the physical quantity to the level of the inevitable residual uncertainty determined by the measurement error. The value of the measurement error depends on the perfection of technical devices, the method of their use, and the conditions of the experiment. *The measurement principle* is a physical phenomenon or a set of physical phenomena underlying the measurement. An example is the measurement of temperature using the thermo-electric effect and other physical phenomena used for conducting experiments, which must be chosen with the aim of achieving the required measurement accuracy. *A measurement experiment* is a scientifically justified experiment for obtaining quantitative information with the required or possible accuracy of determining the measurement results.

Conducting a measurement experiment implies the presence of technical devices that can ensure the required accuracy of obtaining the result. Technical devices participating in the experiment are standardized in terms of accuracy characteristics and are referred to as measuring instruments. *A measuring instrument* is a technical device used in a measurement experiment and having standardized accuracy characteristics. The quantitative information obtained by measurement represents measurement information. *Measurement information* is quantitative information about the property or properties of a material object, phenomenon, or process obtained using measuring instruments as a result of their interaction with the object.

The amount of measurement information is a numerical measure of reducing the uncertainty of the quantitative assessment of object properties. The interaction between the object under study and the measuring instruments in the experiment process implies the presence of *signals*, which serve *as carriers of information.* Important carriers of information include electrical current, voltage, pulses, and other electrical parameters. A *measurement signal* is a signal functionally related to the measured physical quantity with a specified accuracy.

The *measurement method* is a set of techniques for using principles and measuring instruments. Unity of measurement is of great importance in measurement technology—such a state of measurements where their results are expressed in specified units, and the measurement errors are known with a given probability. Unity of measurement allows comparing the results of different experiments conducted under various conditions, performed in many places using different methods and measuring instruments. This is achieved through accurate reproduction and storage of established units of physical quantity and transmitting their sizes to the applied measuring instruments. These issues constitute the subject of metrology.

Metrology is the science of measurement methods and means to ensure unity of measurements and achieve the required accuracy. A *measure* is intended for reproducing the physical quantity of a given size. Legislative metrology is a branch of metrology that includes interconnected and interdependent complexes of rules, requirements, and norms, as well as other issues requiring regulation and control by the state, aimed at ensuring unity of measurements and uniformity of measuring instruments.

In accordance with the above, characteristics of measuring instruments that determine the measurement accuracy through their use are called *metrological characteristics of measuring instruments*. Metrological characteristics are necessarily standardized and regulated to ensure unity of measurements. *Control* is the process of establishing conformity between the state, property of the object under control, and the specified norm. The control results in a judgment about the object's condition.

Let's briefly consider the main concepts and definitions used in technical diagnostics.

The object of technical diagnostics is a specific product or its individual components, the actual technical condition of which needs to be determined. The *diagnostic process* involves determining the technical condition of this object with a certain known accuracy. The result of the diagnostic process is establishing the type of *technical condition* of the object (operational or non-operational), classification, and localization of defects.

A product that fully meets all the requirements of the technical documentation is called *operational*. The occurrence of defects leads to a violation of this correspondence, as a result of which the product transitions into a non-operational state. The task of diagnostics is to timely identify defects and find their location in the object, which will allow restoring the violated correspondence of the object to the technical requirements.

The technical condition of any object can be characterized by a set of numerical parameters. From this set, a subset is distinguished, the elements of which are called *diagnostic parameters*. The informativeness of the parameter is the main criterion for selection. In other words, diagnostic parameters should carry sufficient information about the actual technical condition of the diagnostic object. The determination of diagnostic parameters is possible only through the measurement and processing of physical processes occurring in the diagnostic object. Therefore, another important requirement is the availability of selected parameters for measurement with sufficient accuracy.

Based on the specific values of diagnostic parameters obtained during the diagnostic process, a decision is made about the technical condition of the object, and the identification of a specific type of defect from the set of specified ones (classification) and determination of its location (*localization*) can also be carried out. Diagnostic decision-making rules are used for this purpose.

Diagnosis is carried out using special technical means, such as diagnostic devices or diagnostic AI systems, which implement a specific diagnostic method.

Methods of technical diagnostics can be divided into two broad classes: test and functional diagnostics. In the first case, special test stimuli are applied to the diagnostic object, and its reaction is measured. Often, the use of this type of diagnostics is associated with taking the diagnosed equipment out of operation for the entire period of its testing.

In functional diagnostic systems, working stimuli are used to determine the technical condition of the object—i.e., the processes that occur in the object itself during its operation. In this case, diagnostics can be carried out without taking the equipment out of operation, which is particularly important for the power industry.

Following the works, the diagnostic process involves a series of operations. One of the main results of these operations is obtaining and analyzing the reaction of the diagnostic object to test or working stimuli, which represent *diagnostic signals*. These signals are measured by diagnostic means at selected control points. Based on the measured signals, numerical values of diagnostic parameters are obtained according to a certain algorithm. Using diagnostic decision-making rules, a decision is made about the actual technical condition of the diagnostic object. *Automated diagnostic IMS* can perform all stages of the diagnostic process independently. In *automated equipment,* some functions are transferred to the operator.

It is also important to note that the issues of technical diagnostics of electrical equipment from the perspective of practical use are most fully covered in the literature. This work provides a general classification of various methods and means of technical diagnostics. Additionally, specific issues related to diagnosing thermal and electromechanical equipment, turbo-generators, and nuclear power plant reactors are discussed.

Now, let's examine the aspects related to the development of diagnostic methods and systems for electrical equipment.

1.5 Regulation of EE Diagnostics

Building diagnostic IMS for EE takes into account a number of features associated with the presence of strong electrical and magnetic fields, high electrical and mechanical voltages, vibrations, temperature, etc., accompanying its operation. This means considering the specifics of the equipment's operation, which is the object of diagnostics. Additionally, work must be conducted based on existing standards. Let's start this subsection by addressing the latter issue.

Any monitoring, control, and diagnostic IMS for EE must be developed in accordance with existing regulatory documents. Currently, there are more than 3000 standards in the field of power industry in the European Union. In Ukraine, work is underway to harmonize the national standardization base with the European one. As of 2012, about 1000 standards had been introduced, and approximately two hundred were harmonized. However, a significant number of standards in the field of technical diagnostics and ensuring the reliability of electrical equipment remain unharmonized.

Let's examine the requirements of existing standards and other regulatory documents related to conducting activities to ensure the reliable and safe operation of electrical equipment, such as maintenance, as well as the development and implementation of condition monitoring and technical diagnostic systems.

1.5.1 EE's Technical Maintenance

The system of scheduled preventive maintenance includes maintenance, current and major repairs, preventive and post-repair tests. Before repairs, EM tests are always carried out to detect and subsequently eliminate defects.

The system of planned preventive maintenance includes technical servicing, current and major repairs, preventive and post-repair tests. Prior to repair, tests are always conducted on electrical machinery (EM) to detect and subsequently eliminate defects.

Typical Technical Maintenance Tasks Include

1. *Daily observation*: ensuring compliance with operating rules and manufacturer's instructions (checking load, temperature of individual EM components, coolant temperature in closed cooling systems, lubrication in bearings, absence of abnormal noises and vibrations, excessive sparking at the commutator and contact rings, etc.); daily monitoring of grounding integrity; adherence to safety rules for personnel working on electrical equipment; emergency shutdown of EMs; minor repairs carried out during breaks in the operation of main technological equipment, which do not require special EM shutdown (tightening contacts and fastenings, brush replacement, traverse adjustment, adjustment of starting and regulating equipment, and protection system, cleaning accessible machine parts, etc.); participation in acceptance tests after installation, repair, and commissioning of EMs and their control and protection systems; scheduled inspections of operating EMs according to the schedule approved by the chief engineer, with completion of inspection forms.
2. *Tasks during current repairs*: conducting technical maintenance operations; disconnecting from power supply and drive mechanism (motor); cleaning external surfaces from contamination; disassembly of EMs to the extent necessary for repair; inspection of bearing condition, flushing, and replacement of roller bearings if clearances exceed permissible limits, inspection and repair of forced lubrication systems, oil replacement; inspection, cleaning, and repair

of fan mounting, inspection and repair of forced ventilation system; inspection, cleaning, and compressed air blowing of windings, collector, ventilation ducts; inspection of the condition and reliability of winding end fastening, defect correction; repair of local insulation damage, winding drying, coating of winding end faces with insulating varnishes; inspection and tightening of fastenings and contacts with replacement of defective fastening elements; inspection and adjustment of brush holders, traverses, short-circuiting devices, brush lifting mechanism; collector and contact ring cleaning and grinding, collector undercutting; inspection of the condition and correctness of markings of winding end leads and terminal blocks with necessary repairs; replacement of flange gaskets and seals; inspection of explosion-proof EM tightness; EM assembly and protective grounding verification; EM connection to the power supply and idle and loaded operation testing; rectification of paint damage; acceptance testing and documentation for EM commissioning.

3. *Tasks during major repairs*: conducting current repair operations; checking rotor axial runout and radial clearances of journal bearings with subsequent replacement of inserts; replacement of roller bearings; complete disassembly of EMs with cleaning and flushing of all mechanical parts; replacement of defective windings (including repair of short-circuited windings), cleaning and blowing of windings, impregnation and drying of windings, coating of winding end faces with insulating varnishes and enamels; collector, contact ring, and brush assembly repair (up to their replacement with new ones); stator and rotor core repair, including partial sheet metal replacement; restoration of core pressing; bearing shield repair, housing repair, restoring seat dimensions; shaft repair; repair or replacement of fan; replacement of defective slot wedges, insulation parts; winding end leads marking; EM assembly and painting; acceptance testing and documentation for EM commissioning.

Acceptance testing of each machine and acceptance of the main (experimental) machines are conducted at the manufacturer's stand.

In accordance with existing standards (GOST 183-74, GOST 533-85, GOST 5616-81, GOST 609-84), acceptance tests of each EM include: insulation resistance measurements of windings relative to the frame and between windings, insulation resistance of embedded temperature sensors, windings at direct current in practically cold condition, resistance thermometer at direct current in practically cold condition; insulation resistance tests of windings relative to the frame and between windings for electrical strength; determination of stable short-circuit characteristics (for hydrogenerators at the installation site), no-load operation (for hydrogenerators at the installation site); tests at increased rotation frequency (for turbogenerators); measurement of bearing insulation resistance, oil temperature in bearings (for hydrogenerators at the installation site); inspection of shaft seals assembly and determination of air leakage at excess pressure not less than nominal hydrogen pressure (for EMs with hydrogen cooling).

The acceptance tests for the main (experimental) samples (for hydro-generators in the installation site) additionally include: testing for short-term overload current;

determination of efficiency; heating tests; determination of the distortion coefficient of the sinusoidal curve of voltage, inductive resistances, and time constants of windings; tests for short-circuit impulse current, for heating; determination of vibrations, nominal excitation current, and adjustment characteristics; noise level measurement; checking the operation of the gas-oil system of hydrogen cooling and determination of hydrogen leakage (for electric machines with hydrogen cooling); checking the liquid cooling system (for electric machines with liquid cooling).

Measurement of the insulation resistance of the stator winding relative to the machine casing and between the windings is carried out using a megohmmeter of at least 1000 V in practically cold condition, at which the winding temperatures take the ambient temperature. When measuring the insulation resistance of the generator windings with direct water cooling, the megohmmeter screen output is connected to the water collector, from which the external water system is disconnected. The insulation resistance is determined sequentially for each branch of the stator winding, while the other branches are connected to the machine casing. When determining the absolute value of the insulation resistance, measurements are made no less than 60 s after applying voltage to the insulation. After measuring the insulation resistance of individual parts of the winding, they are discharged to the generator casing.

Resistance measurement of windings with direct current is conducted in a cold state before the generator drying process using a voltmeter and ammeter, utilizing magnetoelectric instruments with an accuracy class of no less than 0.5. Readings on the instruments are taken simultaneously at constant values determined. Winding resistances are determined as the average value from at least three measurements conducted at different current values. The accuracy of measurements largely depends on the quality of contacts at the points of connection of measuring instruments, with the voltmeter connection recommended to be made separately from the current circuits.

Resistance measurements of resistance thermometers with direct current are conducted at ambient temperature using a voltmeter and ammeter method with a measurement error of no more than 0.5%. Measurements are recommended to be conducted directly on the terminal assembly.

At the installation site, all turbo, hydro-generators, and large synchronous machines undergo acceptance tests, which in addition to acceptance tests at the manufacturer's stand include heating tests; bearing vibration measurements; checking the operation of the gas-oil system for hydrogen cooling and determination of hydrogen leakage (for machines with hydrogen cooling); checking the operation of the liquid cooling system (for machines with liquid cooling).

At the installation site, tests are also conducted on hydro-generators and other large-sized synchronous machines, the testing of which at the manufacturer's stand is either not possible or requires significant costs to conduct.

Heating tests. In hydro-generators, the temperature of the stator winding and active steel is measured using resistance thermometers. The winding temperature is determined by the resistance thermometer readings placed between the rods (sides of the coils) of the stator winding. The temperature of the active steel is determined by the resistance thermometer readings placed at the bottom of the slot. According to the requirements of GOST 5616-81E, for measuring the temperature of the winding in hydro-generators exceeding 10 MVA, 12 resistance thermometers should be installed in one or two parallel branches of the winding, and two thermometers per parallel branch for more than two branches. For measuring the temperature of the stator core, no fewer than six thermometers should be installed in a hydro-generator. For generators with water cooling, stator winding thermometers (or other temperature indicators) should be installed at the end of each parallel circuit of the cooling system.

To measure the air temperature at the outlet of each cooler section, one thermometer is installed, and in the hot air zone, two thermometers are installed per generator.

According to GOST 5616-89, temperature of the bearing segments and bearings is also measured in hydro-generators using thermometers (two thermometers per segment) and oil temperature in the pedestal bath and each bearing (according to the readings of two thermometers).

For synchronous machines, in accordance with the requirements of current standards, bearing vibrations of machines are measured. Vibration measurements (vibration displacement or effective value of vibration velocity) are carried out on the upper bearing caps in the vertical direction and at the connector in the transverse and axial directions.

For turbo-generators, the effective value of vibration velocity should not exceed 4.5 mm/s for all operating modes.

In hydro-generators, vibration is measured in the horizontal plane of the crossbars. According to GOST 5616-89, with a rotation frequency of 100 rpm, the allowable vibration is 180 μm; from 100 to 187.5 rpm—150 μm; up to 375 rpm—100 μm; up to 750 rpm—70 μm.

On main (experimental) samples of turbo and hydro-generators, vibration of the core and end parts of the stator winding is measured. The allowable vibration (amplitude of oscillations) for the stator core of a turbo-generator is 60 μm, and for a hydro-generator—30 μm. Vibration (amplitude of oscillations) for the end parts of the stator winding should not exceed 150 μm.

Vibration of contact rings in turbo-generators should not exceed 200 μm.

1.5.2 General Requirements for Technical Diagnostics Systems

Information about the condition of operating equipment is contained in various signals measured during its operation. One of the indicators is the temperature of various parts of the EE. For EO with rotating components, vibration is the most important diagnostic signal. Therefore, the analysis of vibrations and temperatures of individual components forms the basis of EE technical diagnostics. Such research direction has expanded the capabilities of existing non-destructive testing methods, enabling the solution of practical long-term equipment condition forecasting tasks that contain rotating components.

It should be noted that technical diagnostic systems in Ukraine must be developed and put into operation in accordance with the Laws of Ukraine "On Electric Power Industry," "On Environmental Protection," "On Metrology and Metrological Activity," "On Information," as well as the industry document of the Ministry of Fuel and Energy of Ukraine "Technical Operation of Electric Stations and Networks. Rules".

Recently, a series of standards related to the diagnostics of electrical equipment have been adopted by the Interstate Council for Standardization, Metrology, and Certification, which includes Ukraine, corresponding to the international standards of ISO (International Standard Organization). Let's consider some of them that should be used in the creation of modern diagnostic equipment for EE.

Certification, Installation, and Application of Vibration Sensors

In modern vibration diagnostics systems, it is advisable to use accelerometers as vibration sensors. Definitions of accelerometers of various types, their structural elements, main parameters, and characteristics are given in GOST 18955-73. According to this standard, *an accelerometer is designed to measure translational accelerations.* By integrating vibrational acceleration, vibrational velocity of the investigated component can be obtained, and by integrating vibrational velocity, vibrational displacement can be obtained.

The attachment of accelerometers, according to DSTU ISO 5348:2009, is allowed using: studs; methyl cyanoacrylate adhesive; beeswax; double-sided adhesive tape; vacuum mounting; magnet; probe.

The characteristics of accelerometers are determined by DSTU ISO 8042:2008, which specifies the requirements for accelerometer calibration and accounts for the influence of external factors on the sensor.

The effect of variable electromagnetic fields on the output voltage of the sensor, for example, in the form of equivalent vibration due to a magnetic field of a specified intensity at a specified frequency (e.g., 50 Hz), in the direction where this influence is maximum, is established by DSTU ISO 2954:2005. The frequency range in which the developed device should measure vibration is also defined there—10...1000 Hz, and the calibration frequency is determined—80 Hz.

Development of Vibration Diagnostic Equipment

The developed vibration diagnostic equipment must comply with GOST 25275-82, which applies to devices for measuring vibrations of machine housings and bearing supports of rotating machines with a rotation frequency from 10 to 200 rpm, as well as to elements of vibration measuring instruments.

If the vibration measuring instrument contains electronic octave or third-octave passive or active filters, then GOST 17168-82 also applies to them. The methodology for testing such filters is presented in GOST 8.553-88 (does not apply to narrowband, broadband, and corrective filters).

Methods of Vibration Signal Analysis and Decision Rule Construction

The selection and use of statistical methods for analyzing vibration signals are regulated by DSTU ISO/TR10017:2005, which includes paragraph 4.13 on time series analysis.

The construction of decision rules should be carried out taking into account the group of standards DSTU ISO 11843 (parts 1 to 3); DSTU ISO 16269-6:2008, which regulates the determination of confidence intervals, as well as DSTU ISO 11453-6:2004, which defines the methodology for hypothesis testing.

If the developed diagnostic system involves forecasting information parameters, it is necessary to consider the provisions of DSTU ISO 16269-8:2008, which deals with the determination of forecast intervals.

General Requirements for EM Vibration Diagnostic Systems

Analysis of existing standards of Ukraine, interstate, European, and international standards in the field of ensuring the reliability of EO provides grounds to formulate the following requirements that modern competitive vibration diagnostic systems for EO should meet.

Measurement Ranges:

- vibration velocity (minimum): 30 mm/s;
- vibration displacement (minimum): 350 μm.

Frequency Range:

- vibration measurement range (not worse): 10…2500 Hz;
- accelerometer calibration frequency: 79.6 Hz.

Components on which vibrations need to be measured:

- for synchronous machines: vibration on upper bearing caps in vertical direction and at the connector in transverse and axial directions;
- for hydro-generators: vibration in the horizontal plane of the crossbars;
- for turbo-generators: vibration of contact rings;
- for turbo and hydro-generators: vibration of the stator core and end parts of the stator winding.

Attachment of accelerometers (DSTU ISO 5348:2009) using:

- studs;
- methyl cyanoacrylate adhesive;
- beeswax;
- double-sided adhesive tape;
- vacuum mounting;
- magnet;
- manual probe.

Measurement accuracy—according to GOST 25275-82/ST SEV 3173-81. The electrical part control device must have an electrical calibration device, which should output a harmonic signal with a frequency of 79.6 Hz; the error of electrical calibration of vibration measuring instruments should not exceed 5%; the main measurement error of harmonic vibration within the working part of the scale should not exceed 10%.

Considering the above main provisions of standards regarding equipment for measuring and diagnosing electrical equipment, let's move on to the discussion of the general structure and features of building an IMS diagnostic.

1.5.3 Structure and Features of Modern IMSs for EE Diagnostics

From the above list of diagnostic parameters and signals, it follows that the latter have sufficiently wide dynamic and frequency ranges. Considering this, modern IMS for electrical machine diagnostics should be universal, i.e., capable of measuring, recording, and processing diagnostic signals with different physical nature, as well as dynamic and frequency ranges. Let's briefly discuss these issues.

The generalized structural diagram of a modern multichannel IMS for functional diagnostics is shown in Fig. 1.2. It consists of:

1. Measurement transducers designed to measure various physical processes accompanying the operation of the diagnosed object.
2. Analog part intended for preliminary processing of the measured information (impedance matching, amplification, filtering, etc.).
3. Interface between the analog and digital parts, the role of which is to convert measured signals into digital form and transmit them to the digital part of the IDS.
4. Digital part intended for subsequent information processing (e.g., digital filtering) and analysis of the object's condition according to a specified algorithm, defect detection, and classification.

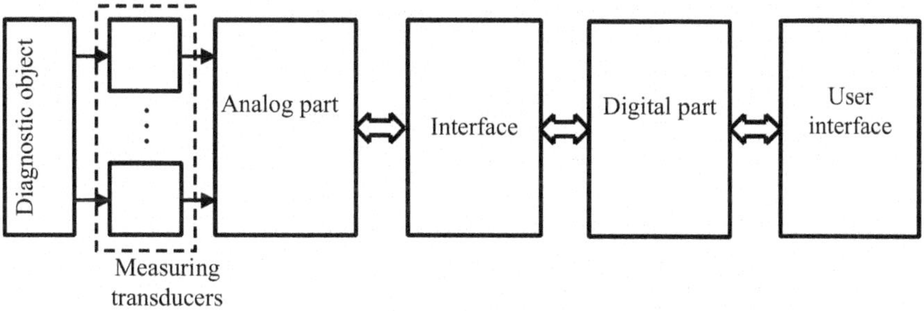

Fig. 1.2 Generalized structural diagram of diagnostics IMS

5. User interface, through which the operator controls the system's operation and receives information about the object's condition.

The digital component of a diagnostic IMS in most cases represents a personal computer operating under specialized software. In IMS for territorially distributed objects, or when access to the nodes being diagnosed is complicated (such as constantly rotating blades of wind turbines), the use of special information transmission channels from measuring transducers over significant distances (via radio channels, optical fiber lines, etc.) may be possible.

The development of diagnostic tools for a specific type of equipment is associated with solving the following main tasks:

- identification of possible defect classes (the most important or most frequently occurring ones) that need to be detected;
- selection of diagnostic signals available for measurement and control points on the object under study;
- development of a mathematical model of the diagnostic object, the analysis of which allows justifying possible diagnostic parameters;
- development of algorithms for obtaining numerical values of selected diagnostic parameters;
- construction of decision rules for detecting and classifying defects;
- creation of technical means implementing certain stages of the diagnostic process: from measuring selected diagnostic signals to making diagnostic decisions.

For the first task, statistical data on failures is mainly required. To solve the second task, it may be necessary to make constructive changes to the diagnostic object to ensure the attachment of corresponding measuring transducers. The solution to the third and fourth tasks is interrelated—namely, the algorithms for obtaining numerical values of selected diagnostic parameters depend on the developed mathematical model of diagnostic signals measured on the diagnostic object.

While solving the fifth task, the determining factor is the type of developed mathematical model (deterministic or statistical). Depending on the chosen model, decision rules are established, allowing the determination and classification of types of defects in the nodes of the diagnosed equipment.

The solution to the last task, related to the development of the IMS itself, is based on the results of solving the above five tasks. Primarily, this concerns various types of software used in the developed IMS.

1.6 Features of Applying Deterministic and Statistical Methods in Diagnostics

As shown by the review of works by domestic and foreign authors, currently, IMS and devices based on both deterministic and statistical methods are used in solving tasks of control and diagnostics of industrial equipment.

While using deterministic methods, a deterministic time function is chosen as the initial mathematical model of information signals coming from sensors installed on the studied node. In this case, amplitude-frequency or phase-frequency parameters of signals, characterizing the operation of the studied nodes of industrial equipment, are most often used as informative. For example, when solving the problem of diagnosing technical objects, the use of deterministic methods mainly boils down to theoretically determining possible diagnostic features and comparing them with quantitative estimates of these features obtained from processing experimental data. If the latter significantly differ from the theoretical results, a conclusion is made about the presence of a defect in the studied node. Essentially, these methods are applicable when the results of all observations under the same initial conditions are completely identical, and if the same parameter (feature) is measured under the same conditions. This situation is either highly idealized or occurs with very low accuracy of the measuring instruments used, when random effects on measurements are not perceived by the measuring instruments. Therefore, in the deterministic approach, there is no need for multiple measurements since they are all the same, and a conclusion about the technical condition of the node can be made based on a single measurement. However, there is always the possibility, by increasing the accuracy of each individual measurement, to detect the non-repeatability of measurement results in the described situation. Then the question arises of what to consider as the result of the conducted measurement from the obtained series of numbers. Therefore, the application of deterministic methods cannot be considered satisfactory and justified since the overwhelming majority of the measured processes occurring in the diagnosed nodes of electrical equipment are random in nature, i.e., their realizations change from observation to observation. Thus, the possibility of obtaining practically reliable information about the technical condition of diagnosed nodes based on a single observation is highly doubtful.

Non-random or deterministic signals can be conveniently regarded as a special (degenerate) case of random processes provided that the variance of such processes tends to zero. This way of reasoning is sometimes encountered both in theoretical and applied scientific works.

While diagnosing using IMS based on statistical methods, it creates the possibility to recommend an algorithm, based on a series of results from a specific measurement experiment, for computing the best probabilistic approximation of the measured parameter. Additionally, when using such IMS, the measure of incorrect decisions about the technical condition of diagnosed nodes is taken into account, and there is also the possibility to estimate the average number of possible incorrect decisions, their dispersion, etc. The main parameters used as diagnostic features when applying IMS based on statistical methods are the mathematical expectation, variance of the investigated signal realizations, and from functional characteristics—the correlation function, energy spectrum, and one-dimensional distribution function. Some issues related to the theoretical justification and obtaining quantitative estimates through statistical processing of experimental data are discussed below.

An integral part of technical diagnostic IMS is the diagnostic object itself. To confidently conduct the diagnosis of a particular node, on the one hand, the researcher needs to have a mathematical model that would sufficiently accurately describe the measured processes occurring in the studied nodes based on the physics of phenomena. On the other hand, there is a need for IMS built taking into account the a priori data on the information signals that can be obtained as a result of the analysis of the mathematical model of the studied processes. In chronological terms, the model is primary, and the methods (deterministic or statistical) underlying the used IMS are secondary, as they are developed with respect to the chosen model.

Analysis of works by domestic and foreign authors has shown that the use of statistical methods for processing information signals is more justified compared to deterministic methods since many physical processes occurring in the diagnosed nodes of electrical equipment are random in nature.

Based on the analysis of a whole range of scientific works, it can be argued that when creating TDIIS for industrial equipment diagnostics, it is necessary to use statistical methods for processing information signals, i.e., such IMS should be based on a probabilistic mathematical model. Additionally, when creating such IMS, it is necessary to proceed from the fact that diagnosing the technical condition of nodes of industrial equipment is a complex task, which includes:

- creation mathematical probabilistic models of information signals (processes) occurring in the diagnosed nodes;
- theoretical justification of possible diagnostic features;
- development of software and hardware for diagnostics IMS;
- selection of diagnostic spaces and formation of training sets in these spaces;
- construction of decision rules for diagnosis and classification of specific types of defects in nodes of industrial equipment.

Based on the mathematical models of the studied processes, a IMS for statistical diagnostics of industrial equipment is constructed, which should include:

- basic software and hardware;
- auxiliary means of support and management of the operation of basic means.

The main software and hardware tools included in the composition of IMS for statistical diagnostics are intended for measurement, processing, analysis, and display of information about the condition of the diagnostic object. They are sufficiently detailed in numerous literary sources, for example [15, 122].

The choice of means of support and management of IMS, including information-reference, algorithmic (typical and special programs for computers), and metrological support, is carried out taking into account the diagnostic objects, the a priori information about which is provided by their mathematical model.

Algorithmic support determines the order of operation of the basic software and hardware tools.

Metrological support includes the methodology of periodic verification of the main metrological characteristics of IMS. The determining moment for the development of the specified methodology is the choice of the class of test signals, which is determined by the mathematical model of the studied processes.

The development of information-reference support can conditionally be reduced to three main stages:

- determining quantitative estimates of the boundaries of diagnostic features corresponding to various technical conditions of diagnosed nodes of industrial equipment;
- selection of diagnostic spaces and formation of training sets corresponding to specific defects in diagnosed nodes;
- construction of decision rules for diagnosis and classification of specific types of defects in nodes of diagnosed equipment.

The solution to the entire complex of tasks related to the construction of IMS for industrial equipment diagnostics is inseparably linked to the chosen mathematical model of physical processes, the realizations of which are measured and studied in the process of diagnostics. Therefore, the development and study of mathematical models of these processes are primary when creating any technical diagnostic IMS.

Both in the case of deterministic and statistical models, the mathematical model of the diagnostic object can be represented as follows:

$$\Xi \stackrel{\Delta}{=} \psi[\boldsymbol{H}, \Phi, t]; t \in \boldsymbol{T}, \tag{1.1}$$

where $\boldsymbol{\Phi} = \{\phi_1(\tau, t), \phi_2(\tau, t), \ldots, \phi_n(\tau, t)\}$ is an n-dimensional vector characterizing the internal state of the diagnosed object; $\boldsymbol{H} = \{\eta_1(t), \eta_2(t), \ldots, \eta_n(t)\}$ is a vector of input stimuli; $\boldsymbol{\Xi} = \left\{\xi_1(t), \xi_2(t), \ldots, \xi_j(t)\right\}$ is a j-dimensional vector describing the response of the diagnosed object to the input stimuli $\boldsymbol{H}$.

The functions $\phi_k(\tau, t)$, $k = \overline{1, n}$, $\tau \in (-\infty, \infty)$ forming the vector $\boldsymbol{\Phi}$ are usually assumed to be non-random and dependent on a known set of numerical parameters. Thus, it is assumed that at each moment in time $t \in \boldsymbol{T}$, the internal state of the diagnosed object is fully described by some set of numerical parameters. The appearance of certain defects in the diagnosed object causes changes in its internal state, i.e., changes in the parameters included in the functions, forming the vector $\boldsymbol{\Phi}$. This, in turn, leads to changes in the nature of the response of the diagnosed object, which is described by the vector $\boldsymbol{\Xi}$, to the same input stimulus defined by the vector $\boldsymbol{H}$. Thus, by measuring and analyzing the components of the vector $\boldsymbol{\Xi}$, it is possible to conduct the diagnosis of the investigated object.

The theoretical investigation of model (1.1) allows for the justification of certain diagnostic features of the technical condition of the diagnosed equipment. By diagnostic features, we mean parameters characterizing model (1.1) that allow diagnosing the technical condition of the investigated object with a certain reliability. These diagnostic parameters can be encapsulated in vectors $\mathbf{\Phi}$ or $\mathbf{H}$ and are indirectly measured through the vector $\boldsymbol{\Xi}$ related to them by expression (1.1). In fact, the diagnostic process boils down to measuring parameters characterizing vectors $\mathbf{\Phi}$ and $\mathbf{H}$ through the vector $\boldsymbol{\Xi}$, followed by decision-making based on pre-learning data and constructed decision rules.

In addition to the theoretical justification of diagnostic features based on the analysis of model (1.1), the question arises as to which of the components included in (1.1) to measure. For instance, it is quite challenging to measure the components of the generating process characterized by the vector **H**. Similarly, it is difficult to measure the parameters characterizing the internal state of the system and included in the vector **Φ**. The most accessible in terms of conducting diagnostic measurements are the components of the vector Ξ. By components of this vector, we mean the realizations of diagnostic signals (physical processes) measured by corresponding sensors.

Models of non-random signals are generally simpler than models of noise and are often represented as a sum or product of elementary (mostly harmonic) functions, so we won't delve into them in detail. Here are some examples:

(1) $x(t) = u \cos(\omega t + \phi)$—simple harmonic oscillation defined over one of the continuous time intervals $t \in (-\infty, \infty)$ or $t \in [0, \infty)$ or $t \in [t_0, t_1]$. Here u is a positive numerical parameter called amplitude, ϕ is a numerical parameter called initial phase, and ω is the angular frequency $\omega = 2\pi f$.
(2) $x(t) = u \cdot [1 + mf(t)] \cdot \cos \omega_H t$—amplitude-modulated oscillation, $0 \le m \le 1$, m is the modulation depth, ω_H is the carrier frequency, $|f(t)| \le 1$, ω_M is the modulation frequency, usually $\omega_M << \omega_H$.
(3) $x(t) = e^{-\alpha t} \cos(\omega t + \phi)$—decaying rhythmic process (non-periodic).

Models of random diagnostic signals. Addressing questions about building models of signals measured in the investigated units based on random processes, let's briefly consider the following example.

The determining factor in constructing a mathematical model of a technical object as a diagnostic object in the form of a multidimensional vector (1.1) is the choice of the class of processes characterizing the action vector **H**, as well as the determination of the vector of eigenfunctions **Φ** characterizing the diagnosed object itself. In this case, one starts from the characteristics and properties of physical processes accompanying the operation of a specific diagnosed unit. Many models of such processes are based on the fact that at a measurement point at any fixed moment in time, the studied process can be regarded as an algebraic sum of a large number of random variables characterizing the energy coming from various random sources located in space at some distance from this point. Such models can be used, for example, in describing noise and vibration processes and fields, acoustic emission processes occurring in a solid body under mechanical loading, or processes of occurrence of electrical discharges in insulation under high voltage.

In the simplest one-dimensional case, the task of constructing a mathematical model of the investigated physical process can be reduced to the classical problem of analyzing the response of a system with an impulse response function $h(t)$ when subjected to inputs of a random process $N(t)$, which is commonly referred to as the excitation process in systems theory (see Fig. 1.3). When solving such problems, it is assumed that the characteristics of the excitation process are known. Analytically, the solution to the problem under consideration can be expressed in the form of a

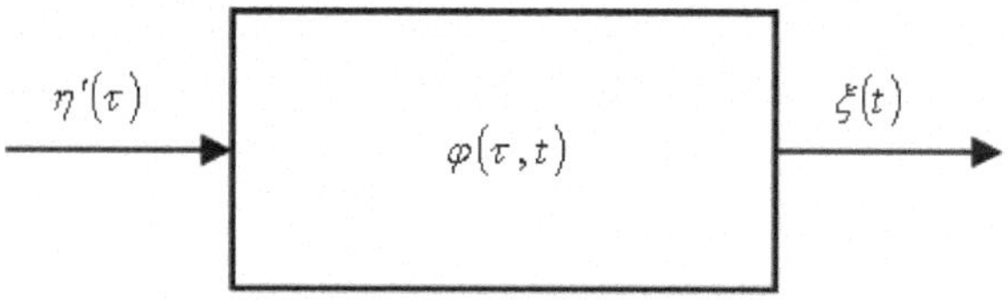

Fig. 1.3 Construction of a mathematical model forming a filter by the method

stochastic integral representation given by:

$$\xi(t) = \int_0^\infty \phi(\tau,\ t) d\eta(\tau);\ t \in T, \tag{1.2}$$

where $\phi(\tau,\ t) \in L_2(-\infty,\ \infty)$ is the response at each fixed time $t \in (-\infty,\ \infty)$.

Expression (1.2) represents a one-dimensional mathematical model of the process measured during the diagnosis of a certain component, formulated as a stochastic integral representation.

In this representation, as noted above, $\eta'(\tau)$ is the generating process, typically chosen to be a process for which the complete probabilistic characteristics are known, namely the distribution function $F(x)$ or the characteristic function $f(t,\ u)$, $\phi(\tau,\ t)$ is the impulse response function, characterizing the diagnosed object (in mathematics, this function is an integrable function τ with a finite square, meaning it characterizes a system with finite power or a physically realizable system) $\xi(t)$ is the measured random process.

Unlike the deterministic models discussed earlier, obtaining estimates of random diagnostic signals requires the use of the mathematical apparatus of probability theory and mathematical statistics. In this case, various numerical and functional characteristics of the measured random processes can serve as the main parameters and characteristics, namely:

- estimates of initial and central moments (mean, variance, skewness and kurtosis coefficients, and in some cases estimates of central moments above the fourth order);
- estimates of spectral correlation functions;
- estimates of probability density functions or distributions.

Additionally, before conducting statistical processing of the measured diagnostic signals, it is necessary to perform their preliminary verification for stationarity, ergodicity, and to establish to which type of probability distribution the investigated diagnostic signals belong. Various types of statistical criteria are used for this purpose.

The main issues related to the use of probabilistic methods and models for studying random diagnostic signals are discussed in subsequent sections of this book.

Based on the measured parameters of diagnostic signals, it is possible not only to assess the technical condition of the diagnosed object but also to predict possible failures during the further operation of the investigated object.

Let's briefly consider these issues.

1.7 Methods of Forecasting Failures of EE

One of the most important characteristics of any technical object is its technical resource, i.e., the operation time of this object from the beginning of its operation or after repair until it reaches its limit state, determined by normative-technical documentation. During the direct operation of the object during the period determined by the technical resource, the possibility of diagnosing (functional or test) the state of this object (product) is envisaged. The measurements obtained as a result of evaluating the diagnostic features (parameters) of the monitored signals allow conclusions to be drawn about the technical condition of the diagnosed object at a fixed point in time. Having data on the dynamics of changes in the estimates of diagnostic features over some period of time of the operation of the studied object, it is possible to forecast the possible changes in these parameters in the future. This approach allows forecasting the expected number of product failures in the future or making a forecast of the residual resource of the studied product.

1.7.1 Relationship Between Diagnostic and Forecasting Models in Reliability Assurance Tasks

As rightly noted in [131–133], the task of predicting the technical condition of an object during its operation is closely related to the tasks of technical diagnostics. Assessing the area covered by technical diagnostics, three types of tasks for determining the technical condition of objects are considered.

The first type of task is the actual diagnostic task, i.e., determining the technical condition of the investigated object.

The second type of task is devoted to considering issues of *forecasting* or predicting (with some probability) the possible technical condition of the investigated object at a future moment (interval) in time.

The third type includes tasks of a so-called technical *genesis*, the main goal of which is to determine the technical condition of the investigated object at some point in the past.

Below, we will only focus on the consideration of diagnostic and forecasting tasks for the technical condition of objects since solving the technical genesis task is somewhat specific. For example, the result of solving the technical genesis task of an object is related to determining its state at a past moment in the operation time. Such tasks arise when examining (expertise) accidents of an object to identify possible root causes that led to the accident.

Solving the forecasting task is usually associated with determining the technical resource of an object or predicting its failure by a certain point in time in the future. It is this latter circumstance that is the determining factor for solving the task of technical maintenance or repair of the studied object based on its actual condition (rather than on scheduled dates).

As with solving the task of diagnosing a technical object, as well as forecasting its condition in the future, the primary role is assigned to building a mathematical model. In this case, the mathematical models used to solve these tasks for the studied objects are quite different. Thus, solving the task of diagnosing is based on a mathematical model (deterministic or statistical) characterizing either the object itself or some physical process (diagnostic signal) accompanying its operation. In the case of solving the forecasting task, the researcher usually deals with a mathematical model describing the evolution of observed parameters over a certain time interval.

To build mathematical models for diagnosing and forecasting the condition of a certain object, information about its technical condition at various moments (periods) of time during its operation is used: t_0—the present moment in time; T_1—the past period of time; T_2—the future period of time. Using this approach in [75–77], four main types of mathematical models are constructed and considered, providing solutions to both diagnostic and forecasting tasks.

Individual instances of diagnostic or forecasting objects will be denoted by the symbol s_j; s_0—one specific studied instance of the object; S_1– a group of k instances of the object undergoing examination to obtain prior information about their technical condition; S_2—a group of m instances of the object, the technical condition of which (present or future) is determined based on the obtained prior information about the technical condition of instance s_0 or the group of instances S_1.

Prior information required for diagnosis or prognosis consists of various data on the technical condition of one s_0 or a group of S_1 object instances. This data is obtained either as a result of a single measurement at time t_0 or multiple measurements over a period of time T_1 diagnosing specific object instances.

Denoting the data on the technical condition by the symbol e, let's highlight four volumes of received prior information:

- $e(S_1, T_1)$—group of S_1 object instances diagnosed multiple times over a period of time T_1;
- $e(s_0, T_1)$—instance s_0 of the object diagnosed multiple times over a period T_1;
- $e(S_1, t_0)$—group of S_1 object instances diagnosed once at time t_0;
- $e(s_0, t_0)$—instance s_0 of the object diagnosed once at time t_0.

The first situation corresponds to obtaining the largest, and the fourth to the smallest volume of prior information. The second and third situations occupy an intermediate position in terms of information volume, but these situations are not comparable to each other.

Similarly, four types of processing prior information can be distinguished to determine:

- $e(S_2, T_2)$—the technical condition of a group of S_2 object instances in the future period T_2;
- $e(s_0, T_2)$—the technical condition of one instance s_0 of the object in the future period T_2;
- $e(S_2, t_0)$—the technical condition of a group of S_2 object instances in the current period t_0;
- $e(s_0, t_0)$—the technical condition of one instance s_0 of the object in the current period t_0.

As noted in [13], of the specified types of processing prior information, the first two correspond to forecasting tasks, and the last two correspond to diagnostic tasks.

When setting a forecasting task, it is assumed that a set of parameters $\{\xi_l, l = 1, 2, \ldots, n\}$, characterizing the studied object (vector $\overline{\xi}$), is chosen for measurement, which is assumed to allow (using certain forecasting means implementing the forecasting algorithm) to predict the future technical condition of the object e. These parameters are called forecasting. It should be noted that the quantitative estimates of these parameters depend significantly on a vector $\overline{X}$ characterizing irreversible degradation processes of the physical–chemical properties of the diagnosed nodes of the studied object.

During measurements in real conditions, disturbances are imposed on the values of forecasting parameters described by the components of the vector $\overline{\xi}$, due to the presence of electromagnetic, mechanical, thermal, and other fields. In addition, it is assumed that the measured values of the vector $\overline{\xi}$ components depend significantly on the methods applied and the class of accuracy of the measuring means, which are characterized by the vector $\overline{Z}$). The forecasting results may be affected by forecasting errors (vector $\overline{W}$). Thus, the future technical condition of the object e depends on several random vector arguments:

$$e \stackrel{\Delta}{=} f(\overline{X}, \overline{Y}, \overline{Z}, \overline{W}). \tag{1.3}$$

Thus, as a result of measurement, instead of the true values of the vector $\overline{\xi}$, we obtain the vector $\overline{\xi}^{\bullet} \neq \overline{\xi}$, and expression (1.3) essentially represents a forecasting process model. The probabilistic nature of this model is determined by the fact that the arguments $\overline{X}$, $\overline{Y}$, $\overline{Z}$ and $\overline{W}$ are inherently random functions. Obtaining the dependence (1.3) in explicit analytical form for complex objects is practically impossible. Therefore, various simplification techniques are used for both the model itself and its processing procedures. Among these techniques is the decomposition of the overall forecasting task into two independent tasks—the task of measuring forecasting parameters when working with a model of the form

$$\overline{\xi_l^{\bullet}} \stackrel{\Delta}{=} \phi_l(\overline{X}, \overline{Y}, \ldots, \overline{Z}); \quad l = 1, 2, \ldots, n, \tag{1.4}$$

and the task of obtaining a forecast (forecasting result) based on a model of the form

$$e \stackrel{\Delta}{=} \psi\left(\overline{\xi}^{\bullet}, \overline{W}\right). \tag{1.5}$$

The simplest would be an explicit analytical model of the form

$$e \stackrel{\Delta}{=} f_x\left(\overline{X}\right). \tag{1.6}$$

In the last expression, there is no dependence of the future technical state of the object on random disturbances and measurement errors. To reduce the dependence of the final forecasting results on random functions defined by vectors $\overline{Y}$, $\overline{Z}$ and $\overline{W}$,, various smoothing methods are usually applied.

Below, as an example, let's consider the use of statistical spline functions for establishing the relationship as described in (1.6).

1.7.2 Using Statistical Spline Function Methods for Failure Forecasting of EE

The main idea of forecasting failures in electrical equipment using statistical spline functions is that certain numerical characteristics of physical processes occurring in the nodes of the equipment change when faults and defects occur, allowing them to be detected and identified. Regular monitoring of the gradual change of these parameters over time can provide important information about defect development trends and enable forecasting the potential occurrence of failure [4].

Let's briefly consider the forecasting of failures of individual components of electrical equipment using statistical spline functions. We will only consider so-called gradual failures caused by changes in diagnostic object parameters due to material aging, accumulation of micro-defects, etc. Random failures caused by unpredictable factors will not be considered.

To diagnose a specific component of electrical equipment, it is necessary to select a certain numerical diagnostic parameter (or a set of diagnostic parameters) and a corresponding criterion according to which a decision is made about the technical state of this component. These issues have been extensively discussed in a number of works, for example [75–77]. To forecast the potential failure of the selected component, it is additionally necessary to have statistics of diagnostic parameter values over a certain period of time, which we will call the observation interval [5, 59, 61, 63, 69].

Let's assume that the studied equipment (i.e., the diagnostic object) operates under identical conditions throughout the entire observation interval $[a, \ b]$, namely, the operating mode does not significantly change, no repair work is carried out on the examined components, and planned maintenance activities do not substantially change the operating conditions of these components. In this case, it can be assumed that the diagnostic parameter changes over time t according to some functional dependency $y = f(t, \ A)$, where A is a deterministic vector of unknown real parameters linearly included in $f(t, \ A)$.

As a result of measurements conducted during the observation interval, a sequence of values y_i, $i = \overline{1, N}$ is obtained, corresponding to the values of the argument $t_i \in [a, b]$, $i = \overline{1, N}$. Let's assume that the observations are distributed according to a Gaussian law with equal variances ($Dy_i = \sigma^2$, $i = \overline{1, N}$, D), and they are uncorrelated. This follows from the fact that measurements are made by the same measuring instrument with the same accuracy at independent time points.

Let's represent the sought functional dependency as:

$$My_i = \sum_{k=0}^{r} x_{ik} a_k = f(t_i, \mathbf{A}); i = \overline{1, N}, \tag{1.7}$$

where M is the operator of mathematical expectation; $\mathbf{A} = (a_0, a_1, \ldots, a_r)$ are unknown parameters; $\mathbf{X} = (x_{ik})$, $i = \overline{1, N}$, $k = \overline{0, r}$ are the so-called design matrices, consisting of predefined elements that functionally depend on t_i (nonlinear dependence is not excluded).

If we consider $Y = (y_i)$, $i = \overline{1, N}$,and $\mathbf{A}$ as column matrices, then expression (1.7) can be rewritten in matrix form: $M\mathbf{Y} = \mathbf{XA}$. Thus, the task of reconstructing the dependency $y = f(t, \mathbf{A})$ is reduced to finding statistical estimates of the elements of the vector $\mathbf{A}$ based on the observations y_i, $i = \overline{1, N}$. Arbitrary selection of elements of the design matrix is allowed, which enables seeking an optimal solution in some sense.

It should be noted that the form of the design matrix determines the class of recoverable functions. Let's consider solving this problem in the class of spline functions [4], and statistical estimates of unknown parameters will be constructed using the method of least squares [104].

Typically, a spline function is "glued" from pieces of various functions according to a certain scheme. Polynomial splines consist of segments of different polynomials in such a way as to ensure the necessary smoothness of the resulting spline. When interpolating using splines on a segment of the $[a, b]$, a grid is defined.

$$\Delta_r = \left\{ t_j \right\}_{j=0}^{r}, a = t_0 < t_1 < \ldots < t_r = b. \tag{1.8}$$

The function $S_m(t) = S_{m,k}(t, \Delta_r)$ is called a polynomial spline of degree m with k $1 \le k \le m$ nodes (1.8) if:

(a) $S_m(t) \in \mathbf{P}_m$ at $t \in \left[t_j, t_{j+1}\right] j = \overline{0, r-1}$,
(b) $S_m(t) \in \mathbf{C}^{m-k}[a, b]$,

where $\mathbf{P}_m$—the set of valid polynomials of degree at most m, $\mathbf{C}^k[a, b]$—the set of continuous functions $[a, b]$ with continuous k-derivative.

In the conditions of the formulated problem on the interval $[a, b]$ with a given r, we need to find such a grid Δ_r that the spline constructed on it provides the optimal statistical estimate of the vector $\mathbf{A}$ according to the method of least squares. The elements of this vector will be considered as ordinates at the nodes of the interpolating spline of the 1st degree. According to the method of least squares, it is necessary to minimize the expression:

$$(\boldsymbol{Y}-\boldsymbol{XA})*(\boldsymbol{Y}-\boldsymbol{XA})=\sum_{i=1}^{N}\left(y_i-\sum_{j=r}^{r}a_jx_{ij}\right)^2. \tag{1.9}$$

The vector **A** is sought according to the expression.

$$\boldsymbol{A}=(\boldsymbol{X}*\boldsymbol{X})^{-1}\boldsymbol{X}*\boldsymbol{Y}. \tag{1.10}$$

The confidence interval with a probability β for the found estimates **A** is calculated using the formula

$$I_\beta^{(j)}=\left[a_j\mp\gamma_\beta\sqrt{\left\{(\boldsymbol{X}*\boldsymbol{X})^{-1}\right\}_{jj}\frac{d}{N-r-1}}\right], \tag{1.11}$$

where γ_β is a value that satisfies the relation $\mathrm{P}\left\{|t_{N-r-1}|\leq\gamma_\beta\right\}=1-\beta$ if the random variable t_{N-r-1} follows a Student's $N-r-1$-distribution with degrees of freedom; d is the sum of squares of deviations of observations from the values of the obtained spline at the corresponding points.

Relation (1.11) allows constructing confidence intervals at each spline node, and over thentire interval—a confidence corridor at some specified confidence level α.

To make a forecast using a statistical spline, an additional node is introduced to the set of spline nodes, whose abscissa corresponds to the forecast interval. By means of computational iteration, a grid is selected that satisfies Eq. (1.10) and simultaneously minimizes (1.9) over the set of possible non-uniform grids and the width of the confidence interval at the forecast node.

As a result, we obtain the expected value and a confidence interval for the selected diagnostic parameter at the end of the forecast interval.

The boundaries of the confidence corridor are formed by linearly interpolating the upper and lower bounds of the confidence intervals at all nodes of the obtained spline, including the forecast node.

Examples of practical application of the method of statistical spline functions for determining the possible number of failures of the investigated nodes of electrical equipment are discussed in Chap. 8 of this book.

References

1. Lin, J.G., Xie, F.C., Wei, B.C.: Statistical diagnostics for skew-t-normal nonlinear models. Commun. Stat. Simul. Comput. **38**(10), 2096–2110 (2009)
2. Pan, J.X., Fang, K.T.: Growth Curve Models and Statistical Diagnostics. Springer Science & Business Media (2002)

3. Garay, A.M., Lachos, V.H., Labra, F.V., Ortega, E.M.: Statistical diagnostics for nonlinear regression models based on scale mixtures of skew-normal distributions. J. Stat. Comput. Simul. **84**(8), 1761–1778 (2014)
4. Babak, S., Babak, V., Zaporozhets, A., Sverdlova, A.: Method of statistical spline functions for solving problems of data approximation and prediction of objects state. In: Proceedings of the Second International Workshop on Computer Modeling and Intelligent Systems (CMIS-2019), Zaporizhzhia, Ukraine, April, pp. 15–19 (2019, April)
5. Babak, V., Eremenko, V., Zaporozhets, A.: Research of diagnostic parameters of composite materials using Johnson distribution. Int. J. Comput. **18**(4), 483–494 (2019)
6. Myslovycn, M.V.: Scientific Foundations of the Theory of Information-Measuring Systems for Statistical Diagnostics of Electrical Equipment (1995)
7. Zaporozhets, A.A., Eremenko, V.S., Serhiienko, R.V., Ivanov, S.A.: Development of an intelligent system for diagnosing the technical condition of the heat power equipment. In: 2018 IEEE 13th International Scientific and Technical Conference on Computer Sciences and Information Technologies (CSIT), vol. 1, pp. 48–51. IEEE (2018, September)
8. Zaporozhets, A., Babak, V., Sverdlova, A., Isaienko, V., Babikova, K.: Development of a system for diagnosing heat power equipment based on IEEE 802.11 s. In: Systems, Decision and Control in Energy II, pp. 141–151. Springer International Publishing, Cham (2021)
9. Babak, V., Mokiychuk, V., Zaporozhets, A., Redko, O.: Improving the efficiency of fuel combustion with regard to the uncertainty of measuring oxygen concentration. Eastern Eur. J. Enterp. Technol. **6**(8), 84 (2016)
10. Zaporozhets, A.O., Redko, O.O., Babak, V.P., Eremenko, V.S., Mokiychuk, V.M.: Method of indirect measurement of oxygen concentration in the air. Naukovyi Visnyk Natsionalnoho Hirnychoho Universytetu **5**, 105–114 (2018)
11. Zaporozhets, A.O.: Methods and means for the control of the fuel combustion process. In: Control of Fuel Combustion in Boilers, pp. 1–33 (2020)
12. Zaporozhets, A.O.: Experimental research of a computer system for the control of the fuel combustion process. In: Control Fuel Combustion in Boilers, pp. 89–123 (2020)
13. Babak, V., Scherbak, L., Kuts, Y., Zaporozhets, A.: Information and measurement technologies for solving problems of energy informatics. In: The 1st International Workshop on Information Technologies: Theoretical and Applied Problems 2021. CEUR Workshop Proceedings, vol. 3039, pp. 24–31 (2021, December)
14. Sverdlova, A., Zaporozhets, A.: Predicting anomaly conditions of energy equipment using neural networks. In: E3S Web of Conferences, vol. 280, p. 09005 (2021, June)
15. Babak, V.P., Babak, S.V., Myslovych, M.V., Zaporozhets, A.O., Zvaritch, V.M.: Simulation and software for diagnostic systems. In: Diagnostic Systems for Energy Equipments, pp. 71–90 (2020)
16. Zaporozhets, A., Babak, V., Starenkiy, V., Gryb, O., Karpaliuk, I., Luka, O.: Instruments for identification of corona discharge presence by spectral characteristics of acoustic radiation. In: Detection of Corona Discharge in Electric Networks, pp. 113–136. Springer Nature Switzerland, Cham (2023)
17. Bohachev, I., Babak, V., Zaporozhets, A.: Novel small-aperture transducers based on magnetostrictive effect for diagnostic systems. Tech. Electrodyn. **3**, 69–78 (2022)
18. Babak, V.P., Babak, S.V., Myslovych, M.V., Zaporozhets, A.O., Zvaritch, V.M., Babak, V.P., Zvaritch, V.M., et al.: Technical provision of diagnostic systems. In: Diagnostic Systems for Energy Equipments, pp. 91–133 (2020)
19. Kondratenko, I., Vasetsky, Y., Zaporozhets, A.: Electromagnetic systems with iron-free inductors for induction heating of moving strip in transverse magnetic field. In: Interactions Between Electromagnetic Field and Moving Conducting Strip, pp. 91–119. Springer Nature Switzerland, Cham (2023)
20. Zaporozhets, A., Babak, V., Popov, O., Scherbak, L., Kuts, Y.: Methodological support of air pollution monitoring system. In: Systems, Decision and Control in Energy V, pp. 713–737. Springer Nature Switzerland, Cham (2023)

21. Zabulonov, Y., Popov, O., Skurativskyi, S., Hromova, I., Zaporozhets, A.: Remote gamma ray mapping of ground surface: mathematical issues. In: Systems, Decision and Control in Energy IV: Volume II. Nuclear and Environmental Safety, pp. 173–184. Springer Nature Switzerland, Cham (2023)
22. Babak, V.P., Babak, S.V., Eremenko, V.S., Kuts, Y.V., Myslovych, M.V., Scherbak, L.M., Zaporozhets, A.O., et al.: Models and measures for atmospheric pollution monitoring. In: Models and Measures in Measurements and Monitoring, pp. 227–266 (2021)
23. Babak, V.P., Babak, S.V., Eremenko, V.S., Kuts, Y.V., Myslovych, M.V., Scherbak, L.M., Zaporozhets, A.O., et al.: Monitoring the air pollution with UAVs. In: Models and Measures in Measurements and Monitoring, pp. 191–225 (2021)
24. Gulevets, D., Zaporozhets, A., Isaienko, V., Babikova, K.: Ecological and energy analysis of the green areas and the surface layer of atmospheric air in the districts of the Kyiv City. In: Environmental Control for Ensuring Cities Safety, pp. 73–91 (2021)
25. Zaporozhets, A.: Overview of quadrocopters for energy and ecological monitoring. In: Systems, Decision and Control in Energy I, pp. 15–36. Springer International Publishing, Cham (2020)
26. Zaporozhets, A., Babak, V., Sverdlova, A., Scherbak, L., Kuts, Y.: Review of the state of air pollution by energy objects in Ukraine. System Research in Energy **2**(71), 42–52 (2022)
27. Khomenko, O., Tsendjav, L., Kononenko, M., Janchiv, B.: Nuclear-and-fuel power industry of Ukraine: production, science, education. In: Mining of Mineral Deposits (2017)
28. Sala, D., Bashynska, I., Pavlova, O., Pavlov, K., Chorna, N., Chornyi, R.: Investment and innovation activity of renewable energy sources in the electric power industry in the south-eastern region of Ukraine. Energies **16**(5), 2363 (2023)
29. Babenko, V.A., Jenkovszky, L.L., Pavlovych, V.N.: Nuclear power industry: tendencies in the world and Ukraine. Phys. Part. Nucl. **38**, 795–826 (2007)
30. Vivcharenko, O.: Development of coal industry of Ukraine in the context of contemporary challenges. In: Geomechanical Processes During Underground Mining, pp. 1–5 (2012)
31. Chernyavskyy, M., Dunayevska, N., Novoseltsev, O., Eutukhova, T.: Comparative assessment of electricity supply costs from different sources in baseload and regulated modes. Syst. Res. Energy **2**(77), 4–15 (2024)
32. Denysov, V., Eutukhova, T.: Dynamic models for developing reference scenarios of energy system in the low-carbon transition. Syst. Res. Energy **1**(76), 17–26 (2024)
33. Stanytsina, V., Nechaieva, T., Trokhaniak, V., Horskyi, V., Teslenko, O.: Electricity and heat supply technologies for increasing the energy independence of certain territorial communities. Syst. Res. Energy **4**(75), 32–44 (2023)
34. Kostenko, G., Zaporozhets, A.: Enhancing of the power system resilience through the application of micro power systems (microgrid) with renewable distributed generation. Syst. Res. Energy **3**(74), 25–38 (2023)
35. Tsagkaris, C., Matiashova, L., Isayeva, A.: The impacts of putting the Zaporizhzhia nuclear power plant in the line of fire. Public Health Chall. **1**(4), e26 (2022)
36. Molhem, A.: Possible accidents scenarios in Zaporizhzhia nuclear power plant. Open Access Libr. J. **10**(2), 1–19 (2023)
37. Roecker, S.: Nuclear power plants under attack: the legacy of Zaporizhzhia. Arms Control Today **53**(3), 6–10 (2023)
38. Davis, A.: Playing with fire': implications of Russia's occupation of Chornobyl and Zaporizhzhia nuclear power plants for the global nuclear order. St Antony's Int. Rev. **18**(1), 137–149 (2023)
39. Kurando, M.: Nuclear security in conflict zones: the dangerous case of Zaporizhzhia. Int. J. Nucl. Secur. **8**(2), 10 (2023)
40. Abdullayev, A., Kulish, G., Slyeptsov, O., Slyeptsov, S., Aleshin, Y., Sparrow, S., Latorre, R., et al.: Westinghouse fuel assemblies performance after operation in South-Ukraine NPP mixed core (2013)
41. Yastrebenetsky, M., Rozen, Y., Klevtsov, A., Gashev, M.: Strategy of NPP I&C systems' modernization in Ukraine. In: Nuclear and Radiation Safety (2012)

42. Becker, O., Lorenz, P., Wallner, A.: Critical review of EU nuclear stress tests in Bulgaria, Hungary, Romania and Ukraine (2012)
43. Matiichuk, L.: Structure and functioning system of the fuel and energy complex of Ukraine. Innov. Sustain. **№ 4**, 109–118 (2022)
44. Tulchinskiy, R., Butko, M., Tulchynska, S., Khudolei, V.: Ukraine's electricity system striving for European energy integration. Polityka Energetyczna **23**, 93–110 (2020)
45. Kyrylenko, O.V., Pavlovsky, V.V., Steliuk, A.O., Lenga, O.V., Vyshnevskyi, M.V.: Simulation of the normal and emergency operation of interconnected power system of Ukraine for frequency stability study. In: Technical Electrodynamics (2017)
46. Kuznietsova, K.O.: Preconditions and perspectives of Ukrainian energy market development. Econ. Proces. Manage. **1** (2014)
47. Denysov, V.A.: Evaluation of the parameters of optimized coating of load graphs in the modes of synchronous operation of the power systems of Ukraine and Poland for characteristic days on the example of the actual values of the established power and load graphs. Syst. Res. Energy **1–2**(68–69), 50–57 (2022)
48. Buratynskyi, I., Nechaieva, T.: Modeling of the combined operation of a solar photovoltaic power plant and a system of electric energy storage. Probl. General Energy **3**(62), 30–36 (2020)
49. Kulyk, M., Nechaieva, T., Zgurovets, O., Shulzhenko, S., Maistrenko, N.: Comparative analysis of energy-economic indicators of renewable technologies in market conditions and fixed pricing on the example of the power system of Ukraine. In Systems, Decision and Control in Energy IV: Volume I. Modern Power Systems and Clean Energy, pp. 433–449. Springer Nature Switzerland, Cham (2023)
50. Rudniev, Y., Romanchenko, J.: Comprehensive condition analysis and perspectives of energy development in Ukraine according to the Smart Grid concept. Power Eng. Econ. Tech. Ecol. **1** (2024)
51. Khalatov, AA: Energy sector of Ukraine: modern state and nearest prospects. Visn. Nac. Acad. Nauk Ukr **6**, 53–61 (2016)
52. Chernousenko, O.Yu., Nikulenkova, T.V., Nikulenkov, A.H.: State Assessment of Ukrainian and other countries power equipments. Bull. NTU "KhPI". Ser.: Power Heat Eng. Proces. Equip. **8**(1180), 22–27 (2016). ISSN 2078–774X, https://doi.org/10.20998/2078-774X.2016.08.03
53. Kyrylenko, O.V., Basok, B.I., Baseyev, Y., Blinov, I.V.: Power industry of Ukraine and realities of the global warming. Tech. Electrodyn. **3**, 52–61 (2020)
54. Čepin, M., Čepin, M.: Reliability and performance indicators of power plants. In: Assessment of Power System Reliability: Methods and Applications, pp. 197–214 (2011)
55. Maalla, A., Lin, X.L., Zhuang, C.Y.: Direct current transmission based on reliability indicators system. In: 2020 IEEE 5th Information Technology and Mechatronics Engineering Conference (ITOEC), pp. 947–951. IEEE (2020, June)
56. Heylen, E., Deconinck, G., Van Hertem, D.: Review and classification of reliability indicators for power systems with a high share of renewable energy sources. Renew. Sustain. Energy Rev. **97**, 554–568 (2018)
57. Blinov, I., Zaitsev, I., Parus, E., Bereznychenko, V.: Faults Indicators applying for smart monitoring system for improving reliability electric power distribution. In: Power Systems Research and Operation: Selected Problems II, pp. 235–256. Springer International Publishing, Cham (2022)
58. Kuric, I., Klačková, I., Nikitin, Y.R., Zajačko, I., Císar, M., Tucki, K.: Analysis of diagnostic methods and energy of production systems drives. Processes **9**(5), 843 (2021)
59. Eremenko, V., Zaporozhets, A., Isaienko, V., Babikova, K.: Application of wavelet transform for determining diagnostic signs. In: Proceedings of the 15th International Conference on ICT in Education, Research and Industrial Applications (2019, June)
60. Babak, V.P., Babak, S.V., Eremenko, V.S., Kuts, Y.V., Myslovych, M.V., Scherbak, L.M., Zaporozhets, A.O., et al.: Models and measures for standardless measurements of the composite materials characteristics. In: Models and Measures in Measurements and Monitoring, pp. 157–190 (2021)

61. Zaporozhets, A., Eremenko, V., Babak, V., Isaienko, V., Babikova, K.: Using Hilbert transform in diagnostic of composite materials by impedance method. Periodica polytechnica Electr. Eng. Comput. Sci. **64**(4), 334–342 (2020)
62. Zaporozhets, A.: Analysis of control system of fuel combustion in boilers with oxygen sensor. Periodica Polytechnica Mechan. Eng. **63**(4), 241–248 (2018)
63. Eremenko, V., Babak, V., Zaporozhets, A.: Method of reference signals creating in non-destructive testing based on low-speed impact. Tekhnichna Elektrodynamika **4**, 070 (2021)
64. Baranski, M., Decner, A., Polak, A.: Selected diagnostic methods of electrical machines operating in industrial conditions. IEEE Trans. Dielectr. Electr. Insul. **21**(5), 2047–2054 (2014)
65. Contreras-Valdes, A., Amezquita-Sanchez, J.P., Granados-Lieberman, D., Valtierra-Rodriguez, M.: Predictive data mining techniques for fault diagnosis of electric equipment: A review. Appl. Sci. **10**(3), 950 (2020)
66. Babak, V., Zaporozhets, A., Zvaritch, V., Scherbak, L., Myslovych, M., Kuts, Y.: Models and measures in theory and practice of manufacturing processes. IFAC-PapersOnLine **55**(10), 1956–1961 (2022)
67. Zaporozhets, A., Eremenko, V., Isaenko, V., Babikova, K.: Approach for creating reference signals for detecting defects in diagnosing of composite materials. In: Conference on Computer Science and Information Technologies, pp. 154–172. Springer International Publishing, Cham (2019, September)
68. Babak, V., Bohachev, I., Zaporozhets, A., Khaidurov, V., Havrysh, V., Kalinichenko, A.: Some features of modeling ultrasound propagation in non-destructive control of metal structures based on the magnetostrictive effect. Electronics **12**(3), 477 (2023)
69. Farooq, U., Myler, P.: Flat nose low velocity drop-weight impact response of carbon fibre composites using non-destructive damage detection techniques. Open Eng. **5**(1) (2015)
70. Dwivedi, S.K., Vishwakarma, M., Soni, A.: Advances and researches on non destructive testing: a review. Mater. Today Proc. **5**(2), 3690–3698 (2018)
71. Blitz, J.: Electrical and Magnetic Methods of Non-destructive Testing, vol. 3. Springer Science & Business Media (1997)
72. Antonino-Daviu, J., Quijano-López, A., Climente-Alarcon, V., Garín-Abellán, C.: Reliable detection of rotor winding asymmetries in wound rotor induction motors via integral current analysis. IEEE Trans. Ind. Appl. **53**(3), 2040–2048 (2017)
73. Siddique, A., Yadava, G.S., Singh, B.: A review of stator fault monitoring techniques of induction motors. IEEE Trans. Energy Convers. **20**(1), 106–114 (2005)
74. Ferreira, F.J., Baoming, G., de Almeida, A.T.: Reliability and operation of high-efficiency induction motors. In: 2015 IEEE/IAS 51st Industrial & Commercial Power Systems Technical Conference (I&CPS), pp. 1–13. IEEE (2015, May)
75. Babak, V.P., Babak, S.V., Myslovych, M.V., Zaporozhets, A.O., Zvaritch, V.M.: Principles of construction of systems for diagnosing the energy equipment. In: Diagnostic Systems for Energy Equipments, pp. 1–22 (2020)
76. Babak, V.P., Babak, S.V., Eremenko, V.S., Kuts, Y.V., Myslovych, M.V., Scherbak, L.M., Zaporozhets, A.O., et al.: Problems and features of measurements. In: Models and Measures in Measurements and Monitoring, pp. 1–31 (2021)
77. Babak, V.P., Babak, S.V., Eremenko, V.S., Kuts, Y.V., Myslovych, M.V., Scherbak, L.M., Zaporozhets, A.O., et al.: Models and measures for the diagnosis of electric power equipment. In: Models and Measures in Measurements and Monitoring, pp. 99–126 (2021)
78. Babak, V., Zaporozhets, A., Kovtun, S., Myslovych, M., Kuts, Y., Scherbak, L.: Information support for identification of the technical state of electric power facilities. In: Systems, Decision and Control in Energy V, pp. 129–153. Springer Nature Switzerland, Cham (2023)
79. Cruz, S.M., Cardoso, A.M.: Stator winding fault diagnosis in three-phase synchronous and asynchronous motors, by the extended Park's vector approach. IEEE Trans. Ind. Appl. **37**(5), 1227–1233 (2001)
80. Casimir, R., Bouteleux, E., Yahoui, H., Clerc, G., Henao, H., Delmotte, C., Faucher, J., et al.: Comparison of modelling methods and of diagnostic of asynchronous motor in case of defects. In: 9th IEEE International Power Electronics Congress, 2004. CIEP 2004, pp. 101–108. IEEE (2004, October)

81. Capolino, G.A., Antonino-Daviu, J.A., Riera-Guasp, M.: Modern diagnostics techniques for electrical machines, power electronics, and drives. IEEE Trans. Industr. Electron. **62**(3), 1738–1745 (2015)
82. Kostenko, G., Zaporozhets, A.: Transition from electric vehicles to energy storage: review on targeted lithium-ion battery diagnostics. Energies **17**(20), 5132 (2024)
83. Bucci, G., Ciancetta, F., Fiorucci, E.: Apparatus for online continuous diagnosis of induction motors based on the SFRA technique. IEEE Trans. Instrum. Meas. **69**(7), 4134–4144 (2019)
84. Venugopal, G., Udayakumar, A.K., Balashanmugham, A., Houran, M.A., Alsaif, F., Elavarasan, R.M., Alsharif, M.H., et al.: Fault identification and classification of asynchronous motor drive using optimization approach with improved reliability. Energies **16**(6), 2660 (2023)
85. Kouakeuo, S.H.N., Solignac, A., Sabariego, R.V., Morel, L., Raulet, M.A., Toutsop, B., Ducharne, B., et al.: Internal characterization of magnetic cores, comparison to finite element simulations: a route for dimensioning and condition monitoring. IEEE Trans. Instrum. Meas. **71**, 1–10 (2022)
86. Gawrylczyk, K.M., Trela, K.: Frequency response modeling of transformer windings utilizing the equivalent parameters of a laminated core. Energies **12**(12), 2371 (2019)
87. Abeywickrama, N., Serdyuk, Y.V., Gubanski, S.M.: High-frequency modeling of power transformers for use in frequency response analysis (FRA). IEEE Trans. Power Delivery **23**(4), 2042–2049 (2008)
88. Babak, V.P., Babak, S.V., Myslovych, M.V., Zaporozhets, A.O., Zvaritch, V.M.: Diagnostic Systems for Energy Equipments, p. 281. Springer International Publishing (2020)
89. Song, L., Wang, H., Chen, P.: Vibration-based intelligent fault diagnosis for roller bearings in low-speed rotating machinery. IEEE Trans. Instrum. Meas. **67**(8), 1887–1899 (2018)
90. Cong, F., Chen, J., Dong, G., Pecht, M.: Vibration model of rolling element bearings in a rotor-bearing system for fault diagnosis. J. Sound Vib. **332**(8), 2081–2097 (2013)
91. Huo, Z., Zhang, Y., Francq, P., Shu, L., Huang, J.: Incipient fault diagnosis of roller bearing using optimized wavelet transform based multi-speed vibration signatures. IEEE Access **5**, 19442–19456 (2017)
92. Junsheng, C., Dejie, Y., Yu, Y.: A fault diagnosis approach for roller bearings based on EMD method and AR model. Mech. Syst. Signal Process. **20**(2), 350–362 (2006)
93. Jha, R.K., Swami, P.D.: Fault diagnosis and severity analysis of rolling bearings using vibration image texture enhancement and multiclass support vector machines. Appl. Acoust. **182**, 108243 (2021)
94. Caesarendra, W., Kosasih, B., Tieu, A.K., Zhu, H., Moodie, C.A., Zhu, Q.: Acoustic emission-based condition monitoring methods: review and application for low speed slew bearing. Mech. Syst. Signal Process. **72**, 134–159 (2016)
95. Grosse, C.U., Ohtsu, M., Aggelis, D.G., Shiotani, T. (eds.): Acoustic Emission Testing: Basics for Research–Applications in Engineering. Springer Nature (2021)
96. Li, C.J., Li, S.Y.: Acoustic emission analysis for bearing condition monitoring. Wear **185**(1–2), 67–74 (1995)
97. Maughan, C.V., Reschovsky, J.M.: Advances in motor and generator rotor health. In: 2010 IEEE International Symposium on Electrical Insulation, pp. 1–4. IEEE (2010, June)
98. Tinta, D., Petrovčič, J., Benko, U., Juričić, Đ, Rakar, A., Žele, M., Stefanovska, A., et al.: A diagnostic system for vacuum cleaner motors. IFAC Proc. Vol. **36**(5), 831–836 (2003)
99. Smit, D., Raghuram, S.: Degradation and Diagnostics of Electrical Machines (2021)
100. Singh, G.K.: Induction machine drive condition monitoring and diagnostic research—a survey. Electric Power Syst. Res. **64**(2), 145–158 (2003)
101. Stone, G.C., Culbert, I., Boulter, E.A., Dhirani, H.: Electrical insulation for rotating machines: design, evaluation, aging, testing, and repair, vol. 83. Wiley (2014)
102. Energy Sector in Ukraine and the World: Forecasts and Challenges. https://www.bdo.ua/en-gb/insights-1/information-materials/2024/energy-sector-in-ukraine-and-the-world-forecasts-and-challenges

103. Prabhu, R.: Rolling bearing diagnostics. In: Proceedings of the Indo-US Symposium on Emerging Trends in Vibration and Noise Engineering, New Delhi, pp. 311–320 (1996)
104. Babak, V.P., Babak, S.V., Eremenko, V.S., Kuts, Y.V., Myslovych, M.V., Scherbak, L.M., Zaporozhets, A.O., et al.: Models of measuring signals and fields. In: Models and Measures in Measurements and Monitoring, pp. 33–59 (2021)
105. Dybała, J., Zimroz, R.: Rolling bearing diagnosing method based on empirical mode decomposition of machine vibration signal. Appl. Acoust. **77**, 195–203 (2014)
106. Unal, M., Sahin, Y., Onat, M., Demetgul, M., Kucuk, H.: Fault diagnosis of rolling bearings using data mining techniques and boosting. J. Dyn. Syst. Meas. Contr. **139**(2), 021003 (2017)
107. Kostek, R., Landowski, B., Muślewski, Ł: Simulation of rolling bearing vibration in diagnostics. J. Vibroeng. **17**(8), 4268–4278 (2015)
108. Babak, V., Zaporozhets, A., Kuts, Y., Myslovych, M., Fryz, M., Scherbak, L.: Models and characteristics of identification of noise stochastic signals of research objects. In: CEUR Workshop Proceedings, vol. 3309, pp. 349–362 (2022, December)
109. Babak, V., Zaporozhets, A., Kuts, Y., Scherbak, L., Eremenko, V.: Application of material measure in measurements: theoretical aspects. In: Systems, Decision and Control in Energy II, pp. 261–269. Springer International Publishing, Cham (2021)
110. Ram, R., Bhatnagar, P.V., Rajesh, M.G., Das, D.: Generation of on-line test signals for nuclear instrumentation for PFBR. In: Proceedings of DAE-BRNS National Symposium on Nuclear Instrumentation-2010 (2010)
111. Bianchi, S., Corsini, A., Rispoli, F., Sheard, A.G.: Detection of aerodynamic noise sources in low-speed axial fans with tip end-plates. Proc. Inst. Mech. Eng. C J. Mech. Eng. Sci. **223**(6), 1379–1392 (2009)
112. Liu, W.Y.: A review on wind turbine noise mechanism and de-noising techniques. Renew. Energy **108**, 311–320 (2017)
113. Janda, M., Vitek, O., Skalka, M.: Noise diagnostic of induction machine. In: The XIX International Conference on Electrical Machines-ICEM 2010, pp. 1–4. IEEE (2010, September)
114. Di Matteo, A., Pirrotta, A.: Path integral method for nonlinear systems under Lévy white noise. ASCE-ASME J. Risk Uncertain. Eng. Syst. Part B: Mechan. Eng. **3**(3), 030905 (2017)
115. Di Paola, M., Alotta, G.: Path integral methods for the probabilistic analysis of nonlinear systems under a white-noise process. ASCE-ASME J. Risk Uncertain. Eng. Syst. Part B: Mechan. Eng. **6**(4), 040801 (2020)
116. Du, Z., Chen, X., Zhang, H., Miao, H., Guo, Y., Yang, B.: Feature identification with compressive measurements for machine fault diagnosis. IEEE Trans. Instrum. Meas. **65**(5), 977–987 (2016)
117. Wu, Y.J., Wang, H.Y.: First-Crossing problem of weakly coupled strongly nonlinear oscillators subject to a weak harmonic excitation and Gaussian white noises. J. Vib. Acoust. **140**(4), 041006 (2018)
118. Myslovitch, M., Ulitko, A.: Use of the Poisson Spectral Function in Acoustic Emission Expert System of the Electrical Engineering Equipment
119. Meyer, H.K.H.: Dielectric Barriers Under Lightning Impulse Stress: Breakdown and Discharge-Dielectric Interaction in Short Non-Uniform Air Gaps (2019)
120. Aliramezani, M., Koch, C.R., Shahbakhti, M.: Modeling, diagnostics, optimization, and control of internal combustion engines via modern machine learning techniques: a review and future directions. Prog. Energy Combust. Sci. **88**, 100967 (2022)
121. Zaporozhets, A.O.: Research of the process of fuel combustion in boilers. In: Control of Fuel Combustion in Boilers, pp. 35–60 (2020)
122. Zaporozhets, A.O.: Hardware and software implementation of modules of the system of the fuel combustion control process. In: Control of Fuel Combustion in Boilers, pp. 61–87 (2020)
123. Zaichenko, S., Erçetin, Ü., Kulish, R., Derevyanko, D., Shalenko, V.: Determination of diagnostic parameters of power plants based on internal combustion engines. Mining Revue **27**(3), 86–92 (2021)
124. Saraiev, O., Saraieva, I., Gritsuk, I.V., Volkov, V., Litikova, O., Belousov, E., Kurnosenko, D., et al.: Automated Diagnostic System for Engine Cylinder-Piston Group (No. 2020–01–2022). SAE Technical Paper (2020)

125. Hrynkiv, A., Aulin, V., Lysenko, S., Rogovskii, I., Titova, L., Zagurskiy, O., Kolosok, I.: Development of a system for determining the informativeness of the diagnosing parameters for a cylinder-piston group in the diesel engine during operation. Eastern-Eur. J. Enterp. Technol. **3**(5–105), 19–29 (2020)
126. Kapłan, M., Klimek, K., Maj, G., Zhuravel, D., Bondar, A., Lemeshchenko-Lagoda, V., Wałowski, G., et al.: Method of evaluation of materials wear of cylinder-piston group of diesel engines in the biodiesel fuel environment. Energies **15**(9), 3416 (2022)
127. Babak, V.P., Babak, S.V., Eremenko, V.S., Kuts, Y.V., Myslovych, M.V., Scherbak, L.M., Zaporozhets, A.O.: Models and measures for measuring random angular quantities. In: Models and Measures in Measurements and Monitoring, pp. 61–97 (2021)
128. Babak, V.P., Babak, S.V., Eremenko, V.S., Kuts, Y.V., Myslovych, M.V., Scherbak, L.M., Zaporozhets, A.O., et al.: Examples of using models and measures on the circle. In: Models and Measures in Measurements and Monitoring, pp. 127–156 (2021)
129. Babak, V., Zaporozhets, A., Kulyk, M., Kuts, Y., Scherbak, L.: Application of discrete Hilbert transform to estimate the characteristics of cyclic signals: information provision. In: Systems, Decision and Control in Energy IV: Volume I. Modern Power Systems and Clean Energy, pp. 93–115. Springer Nature Switzerland, Cham (2023)
130. Babak, V.P., Babak, S.V., Myslovych, M.V., Zaporozhets, A.O., Zvaritch, V.M.: Methods and models for information data analysis. In: Diagnostic Systems for Energy Equipments, pp. 23–70 (2020)
131. Rudnichenko, N., Vychuzhanin, V., Polyvianchuk, A., Mateichyk, V.: Complex technical system condition diagnostics and prediction computerization. In: CEUR Workshop Proceedings. 2608, pp. 42–56. RWTH Aachen University (2020)
132. Zaporozhets, A.O.: Correlation analysis between the components of energy balance and pollutant emissions. Water Air Soil Pollut. **232**, 1–22 (2021)
133. Panda, A., Nahornyi, V., Pandová, I., Harničárová, M., Kušnerová, M., Valíček, J., Kmec, J.: Development of the method for predicting the resource of mechanical systems. Int. J. Adv. Manuf. Technol. **105**, 1563–1571 (2019)

Chapter 2
Linear Random Functions as Models of Diagnostic Signals

Abstract This comprehensive research delves into the mathematical modeling of diagnostic signals in electrical equipment diagnostics using linear random processes (LRPs) and their properties. Beginning with an exploration of stochastic processes with independent increments, the chapter establishes foundational concepts crucial for understanding LRPs. These processes, characterized by their reliance on past and present values, are further examined for their applicability in both continuous and discrete-time scenarios. Special attention is given to the properties of LRPs, including their closure under linear transformations and the simplification of probabilistic analysis through characteristic functions. The chapter extends the discussion to multidimensional random fields, showcasing the versatility of LRPs in modeling complex phenomena across various fields like radiophysics, radio engineering, and optics. Additionally, the exploration of stationarity and homogeneity within random fields elucidates the conditions under which these models can be simplified, further contributing to their practical utility in diagnostics. By providing a fundamental understanding of these processes and demonstrating their applicability through examples, this chapter aims to equip researchers and practitioners in the field of electrical equipment diagnostics with robust mathematical models for diagnostic signals.

Keywords Linear random process · Random proceses · Noise signals · White signal · Hilbert linear processes · Characteristic function · Mathematical expectation · Correlation function

As mentioned above, when developing any diagnostic method, first of all, it should build a mathematical model of the diagnostic object. Since the determination of the state of an object is carried out by measuring certain signals generated by the corresponding sensors, the main efforts are directed to the development of mathematical models of such signals. In statistical diagnostics, the apparatus of the theory of random processes [1–4] is used to mathematically describe diagnostic signals.

Often the processes under study are characterized by time invariance in the sense that their main parameters remain unchanged throughout the entire duration of the diagnostic experiment. In such cases, for a mathematical description of diagnostic

V. Babak et al., *Statistical Diagnostics of Electric Power Equipment*, Studies in Systems, Decision and Control 573, https://doi.org/10.1007/978-3-031-76253-6_2

signals, it is natural to use mathematical models of stationary random processes [5–8]. At the same time, in many practical cases, an EE operating in a cyclic mode is considered. The signals that are measured in the process of diagnosing such equipment are characterized by some repeatability over time. This repeatability should be taken into account in the mathematical model. Therefore, there is a need to consider classes of non-stationary random processes [5, 6, 9] whose parameters change cyclically in time.

In this chapter, we will consider various classes of random processes that are used as mathematical models of information signals in the statistical diagnostics of EE.

2.1 Infinitely Divisible Distribution Laws and Random Processes

In many cases, a diagnostic signal can be considered as the sum of a large number of random variables that characterize the influence of various factors on the operation of the equipment under study. Thus, there is a need to consider the properties of sums of independent random variables of the form

$$\xi_n = \zeta_1 + \zeta_2 + \ldots + \zeta_n, \tag{2.1}$$

where $\zeta_j, j = \overline{1,n}$ may depend on time.

In this case, we are interested in the conditions under which the sequence of sums ξ_n converges in a certain probabilistic sense to some boundary quantity ξ_n, as well as the class of all those random variables (that is, the laws of their distribution) to which sums of the form (2.1) can converge.

In probability theory, a random variable that can be the limit of sums of the form (2.1) is called infinitely divisible, and its distribution function is called an infinitely divisible distribution function [10]. The same applies to the characteristic function (CF) [11, 12].

Generally speaking, without additional restrictions on random variables ζ_j distribution of a limiting random variable ξ may be arbitrary. This is easy to show if you accept $\zeta_j \equiv 0, j > 1$. Then the distribution of the quantity ξ will coincide with the distribution ζ_1, which can be chosen arbitrarily. In order to avoid such trivial cases, it is assumed that with growth n there is a uniform decrease in the probabilistic sense of all terms of the sum (2.1), thus, the condition of uniform asymptotic smallness of the quantities is imposed ζ_j. Moreover, the convergence of sums (2.1) to the limit ξ understood in the sense of convergence in distribution, that is, as the convergence of a sequence of distribution functions for partial sums ξ_n to the limiting distribution function for a random quantity ξ.

Let $F(x)$ – distribution function of a random quantity ξ, a $f(u)$ – its CF. Distribution function $F(x)$ called *infinitely divisible in a generalized sense* [11], if for everyone $n \geq$

1 function $F(x)$ can be represented as a convolution of distribution functions $F_1(x)$, $F_2(x)$, ..., $F_n(x)$, that is $F(x) = F_1(x)\#F_2(x)\# \cdots \#F_n(x)$.

For CF it means that

$$f(u) = \prod_j f_j(u),$$

where $\left|1 - f_j(u)\right| \leq \varepsilon$ at $|u| \leq 1/\varepsilon$, $1 \leq j \leq n$.

For random variables, the condition of uniform asymptotic smallness has the following form:

$$\lim_{n\to\infty} \max_{1\leq j\leq n} \mathrm{P}\left\{\zeta_j \geq \varepsilon\right\} = 0,\ \varepsilon > 0.$$

Sometimes another condition is imposed on the terms in (2.1), namely, they require that all random variables ζ_j have the same distribution. Related to this is the following definition: distribution function $F(x)$ called *infinitely divisible* [11], if for everyone $n \geq 1$ function $F(x)$ can be represented as a convolution, provided that $F_1(x) \equiv F_2(x) \equiv \cdots \equiv F_n(x)$. In this case $f(u) = (f_n(u))^n$.

In [13, 14], a theorem was proven that states that every infinitely divisible distribution law is an infinitely divisible distribution law in the generalized sense and vice versa. Thus, the two previous definitions are equivalent, that is, both under the condition of uniform asymptotic smallness of the terms and under the condition of the same distribution, the same class of limit distributions is obtained.

Infinitely divisible distributions have the following properties [15–17]:

1. The CF of the infinitely divisible law does not vanish anywhere.
2. The distribution function of the sum of independent random variables that have infinitely divisible distribution laws is also infinitely divisible.
3. The distribution function, which is limiting (in the sense of convergence in the main) for a sequence of infinitely divisible distribution functions, is itself infinitely divisible.

The CF of infinitely divisible laws can be represented in a general form - the so-called canonical representation. The formulas of Kolmogorov, Levy-Khinchin and Levy are known [18–23]. The first formula is valid only for infinitely divisible random variables with finite variance (Hilbert), while the other two formulas remain valid for the general case [24–26]. The general form of the logarithm of the CF of an infinitely divisible distribution, provided that its variance is finite, in Kolmogorov form [27–30]:

$$\ln f(u) = imu + \int_{-\infty}^{\infty} \left(e^{iux} - 1 - iux\right)\frac{dK(x)}{x^2}, \tag{2.2}$$

where m—real parameter coinciding with the mathematical expectation of a random variable ξ; $K(x)$—non-decreasing left continuous bounded function, $K(-\infty) = 0$;

integrand value at $x = 0$ is taken equal to $-u^2/2$. Here and below $\ln f(u)$ the main meaning is understood; at the point $u = 0$ value $\ln f(u)$ goes to zero.

Representation (2.2) is completely specified by the parameter m and function $K(x)$. If CF $f(u)$ is known, then the value is determined unambiguously as follows:

$$m = (-i)\frac{d}{du}\ln f(u)\bigg|_{u=0}. \tag{2.3}$$

Function $K(x)$, included in formula (2.2), can be defined as the Fourier-Stieltjes transform [31–33] from the relation

$$\frac{d^2}{du^2}\ln f(u) = -\int_{-\infty}^{\infty} e^{iux}dK(x),$$

from which we get

$$\hat{K}(x) = \frac{1}{2\pi}\lim_{y\to\infty}\lim_{c\to\infty}\int_{-c}^{c}\frac{e^{-iuy}+e^{-iux}}{iu}\frac{d^2}{du^2}\ln f(u)du, \tag{2.4}$$

where $\hat{K}(x) = \frac{1}{2}[K(x+0) + K(x-0)]$.

At all points of continuity of the function $K(x)$ values $\hat{K}(x) = K(x)$, so the function $K(x)$ is determined unambiguously. At break points the value $K(x)$ may differ from $\hat{K}(x)$ but this does not affect the value of the integral (2.2) (due to the properties of the Stieltjes integrals [34–37]).

Function $K(x)$ is called the Poisson spectrum of jumps in Kolmogorov form [27–30]. Its physical meaning is as follows. As is known from the properties CF random variable variance

$$\mathrm{D}\xi = -\frac{d^2}{du^2}\ln f(u)\bigg|_{u=0}.$$

Using formula (2.2), we obtain

$$\mathrm{D}\xi = \int_{-\infty}^{\infty} dK(x) = K(+\infty) - K(-\infty) = K(+\infty),$$

that is, a complete variation of the function $K(x)$ equal to the variance of the random variable ξ. Thus, $K(x)$ is the distribution function up to a non-negative numerical factor equal to $\mathrm{D}\xi$.

Since in the practice of technical diagnostics only Hilbert random processes are encountered [24–26], within the framework of this work we will limit ourselves to considering formula (2.2). Detailed information about the Levy and Levy-Khinchin formulas can be found in special leiterature.

The class of infinitely divisible laws includes a number of distribution laws that have received wide practical application, in particular, normal, Poisson, log-normal, Cauchy, gamma distribution and others [38–43]. A degenerate distribution (a non-random quantity or function whose variance is zero) is also infinitely divisible, which allows one to consider additive signal-to-noise models [44–46]. This class does not include uniform, triangular, or, in general, all distributions that have finite density. In [11] the parameters of the canonical representations of the main infinitely divisible distributions are given, which will be used in the subsequent presentation.

Random value ξ called *infinitely divisible* [11], if for everyone $n \geq 1$ one can find such independent random variables $\eta_1, \ldots, \eta_n$ (either identically distributed or satisfying the condition of uniform asymptotic smallness), that $\xi \overset{d}{=} \eta_1 + \eta_2 + \cdots + \eta_n$.

Here's the notation $\xi \overset{d}{=} \eta$ means that random variables ξ and η coincide in distribution, that is $F_\xi(x) = F_\eta(x), x \in \mathbf{R}$.

In [11] it was proven that a random variable ξ can be a limit on the distribution of sums of the form (2.1) if and only if ξ is infinitely divisible. Thus, the class of limit distributions for sums of the form (2.1) and the class of infinitely divisible distributions coincide.

A separable random process $\xi(t)$ is called *infinitely divisible* if all its finite-dimensional distribution functions are infinitely divisible. Examples of infinitely divisible random processes can be a continuous process in the mean square sense with independent increments, as well as a linear random process, which will be discussed in detail below.

2.2 White Noise, Random Processes and Fields

When constructing mathematical models of many phenomena and processes, including in electrical circuits of the IMS, processes such as white noise play an important role [47–51]. In the literature on radiophysics and radio engineering, white noise is understood as a random process with a constant spectral density intensity at all frequencies [52–54]. In this case, it is implicitly assumed that it is stationary. In some works, for example [55, 56], white noise is introduced as a generalized derivative of a process with independent increments, that is, as a process with independent values. In the stationary case, such a definition turns out to be equivalent to that indicated above, but opens up new possibilities when considering non-stationary processes. In particular, we can consider periodic white noises, which are used in constructing mathematical probabilistic models of various processes occurring in power equipment and characterized by cyclicity.

Processes with independent increments have been studied quite fully at the current time. In fact, it was from the study of these processes that the development of the theory of random processes began. First, the Wiener process (the process of Brownian motion) was considered, and then more general processes with independent increments. In a complete form, the analysis of processes with independent increments was carried out in the works of B. de Finetti (1929), A.M. Kolmogorov (1932) and P. Levy (1934). The results of studies of processes with independent increments are quite closely related to the theory of infinitely divisible distributions discussed in the previous subsection.

Process $\{\eta(t),\ t \in \mathbf{T}\}$ is called *a process with independent increments* if, for a fixed $\mathbf{T}$ for all points in time $t_{-m} < t_{-m+1} < \ldots < t_{-1} < t_0 < t_1 < \ldots < t_{n-1} < t_n$ from $\mathbf{T} \subset \mathbf{R}^1$ random variables $\eta(t_{-m+1}) - \eta(t_{-m}),\ \ldots,\ \eta(t_0) - \eta(t_{-1}),\ \eta(t_0),\ \eta(t_1) - \eta(t_0),\ \ldots,\ \eta(t_n) - \eta(t_{n-1})$ independent.

The quantity $\eta(t_0)$ is called the initial value (or state) of the process. Most works provide definitions of processes with independent increments on a finite interval or on a semi-axis, with the point being the beginning. But when solving some problems, in particular, when constructing linear random processes, which will be described below, it becomes necessary to consider such processes on the entire numerical axis $(-\infty,\ \infty)$. In this case, the point t_0 inevitably falls inside the time interval, which leads to a more complex, but also more general definition.

Process increment $\eta(t)$ over a period of time $[\tau_1,\ \tau_2]$, $\tau_1,\ \tau_2 \in \mathbf{T}$, let's denote $\Delta\eta(\tau_1,\ \tau_2) = \eta(\tau_2) - \eta(\tau_1)$. We denote the one-dimensional $\eta(t)$ CF of a process by $f_\eta(u;\, t)$, and the C $\Delta\eta(s,\ t)$ F of its increments $\Delta\eta(s,\ t)$ by $f_{\Delta\eta}(u;\, s, t)$. Obviously, for any time $\tau_1 < \tau_2 < \tau_3$, $\tau_1,\ \tau_2,\ \tau_3 \in \mathbf{T}$ with probability 1 the relation is satisfied

$$\Delta\eta(\tau_1, \tau_3) = \Delta\eta(\tau_1, \tau_2) + \Delta\eta(\tau_2, \tau_3). \tag{2.5}$$

In a more general case,

$$\Delta\eta(t_n, t_{-m}) = \sum_{j=1}^{n} \Delta\eta\big(t_{j-1}, t_j\big) + \sum_{j=1}^{m} \Delta\eta\big(t_{-j}, t_{-j+1}\big). \tag{2.6}$$

This shows that the increments of the process on intervals containing the point t_0 do not depend on the initial value of the process $\eta(t_0)$. On intervals in which the point t_0 is extreme (left or right), they are independent by definition. Therefore, the value $\eta(t_0)$ can be chosen arbitrarily, even non-randomly. It is often convenient to put $\eta(t_0) = 0$ with probability 1. In this case, the CF of the process $\eta(t)$ $t = t_0$ degenerates, that is, $f_\eta(u;\, t_0) \equiv 1$. In the future we will always consider this assumption to be fair.

Next, taking into account the notation of increments, we have:

$$f_{\Delta\eta}(u;\, s, t) = \overline{f_{\Delta\eta}(u;\, t, s)},\ t, s \in \mathbf{T}, \tag{2.7}$$

and from the definition of the process and from expression (2.5) we obtain

$$f_{\Delta\eta}(u;\, \tau_1, \tau_3) = f_{\Delta\eta}(u;\, \tau_1, \tau_2) f_{\Delta\eta}(u;\, \tau_2, \tau_3). \tag{2.8}$$

The connection between the CF $f_\eta(u;\ t)$ of the process values and the CF $f_{\Delta\eta}(u;\ s,\ t)$, $s < t$ of its increments is expressed as follows:

$$f_\eta(u;\, t_n) = \prod_{j=1}^{n} f_{\Delta\eta}\big(u;\, t_{j-1}, t_j\big);$$
$$t_1, \ldots, t_n > t_0, t_{-1}, \ldots, t_{-m} < t_0,\ m, n = 1, 2, \ldots \tag{2.9}$$

An important fact is that the one-dimensional distribution function of a stochastically continuous random process with independent increments is infinitely divisible, which is proven. Therefore, the CF logarithm of such a process (provided that it is Hilbertian) can be written in Kolmogorov form [11]:

$$\ln f_\eta(u;\, t) = im(t)u + \int_{-\infty}^{\infty} \big(e^{iux} - 1 - iux\big)\frac{d_x K(x;\, t)}{x^2}, \tag{2.10}$$

where $m(t)$, and $K(x, t)$, $x \in (-\infty, \infty)$ have the same meaning as m and $K(x)$ in formula (2.2), but additionally depend on t.

As P. Levy showed, any random process with independent increments can be decomposed into three stochastically independent components:

$$\eta(t) = a_0(t) + \eta_d(t) + \eta_c(t), \tag{2.11}$$

where $a_0(t)$—non-random centering function; $\eta_d(t)$– component with fixed discontinuities; $\eta_c(t)$—stochastically continuous component.

The listed components are not necessarily present at the same time. The terms $\eta_c(t)$ and $\eta_d(t)$ in (2.11) are called processes of Gaussian and Poisson types, respectively. All implementations of processes of the first type with probability 1 are absolutely continuous functions of time. Poisson-type processes, on the contrary, have stepwise implementations. In Kolmogorov's representation, $m(t)$ is the mathematical expectation of the process, and the spectral jump function contains information about both of these components simultaneously. The dispersion of the Gaussian component is determined by the magnitude of the jump $K(x, t)$ at zero. In general, $K(x, t)$ makes sense to distribute the power of jumps.

Let us now consider two partial cases that are important for practice. To record process increments over an interval of length $h > 0$, we use a difference operator, which is more convenient for subsequent presentation than the notation adopted above. Thus, we denote

$$\Delta^h \eta(t) = \Delta\eta(t,\ t + h) = \eta(t + h) - \eta(t). \tag{2.12}$$

A process with independent increments $\eta(t)$ is called homogeneous if the distribution of its increments $\Delta^h\eta(t)$, $h > 0$ does not depend on t. Otherwise, the process $\eta(t)$ is called heterogeneous.

For a homogeneous process with independent increments $m(t) \equiv m = const$ and $K(x,t) = K(x)$, $x \in (-\infty, \infty)$. In this case, expression (2.10) simplifies and takes the form

$$\ln f_\eta(u;\ t) = |t| \left(imu + \int_{-\infty}^{\infty} \left(e^{iux} - 1 - iux\right) \frac{d_x K(x)}{x^2} \right). \tag{2.13}$$

An inhomogeneous process with independent increments $\eta(t)$ is called a process with independent periodic increments [57, 58] if, for an arbitrary number $h > 0$, there exists a number $T > 0$ called the period of increments such that the random vectors $\left(\Delta^h \eta(\tau_1),\ \ldots,\ \Delta^h \eta(\tau_n)\right)$ and $\left(\Delta^h \eta(\tau_1 + T),\ \ldots,\ \Delta^h \eta(\tau_n + T)\right)$ are stochastically equivalent in the broad sense (i.e., distributionally).

Note that in the special case when $h = T$, where T is the period of process increments $\eta(t)$, the increments $\Delta^h \eta(t)$ are homogeneous, that is, they have an arbitrary period, and are centered.

Processes with independent increments can also be classified according to the type of distribution of increments: Wiener, Poisson, generalized Poisson, gamma process and others [59–64].

Homogeneous stochastically continuous process with independent increments distributed according to a normal law with parameters $m = 0$ and $\mathrm{D}\eta(t) = \sigma^2 |t|$, is called a Wiener process and is denoted by. For the first time, such a process was considered when modeling Brownian motion. The mathematical expectation and dispersion of the Brownian motion process $\eta(t)$, $\eta(0) = 0$, are determined by the expressions

$$\mathrm{M}\eta(t) = mt,\ \mathrm{D}\eta(t) = \sigma^2 |t|.$$

Moreover, the parameters m and σ can be arbitrary.

A stochastic continuous process with independent increments, distributed according to Poisson's law with parameter $a(t)$, is called a Poisson process and is denoted by $\pi(t)$. The function $a(t)$ must be continuous, which follows from the stochastic continuity of the process.

Processes with independent increments can be considered both in continuous and discrete time. Both of these cases find application in statistical diagnostics.

Along with processes with independent increments, processes with uncorrelated increments are also considered. The last class of processes is broader; processes with independent increments constitute its subclass.

In addition to their independent use as mathematical models of various phenomena, processes with independent increments are of great importance in the construction of such important classes of random processes as white noise and linear random processes.

For the first time, white noise as a generalized derivative of a process with independent increments was considered by K. Ito (1954) and I. M. Gelfand (1955). Such processes are called generalized.

In [11], white noise $\zeta(t)$ is understood as the generalized derivative of a homogeneous process with uncorrelated increments $\eta(t)$. The process $\eta(t)$ is called generating for $\zeta(t)$. The process $\zeta(t)$ obtained in this way is stationary in the broad sense.

An inhomogeneous process with independent (or uncorrelated) increments can also be used as a generator. In this case, non-stationary white noises are obtained, among which periodic noises introduced in the robot are of great practical importance [68].

By *white noise* $\zeta(t)$ we mean a real random process with independent values (white noise in the strict sense) or with uncorrelated values (white noise in the broad sense), which is a generalized derivative of some process with independent (or, accordingly, uncorrelated) increments.some process with independent (or, accordingly, uncorrelated) increments $\eta(t)$. In this case, time can change both continuously and discretely. In the latter case, we obtain a random sequence - discrete white noise. Using a homogeneous process as a generator, we obtain stationary white noise. If the process increments $\eta(t)$ are periodic with a T period, then the white noise will be periodic with the same period. According to the type of distribution of process values, white noise can be classified into Gaussian, Poisson, gamma and others. Such white noises are generalized derivatives of the corresponding processes with independent increments.

Stationary white noise with continuous time $\{\zeta(t), t \in (-\infty, \ \infty), \mathrm{M}\zeta(t) = 0\}$ has a correlation function [11, 69–72]

$$\mathrm{M}\zeta(t)\zeta(t+\tau) = R(\tau) = \mathrm{k}_2\delta(\tau),$$

where $\kappa_2 > 0$ is the noise intensity; $\delta(\tau)$ is Dirac delta function [73–75].

The variance of continuous-time white noise is infinite, meaning it is not Hilbert noise. The spectral density of such a process $S(\omega) = \kappa_2$ is constant for all frequency values. It follows from this that white noise with continuous time is not physically realizable, that is, it does not exist in the usual sense.

Discrete-time stationary white noise is a sequence $\{\zeta_t, t = 0, \pm 1, \pm 2, \ldots\}$ independent (or uncorrelated) random variables whose mathematical expectations are equal to zero, $\mathrm{M}\zeta_t = 0$, and the correlation function has the form

$$\mathrm{M}\zeta_k\zeta_{k+\tau} = R_\tau = \kappa_2\delta_\tau,$$

where $\kappa_2 > 0$ is noise intensity; δ_τ is discrete delta function:

$$\delta_\tau = \begin{cases} 0 \text{ at } \tau \neq 0, \\ 1 \text{ at } \tau = 0. \end{cases}$$

In this case, the dispersion of the process is finite and coincides with its intensity κ_2. Spectral density of such process is

$$S(\omega) = \begin{cases} 0 \ \text{at} \ \omega < -\pi \ \text{або} \ \omega \geq \pi, \\ k_2 \quad \text{at} \ -\pi \leq \omega < \pi. \end{cases}$$

Thus, discrete-time white noise is physically realizable.

Since white noise is a generalized derivative of a process with independent increments, it is itself infinitely divisible. As a result, the class of white noise considered in this work is narrower than the class of all processes with independent (uncorrelated) values. But the unlimited divisibility of such processes makes it possible to perform linear operations on them without going beyond the class, which is important for their application.

2.3 Linear Random Processes and Fields

The class of infinitely divisible processes includes the class of linear random processes (LRP) [11]. From an applied point of view, LRP can be considered as the result of filtering white noise by some linear system, which allows it to be used as a model of signals in radio circuits. The response of some nonlinear systems can also be represented as an LRP. Therefore, LRPs are widely used in solving problems of signal extraction and classification not only in technical diagnostics, but also in other fields of science [76–79].

The theory of LRP is based on fundamental results obtained by A.P. Kolmogorov, D. Dub, B.V. Gnedenko. The construction of such mathematical models is based on the so-called constructive methods for specifying random processes. LRPs of Poisson type in the sense of expression (2.11) were developed in the works of R. Lugannani and D. Thomas [80].

Fundamental results on the theory of LRP were obtained in the works of B.G. Marchenko [13, 14]. He proved the theorem that establishes a connection between the construction of the CF of infinitely divisible distribution laws and the construction of LRP parameters.

For Hilbert linear processes (that is, processes with finite dispersion, the CF of which can be represented in Kolmogorov's canonical form), studies of their properties, as well as the possibility of their use for classifying signals in various applied areas, were carried out in [24–26].

Since modern IMS contain microprocessors or personal computers, discrete-time processes are increasingly used [81, 82]. Discrete-time LRPs were apparently first considered in [83]. Subsequently, the theory of LRP with discrete time was developed in [81, 84]. Recently, a number of robots have appeared in which LRPs with continuous time are represented as autoregressive-moving average processes with continuous time (they are designated as continuous time ARMA process – CARMA). The generating process is the so-called Levy process, that is, homogeneous infinitely divisible processes with independent stationary increments without a Gaussian component [85–87].

A linear (in the narrow sense) random process is a stochastic integral of the form [11]

$$\xi(t) = \int_{-\infty}^{\infty} \varphi(\tau, t) d\eta(\tau), t \in \mathbf{T}, \mathbf{T} \subset \mathbf{R}, \tag{2.14}$$

where $\varphi(\tau, t)$, $\tau \in (-\infty, \infty)$, $t \in \mathbf{T}$– real non-random numerical function (kernel (2.14)), such that $\int_{-\infty}^{\infty} |\varphi(\tau, t)|^p d\tau < \infty$ evenly t across $p = 1, 2$; $\{\eta(\tau), \mathrm{P}\{\eta(0) = 0\} = 1, \tau \in (-\infty, \infty)\}$– stochastically continuous random process with independent increments.

A parametric set $\mathbf{T}$ can be defined as an interval on a number line; then we speak of a linear random process of continuous time. A $\mathbf{T}$ set can also represent some countable or finite subset of the number line, in particular $\mathbf{T} = \Delta_N = \{t_n : t_n = n\Delta t\}_{n=0}^{N}$. In the latter case (2.14) is a discrete-time process.

The deterministic function $\varphi(\tau, t)$ is called the kernel of the integral representation (2.14), and the random process $\eta'(\tau)$, which is a generalized derivative of the process with independent increments $\eta(\tau)$, is called the generating one. Thus, the LRP can be thought of as the response of a linear impulse $\varphi(\tau, t)$ response filter to white noise $\eta'(\tau)$.

If the process $\eta(\tau)$ is a process with uncorrelated increments, then (2.14) is called linear in the broad sense.

Considering that diagnostic signals are generated by a real physical system $\xi(t)$, the process must have finite values of energy characteristics. Therefore, we will assume that the Hilbert process $\xi(t)$ that is, the condition $\mathrm{M}\left[|\xi(t)|^2\right] < \infty$ is satisfied. In order for process (2.14) to be Hilbert, it is necessary that the dispersion of the $\eta(\tau)$ process increments be finite.

Representation (2.14) can be accepted as a mathematical model of diagnostic signals for use in measurement and diagnostic problems, since the properties of the LRP of the form (2.14) are determined by the characteristics of the kernel $\varphi(\tau, t)$ and function $\eta(\tau)$, which, in most cases important for practice, can be uniquely determined by a given process $\xi(t)$. In addition, the class of processes of the form (2.14) is closed under linear transformations, which in this case are reduced to the corresponding linear operations on non-random kernels $\varphi(\tau, t)$, which will be shown below.

For a linear, in the narrow sense, Hilbert random process $\xi(t)$ (2.14), under the assumption that $\int_{-\infty}^{\infty} \varphi^2(\tau, t) d\tau < \infty$, the logarithm of the one-dimensional CF is given by the expression [11]

$$\begin{aligned} \ln f_{\xi}(u; t) &= \ln \mathrm{M}\left[e^{iu\xi(t)}\right] \\ &= iu \int_{-\infty}^{\infty} \varphi(\tau, t) dm(\tau) \end{aligned}$$

$$+\int_{-\infty}^{\infty}\int_{-\infty}^{\infty}\left[e^{iux\varphi(\tau,t)}-1-iux\varphi(\tau,t)\right]\frac{dK_{x\tau}(x,\ \tau)}{x^2}, \tag{2.15}$$

where $\varphi(\tau, t)$ the same as in (2.14), and pair $(m(\tau), K(x, \tau))$, defining the properties of the process $\eta(\tau)$, is defined in (2.10).

Expression (2.15) is not the canonical Kolmogorov form for the one-dimensional CF process $\xi(t)$.

Multidimensional CFs of a process $\xi(t)$ are also infinitely divisible. The logarithm of the n-dimensional CF LRP in Kolmogorov form has the form:

$$\begin{aligned} f_\xi(u_1, \ldots, u_n; t_1, \ldots, t_n) &= i\sum_{k=1}^{n} u_k \int_{-\infty}^{\infty} \varphi(\tau, t_k)dm(\tau) \\ &+ \int_{-\infty}^{\infty}\int_{-\infty}^{\infty}\left[e^{ix\sum_{k=1}^{n} u_k\varphi(\tau,t_k)} - 1 - ix\sum_{k=1}^{n} u_k\varphi(\tau, t_k)\right]\frac{d_{x\tau}K(x,\tau)}{x^2}. \end{aligned} \tag{2.16}$$

The fact that the CF LRP (2.14) can be represented in general form by the expression (2.16) determines the same universality in application terms of the model (2.14), as, for example, the model of Gaussian processes, which, by the way, is a partial case of linear processes. The general form of the CF (2.16) allows for a complete probabilistic analysis of the responses of linear systems: finding cumulants, distribution functions of their responses, studying the distribution of implementation jumps at the input and output of such systems, exploring the connections between the input and output characteristics of linear circuits, etc.

The known properties of CF make it possible to find expressions for the cumulants of the k-th order of the LRP. Taking into account (2.16), we have:

$$\kappa_1\{\xi(t)\} = i\frac{\partial}{\partial u}\ln f_\xi(u;t)\bigg|_{u=o} = \int_{-\infty}^{\infty}\varphi(\tau,t)\,dm(\tau);$$

$$\begin{aligned} \kappa_2\{\xi(t_1),\xi(t_2)\} &= \frac{\partial^2}{\partial u_1\partial u_2}\ln f_\xi(u_1,u_2;t_1,t_2)\bigg|_{u_1=u_2=o} \\ &= \int_{-\infty}^{\infty}\int_{-\infty}^{\infty}\varphi(\tau,t_1)\,\varphi(\tau,t_2)\,d_{x\tau}K(x,\tau); \end{aligned}$$

$$\begin{aligned} \kappa_k\{\xi(t_1),\ldots,\xi(t_k)\} &= (-i)^k\frac{\partial^k}{\partial u_1\ldots\partial u_k}\ln f_\xi(u_1,\ldots,u_k;t_1,\ldots,t_k)\bigg|_{u_1=\ldots=u_k=o} \\ &= (-i)^k\int_{-\infty}^{\infty}\int_{-\infty}^{\infty}x^{k-2}\varphi(\tau,t_1)\ldots\varphi(\tau,t_k)d_{x\tau}K(x,\tau); \end{aligned}$$

$$k = 3, 4, ...,$$

moreover, for the existence of cumulants of the third and higher orders, it is necessary to impose additional conditions of integrability τ over the functions $\varphi^k(\tau, t)$, $k = 3, 4, ...$, for all t.

If we additionally assume that the process $\eta(\tau)$ is a homogeneous process with independent increments, the expression for the logarithm of the one-dimensional CF of the process (2.14) can be written as

$$\ln f_\xi(u; t) = imu \int_{-\infty}^{\infty} \varphi(\tau, t) d\tau + \int_{-\infty}^{\infty} \int_{-\infty}^{\infty} \left[e^{iux\varphi(\tau,t)} - 1 - iux\varphi(\tau, t)\right] \frac{dK(x)}{x^2} d\tau, \tag{2.17}$$

where m and $K(x)$—parameters of the infinitely divisible CF increments of a homogeneous process $\eta(\tau)$ in Kolmogorov form (2.13) [27–30].

The logarithm of the n-dimensional CF for this case has the form:

$$\ln f_\xi(u_1, ..., u_n; t_1, ..., t_n) = \ln \mathbf{M}\left[\exp\left(i \sum_{k=1}^{n} u_k \xi(t_k)\right)\right] = im \sum_{k=1}^{n} u_k \int_{-\infty}^{\infty} \varphi(\tau, t_k) d\tau$$

$$+ \int_{-\infty}^{\infty} \int_{-\infty}^{\infty} \left[e^{ix \sum_{k=1}^{n} u_k \varphi(\tau,\, t_k)} - 1 - ix \sum_{k=1}^{n} u_k \varphi(\tau,\, t_k)\right] \frac{dK(x)}{x^2} d\tau,$$

$$n = 1,\ 2,\ ...; t_1,\ ...,\ t_2 \in T. \tag{2.18}$$

In the case of integrability with τ respect to function $\varphi^k(\tau, t)$, $k = 1,\ 2,\ ...$ for all t and the existence of all cumulants of the process $\eta(\tau)$ at the point $\tau = 1$ mixed cumulants of process values $\xi(t)$ of k-th order in moments of a time $t_1,\ ...,\ t_k$ look like:

$$\kappa_k[\xi(t_1),\ ...,\ \xi(t_k)] = \kappa_k \int_{-\infty}^{\infty} \prod_{j=1}^{k} \varphi(\tau, t_j) d\tau, \quad k = 1,\ 2,\ ..., \tag{2.19}$$

where κ_k—k-th cumulant of a random variable $\eta(1)$.

Taking into account (2.19), the expressions for the mathematical expectation $\mathtt{M}[\xi(t)]$ and the correlation function $R_\xi(t_1, t_2)$ of the LRP (2.14) have the form:

$$\mathtt{M}[\xi(t)] = \kappa_1 \int_{-\infty}^{\infty} \varphi(\tau, t) d\tau, R_\xi(t_1, t_2) = \kappa_2 \int_{-\infty}^{\infty} \varphi(\tau, t_1)\varphi(\tau, t_2) d\tau. \tag{2.20}$$

If in expression (2.14) the function $\varphi(\tau,\ t) = \varphi(t-\tau)$ and $\eta(t)$ is a homogeneous process with independent increments, then the process $\xi(t)$ is stationary. For a stationary real LRP, representation (2.14) can be written as:

$$\xi(t) = \int_{-\infty}^{\infty} \varphi(t-\tau) d\eta(\tau), t \in \mathbf{T}. \tag{2.21}$$

In this case, the expressions for the mathematical expectation and the correlation function take the form:

$$\mathrm{M}[\xi(t)] = \kappa_1 \int_{-\infty}^{\infty} \varphi(x)dx = const, R_\xi(\tau) = \kappa_2 \int_{-\infty}^{\infty} \varphi(x)\varphi(x+\tau)dx. \tag{2.22}$$

Above, we discussed the main properties of the mathematical model of a diagnostic signal in the form of an LRP specified over a continuous time interval. It is advisable to use such a model when researching using analog instruments included in the IMS.

However, modern diagnostic IMSs also include digital information processing tools. Such devices operate not with electrical signals continuously varying in time, but with digital processes – sequences of numbers obtained as a result of time sampling and quantization by the level of real continuous diagnostic signals. Such sequences are also called time series to emphasize their dependence on time. To describe digital signals, it is necessary to have discrete-time models. Naturally, they must remain in close connection with the original models of continuous argument.

We will consider processes with discrete time as specified on an equidistant lattice $\Delta_N = \{t_n : t_n = n\Delta t\}_{n=0}^{N}$ with a sampling step $\Delta t = const, \forall t_n \in \mathbf{T}$.

According to [13, 14], a variant of representing the process (2.14) with the imposition of certain conditions can be the following:

$$\xi(n) = \sum_{m=-\infty}^{\infty} \phi(m,n)\zeta(m), \quad n \in \left(\overline{-\infty,\infty}\right), \tag{2.23}$$

where $\varphi(m,n)$, $m \in \left(\overline{-\infty,\infty}\right)$—deterministic function—the kernel of representation (2.23), for which the condition holds $\sum\limits_{m=-\infty}^{\infty} |\varphi(m,n)|^2 < \infty$ for every fixed $n \in \left(\overline{-\infty,\infty}\right)$; $\zeta(m)$—generating white noise in the narrow sense (a random sequence with independent values), which we will consider Hilbert.

The properties of model (2.14) considered above have corresponding equivalents for the discrete linear process (2.23), with the only difference being that the time argument changes discretely. In particular, for LRP $\xi(n)$ it is possible to write down the representation of one- and multidimensional CFs, which are expressed through CFs $\varphi(m,n)$ and $\zeta(m)$, considering the time argument as discrete and replacing in

(2.23) integration over summation over m. As a result, the mathematical expectation and correlation function of the LRP with discrete time (2.23) have the form:

$$\begin{aligned} &\mathrm{M}[\xi(n)] = \kappa_1 \sum_{m=-\infty}^{\infty} \varphi(m, n), n \in \left(\overline{-\infty, \infty}\right); \\ &R_\xi(n_1, n_2) = \kappa_2 \sum_{m=-\infty}^{\infty} \varphi(m, n_1)\varphi(m, n_2), \quad n_1, n_2 \in \left(\overline{-\infty, \infty}\right), \end{aligned} \tag{2.24}$$

where $\kappa_1 = \mathrm{M}[\zeta(m)]$, $\kappa_2 = \mathrm{D}[\zeta(m)]$.

For a stationary LRP with discrete time, representation (2.23) takes the form:

$$\xi(n) = \sum_{m=-\infty}^{\infty} \varphi(n-m)\zeta(m), n \in \left(\overline{-\infty, \infty}\right), \tag{2.25}$$

and the mathematical expectation and correlation function

$$\mathrm{M}[\xi(n)] = \kappa_1 \sum_{m=-\infty}^{\infty} \phi(m) = const, R_\xi(m) = \mathrm{k}_2 \sum_{k=-\infty}^{\infty} \phi(k)\varphi(k+m). \tag{2.26}$$

Analyzing the integral representation (2.14), it can be shown that LRPs have such properties [13, 14].

1. The sum (difference) of two linear processes.

If $\xi_1(t)$ and $\xi_2(t)$ are linear processes, $\xi_3(t) = \alpha\xi_1(t) + \beta\xi_2(t)$, where α и β—arbitrary real numbers is also a linear process.

2. Convolution of a linear process with a non-random function.

Let $\xi(t)$ be a linear process, $\varphi_0(t) \in L_2(-\infty, \infty)$—non-random function. We obtain the convolution of a process $\xi(t)$ with a function $\varphi_0(t)$ in the form

$$\xi_1(t) = \int_{-\infty}^{\infty} \xi(s)\,\varphi_0(t-s)ds = \int_{-\infty}^{\infty}\int_{-\infty}^{\infty} \varphi(\tau, s)\,\varphi_0(t-s)\,d\eta(\tau)ds = \int_{-\infty}^{\infty} \varphi_1(\tau, t)\,d\eta(\tau), \tag{2.27}$$

where $\varphi_1(\tau, t) = \int\limits_{-\infty}^{\infty} \varphi(\tau, s)\,\varphi_0(t-s)ds$.

If the process $\xi(t)$ is stationary, then

$$\varphi_1(\tau,\ t) = \varphi_1(t-\tau) = \int_{-\infty}^{\infty} \varphi(t-\tau)\varphi_0(t-s)ds = \int_{-\infty}^{\infty} \varphi(s)\varphi_0(t-\tau-s)ds. \tag{2.28}$$

Thus, as a result of the convolution of a linear process $\xi(t)$ with a non-random square-integrable function $\varphi_0(t)$, an LRP is again obtained.

3. Differentiation of linear processes.

If the LRP $\varphi(\tau, t)$ kernel $\xi(t)$ is differentiable with respect to t, then there is a derivative of the LRP defined by the expression

$$\xi'(t) = \int_{-\infty}^{\infty} \frac{\partial}{\partial t}\phi(\tau, t)\, d\eta(\tau).$$

4. Integration of a linear process.

If the LRP $\phi(\tau, t)$ kernel $\xi(t)$ satisfies the condition

$$\int_{-\infty}^{\infty} \left| \int_{-\infty}^{t} \varphi(\tau, s) ds \right|^2 d\tau < \infty,\ t \in T,$$

then the stochastic integral

$$\int_{-\infty}^{t} \xi(s) ds = \int_{-\infty}^{\infty} \int_{-\infty}^{t} \varphi(\tau, s)\, d\eta(\tau) ds$$

exists and is LRP.

Thus, linear operations on LRPs are reduced to the corresponding linear operations on non-random kernels of representation (2.14) and as a result also yield linear processes. This means that the class of LRPs is closed under linear transformations. This feature of LRPs, combined with the existence of explicit expressions for their dimensional CFs, allows for a complete probabilistic analysis of both the processes themselves and the results of their transformations by linear operators. This fact is an absolute advantage of LRPs as models of real physical processes, since it significantly simplifies the analysis of the passage of these processes through the linear paths of diagnostic IMS, even for non-Gaussian cases.

Model (2.14) can be generalized to the multidimensional case if we consider the dependence ξ not on the one-dimensional parameter t, but on some multidimensional vector $\mathbf{x}$. Then $\xi(\mathbf{x})$ is called a random field.

Random fields play an important role in solving many applied problems in radiophysics, radio engineering, optics, measurement theory, statistical theory of turbulence, hydroacoustics, etc. For example, the radiation field of a radio transmitter, the field of temperature, humidity, and dielectric constant in real environments are stochastic in nature [88–95].

The choice of mathematical model and the method of specifying the random field that is analyzed is critical. Often the form of recording a field is symbolic, and all information about the field is contained in the corresponding finite-dimensional distributions or CFs. But for engineering applications, it would be more convenient to include in the symbolic record of a field complete information about the parameters that characterize it, while maintaining the ability to perform appropriate operations on such a symbolic record without losing the meaning of the parameters included in this form of representation. Closely related to this form of specifying random fields is the method of stochastic integral representations [13, 14], which turns out to be effective in solving problems of linear and some nonlinear transformations of various random fields, as well as in statistical modeling. The class of linear random fields, which are both infinitely divisible, can be successfully used to solve these problems. We emphasize that when using the method of integral representations, a unified approach is maintained for the entire class of linear and infinitely divisible random fields, and not just for Gaussian ones.

Let $\xi(\mathbf{x})$ be a random field defined in an n-dimensional space $\mathbf{R}^n$ with a fixed system of rectangular coordinates $\mathbf{x} = (x_1, x_2, \ldots, x_n)$, and let be a point in this space. A field $\xi(\mathbf{x})$ is called Hilbert if its second moment is finite, that is

$$\mathtt{M}\{\xi^2(\mathbf{x})\} < +\infty,$$

where $\mathtt{M}$ is the mathematical expectation operator.

A real Hilbert random field $\xi(\mathbf{x})$ in an n-dimensional space $\mathbf{R}^n$ is called linear if it can be represented in the form

$$\xi(\mathbf{x}) = \lim_{m\to\infty} \sum_{k=1}^{m} \varsigma_{mk}(\mathbf{x}), \tag{2.29}$$

where $\{\varsigma_{mk}(\mathbf{x}),\ k = \overline{1,\ m},\ \ m = 1,\ 2,\ \ldots\}$—an infinite sequence of random fields that are independent in the aggregate for each m, the successive partial sums of which satisfy the condition of uniform asymptotic smallness: for any $\varepsilon > 0$

$$\lim_{m\to\infty} \max_{k} \mathtt{P}\{|\varsigma_{mk}(\mathbf{x})| \geq \varepsilon\} = 0.$$

It can be shown that the one-dimensional distribution function of a linear random field (2.29) is infinitely divisible.

Let the random field $\xi(\mathbf{x})$ in $\mathbf{R}^n$ admit a representation in the form of a stochastic integral

$$\xi(\mathbf{x}) = \int_{\mathbf{R}^n} \varphi(\boldsymbol{\tau};\mathbf{x})d\eta(\boldsymbol{\tau}),$$

where $\varphi(\boldsymbol{\tau};\mathbf{x})$ is a non-random function that is square-integrable $\boldsymbol{\tau}$ over the entire space $\mathbf{R}^n$ at any $\mathbf{x} \in \mathbf{R}^n$; $\eta(\boldsymbol{\tau})$, $\boldsymbol{\tau} = (\tau_1, \tau_2, \ldots, \tau_n)$,—random field with independent increments, such that when $k = \overline{1, n}$ condition is met $\eta(\tau_1, \tau_2, \ldots, \tau_{k-1}, 0, \tau_{k+1}, \ldots, \tau_n) \equiv 0$ with probability 1. Then the field $\xi(\mathbf{x})$ is linear.

The most interesting for practice are random fields in $\mathbf{R}^3$ space additionally depending on the actual t parameter. Such fields are called spatiotemporal and are written in the form $\xi(\mathbf{x}, t)$, where $\mathbf{x} = (x_1, x_2, x_3)$, x_1, x_2, x_3—coordinates in three-dimensional space, and $t \in \mathbf{R}^1$ is a time.

A spatiotemporal linear random field can be represented as

$$\xi(\mathbf{x}, t) = \int_{\mathbf{R}^n} \int_{-\infty}^{\infty} \varphi(\boldsymbol{\tau}, \theta; \mathbf{x}, t) d\eta(\boldsymbol{\tau}, \theta), \tag{2.30}$$

where $\eta(\boldsymbol{\tau}, \theta)$—spatiotemporal random field with independent increments, with probability 1 equal to zero if at least one of the coordinates is equal to zero; $\varphi(\boldsymbol{\tau}, \theta; \mathbf{x}, t)$ non-random function, integrable with a square for $\boldsymbol{\tau}$ and θ for any $\mathbf{x} \in \mathbf{R}^3$ and $t \in \mathbf{R}^1$.

A random field $\xi(\mathbf{x}, t)$ is called stationary (in the broad sense) in time if its mathematical expectation does not depend on t and the correlation function does not depend on the origin of time, and homogeneous (in the broad sense) in space if its mathematical expectation does not depend on and the correlation function does not depends on the choice of the origin of spatial coordinates. Note that the random field (2.30) can be stationary, but inhomogeneous in space, or, conversely, homogeneous, but nonstationary. For a linear random field that is homogeneous in space and stationery in time, expression (2.30) is transformed to the form

$$\xi(\mathbf{x}, t) = \int_{\mathbf{R}^3} \int_{-\infty}^{\infty} \varphi(\mathbf{x} - \boldsymbol{\tau}, t - \theta) d\eta(\boldsymbol{\tau}, \theta).$$

References

1. Porat, B.: Digital Processing of Random Signals: Theory and Methods. Courier Dover Publications (2008)
2. Gray, R.M.: Probability, Random Processes, and Ergodic Properties. Springer Science & Business Media (2009)
3. Wong, E., Hajek, B.: Stochastic Processes in Engineering Systems. Springer Science & Business Media (2012)
4. Yates, R.D., Goodman, D.J.: Probability and stochastic processes: a friendly introduction for electrical and computer engineers. John Wiley & Sons (2014)
5. Javorskyj, I., Kravets, I., Matsko, I., Yuzefovych, R.: Periodically correlated random processes: application in early diagnostics of mechanical systems. Mech. Syst. Signal Process. **83**, 406–438 (2017)

6. Javorskyj, I., Yuzefovych, R., Lychak, O., Trokhym, G., Varyvoda, M.: Methods of periodically non-stationary random processes for vibrations monitoring of rolling bearing with damaged outer race. Digital Signal Processing **145**, 104343 (2024)
7. Babak, V.P., Babak, S.V., Myslovych, M.V., Zaporozhets, A.O., Zvaritch, V.M.: Methods and models for information data analysis. Diagnost. Syst. Energy Equipments, 23–70 (2020)
8. Li, Y., Mulani, S.B., Kapania, R.K., Fei, Q., Wu, S.: Non-stationary random vibration analysis of structures under multiple correlated normal random excitations. J. Sound Vib. **400**, 481–507 (2017)
9. Javorskyj, I., Yuzefovych, R., Lychak, O., Semenov, P., Slyepko, R.: Detection of distributed and localized faults in rotating machines using periodically non-stationary covariance analysis of vibrations. Meas. Sci. Technol. **34**(6), 065102 (2023)
10. Myslovych, M., Hyzhko, Y., Hutorova, M., Ostapchuk, L.: On peculiarities of development of information support for technical diagnostics multilevel systems of electrical equipment. Comput. Probl. Electr. Eng. **9**(2), 34–41 (2019)
11. Babak, V., Zaporozhets, A., Kuts, Y., Myslovych, M., Fryz, M., Scherbak, L.: Models and characteristics of identification of noise stochastic signals of research objects. In: CEUR Workshop Proceedings, vol. 3309, pp. 349–362 (2022)
12. Babak, V., Scherbak, L., Kuts, Y., Zaporozhets, A.: Information and measurement technologies for solving problems of energy informatics. In: The 1st International Workshop on Information Technologies: Theoretical and Applied Problems 2021. CEUR Workshop Proceedings, vol. 3039, pp. 24–31 (2021)
13. Zvarich, V.N., Marchenko, B.G.: Linear autoregressive processes with periodic structures as models of information signals. Radioelectron. Commun. Syst. **54**(7), 367–372 (2011)
14. Guzii, V.I., Marchenko, B.G., Protsenko, L.D.: Transformations providing for detection invariance with changes in the spectral parameters of low-frequency noise. Radioehlektronika **28**, 3–10 (1985)
15. Xu, H., Yu, C., Wang, Y., Cheng, D.: Closure under infinitely divisible distribution roots and the Embrechts–Goldie conjecture. Lithuanian Math. J. 1–14 (2024)
16. Artikis, P.T., Artikis, C.T.: Facilitation of proactive decision making by using models incorporating stochastic integrals. Commun. Math. Appl. **14**(1), 257 (2023)
17. Çekyay, B., Frenk, J.B.G., Javadi, S.: On computing the multivariate poisson probability distribution. Methodol. Comput. Appl. Probab. **25**(3), 70 (2023)
18. Xu, H., Wang, Y., Cheng, D., Yu, C.: On the closure under infinitely divisible distribution roots. Lith. Math. J. **62**(2), 259–287 (2022)
19. Mainardi, F., & Rogosin, S.: The origin of infinitely divisible distributions: from de Finetti's problem to Levy-Khintchine formula. arXiv preprint ArXiv:0801.1910 (2008)
20. Ken Iti, S.: Lévy processes and infinitely divisible distributions, vol. 68). Cambridge university press (1999)
21. Watteel, R.N., Kulperger, R.J.: Nonparametric estimation of the canonical measure for infinitely divisible distributions. J. Stat. Comput. Simul. **73**(7), 525–542 (2003)
22. Fristedt, B., Gray, L., Fristedt, B., Gray, L.: Infinitely divisible distributions as limits. A Modern App. Probabil. Theo. 289–322 (1997)
23. Graczyk, P.: Factorization theorems on symmetric spaces of noncompact type. J. Theor. Probab. **12**, 375–383 (1999)
24. Zaporozhets, A., Eremenko, V., Babak, V., Isaienko, V., Babikova, K.: Using Hilbert transform in diagnostic of composite materials by impedance method. Period. Polytech. Electr. Eng. Comput. Sci. **64**(4), 334–342 (2020)
25. Babak, V., Zaporozhets, A., Kulyk, M., Kuts, Y., Scherbak, L.: Application of discrete Hilbert transform to estimate the characteristics of cyclic signals: information provision. In: Systems, Decision and Control in Energy IV: Volume I. Modern Power Systems and Clean Energy, pp 93–115. Springer Nature Switzerland, Cham (2023)
26. Babak, V., Zaporozhets, A., Kuts, Y., Scherbak, L.: Some features of Hilbert transform and their use in energy informatics. Probl General Energy, 1–2 (2022)

27. Matsiuk, A.V., Pryimak, M.V.: Mathematical model of electroretinogram in the form of linear stochastic process. In: 2005 IEEE Intelligent Data Acquisition and Advanced Computing Systems: Technology and Applications, pp. 663–666. IEEE (2005)
28. Zvaritch, V., Glazkova, E.: Application of linear AR and ARMA processes for simulation of power equipment diagnostic systems information signals. In: 2015 16th International Conference on Computational Problems of Electrical Engineering (CPEE), pp. 259–261. IEEE, (2015)
29. Rosiński, J., Szulga, J.: Product random measures and double stochastic integrals. In: Martingale Theory in Harmonic Analysis and Banach Spaces: Proceedings of the NSF-CBMS Conference Held at the Cleveland State University, Cleveland, Ohio, July 13–17, 1981, pp. 181–199. Springer Berlin Heidelberg, Berlin, Heidelberg (2006)
30. Devroye, L.: Simulating perpetuities. Methodol. Comput. Appl. Probab. **3**(1), 97–115 (2001)
31. Gotovac, V., Helisová, K.: Testing equality in distribution of random convex compact sets via theory of N-distances and its application to assessing similarity of general random sets. (2018). ArXiv preprint arXiv:1801.02090.
32. Gilgen, H.: Fourier representation of a stationary stochastic process, pp. 441–498. Univariate Time Series in Geosciences, Theory and Examples (2006)
33. Babak, V., Zaporozhets, A., Kovtun, S., Myslovych, M., Kuts, Y., Scherbak, L.: Information support for identification of the technical state of electric power facilities. In: Systems, Decision and Control in Energy V, pp. 129–153. Springer Nature Switzerland, Cham (2023)
34. Ma, Y., Ji, D.: Existence of solutions to a system of riemann-liouville fractional differential equations with coupled riemann-stieltjes integrals boundary conditions. Fractal Fract. **6**(10), 543 (2022)
35. Zhao, D., Mao, J.: Positive solutions for a class of nonlinear singular fractional differential systems with riemann-stieltjes coupled integral boundary value conditions. Symmetry **13**(1), 107 (2021)
36. Bonotto, E.M.: Generalized ordinary differential equations in abstract spaces and applications. In: Federson, M., Mesquita, J.G. (eds.). John Wiley & Sons, Incorporated (2021)
37. Zítek, P., Vyhlídal, T.: Steady state observer for fault detection of hereditary systems with weak dynamics. IFAC Proc. Vol. **33**(11), 243–248 (2000)
38. Chaudhary, A.K., Sapkota, L.P., Kumar, V.: Truncated Cauchy power–exponential distribution: theory and Applications. IOSR J. Math. (IOSR-JM), **16**(6), 44–52 (2020)
39. Yağın, F.H., Güldoğan, E., Colak, C.: A web-based software for the calculation of theoretical probability distributions. J. Cognit. Syst. **6**(1), 44–50 (2021)
40. Pham, H.: Basic statistics. In: Springer Handbook of Engineering Statistics, pp. 3–41. Springer, London, London (2023)
41. Reed, W.J.: The normal-Laplace distribution and its relatives. Adv. Distribut. theory, order statistics, and inference, 61–74 (2006)
42. Zubair, M., Tahir, M.H., Cordeiro, G.M., Alzaatreh, A., Ortega, E.M.: The power-Cauchy negative-binomial: properties and regression. J. Statist. Distribut. Appl. **5**, 1–17 (2018)
43. Babak, V., Eremenko, V., Zaporozhets, A.: Research of diagnostic parameters of composite materials using Johnson distribution. Int. J. Comput. **18**(4), 483–494 (2019)
44. Cancho, V.G., Yiqi, B., Fiorucci, J.A., Barriga, G.D., Dey, D.K.: Estimation and influence diagnostics for zero-inflated hyper-Poisson regression model: full Bayesian analysis. Commun. Stat.-Theory Meth. **47**(11), 2741–2759 (2018)
45. Ospina, R., Ferrari, S.L.: A general class of zero-or-one inflated beta regression models. Comput. Stat. Data Anal. **56**(6), 1609–1623 (2012)
46. Ling B, Khonsari M, Hathaway R (2009) Data-driven roller bearing diagnosis using degree of randomness and Laplace test. In: Annual Conference of the PHM Society, vol. 1, No. 1
47. Wang, X., Liu, C., Bi, F., Bi, X., Shao, K.: Fault diagnosis of diesel engine based on adaptive wavelet packets and EEMD-fractal dimension. Mech. Syst. Signal Process. **41**(1–2), 581–597 (2013)
48. Sverdlova, A., & Zaporozhets, A.: Predicting anomaly conditions of energy equipment using neural networks. In: E3S Web of Conferences, vol. 280, p. 09005 (2021)

49. Eremenko, V., Zaporozhets, A., Isaienko, V., & Babikova, K.: Application of wavelet transform for determining diagnostic signs. In: Proceedings of the 15th International Conference on ICT in Education, Research and Industrial Applications. Integration, Harmonization and Knowledge Transfer, vol. 1, pp. 12–15 (2019)
50. Babak, S., Babak, V., Zaporozhets, A., Sverdlova, A.: Method of statistical spline functions for solving problems of data approximation and prediction of objects state. In: Proceedings of the Second International Workshop on Computer Modeling and Intelligent Systems (CMIS-2019), pp. 15–19. Zaporizhzhia, Ukraine, (2019)
51. Eremenko, V., Babak, V., Zaporozhets, A.: Method of reference signals creating in non-destructive testing based on low-speed impact. Tekhnichna Elektrodynamika **4**, 070 (2021)
52. AR ZK, Péter HORV (2010) Modulation schemes for cognitive radio in white spaces. Radioeng. **19**(4), 511
53. Wijenayake, C., Madanayake, A., Kota, J., Bruton, L.: Space-time spectral white spaces in cognitive radio: Theory, algorithms, and circuits. IEEE J. Emerg. Selected Top. Circ. Syst. **3**(4), 640–653 (2013)
54. Lanzisera, S., Zats, D., Pister, K.S.: Radio frequency time-of-flight distance measurement for low-cost wireless sensor localization. IEEE Sens. J. **11**(3), 837–845 (2011)
55. Di Nunno, G., Øksendal, B., Proske, F.: White noise analysis for Lévy processes. J. Funct. Anal. **206**(1), 109–148 (2004)
56. Dybiec, B., Gudowska-Nowak, E., Hänggi, P.: Lévy-Brownian motion on finite intervals: Mean first passage time analysis. Phys. Rev. E **73**(4), 046104 (2006)
57. Fryz, M., Mlynko, B.: Property analysis of conditional linear random process as a mathematical model of cyclostationary signal. In ITTAP, pp. 77–82 (2022)
58. Chojnowska-Michalik, A.: Periodic linear equations with general additive noise in Hilbert spaces. In Stochastic Systems and Optimization: proceedings of the 6th IFIP WG 7.1 Working Conference Warsaw, Poland, September 12–16, 1988, pp. 169–183. Berlin, Heidelberg: Springer Berlin Heidelberg (2005)
59. Suzuki, A.K., Cancho, V.G., Louzada, F.: The Poisson–Inverse-Gaussian regression model with cure rate: a Bayesian approach and its case influence diagnostics. Stat. Pap. **57**(1), 133–159 (2016)
60. Pan, Z., Balakrishnan, N.: Reliability modeling of degradation of products with multiple performance characteristics based on gamma processes. Reliab. Eng. Syst. Saf. **96**(8), 949–957 (2011)
61. Ye, Z.S., Chen, N.: The inverse Gaussian process as a degradation model. Technometrics **56**(3), 302–311 (2014)
62. Ye, Z.S., Xie, M.: Stochastic modelling and analysis of degradation for highly reliable products. Appl. Stoch. Model. Bus. Ind. **31**(1), 16–32 (2015)
63. Babak, V.P., Babak, S.V., Myslovych, M.V., Zaporozhets, A.O., Zvaritch, V.M., Babak, V.P., Zvaritch, V.M.: Simulation and software for diagnostic systems. Diagnostic Systems For Energy Equipments, 71–90 (2020)
64. Bousquet, N., Fouladirad, M., Grall, A., Paroissin, C.: Bayesian gamma processes for optimizing condition-based maintenance under uncertainty. Appl. Stoch. Model. Bus. Ind. **31**(3), 360–379 (2015)
65. Keceli, S., Inui, K., Okamoto, H., Otsuru, N., Kakigi, R.: Auditory sustained field responses to periodic noise. BMC Neurosci. **13**, 1–8 (2012)
66. Li, H., Peng, H., & Shu, N.: Study on suppressing White noises and periodic narrow bandwidth noise of PD by vertically invariant gray-level morphological filter. In: 2010 International Conference on E-Product E-Service and E-Entertainment, pp. 1–3. IEEE (2010)
67. Bohachev, I., Babak, V., Zaporozhets, A.: Novel small-aperture transducers based on magnetostrictive effect for diagnostic systems. Techn. Electrodynam. **3**, 69–78 (2022)
68. Zaporozhets, A., Babak, V., Sverdlova, A., Isaienko, V., Babikova, K.: Development of a system for diagnosing heat power equipment based on IEEE 802.11 s. In: Systems, Decision and Control in Energy II. Springer International Publishing, Cham pp. 141–151 (2021)

69. Guo, S.S.: Nonstationary solutions of nonlinear dynamical systems excited by Gaussian white noise. Nonlinear Dyn. **92**, 613–626 (2018)
70. Zaporozhets, A.O.: Correlation analysis between the components of energy balance and pollutant emissions. Water Air Soil Pollut. **232**, 1–22 (2021)
71. Galleani, L., Cohen, L.: Nonstationary and stationary noise. In: Automatic Target Recognition XVI, vol. 6234, pp. 360–384. SPIE (2006)
72. Fang, T., Zhang, T.S.: Non-stationary mean square response due to uniformly amplitude modulated random excitations. J. Sound Vib. **182**(3), 369–379 (1995)
73. Yu, G., Lin, T., Wang, Z., Li, Y.: Time-reassigned multisynchrosqueezing transform for bearing fault diagnosis of rotating machinery. IEEE Trans. Industr. Electron. **68**(2), 1486–1496 (2020)
74. Antoni, J., Randall, R.B.: A stochastic model for simulation and diagnostics of rolling element bearings with localized faults. J. Vib. Acoust. **125**(3), 282–289 (2003)
75. Dumonteil, E., Courau, T.: Dominance ratio assessment and Monte Carlo criticality simulations: dealing with high dominance ratio systems. Nucl. Technol. **172**(2), 120–131 (2010)
76. Zvaritch, V., Myslovych, M., Gyzhko, Y.: Application of linear random processes to construction of diagnostic system for power engineering equipment. In: IFIP International Conference on Advances in Production Management Systems.Springer International Publishing, Cham, pp. 617–622
77. Fryz, M., Mlynko, B.: Property analysis of multivariate conditional linear random processes in the problems of mathematical modelling of signals. Technol Audit Product Res **3**(2/65), 29–32 (2022)
78. Fryz, M., Scherbak, L., Karpinski, M.P., Mlynko, B. (2021) Characteristic function of conditional linear random process. In ITTAP, pp. 129–135
79. Fryz, M., Mlynko, B.: Properties of stationarity and cyclostationarity of conditional linear random processes. In: 2020 IEEE 15th International Conference on Advanced Trends in Radioelectronics, Telecommunications and Computer Engineering (TCSET) (pp. 166–170). IEEE (2020)
80. Pierre, P.A.: Central limit theorems for conditionally linear random processes. SIAM J. Appl. Math. **20**(3), 449–461 (1971)
81. Pierre, P.: Characteristics of Gaussian random processes by representations in terms of independent random variables. IEEE Trans. Inf. Theory **15**(6), 648–658 (1969)
82. Mann, H.B., Wald, A.: On the statistical treatment of linear stochastic difference equations. Econometrica, J. Economet. Soc. 173–220 (1943)
83. Grenander, U., Rosenblatt, M.: Statistical spectral analysis of time series arising from stationary stochastic processes. Ann Math Stat. 537–558 (1953)
84. Brockwell, P.J., Davis, R.A.: Time series: theory and methods. Springer Science & Business Media (1991)
85. Barndorff-Nielsen, O.E., Mikosch, T., Resnick, S.I. (eds.). Lévy Processes: Theory and Applications. Springer Science & Business Media (2001)
86. Jimenez, J.J.M., Schwartz, S., Vingerhoeds, R., Grabot, B., Salaün, M.: Towards multi-model approaches to predictive maintenance: a systematic literature survey on diagnostics and prognostics. J. Manuf. Syst. **56**, 539–557 (2020)
87. Wang, P., Gao, R.X., Woyczynski, W.A.: Lévy process-based stochastic modeling for machine performance degradation prognosis. IEEE Trans. Industr. Electron. **68**(12), 12760–12770 (2020)
88. Babak, V., Zaporozhets, A., Zvaritch, V., Scherbak, L., Myslovych, M., Kuts, Y.: Models and measures in theory and practice of manufacturing processes. IFAC-PapersOnLine **55**(10), 1956–1961 (2022)
89. Zaporozhets, A., Eremenko, V., Isaenko, V., Babikova, K.: Approach for creating reference signals for detecting defects in diagnosing of composite materials. In: Conference on Computer Science and Information Technologies, pp. 154–172. Springer International Publishing, Cham (2019)
90. Babak, V., Zaporozhets, A., Kuts, Y., Scherbak, L., Eremenko, V.: Application of material measure in measurements: theoretical aspects. In: Systems, Decision and Control in Energy II, pp. 261–269. Springer International Publishing, Cham (2021)

91. Babak, V.P., Babak, S.V., Eremenko, V.S., Kuts, Y.V., Myslovych, M.V., Scherbak, L.M., Zaporozhets, A.O. (2021) Models of measuring signals and fields. Models and Meas Measurements Monit. 33–59
92. Babak, V.P., Babak, S.V., Eremenko, V.S., Kuts, Y.V., Myslovych, M.V., Scherbak, L.M., Zaporozhets, A.O.: Models and measures for the diagnosis of electric power equipment. Models Measures Measurem Monit. pp. 99–126 (2021)
93. Babak, V.P., Babak, S.V., Eremenko, V.S., Kuts, Y.V., Myslovych, M.V., Scherbak, L.M., Zaporozhets, A.O.: Problems and features of measurements. Models Measur Measurem Monitor. pp. 1–31 (2021)
94. Babak, V.P., Babak, S.V., Eremenko, V.S., Kuts, Y.V., Myslovych, M.V., Scherbak, L.M., Zaporozhets, A.O.:. Models and measures for atmospheric pollution monitoring. Models Measures Measurements Monitor. 227–266 (2021)
95. Korban, D., Melnyk, O., Onishchenko, O., Kurdiuk, S., Shevchenko, V., Obniavko, T.: Radar-based detection and recognition methodology of autonomous surface vehicles in challenging marine environment. Sci. J. Silesian Univ. Technol. Series Transport/Zeszyty Naukowe Politechniki Slaskiej. Seria Transport, 122 (2024)

Chapter 3
Stochastic Models of Diagnostic Signals Arising During the Operation of Electrical Equipment

Abstract This chapter presents a comprehensive mathematical framework for the analysis and diagnostic application of vibration and acoustic emission (AE) signals in electrical equipment (EE). Through the development of linear random processes and multi-resonant models, it offers a detailed representation of vibration and AE signals generated by rolling bearings and laminated magnetic circuits within electrical machines. The models are designed to decipher the complex condition of equipment components by analyzing vibrations caused by mechanical interactions and AE resulting from operational stresses. Key contributions include the establishment of a vibration model that simulates the multi-resonant system's response to generating impulses, reflecting the dynamics of rolling bearings, and a model for AE that integrates continuous and discrete signal components. This dual approach allows for a nuanced analysis of EE conditions, particularly focusing on bearing vibrations and laminated magnetic circuit responses to operational defects. The proposed models are centered on the understanding that diagnostic signals can be dissected into probabilistic characteristics, offering novel diagnostic signs through the statistical estimation of signal parameters. These models are crucial for identifying defects and monitoring the condition of EE, with significant implications for enhancing reliability and extending the lifespan of such equipment. The findings propose a methodological advancement in the diagnostic domain, providing a solid foundation for future research into the application of these models in real-world diagnostics and preventive maintenance strategies.

Keywords Vibration diagnostics · Acoustic emission · Electrical equipment · Information-measuring system · Linear random processes · Diagnostic signs · Probabilistic characteristics

Modern theory and practice of measuring the probabilistic characteristics of random processes proceed from the fact that measurements of characteristics should be preceded by the construction of a model of the analyzed process. It is built on the basis of a priori data supplemented by assumptions and can be mathematical and

V. Babak et al., *Statistical Diagnostics of Electric Power Equipment*, Studies in Systems, Decision and Control 573, https://doi.org/10.1007/978-3-031-76253-6_3

physical. In other words, the model of the process under study is primary, and the processing and analysis of experimental data results is secondary [1–5].

This chapter discusses mathematical models of vibration and acoustic emission processes, which can be used as diagnostic signals characterizing information about the technical condition of individual components of electrical equipment (EE). It is also shown below that linear random processes are the most adequate mathematical model that describes these diagnostic signals.

To solve the problem of vibration diagnostics of the technical condition of EE components, it is first of all necessary to have a mathematically reasoned model of the vibration process, which is used as a stochastic carrier of information about the state of these components. To build it, you need to connect the basic mathematical premises with the physics of the occurrence of vibrations in the units of operating electrical equipment.

In this chapter, as examples, the issues of vibration occurrence in rolling bearings and laminated magnetic circuits are considered. It is shown that vibrations in the bearing are generated mainly by some impulse flow that occurs during the operation of the bearing elements. The class of random impulse effects also causes vibrations in the laminated magnetic circuit when a defect appears in the latter, such as fluffing of the outer packages. The starting point for the mathematical formalization of vibrations is to represent them in the form of a response of the mechanical system under study (bearing, laminated magnetic circuit) to the influence of the generating impulse flow generated in the system itself. The white noise described above is used as a model of the generating pulse flux, and a multi-resonant linear system, characterized by a certain set of equivalent coupled circuits, is used as models of the mechanical systems of a bearing and a laminated magnetic circuit. To analyze such systems, the method of electromechanical analogies is used.

Analysis of the mathematical models of vibrations of the bearing and laminated magnetic circuit proposed below made it possible to find the characteristic, correlation functions, and spectral density of the vibrations under study under the assumption that the analyzed mechanical systems are multi-resonant. Based on the obtained probabilistic characteristics of vibrations, in subsequent chapters, diagnostic signs are established and decisive rules are constructed for diagnosing and classifying the type of defect in the nodes under study.

Using the theory of linear random processes (LRP), an acoustic emission (AE) model has been proposed and analyzed, which acts as a diagnostic signal characterizing the degree of loading (including cyclic) of some electrical machine (EM) components. Such units may include the housing and its fastening elements, bearing units, etc.

3.1 Representation of Vibrations of EM Components Using LRP Model

While constructing a mathematical model of vibrations of EM nodes, we assume that the vibration process observed at the point of sensor placement $\xi(t)$—linear system response (Fig. 3.1) on the influence of the generative process, which is a sequence of individual impulses. Each of these pulses is determined by the probabilistic law of occurrence in time and has a certain duration and intensity, the values of which are random. Generative process $\eta'(\tau)$ can be described as a random process with independent values, the distribution of which is infinitely divisible [6–8].

The possible relative arrangement of the introduced pulses occurring at moments in time $\tau = \tau_1, \tau = \tau_2$ is shown in Fig. 3.2. The mechanism of the occurrence of pulses, the shape of which is characteristic of a resonant system, is illustrated in Fig. 3.2, and an aperiodic one – in Fig. 3.3.

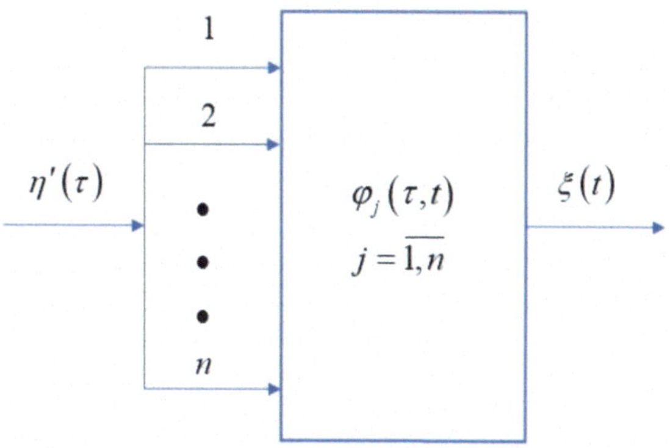

Fig. 3.1 Model of vibrations of EM nodes in the form of a response of a linear system

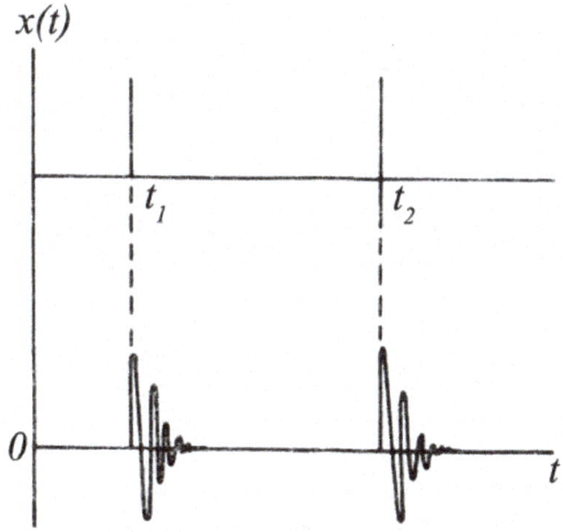

Fig. 3.2 Pulse shape characteristic of a resonant system

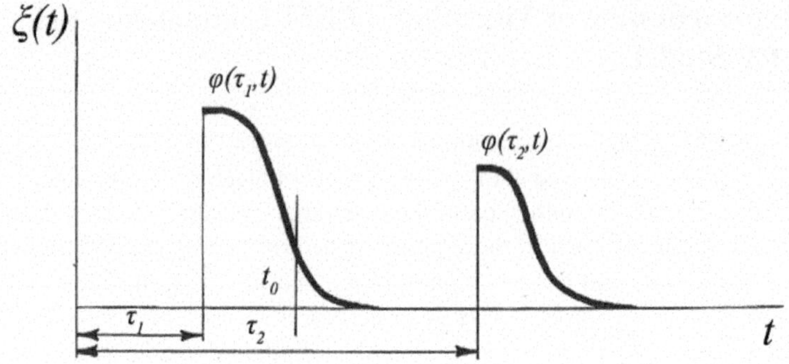

Fig. 3.3 Pulse shape characteristic of an aperiodic system

Let us make some assumptions that are necessary to construct a mathematical model of the generating process $\eta'(t)$.

1. The individual elementary impulses that are formed during the rotation of the bearing are statistically independent, that is, the assumption is fulfilled that the magnitude and nature of the impulse that appears in the interval $\Delta t_2 = t_3 - t_2$, $t_3 > t_2$ does not depend on the magnitude and nature of the impulse, arising on the interval $\Delta t = t_1 - t_0$, $t_2 > t_1 > t_0$.
2. Over a short period $(\tau, \tau + \Delta\tau)$, the probability of the occurrence of an impulse is $\lambda\Delta\tau + 0(\Delta\tau)$, where λ is the probability of the occurrence of more than one impulse.
3. Random function $\tilde{\varphi}(\tau, t)$ describing a single impulse satisfies the condition $\tilde{\varphi}(\tau, t) = \tilde{\lambda}_\tau \varphi(\tau, t)$, where $\varphi(\tau, t)$—non-random function that satisfies the condition of physical realizability $\varphi(\tau, t) \equiv 0$ at $t < \tau$, $\tau \in [0, \infty)$; coefficient $\tilde{\lambda}_\tau$—random stationary function τ with given distribution function $F(x)$, having finite dispersion [9–11].

Due to the assumptions made, we can assume that elementary pulses have characteristics that change in a non-random way, and their factors $\tilde{\lambda}_\tau$ are random and stationary in time. The moments of arrival of pulses are random, the probability of their occurrence depends only on the length of the observed time interval Δt, and does not depend on the moment the observation begins.

Assumption 2 is sufficient to conclude that during time the probability of the appearance of exact impulses

$$P_k = \frac{(\lambda_0 \tau)^k}{k!} e^{-\lambda_0 \tau} \text{ at } \tau,$$

where—λ_0 pulse intensity is equal to the mathematical expectation of the number of pulses per unit time.

Let $\pi_1(\tau)$—generalized Poisson process with independent increments [11–13], specified on a time interval $-\infty < \tau < \infty$ provided that $\mathrm{P}\{\pi_1(0) = 0\} =$

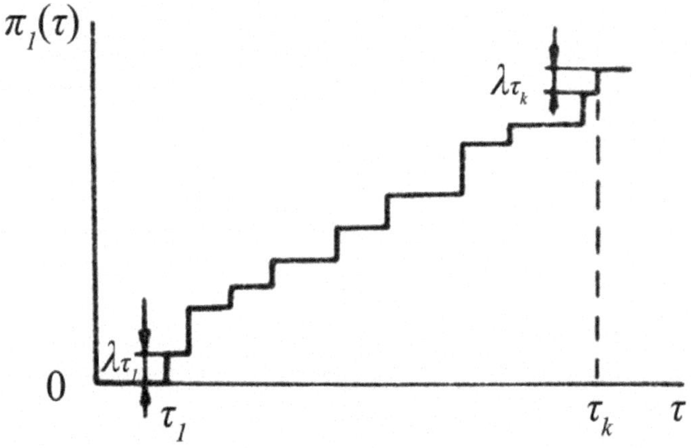

Fig. 3.4 Possible implementation of the process $\pi_1(\tau)$

1. We believe that growth points τ_k the implementation of the process coincide with the moments of the onset of the impulse. At these points in the process, the implementation gets jumps of magnitude $\tilde{\lambda}_{\lambda_k}$(Fig. 3.4). The time intervals between jumps $\pi_1(\tau)$ satisfy Assumption 2.

We believe that the process fulfills the condition of independence of increments over time intervals (τ_3, τ_4), which do not intersect, that is, when $\tau_1 < \tau_2 \leq \tau_3 < \tau_4$ valuables $\pi_1(\tau_2) - \pi_1(\tau_1)$ and $\pi_1(\tau_4) - \pi_1(\tau_3)$ mutually independent. During this time $\tau + \Delta\tau$, the increment of this process is equal to the sum of the jumps during this period the increment of this process is equal to the sum of the jumps during this period

$$\pi_1(\tau + \Delta\tau) - \pi_1(\tau) = \sum_{\tau_k \in (\tau, \tau+\Delta\tau)} \lambda_{\tau_k}.$$

Thus, taking into account the assumptions made, the vibration process at the point where the accelerometer is placed for a fixed point in t time can be represented as

$$\xi(t) = \sum_{k=-\infty}^{\infty} \lambda_{\tau_k}\phi(\tau_k, t). \tag{3.1}$$

Using a generalized Poisson process $\pi_1(\tau)$ we write (3.1) in the form

$$\xi(t) = \int_{-\infty}^{t} \phi(\tau, t)d\pi_1(\tau) = \int_{-\infty}^{\infty} \phi(\tau, t)d\pi_1(\tau), \tag{3.2}$$

because $\varphi(\tau, t) \equiv 0$ at $t < \tau$.

A working rolling bearing can be represented as a mechanical or corresponding electrical multi-resonant multi-input system (see Fig. 3.1) with a vector of pulse transition functions $\left\{\varphi_j(\tau, t), j = \overline{1, n}\right\}$. In this case, the response of the multi-input system under study can be considered as the sum of random processes coming from different interacting elements of the bearing, i.e.

$$\xi(t) = \sum_{j=1}^{n} a_j \xi_j(t),\ t \in (-\infty, \infty), \tag{3.3}$$

where n—certain positive number, which is determined by the design, technological, and operational characteristics of the bearing a_j; weight coefficients that take into account attenuation in the corresponding channel $a_j > 0$;

$$\xi_j(t) = \int_{-\infty}^{\infty} \phi_j(\tau, t) d\eta(\tau), j = \overline{1, n}$$

The generating process $\eta'(\tau)$, the nature of which is largely determined by the technical condition of the rolling bearing, can be specified using CF $f_\eta(u, t)$ [14–19]. Depending on the components that are included in the CF $f_\eta(u, t)$, one can judge the generative process $\eta'(\tau)$. So, if the generating process defines the CF in Kolmogorov form [20–23] and assumes that $K(x) \equiv 0, |x| > 0$, then in this case the generating process $\eta'(\tau)$ acting on the bearing, which we can imagine as a linear system with an impulse transition function $\varphi_j(\tau, t)$, is normal.

If in the expression for the considered CF $K(x) \neq 0$, then it will be the sum of normal (continuous) and impulse random processes. Depending on the type of function selected $K(x)$ the impulse process $\eta'(\tau)$ can be a Poisson process, a gamma process, or some other process with independent increments [24–26]. Similar use of the canonical CF representation $f_\eta(u, t)$ allows, when considering a bearing vibration model, to simultaneously take into account the continuous and impulse components. The latter, in particular, may be due to some bearing defects. Thus, based on the proposed models, the process of vibrations $\xi(t)$ at the point where the accelerometer is located can be considered as the response of a multi-resonant linear system, which is described by a vector of impulse transition functions $\varphi_j(\tau, t)$ when its inputs are affected by a generating process $\eta'(\tau)$. In this case, for a mathematical description of the process $\xi(t)$, you can use the class of linear random processes [27–30].

Since the nature of the vibration process $\xi(t)$ is largely determined by the type of impulse transition function $\varphi_j(\tau, t)$, let us consider in more detail the mechanical model of the bearing and its electrical analogy.

3.2 Mechanical Model of a Rolling Bearing Operating on the Shaft of an EM, and Its Electrical Analogy

To explain the multi-resonance structure of vibrations that arise during the operation of rolling bearings, let us consider a flat simplified model of a single-row bearing that has a rolling element and operates on a horizontally located EM. A simplified flat mechanical diagram of a bearing, in which the outer ring is fixedly fixed on the shaft, and contains the n_r number of rolling elements [31–34], is presented in Fig. 3.5.

In this diagram the parameters $C_k\prime$ and $C_k^{\prime\prime}$, $k = \overline{1, n_r}$ describe the equivalent elasticity of the bearing rings. It is taken into account that the space between these rings, rolling elements, and the separator is filled with grease. Variable r_k, $k = \overline{1, n_r}$—friction, which is introduced into the system due to the movement of rolling elements in the cage and caused by the viscosity of the lubricant. Variable M – added rotor mass EM. Each bearing rolling element has a mass m_k, $k = \overline{1, n_r}$. We roughly assume that the masses of the rolling elements of the new bearing are equal to each other, i.e. $m_1 = m_2 = m_3 = \ldots = m_n$. However, as wear occurs, a noticeable scatter of masses may appear between the masses of the rolling elements, which will cause asymmetry of the bearing elements and lead to the appearance of additional components in the spectral composition of its vibrations.

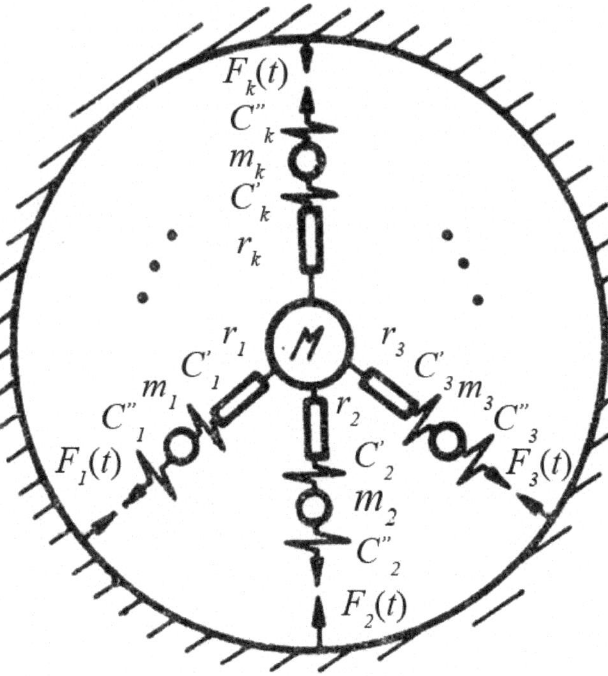

Fig. 3.5 Mechanical diagram of a rolling bearing

Following the method of electromechanical analogies [35, 36], the mechanical bearing system (Fig. 3.5), which contains masses M, m_1, m_2, …, $m_{n_{iii}}$, elastic connections C'_k and friction r_k,$k = \overline{1, n_r}$, transforms into an equivalent electrical circuit with inductances L , L_1 , L_2 , … , L_n, containers,C_k and resistances R_k, $k = \overline{1, n_r}$.

m_k, C_k and r_k do not go directly to L_k, an electrical circuit, and are associated with complex dependencies. Elements L_k, C_k and R_k are connected in a certain way in the equivalent circuit of a rolling bearing and form unique *RLC*-circuits capable of resonating at certain frequencies. So, based on consideration of the simplified mechanical diagram of a rolling bearing, we can conclude that it is a multi-resonant system. This is confirmed experimentally in this chapter, as well as in [17, 18].

Every equivalent force $\{F_k, k = \overline{1, n_r}\}$, which acts on the rolling bodies in the radial direction, can be considered as a projection onto the corresponding radial directions of two force components: the outer ring - the rolling bodies; the inner ring is the rolling elements. On the inner ring, these forces are applied at a distance (the apex of the angle coincides with the center of the rotor) equal to $2\pi / n_r$ radians multiplied by the radius of the ring [37, 38]. In addition, the mass of the rotor at idle uniform speed or rest is distributed over all rolling bodies that are located in the lower half-plane. Thus, when the rotor rotates, each rolling body during the half-cycle of rotation will experience the influence of the rotor mass, which changes according to a sinusoidal law and is "cut off" when the positive half-wave passes. Consequently, each force represents rectified half-cycles of a pure harmonic oscillation (Fig. 3.6), on which vibrations from electromagnetic forces are superimposed, due to the mechanical imbalance of the rotor, due to the operation of the brush-collector unit, as well as due to aerodynamic forces. In addition, anyone $F_k(t)$ is affected by the generating process, which is caused by the waviness of the raceways, the irregularity of the shape of the rolling elements, the tracks of the inner and outer rings, as well as their possible defects (cracks and chips of metal), the presence of wear products in the lubricant, etc.

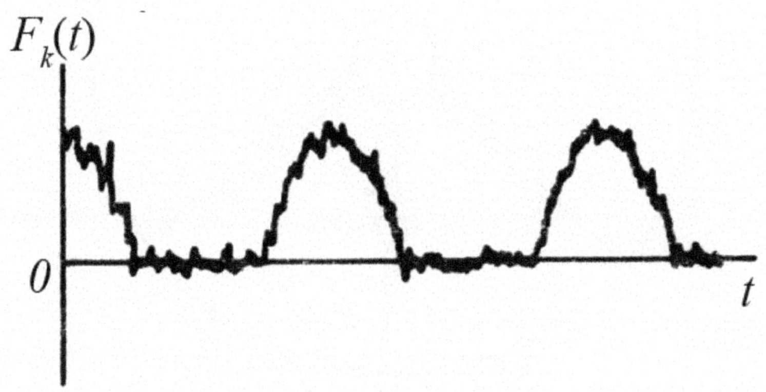

Fig. 3.6 Force acting on the rolling elements of the bearing

As a result, each equivalent force, taking into account the listed factors, can be described as

$$F_k(t) = \frac{1}{2}\overline{a}[\cos(\omega t + \phi_k) + |\cos(\omega t + \phi_k)|] + \eta'(t) + f_k(t), \tag{3.4}$$

where $\overline{a}$ – amplitude equal to the average value of the force that acts on the rolling body in the lower position; φ_k– the initial phase of the corresponding rolling body, if for the first we take $\varphi_1 = 0$, то $\varphi_k = 2\frac{\pi}{n_\omega}(k - 1)$; $\eta'(\tau)$—generative process; $f_k(t)$—a component that includes electromagnetic forces caused by the unbalance of the rotor, as well as vibrations due to the brush-commutator assembly and electrodynamic forces.

The electrical analog of force $F_k(t)$ in the equivalent electrical circuit of a rolling bearing by the methods of electromechanical analogies, equivalent currents can serve $I_k(t)$ or voltage $U_k(t)$. The image of a bearing in the form of an electrical circuit allows the use of well-developed methods of the theoretical foundations of electrical engineering, statistical radio engineering, and probability theory in its analysis. In addition, this approach allows you to simulate the operation of a bearing on a computer.

The main task of diagnosing the technical condition of a bearing is, first of all, to identify a defect in it [39–43]. The occurrence of a defect causes a change in the multi-resonant bearing system and changes the nature of the generating process $\eta'(t)$. This in turn leads to changes in $\xi(t)$ vibrations, which are recorded by the accelerometer. Using the above results, it is possible to construct a mathematical model of the oscillatory process of vibrations, which are caused by the operation of the bearing assembly and are observed at the point where the accelerometer is installed. However, before moving on to considering the issue of constructing mathematical vibrations of a rolling bearing, let us dwell on some issues of developing models of vibrations of a laminated magnetic core of an EM. This is because the class of linear random processes, which is used in the construction of the mentioned mathematical models, is a universal mathematical apparatus that makes it possible to construct and further study many physical processes that accompany the operation of various technical objects.

3.3 Vibration Models of Laminated Magnetic Circuits of EM

While constructing a model of vibrations of a laminated magnetic core [44–46], we will make some assumptions related to the fact that the study of vibrations of continuous elastic systems (which include a laminated magnetic core) as systems with an infinite number of degrees of freedom encounters great difficulties of a fundamental nature. The possibility of solving a mathematical problem about the vibrations of the node under consideration becomes feasible only if significant simplifications are introduced into the calculation. Various methods are known [47–49] for constructing simplified diagrams of devices for calculating vibrations.

As an example, let us consider the formation of vibrations in a laminated magnetic core, which is used as the stator core of a turbogenerator [50–52]. Structurally, this core (like any laminated magnetic circuit) is a set of plates made of sheet magnetostrictive steel. If we do not take into account the thickness of the insulating material between the sheets in comparison with the thickness of the tooth when ensuring the required compaction density, it can be assumed that the laminated magnetic circuit is made of a homogeneous material having the same density throughout the entire volume.

To determine the parameters of the mechanical system that affect the frequency characteristics of the laminated magnetic circuit, we will consider the process of the occurrence of oscillations in the outermost package of the stator core.

The outermost package of the stator core can be simplified as a homogeneous rod. As is known from the theory of vibrations [53, 54], free longitudinal vibrations of a homogeneous rod are described by a second-order linear partial differential equation:

$$\frac{E}{\rho}\frac{\partial^2 u}{\partial x^2} = \frac{\partial^2 u}{\partial t^2}, \tag{3.5}$$

where E—elastic modulus; ρ—density of the rod material; u—longitudinal displacement of the equilibrium point of the cross-section of the rod.

The solution of the linear differential Eq. (3.5) with zero boundary and initial conditions gives a relation for determining the natural frequencies of oscillations of the rod

$$f_n = \frac{n}{2 \cdot l}\sqrt{\frac{E}{\rho}},\ n = 1,\ 2,\ 3,\ \ldots, \tag{3.6}$$

where l—rod length.

Expression (3.6) is valid for the case of a thin rod, i.e. when the transverse dimensions of the rod are small compared to its length. If this condition is not met, a correction factor must be introduced into Eq. (3.5). Taking into account the correction, instead of expression (3.6) for a round rod according to [], one should use the relation

$$f_n = \frac{n}{2 \cdot l}\sqrt{\frac{E}{\rho}}\left[1 - \left(\frac{n \cdot \nu \cdot \pi \cdot r}{2 \cdot l}\right)^2\right], \tag{3.7}$$

where ν – Poisson's ratio, r—radius of the rod.

Expression (3.7) $(\pi\ r)^2$ can be represented as $S\pi$, where S—the cross-sectional area of the rod. Expression (3.7) can also be used for rods of a shape other than cylindrical, if the cross-section of the rod has a slight deviation from the circle.

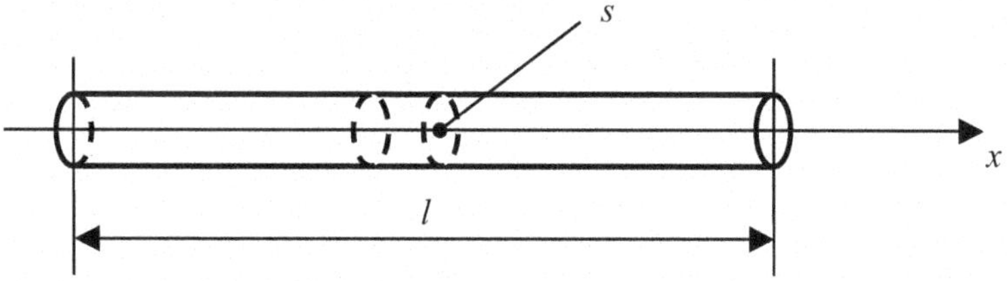

Fig. 3.7 Model of a rod to describe the process of propagation of longitudinal vibrations

The propagation of free longitudinal vibrations described by Eq. (3.5) in a homogeneous rod is illustrated in Fig. 3.7.

The outer package of the stator core of a turbogenerator can be considered a linear mechanical system if we assume that the density of the material is the same throughout the entire volume, and then its vibrations will be described by a linear equation of the form (3.7).

As already mentioned, during the operation of the turbogenerator, a gradual weakening of the pressing of the active steel occurs, and, as a result, the separation of the tooth plates from the laminated core mass occurs. In this case, the natural frequency of oscillation of the tooth plate is determined by the length of the part of the tooth plate that has completely lost contact with the laminated core mass, and the boundary conditions depending on the nature of the fastening of the tooth plate, i.e. the ability to move that part of the tooth plate where the pressing pressure has been maintained and a fixed gap has not yet formed between the plates and the core array.

The most common is the mechanical model of the tooth plate as a cantilevered rod. In works [17, 18], a schematic picture of the magnetic field of an individual plate in the toothed zone of the outermost package is considered. Similar expressions were also obtained there for the forces when the induction in the core changes according to a harmonic law. However, as practice shows, the oscillations of the plate under the influence of the force arising as a result of this induction cannot be considered strictly deterministic. Therefore, we will substantiate the case when in the oscillations of the outer plate, in addition to the deterministic component, there is also a random one.

The works [17, 18] consider a flat schematic picture of the magnetic field of an individual plate in the tooth zone, as well as simplified diagrams of the penetration of an axial flow into the end zone of the core for two states of the outermost package: for compressed and fluffy packages, i.e. for a condition when a gap forms between the outer and other plates. For a more visual representation of the distribution of forces acting on a separate plate in the end zone when fluff appears, this work provides three-dimensional diagrams of the magnetic field pattern in the tooth zone. Figure 3.8 shows a volumetric schematic diagram of the magnetic field of an individual plate in the toothed zone of the outermost package, and Figs. 3.9 and 3.10 show simplified volumetric diagrams of the penetration of the axial flow into the end zone of the core for two states of the outer package—tightly compressed (Fig. 3.9) and fluffy (Fig. 3.10).

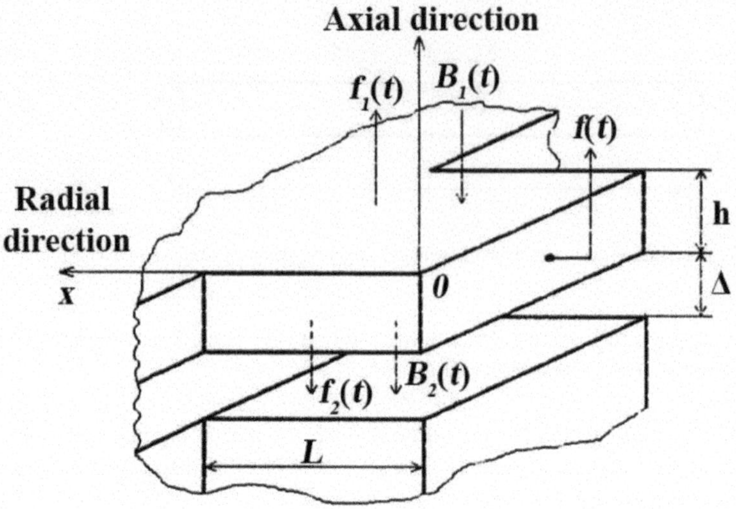

Fig. 3.8 Schematic picture of the volumetric magnetic field of an individual plate in the toothed zone of the outer core package

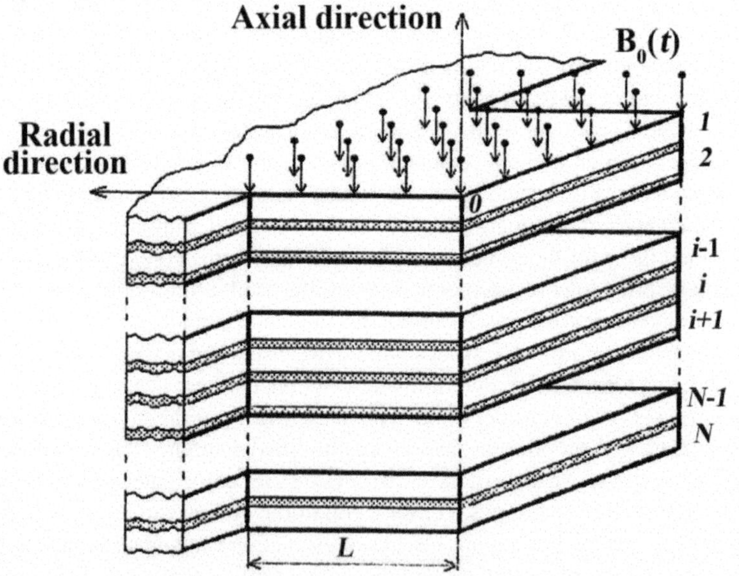

Fig. 3.9 Scheme of penetration of axial flow into the end zone of the core for a tightly compressed outermost

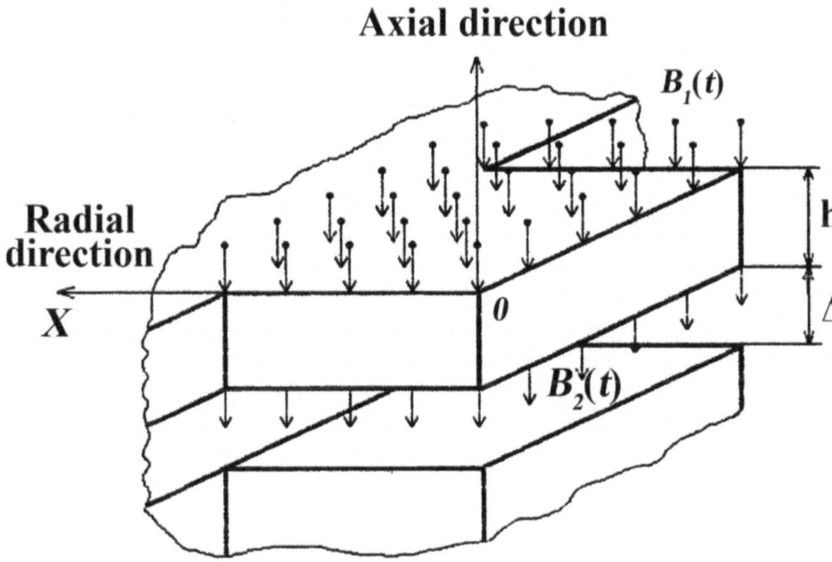

Fig. 3.10 Scheme of axial flow penetration into the end zone of the core for a fluffy outer package

In the given figures *h*—plate thickness; Δ—size of the gap that forms between the outer plate and the rest of the iron package as a result of fluffing; *x*—radial coordinate of the plate for which tight contact with the rest of the stacked magnetic circuit has been lost; $f_1(t)$—force arising due to induction $B_1(t)$ entering into a separate tooth plate: $f_2(t)$—force arising from induction leaving the plate; $f_0(t)$—resultant force acting on the tooth plate; $B_0(t)$—axial induction in the end plane for a compressed package.

Based on these results, the force $f(t)$ is determined by the relation

$$f(t) = f_1(t) - f_2(t) = \frac{1}{2\mu_0}\left[B_1^2(t) - B_2^2(t)\right], \tag{3.8}$$

where $\mu_0 = 1.27 \cdot 10^{-6}$ T·m/A—magnetic permeability of air, i.e. layer dividing the space between the plates.

It was noted above that the deterministic component $f(t)$, determined according to (3.8), is also influenced by the random component. The latter arises as a result of the mutual collision of the fluffy outer plate (or plates) with the remaining compressed mass of the magnetic circuit core. In turn, these mutual collisions of the fluffy plates are caused mainly by vibration waves that arise in a working EM due to vibrations of bearing units, possible imbalance of the rotor, as well as vibrations of the brush-collector unit and aerodynamic noise. Thus, the resulting force acting on the end plate in a fluff bag can be described as follows:

$$F(t) = f(t) + \eta'(t), \tag{3.9}$$

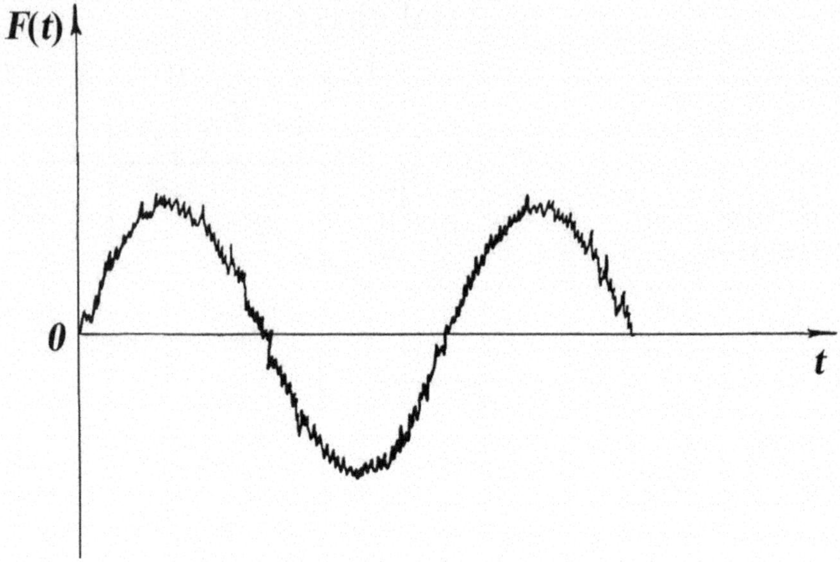

Fig. 3.11 Total force acting on the outer plate of the magnetic core

where $f\ (t)$ – force acting on the outer plate as a result of a harmonic change in induction and determined according to (4.8), $\eta\prime(t)$ – impact that causes random vibrations of the outer plates of the magnetic core.

The harmonic component and the random pulses additively superimposed on it are illustrated schematically in Fig. 3.11.

Next, we will consider the component $\eta\prime(t)$, included in relation (3.9), since by measuring the random process generated by this component, it is possible to diagnose a fuzz-type defect in a laminated magnetic circuit.

So, the component $\eta\prime(t)$ is a flow of impulses random in time and magnitude, which is called the generating process above. We will also assume that the indicated random impulses are independent and satisfy the condition of uniform infinite smallness []. As is known, such a physical process is most adequately described by a mathematical model based on the theory of random processes with independent values, the distribution of which is infinitely divisible [55–57].

Let us draw preliminary conclusions regarding the representation of the stator magnetic circuit as a mechanical system, which is affected by the considered generating process. The laminated magnetic core of the stator of a turbogenerator is a mechanical oscillatory system capable of resonating at certain frequencies under the influence of the generating process. The latter is due to both the design of the magnetic core itself and the entire turbogenerator as a whole. As is known [58–60], an oscillatory system can be represented by one or a set of linear links of the first and second orders. Consequently, to analyze such a representation of a magnetic circuit, one can use the well-developed apparatus of the theory of linear systems [61, 62]. In this case, the vibrations recorded at the point where the vibration sensor is placed represent the response of the linear system to the influence of the generating process.

Having made preliminary assumptions regarding the formation of vibrations in the stator of a turbogenerator, let us proceed directly to the choice of a mathematical apparatus for describing the processes discussed above.

Using physical premises, as well as the mathematical apparatus of the theory of linear random processes [9, 11, 63–65], we will construct a mathematical representation of the generating process $\eta'(t)$. Let us also consider the representation of vibrations of the electromagnetic system in the form of a random process $\xi(t)$ as a response of a linear system, with the use of which the laminated magnetic circuit of the stator core of a turbogenerator is represented, to the influence of the generating process $\eta'(t)$. When constructing a mathematical model of vibrations of the stator of a turbogenerator, we will assume that the vibration process $\xi(t)$ observed at the point of placement of the vibration sensor is the result of the impact on the magnetic circuit of a large number of individual pulses, each of which is determined by the probabilistic law of occurrence in time and has a certain duration and energy. The value of the latter is random. Since the stator core of a turbogenerator consists of various components (laminated magnetic core, bandage rings, coupling prisms, etc.), each of which is characterized by its vibration frequency, we will assume that the core under study can be considered as a multi-resonant multi-input linear system of the second order. The following paragraph is devoted to the construction of the above-mentioned models.

3.4 Mathematical Model of Linear RLC-Noise in Describing Vibrations of Rolling Bearings and Laminated Magnetic Circuits of Electrical Machines

As shown in the previous paragraph, in the general case, the rolling bearing and the laminated magnetic circuit of the stator core can be considered as a multi-input second-order oscillatory system [66–68], characterized by a certain vector of impulse transition functions with components

$$\phi_j(\tau) = \frac{\varpi_j^2}{\psi_j} e^{-\beta_j \tau} \sin(\psi_j \tau) U(\tau), \tag{3.10}$$

where $\psi_j = \sqrt{\omega_j^2 - \beta_j^2}$; $\beta_j = \frac{R_j}{2L_j} > 0$; $\omega_j = \frac{1}{\sqrt{L_j C_j}}$; $\omega_j > \beta_j$.

The choice of a pulsed transition function of the form (3.10) is because the simplest circuit in the oscillatory mode has such a characteristic, as well as the satisfactory agreement of the theoretical conclusions with the experimental simulation results. However, the assumption that a linear mechanical system is characterized only by second-order links of the form (3.10) operating in an oscillatory mode is not always consistent with practice.

Thus, in some spectrograms of vibrations of bearings [69–71] and laminated magnetic circuits, a characteristic rise is sometimes observed in the region of frequencies close to zero, which can be explained by the presence of propagation paths from the source of white noise to the point of recording the vibrogram with high mechanical attenuation, i.e. aperiodic propagation channel with weakly expressed resonance properties. It can be described either by a first-order linear system model with an exponential impulse transition function (low-pass filter) [] or by a second-order aperiodic element. To preserve the universality of the approach, we will focus on the second case, since it includes the aperiodic case as a limit, because in this case the impulse transition function can be obtained as a limit (3.10) at $\omega_j \rightarrow \beta_j$ as

$$\lim_{\omega_j \rightarrow \beta_j} \phi_j(\tau) = \lim_{\psi \rightarrow 0} \omega_j^2 \tau e^{-\beta_j \tau} \frac{\sin \psi_j \tau}{\psi_j \tau} U(\tau) = \beta_j^2 \tau e^{-\beta_j \tau} U(\tau), \tag{3.11}$$

it can be considered as the case of sequential inclusion of two first-order links, i.e. chains of two identical low-pass filters [], with exponential transition functions.

There are also propagation channels described by second-order links with significant attenuation, i.e. It may happen that for some channels $\omega_j < \beta_j$ These are also aperiodic channels and for them, the pulse transition function is formally determined by relation (3.10), if we assume that ψ_j, also takes complex values. In this case, is a real function and is obtained by a formal change in (3.10) φ_j on ψ_j, where $\psi_j = \sqrt{\beta_j^2 - \omega_j^2}$:

$$\begin{aligned} \phi_j(\tau) &= \frac{\omega_j^2}{i\tilde{\psi}_j} e^{-\beta_j \tau} \sin\left(i\tilde{\psi}_j \tau\right) = \frac{\omega_j}{\tilde{\psi}_j} e^{-\beta_j \tau} sh\left(\tilde{\psi}_j\right) U(\tau) \\ &= \frac{\omega_j^2}{2\tilde{\psi}_j} \left[e^{-(\beta_j - \tilde{\psi}_j)\tau} - e^{-(\beta_j + \tilde{\psi}_j)\tau}\right] U(\tau). \end{aligned} \tag{3.12}$$

Function (3.12) is the difference between two exponential pulse transition functions and can be interpreted as a characteristic of two parallel-operating chains, the responses of which are subtracted. In the future, we will limit ourselves to detailed calculations only for links with characteristics (3.10), since, if necessary, aperiodic cases can be analyzed taking into account the reservations made above.

A probabilistic analysis [72–74] of the proposed mathematical model was carried out under the assumption that each of the inputs of a multi-input system (Fig. 3.1), represented by a vector of impulse transition functions (3.10), is affected by a process that can be considered as an additive imposition of a large number of independent impulses arising at random times. As was shown above, such an influencing physical process, within the framework of the formulated assumptions, can be described with sufficient grounds by models of the white noise type.

Vibrations of rolling bearings at the observation point on the EM bearing shield or vibrations of a laminated magnetic circuit when placing an accelerometer on a pressure plate in a steady state can be represented as a sum (3.3) of random processes—system responses (Fig. 3.1) for each of the propagation channels, described by a vector with components (3.10) when the input of each channel is affected by a random process such as white noise. In the general case, the components can be stationary stochastically coupled. In this case, each response component associated with a specific propagation channel and included as a term in (3.3) is represented in the steady state as a linear process

$$\xi_j(t) = \int_{-\infty}^{\infty} \phi_j(t-\tau)d\eta_j(\tau), \quad j = \overline{1,n}, \tag{3.13}$$

where each $\phi_j(\tau)$, $\tau \in (-\infty, \infty)$, is determined by (3.10) taking into account (3.11) for individual components, and $\left\{\eta_j(\tau), j = \overline{1,n}\right\}$ is a vector of generating processes with independent increments, the derivatives of which affect the system inputs (Fig. 3.1). In this case, it is assumed that the components of the vector of the generating process are either stochastically connected, or this connection is so strong that they can be considered as stochastically equivalent.

Thus, the division of the total vibration process at the recording point into components is carried out per the resonant properties of each channel and the oscillatory system of the diagnosed unit as a whole.

The correlation function [75–78] of the process (3.3) in a steady state, taking into account (3.13), is written as

$$R(s) = \sum_{k,j=1}^{n} a_k a_j h_{kj}(s), s \in (-\infty, \infty), \tag{3.14}$$

where $h_{kj}(s)$—mutual correlation function of the k-th and j-th propagation channels. Note that (3.14) is an autocorrelation function of the stationary process (3.3), therefore it is an even function, i.e. $R(s) = R(-s)$. Hence

$$R(s) = R(|s|) = \sum_{k,j=1}^{n} a_k a_j h_{kj}(|s|), s \in (-\infty, \infty), \text{ where } h_{kj}(|s|)$$

$$= \int_{-\infty}^{\infty} \phi_k(\tau)\phi_j(\tau + |s|)d\tau$$

Having performed the integration in the last expression, we get

$$h_{kj}(|s|) = \frac{k_{2kj}}{2}\frac{\omega_k^2\omega_j^2}{\psi_k\psi_j}e^{-\beta_j|s|}\left[a_{kj}\cos\psi_j s + b_{kj}\sin\psi_j|s|\right], \tag{3.15}$$

where for all $k, j = \overline{1, n}$, k_{2kj}—mixed second semi-invariant of random variables $\eta_k(1)$ and $\eta_j(1)$, $k_{2kj} = k_2[\eta_k(1)\eta_j(1)]$, which at $k = j$ goes into the usual variance of the corresponding random variable,

$$\begin{aligned} a_{kj} &= \frac{\beta_{kj}}{\beta_{kj}^2 + \tilde{\psi}_{kj}^2} - \frac{\beta_{kj}}{\beta_{kj}^2 + \psi_{kj}^2} \geq 0; \\ b_{kj} &= \frac{\tilde{\psi}_{kj}}{\beta_{kj}^2 + \tilde{\psi}_{kj}^2} + \frac{\psi_{kj}}{\beta_{kj}^2 + \psi_{kj}^2}; \ \beta_{kj} = \beta_k + \beta_j; \\ \psi_{kj} &= \psi_k + \psi_j; \ \tilde{\psi}_{kj} = \psi_k - \psi_j \end{aligned} \tag{3.16}$$

It should be noted that when $\psi_j \to 0 \psi_j \to 0$

$$\lim_{\psi_j \to 0} h_{kj}(|s|) = k_{2kj} \frac{(\omega_k \beta_j)^2}{\beta_{kj}^2 + \psi_k^2} |s| \, e^{-\beta_j |s|}. \tag{3.17}$$

Thus, (3.14) taking into account (3.15) takes the form

$$R(s) = \sum_{k,j=1}^{n} \frac{1}{2} a_k a_j k_{2kj} \frac{(\omega_k \omega_j)^2}{\psi_k \psi_j} e^{-\beta_j |s|} \left[a_{kj} \cos \psi_j s + b_{kj} \sin \psi_j |s| \right], \ s \in (-\infty, \infty). \tag{3.18}$$

If we introduce the notation

$$A_{jn} = \frac{a_j \omega_j^2}{2\psi_j} \sum_{k=1}^{n} \frac{a_k \omega_k^2 a_{kj}}{\psi_k} k_{2kj} \geq 0; \ B_{jn} = \frac{a_j \omega_j^2}{2\psi_j} \sum_{k=1}^{n} \frac{a_k \omega_k^2 b_{kj}}{\psi_k} k_{2kj}, \tag{3.19}$$

then we can finally represent the correlation function of the process (3.3) in the form

$$R(s) = \sum_{j=1}^{n} e^{-\beta_j |s|} \left[A_{jn} \cos \psi_j s + B_{jn} \sin \psi_j |s| \right], \tag{3.20}$$

at all $s \in (-\infty, \infty)$.

In (3.20) the vector components ψ_j, $j = \overline{1, n}$, we will call resonant frequencies, since they determine the positions of the spectrum maxima, and the components of the vector β_j, are attenuation coefficients. Although the last expression was obtained for the case $\psi_j > 0$, $j = \overline{1, n}$, however, it remains valid for any number of resonant frequencies close to zero and, by continuity, can be extended to the case if, at some j frequency $\psi_j \to 0$. This justifies the name of process (3.3) as a multiresonance model.

At $s = 0$ from (3.20) we have the dispersion of the process (3.3)

$$R(0) = \sum_{j=1}^{n} A_{jn} = \sum_{j,k=1}^{n} \frac{a_j a_k (\omega_j \omega_k)^2}{2\psi_j \psi_k} a_{kj} k_{2kj}. \tag{3.21}$$

Expression (3.20) can be represented as a sum of exponential-sine components

$$R(s) = \sum_{j=1}^{n} e^{-\beta_j |s|} C_{jn} \sin(\psi_j |s| + \Phi_{jn}), \tag{3.22}$$

where $C_{jn} = \sqrt{A_{jn}^2 + B_{jn}^2}$, $C_{jn} = \sqrt{A_{jn}^2 + B_{jn}^2}$. But since $A_{jn} \geq 0$, it is always $0 \leq \Phi_{jn} \leq \pi$.

Thus, the autocorrelation function of process (3.3) is completely determined by the parameters a_j, ψ_j and β_j, $j = \overline{1, n}$, which can be used as diagnostic signs when analyzing vibrations of EM rolling bearings within the framework of the correlation theory.

Process (3.3) is stationary and Hilbert, $R(0) < \infty$, therefore, there is a spectral density for it, which is defined as the Fourier cosine transform, according to [79–82], taking into account (3.20) in the form

$$S(\omega) = 2 \sum_{j=1}^{n} \frac{A_{jn}\beta_j \left(\omega_j^2 + \omega^2\right) + B_{jn}\psi_j \left(\omega_j^2 - \omega^2\right)}{\left(\omega_j^2 + \omega^2\right)^2 - 4\omega^2 \psi_j^2}. \tag{3.23}$$

Let us briefly consider the one-dimensional CF of the process (3.3). Under the assumption that all components of the vector of the generating process are stochastically equivalent, we can immediately write down the general form of its logarithm

$$\begin{aligned} \ln f_{\xi}(u) = ium \sum_{j=1}^{n} a_{jn} + \int_{-\infty}^{\infty} \int_{-\infty}^{\infty} \Bigg[\exp\Bigg(iux \sum_{j=1}^{n} a_{jn} \phi_j(\tau) \Bigg) - \\ -1 - iux \sum_{j=1}^{n} a_{jn} \phi_j(\tau) \Bigg] \frac{dK(x)}{x^2} d\tau, \end{aligned} \tag{3.24}$$

where $\varphi_j(t)$, $j = \overline{1, n}$ determined by (3.2), m—mathematical expectation of the process (3.3), $K(x)$—so-called Poisson spectrum of jumps in Kolmogorov form [20–23].

Expression (3.24) allows one to calculate the values of the probability distribution of process (3.3) using the known characteristics of the generating process using the Fourier-Stieltjes transform. The appearance of certain defects in a bearing is equivalent to a change in parameter values R_j, L_j and C_j in the proposed mathematical model (3.3), which in turn changes the initial and central moments, as well as the nature of the curve of the probability distribution function obtained based on (3.24).

Thus, to diagnose rolling bearings and laminated magnetic cores of EM based on their vibration, the following diagnostic features can be used: attenuation coefficients β_j, $j = \overline{1, n}$; options ψ_j, $j = \overline{1, n}$; magnitudes of initial and central moments; and the nature of the probability distribution function curve.

To obtain estimates of these diagnostic signs, special software is intended, which is part of the IMS for vibration diagnostics of EM.

3.5 Linear Model of AE Signals Accompanying the Operation of Electrical Equipment Units

The most important information about the technical condition of an electric power facility can be obtained using the physical process of AE as a diagnostic signal [83–86]. Particular interest in the AE process is due to its remarkable ability to provide operating personnel with advanced information about the appearance of critical loads in the diagnosed object. This information can be obtained by using a parameter as a diagnostic sign λ, which is characterized by the intensity of the appearance of AE pulses over an observed fixed period Δt.

In this subsection, the results of the development and justification of a general linear model of the AE signal will be used and specific examples of acoustic and electrical AE signals will be considered.

Let us proceed directly to the construction of a refined mathematical model of the AE process. In this case, we will be based on modern results obtained in the field of LRPs and their CFs [9], which allow us to carry out a complete probabilistic analysis, i.e. analyze the distribution of AE processes. It should be immediately noted that it is not possible to solve this problem using other approaches.

Next, we will talk about models of either the acoustic AE signal, that is, the physical process that is recorded on the surface of the object under study, or the electrical AE signal, that is, the response of the electroacoustic transducer, which is subject to processing in the IMS. If there is no need to specify the modelling object, the term "AE signal" will be used.

Studies of the mechanisms of AE, which accompanies such phenomena in solids as plastic deformation by twinning and grain-boundary sliding, the emergence and development of cracks, corrosion, friction, and the like, give reason to believe that the physical processes, that generate continuous and discrete AE, exist together []. However, in well-known works on mathematical modeling of AE signals, this fact is not always taken into account.

Conclusions can be drawn about the nature and some features of AE signals based on the essence of the phenomenon. First of all, it should be noted the qualitative difference between continuous and discrete AE signals. If the first is to some extent similar to thermal noise, then the second is characterized by abrupt changes in its parameters, that is, it has a pulsed nature. Some researchers [87–89] consider the mechanisms of continuous and discrete AE to be similar, but the characteristics of material destruction, the type of AE sources, their intensity and other indicators significantly affect the integral properties of the resulting acoustic field, which is manifested in the dual nature of AE signals.

One remark should be made here, which concerns the terminology traditionally accepted in AE theory. Most definitions are formulated with an "engineering approach" and if the physical interpretation of some terms is satisfactory, then in the mathematical analysis of the phenomenon their application leads to contradictions. For example, the terms "continuous" and "discrete" in mathematics have a very specific content, which far from coincides with what is accepted when defining continuous and discrete AE. Since this section is largely devoted to the issues of mathematical description of the phenomenon, to avoid the above-mentioned contradictions, in what follows, by continuous emission we will understand the generation of a fluctuation AE signal by a source, continuous in time, and by discrete emission, we will understand the activation of the source at certain moments in time, as a result of which an acoustic pulse is emitted.

The use of models that describe only one type of emission, especially in problems of technical diagnostics and control, can lead to a significant decrease in the reliability of the results obtained. For this reason, this work sets the task of developing and analyzing a mathematical model of a general AE signal, which contains both continuous and discrete components. The solution to this problem was obtained based on the LRP theory [9]. Let's look at this issue in more detail.

Let's start with the discrete (pulse) component, which we denote $\xi_I(t),\ \ t \in (-\infty, \infty)$. Let us consider the construction of an impulsive random process.

Taking into account [9], we will make the following physical assumptions for constructing mathematical models of AE signals.

1. In the diagnosed unit, under the influence of load, a large number of micro defects in the structure of the material are formed, ultimately leading to the appearance of macroscopic cracks. In this case, the condition of independence of both the place of occurrence of the defect and its moment is assumed.
2. The appearance of each microdefect is accompanied by the emission of an energy pulse, the sequence of which forms a certain pulse flow. The time of occurrence and amplitude of each pulse in this stream are random.
3. The AE signal under study is considered as a superposition of the response at the output of a system consisting of two linear links: the body of the node being diagnosed – the AE sensor. It is assumed that this system is affected by a pulsed AE flow, which occurs directly in the body of the diagnosed object when it is loaded.
4. Individual elementary AE pulses formed during the transition of each dislocation in the material are stochastically independent, i.e. the assumption is fulfilled that the magnitude and nature of the impulse that appears in the interval $\Delta t_2 = t_3 - t_2,\ \ t_3 > t_2$ do not depend on the magnitude and nature of the impulse arising in the interval $\Delta t = t_1 - t_0,\ \ t_2 > t_1 > t_0$.

5. In a short period, $(t,\ t + \Delta t)$ the probability of an impulse occurring – $\lambda\,\Delta t + o\,(\Delta t)$, where λ – some constant, and $o(\Delta t)$ – probability of occurrence of more than one pulse.
6. A random function $\tilde{\varphi}\,(\tau,\ t)$ describing an individual impulse satisfies the relation $\tilde{\varphi} = \rho_\tau\,\varphi(\tau,\ t)$, where $\varphi(\tau,\ t)$ – non-random function that satisfies the condition of physical realizability $\varphi(\tau,\ t) \equiv 0$ at $t \prec \tau,\ \tau \in [0,\infty)$; coefficient ρ_τ– random stationary function τ with a given distribution function $F(x)$ having finite variance.

Based on the assumptions made, we can assume that elementary pulses $\varphi\,(\tau,\ t)$ have characteristics that change in a non-random manner, and their factors ρ_τ are random and stationary in time.

Assumptions 4–6 are sufficient to conclude that over time the probability of the appearance of exact pulses can be determined by the following relation:

$$P(t) = \frac{(\lambda\,t)^k}{k!}\,e^{-\lambda\,t};\ k = 0,\ 1,\ 2,\ \ldots,\ \text{at } t > 0$$

where λ—intensity of occurrence of AE pulses.

The calculations carried out provide grounds for using Poisson impulse random processes as a mathematical model of the impulse component of the AE process. Based on the assumptions made above for a Poisson pulsed random process, let us consider the process of occurrence of AE pulses in the body of the diagnosed object. Taking into account the results of works [83, 84], we will dwell on the issues of formation and amplitude characteristics of the total acoustic pulse arising as a result of the movement of a cluster of dislocations.

Suppose, it is a Poisson pulse process with independent increments [90–93], specified over a time interval $-\infty < \tau < \infty$ given that $\mathrm{P}\{\pi_I(0) = 0\} = 1$. We will also assume that the growth points τ_n of the process realizations $\pi_I\,(\tau)$ coincide with the moments of the beginning of the elementary impulse. At these points of the process $\pi_I\,(\tau)$ implementation, jumps of magnitude $\rho_{\tau_i} i = \overline{1,\ k}$., are obtained. In accordance with the above assumption 4, we believe that for the process $\pi_I\,(\tau)$ the condition of independence of increments on non-overlapping time intervals is satisfied. In other words, for the selected time intervals $(\tau_1,\ \tau_2)$ and $(\tau_3,\ \tau_4)$, given that $\tau_1 < \tau_2 \le \tau_3 < \tau_4$ variables $\pi_I(\tau_4) - \pi_I(\tau_3)$ are mutually independent. During the observation $(\tau_1,\ \tau_i),\ i \in \overline{1,\ n}$ process $\pi_I\,(\tau)$ will receive an increment equal to the sum of jumps ρ_{τ_i} on the observation interval T_n:

$$\rho_{o\tau} = \pi_I(\tau_i) - \pi_I(\tau_1) = \sum_{j=1}^{n} \rho_{\tau_i},\ j = \overline{1,\ n}. \tag{3.25}$$

Taking into account the calculations carried out, the AE process at a certain point on the surface of the diagnosed object (subject to physical feasibility) can be represented as follows:

$$\xi(t) = \sum_{j=1}^{n} \rho_{\tau_i} \phi(t - t_i), t \in (-\infty, \infty), i = 1, 2, \ldots \quad (3.26)$$

Regarding the function $\varphi(\tau, t)$ included in (3.26), the following is usually assumed. Taking into account the short pulse duration in AE processes ($\Delta t < 10^{-4} s$), the latter can be described by δ-functions [94, 95]. In this case, information about the moments of occurrence of AE pulses and their magnitudes is preserved. Expression (3.26) in this case will be written as follows:

$$\xi(t) = \sum_{j=1}^{n} \rho_{\tau_i} \delta(t - t_i), \quad (3.27)$$

where $\delta(s)$, $s = t - t_i, s \in (-\infty, \infty)$ is the delta function.

Expression (3.27) is consistent with similar expressions that were proposed as possible options for describing the AE process. However, it should be noted that the authors of [83] proposed and used these models without proper theoretical and experimental justification. Probabilistic approaches for the mathematical description of the AE signal using pulse models were proposed in [5, 96, 97]. At the same time, based on the proposed models, analytical dependencies were not obtained that would allow obtaining probabilistic characteristics of AE processes, based on which it would be possible to theoretically substantiate possible diagnostic signs that would allow determining the technical condition of diagnostic objects based on the results of measuring the parameters and characteristics of AE processes.

If we assume that random variables ρ_{τ_i} correspond to the condition of infinite uniform smallness, then we can argue that any process of this type that has finite power belongs to the class of infinitely divisible random processes. Taking into account the hypothesis about the Poisson nature of the flow of AE pulses, the pulse component $\xi_I(t)$ can be represented as the following stochastic integral:

$$\xi_I(t) = \int_{-\infty}^{\infty} \phi_b(\tau, t) d\pi_I(\tau), \quad (3.28)$$

where $\{\pi_I(\tau), \pi_I(0) = 0, \tau \in (-\infty, \infty)\}$ is Hilbert stochastically continuous generalized Poisson process with independent increments. In this case, random variables in construction (3.26) coincide, respectively, with the magnitude and moment of the i-th jump of the process $\pi_I(\tau)$ in (3.28). The process $\xi_I(t)$ is an LRP (it is also called a pulsed LRP [9]) and has the properties discussed above. In particular, its CF can be represented in the Kolmogorov and Levy forms [20–23, 98].

Note that the condition of uniform infinite smallness, which is imposed on random variables ρ_{τ_i} when justifying (3.28), can be considered fulfilled when the contribution of each individual source of acoustic vibrations to the total AE field is infinitesimal in a probabilistic sense. This occurs, for example, at the stage of deformation of solids.

To describe the continuous AE signal we use the Gaussian model. Taking into account the fact that Gaussian random processes admit an integral representation [99–103], we write

$$\xi_c(t) = \int_{-\infty}^{\infty} \phi_c(\tau, t) dw(\tau), \tag{3.29}$$

where $\varphi_c(\tau, t)$ is a function that has the properties of the kernel of the integral representation (2.14); $\{w(\tau),\ w(0) = 0,\ \tau \in (-\infty, \infty)\}$ is Hilbert, in general, inhomogeneous Gaussian process with independent increments. Thus, the process $\xi_c(t)$ is also LRP. As is known, expression (3.29) describes the response of a linear filter with a pulsed transition function $\varphi_c(\tau, t)$ to the action of Gaussian white noise.

Taking into account the properties of random processes with independent increments and the Poisson spectral function [18], we can represent a general AE signal model as the sum of processes (3.28) and (3.29)

$$\xi(t) = \nu_1 \xi_c(t) + \nu_2 \xi_I(t), \tag{3.30}$$

where ν_1 and ν_2 are weighting coefficients characterizing the contribution of the continuous (3.29) and pulsed (3.28) components to the AE signal of general form (3.30). It can be assumed that the magnitude of the coefficients ν_2 depends on the physical and mechanical properties of the material of the diagnosed object in which the AE signal occurs. It should be emphasized that according to [9], the process $\xi(t)$ in (3.30) is linear.

Let's find the probabilistic characteristics of the process $\xi(t)$.

Expression for the n-dimensional CF continuous component $\xi_c(t)$ in the moment of time $t_1, ..., t_n \in T$ taking into account [9], $L(x, \tau) \equiv 0$, $\varphi(\tau, t) \equiv \varphi_c(\tau, t)$, is defined as follows:

$$\begin{aligned} f_{\xi_c}(u_1, ..., u_n; t_1, ..., t_n) = \exp\Bigg[& i \sum_{k=1}^{n} u_k \int_{-\infty}^{\infty} \phi(\tau, t_k)\, d\mu_w(\tau) \\ & - \frac{1}{2} \sum_{k,j=1}^{n} u_k u_j \int_{-\infty}^{\infty} \phi(\tau, t_k)\, \phi(\tau, t_j)\, dD_w(\tau) \Bigg], \end{aligned} \tag{3.31}$$

where $\mu_w(\tau)$, $D_w(\tau) > 0$ is non-random real functions, uniquely determined by the process $w(\tau)$.

Expression for the impulse component $\xi_I(t)$ n-dimensional CF at moments $t_1, ..., t_n \in T$ is defined this way:

$$f_{\xi_I}(u_1,\ldots,u_n;t_1,\ldots,t_n)=\exp\left\{i\sum_{k=1}^{n}u_k\int_{-\infty}^{\infty}\phi_b(\tau,t_k)d\mu_I(\tau)\right.$$
$$\left.+\int_{-\infty}^{\infty}\int_{-\infty}^{\infty}\left[\exp\left(ix\sum_{k=1}^{n}u_k\phi_b(\tau,t_k)\right)-1-\frac{ix}{1+x^2}\sum_{k=1}^{n}u_k\phi_b(\tau,t_k)\right]dL(x,\tau)\right\},\tag{3.32}$$

where $\mu_I(\tau)$ is a non-random real function, uniquely determined by the process $\pi_I(\tau)$; $L(x,SYMBOL 116 \backslash f "Symbol" \backslash s 14t)$ is Poisson spectrum of process jumps $\pi_I(\tau)$.

Taking into account (3.30) and the fact that the processes $\xi_I(t)$ are stochastically independent, the n-dimensional CF of the process $\xi(t)$ is determined as follows:

$$f_\xi(u_1,\ldots,u_n;t_1,\ldots,t_n)=f_{\xi_c}(u_1,\ldots,u_n;t_1,\ldots,t_n)\cdot f_{\xi_I}(u_1,\ldots,u_n;t_1,\ldots,t_n).\tag{3.33}$$

Taking into account (3.31) and (3.32), expression (3.33) comprehensively reveals the structure of the process $\xi(t)$, since it allows us to distinguish its continuous and impulse components.

When the separability condition is satisfied, the process $\xi(t)$ is completely described by a sequence of its n-dimensional CFs of the form (3.33), which makes it possible to find its numerical and functional characteristics. Thus, analytical expressions for the mathematical expectation, correlation function, and process dispersion $\xi(t)$ are determined accordingly

$$M\xi(\tau)=-i\frac{\partial\ \ln f_\xi(u;t)}{\partial u}\bigg|_{u=0}$$
$$=\int_{-\infty}^{\infty}\phi_c(\tau,t)d\mu_c(\tau)+\int_{-\infty}^{\infty}\phi_b(\tau,t)d\mu_b(\tau)+\int_{-\infty}^{\infty}\int_{-\infty}^{\infty}\frac{x^2}{1+x^2}\phi_b(\tau,t)dL(x,\tau);$$
$$R_\xi(t_1,t_2)=\frac{\partial^2\ln f_\xi(u_1,u_2;t_1,t_2)}{\partial u_1\partial u_2}\bigg|_{u_1=u_2=0}$$
$$=\int_{-\infty}^{\infty}\phi_c(\tau,t_1)\phi_c(\tau,t_2)dD(\tau)+\int_{-\infty}^{\infty}\int_{-\infty}^{\infty}x^2\phi_b(\tau,t_1)\phi_b(\tau,t_2)dL(x,\tau),$$

where at $t_1=t_2=t$

$$D\xi(\tau)=\int_{-\infty}^{\infty}\phi_c(\tau,t)dD(\tau)+\int_{-\infty}^{\infty}x^2\varphi_b(\tau,t)dL(x,\tau).$$

Based on the known properties of CF [9], it is possible to perform a complete probabilistic analysis of the linear model, namely, to find semi-invariants, the function, and the distribution density. Thus, mixed semi-invariants of an nth-order process $\xi(t)$ at moments $t_1, ...t_n \in R$ are determined by the expression

$$k_n\{\xi(t_1), ..., \xi(t_n)\} = \int_{-\infty}^{\infty} \prod_{k=1}^{n} \varphi(t_k, \tau)\, dk_{\pi n}(\tau),$$

where $dk_{\pi n}(\tau)$ is the cumulant measure of jumps of a generalized Poisson process $\pi_1(\tau)$, for which the relation holds $\kappa_{\pi n}(\tau) = \kappa[\pi_1(1)] \cdot \tau$ in case if the process $\pi_1(\tau)$ is homogeneous; $\kappa[\pi_1(1)]$—*n*-th semi-invariant of a random variable $\pi_1(1)$; $\varphi(t, \tau) \in L_n(-\infty, \infty)$ for every *t*.

Along with the CF, a complete probabilistic characteristic of the process $\xi(t)$ is its distribution function, which is uniquely determined by the CF using the Fourier-Stieltjes transform. If we assume that the process is $\xi(t)$ stationary, then according to [9] its one-dimensional probability distribution function will be written as follows:

$$F(x) = \frac{1}{2} + \frac{1}{\pi}\left[\int_0^{\infty} \frac{R(u)}{u} \cos A(u) \sin ux\, du - \int_0^{\infty} \frac{R(u)}{u} \sin A(u) \cos ux\, du\right],$$

or

$$F(x) = \frac{1}{2} + \frac{1}{\pi}\int_0^{\infty} \frac{R(u)}{u} \sin(ux - A(u))\, du,\ x \in [0, \infty). \tag{3.34}$$

If the function $F(x)$ is continuous, then for the process $\xi(t)$, according to [9], we can write an expression for its probability distribution density

$$p(x) = \frac{2}{\pi}\int_0^{\infty} R(u) \sin A(u) \sin uxdu, \tag{3.35}$$

where $R(u)$, $A(u)$—respectively, the module and argument of the CF process $\xi(t)$, which are determined following the expressions

$$R(t, u) = \exp\left\{-\frac{1}{2}\int_{-\infty}^{\infty} \ln\left[1 + u^2\alpha^2\phi^2(\tau, t)\right] dt\right\}, A(t, u)$$

$$= \int_{-\infty}^{\infty} arctg\,[u\,\alpha\,\phi(\tau, t)]\, d\tau,$$

where $\alpha > 0$.

By expressions (3.34) and (3.35), the dependencies of functions and probability distribution densities are constructed. To do this, it is necessary to specify a specific class of influences and select a well-defined function $\varphi(\tau, t)$ that most accurately takes into account the physical properties of a real diagnostic object. The solution to these problems is considered in sufficient detail in [9, 10]. Therefore, below we will dwell on the consideration of some probabilistic characteristics used in the study of the AE process.

Following [9], for the generating process $\{\pi_1(\tau), \pi_1(0) = 0, \tau \in R\}$ CF can be represented in the form

$$f(u, t) = \exp\left\{\int_{-\infty}^{\infty}\int_{-\infty}^{\infty}\left(e^{iux\phi(\tau,t)} - 1\right) d F_1(x)\, d\lambda(\tau)\right\}, x \in R, \quad (3.36)$$

where $F_1(x)$ is the distribution function of increments of the generating process; $\pi_1{}'(\tau)$; $\lambda(t)$ is a continuous non-decreasing function, the value of which at the moment t is equal to the mathematical expectation of the number of processes jumps $\pi_1{}'(\tau)$ on the interval $[0, t)$, which is a parameter of the Poisson distribution.

Concepts about the pulsed nature of AE and the unsatisfactory results of using the Gaussian model to describe AE signals in cases, where the latter have a pronounced pulse character, have led to the need to construct new mathematical models.

Considering the AE signal as a stream of events, each of which consists of a single AE act, the authors of [104–107] use impulse models in their studies. First of all, it is a Poisson impulse process.

In works [83, 84], the shape and amplitude characteristics of AE pulses were studied in sufficient detail. According to these works, the main form of an elementary AE signal can be an elementary impulse of a relaxation (Fig. 12a) or acceleration type (Fig. 12b). The type of a particular AE pulse is determined by the physical and mechanical properties of the diagnostic object, as well as the nature of its deformation.

Taking into account the results of works [17, 18], we will briefly dwell on the presentation of issues related to determining the shape and amplitude of the total acoustic pulse, which is considered a result of the mutual superposition of elementary micro pulses, taking into account the non-simultaneity of their radiation (Fig. 3.13). In this case, we will be based on the description of the physical and mechanical aspects of the formation and movement of dislocations in the body of the diagnosed object during loading [108–111].

Under the assumption that clusters consisting of dislocations (according to [], for metals $n \approx 10, \ldots, 10^2$) cross a fixed boundary in the body of an object, emitting a short exponential pulse when each dislocation crosses this boundary ρ_{τ_i}, $i = \overline{1, k}$. In the future, we will assume that these pulses have a relaxation form (Fig. 12a). The total acoustic pulse recorded by the piezoelectric transducer can be considered as a superposition of the specified short exponential pulses ρ_{τ_i}.

The results presented in this paragraph create a theoretical basis for the mathematical modeling of AE signals that accompany the occurrence of various physical

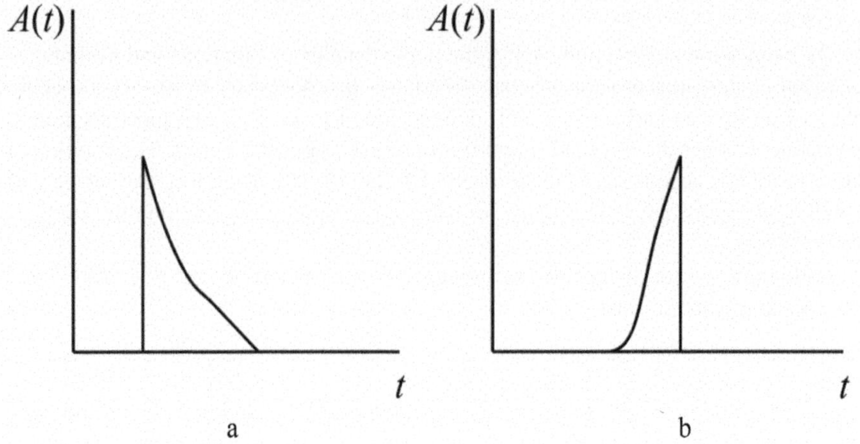

Fig. 3.12 Forms of elementary impulses of relaxation (**a**) and acceleration (**b**) types

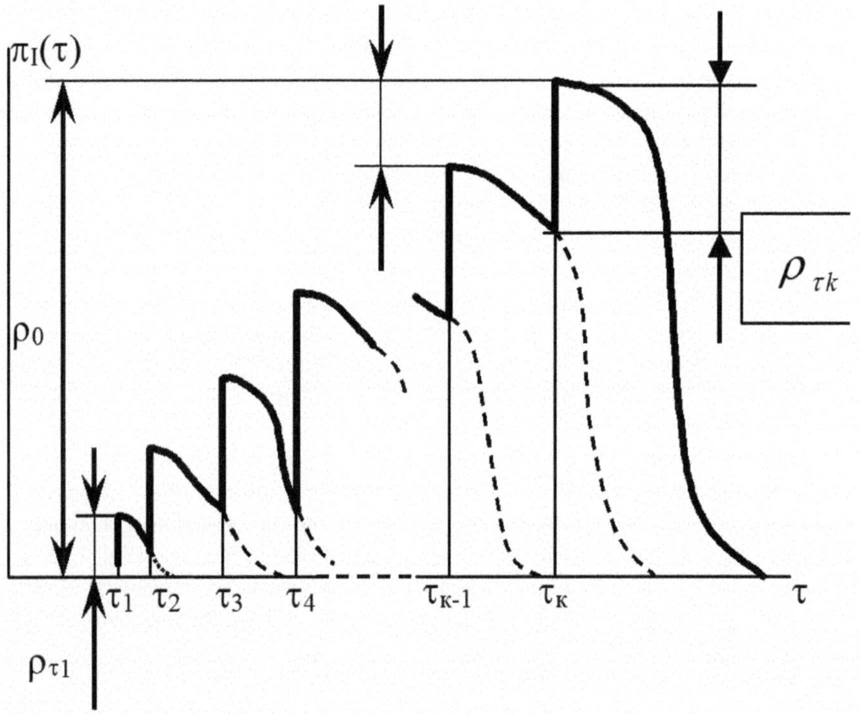

Fig. 3.13 Illustration of mutual superposition of elementary micropulses

processes in solids. Model (2.29) allows one to take into account the properties of the emission source, namely: the nature of the radiation (continuous, pulsed), its nonstationarity, intensity, distribution of elementary AE pulses in amplitude and time, as well as the properties of the propagation medium of the acoustic wave and the AE sensor.

In addition, the proposed mathematical model of AE signals arising during the operation of electrical equipment components makes it possible to theoretically substantiate possible diagnostic signs that can be used when conducting acoustic emission diagnostics of these components. Next, we will dwell on some issues of justification for the choice of diagnostic signs when conducting vibration diagnostics of rolling bearings, laminated magnetic circuits, as well as acoustic emission diagnostics of EM.

3.6 Diagnostic Signs Based on a Statistical Approach to the Diagnosis of Electrical Equipment

While establishing diagnostic characteristics, the main role is played by the choice and results of the study of mathematical probabilistic models of physical processes that arise in the diagnosed components of electrical equipment. The success of diagnostics largely depends on how well its choice is made, and how well this model agrees and is confirmed by experimental data. Let us dwell on the consideration of these issues, choosing as an example diagnostic signals in the form of vibrations of EM rolling bearings, as well as acoustic emission signals that arise in electrical equipment components when they are mechanically or electrodynamically loaded.

3.6.1 Rationale of Diagnostic Signs Based on the Probabilistic Characteristics of Vibrations of EM Rolling Bearings

From the analysis of the mathematical model of bearing vibrations, it follows that the most accessible and at the same time sufficiently complete information for solving diagnostic problems contains the correlation function, power spectral density, and one-dimensional vibration probability distribution density. Based on the results of consideration of these functions, resonance frequencies, extrema of the correlation function, power moments of vibrations, and other parameters are determined that can be used as diagnostic signs. Based on these features, training sets known in the theory of pattern recognition [112–115] are subsequently constructed. The latter are understood as sets of measured parameters formulated in a certain way and divided into subsets. But first of all, attenuation coefficients and frequency parameters, which are directly included in (3.20) and (3.23), can be used as diagnostic signs.

Let us dwell on the issue of statistical assessment of these parameters.

The first stage in processing registered vibrograms is testing the hypothesis for stationarity, which is carried out based on statistical methods and criteria. Depending on the results of this check, the type of averaging is selected, and only after that their correlation and spectral analyses are carried out to obtain empirical correlation functions and power spectral densities. Based on the statistical characteristics of correlation and spectral functions, the values of resonant frequencies and attenuation coefficients are determined, which can act as diagnostic signs.

As already noted, the correlation function of vibrations of EM rolling bearings is a segment of an exponential-sine series of the form (3.22). The task is posed of how, given $R(s)$, to find the representation parameters (3.22) or (3.20), which are diagnostic features.

As a rough approximation that is simply determined, for resonant frequencies one can take the frequencies at which resonant peaks are observed in the spectral density estimate (3.23). This can be done when there are many strongly pronounced resonance peaks. A more accurate but cumbersome way to solve this problem can be implemented using a computer by solving a system of nonlinear algebraic equations that are obtained for different values of s from (3.20) or (3.22) or for different values ω from (3.23). It is convenient to choose extremum points or zeros of the original functions as such values. There are also other ways to solve this problem.

In particular, if the spectral density of vibration power has only one strongly pronounced resonance, provided that the others are practically insignificant, i.e. in (3.20) or (3.22) we can reasonably accept $n = 1$, then we can represent (3.20) in the form

$$R(s) = R(0)\hat{R}(s), \tag{3.37}$$

where

$$R(0) = a_1^2 k_2 \omega_1^2 / 4\beta_1, \tag{3.38}$$

$\hat{R}(s)$ – normalized correlation function,

$$\hat{R}(s) = e^{-\beta_1|s|}\left(\cos\psi_1 s + \frac{\beta_1}{\psi_1}\sin\psi_1|s|\right). \tag{3.39}$$

Figure 3.14 shows the normalized correlation function for $s \geq 0$ for the case of one resonance. The extrema points of function (3.39) are easily found by differentiation. There are an infinite number of them. We will denote them by $s_j, j = 1, 2, 3, \ldots,$ and $s_j = \pi j/\psi_1$ Therefore, as a statistical estimate ψ_1, you can choose a linear estimate from $\tilde{s}_j$, which are observed in the form

$$\tilde{\psi}_1 = \frac{1}{n}\sum_{j=1}^{n}\frac{\pi j}{\tilde{s}_j}, \tag{3.40}$$

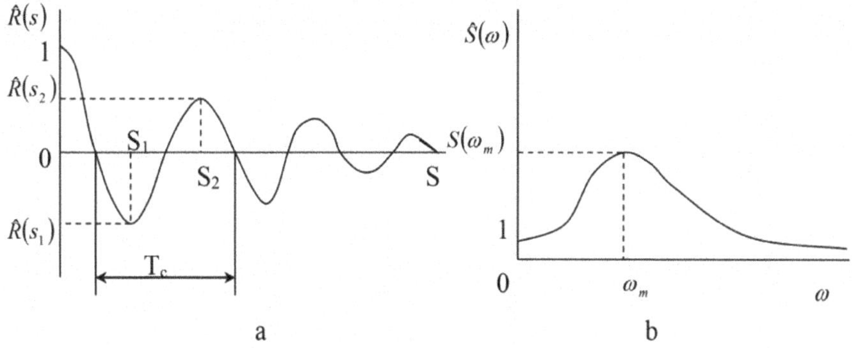

Fig. 3.14 Normalized correlation function and spectral density

where n – random sample size $\left\{\tilde{s}_j, j = \overline{1, n}\right\}$.

To construct an estimate of the attenuation coefficient β_1 you can do this. From (3.39) it is clear that $\hat{R}\left(s_j\right) = (-1)^j e^{-\beta_1 s_j}$ at $s_j > 0, j = 1, 2, \ldots$

Therefore, we take the arithmetic mean as a statistical estimate of β_1

$$\tilde{\beta}_1 = -\frac{1}{n}\sum_{j=1}^{n}\frac{\ln\left[(-1)^j \hat{R}\left(s_j\right)\right]}{\tilde{s}_j}, \tag{3.41}$$

where $\hat{R}\left(s_j\right), j = \overline{1, n}$ are estimates of extreme values of the normalized correlation function (3.39).

However, we must take into account that in practice we are dealing with a time series. Therefore, when estimating these parameters, an additional methodological error appears, which is caused by discretization and which becomes smaller as the discretization step decreases.

Power spectral density at $n = 1$ by (3.23)

$$S(\omega) = S(0)\hat{S}(\omega), \tag{3.42}$$

where $S(0) = a_1^2 k_2$ and normalized spectral density

$$\hat{S}(\omega) = \frac{\omega_1^4}{\left(\omega_1^2 + \omega^2\right)^2 - 4\omega^2 \psi_1^2}. \tag{3.43}$$

The nature of the change in this function is shown in Fig. 3.14b. At $2\psi_1^2 > \omega_1$ it reaches at the point

$$\omega_m = \sqrt{2\psi_1^2 - \omega_1^2} = \sqrt{\omega_1^2 - 2\beta_1^2} \tag{3.44}$$

its maximum value, equal to

$$\hat{S}(\omega_m) = \frac{\omega_1^4}{4\psi_1^2\beta_1^2}. \tag{3.45}$$

In the oscillatory mode at low attenuation, i.e. at $\beta_1 \to 0$ we have $\psi_1 \to 0$ and $\omega_m \to \psi_1$, and $\hat{S}(\omega_m) \to \infty$.

If you construct a statistical estimate of the spectral function, you can estimate the value of the parameters from it ψ_1 and β_1, based on expressions (3.44) and (3.45), also take into account that $\omega_1^2 = \psi_1^2 + \beta_1^2$. To do this it will be necessary to evaluate ω_m and $\hat{S}(\omega_m)$, also solve a system of two equations.

As already noted, the exact solution to the statistical estimation problem β_j and $\psi_j, j = \overline{1, n}$ very bulky. But some assumptions are justified from a practical point of view that makes it possible to simplify it. Thus, assuming that the generating processes for different resonant frequencies are uncorrelated, i.e. $k_{2kj} = 0$ at $k \neq j$, the expression for the correlation function (3.20) takes the form

$$R(s) = \sum_{j=1}^{n} \frac{a_j^2 k_{2jj} \omega_j^2}{4\beta_j} e^{-\beta_j |s|} \left(\cos \psi_j s + \frac{\beta_j}{\psi_j} \sin \psi_j |s| \right), \tag{3.46}$$

and at $s \geq 0$

$$R'(s) = -\sum_{j=1}^{n} \frac{a_1^2 k_{2jj} \omega_j^4}{\psi_j} e^{-\beta_j s} \sin \psi_j s. \tag{3.47}$$

In this case, following (3.23), the power spectral density has the form

$$S(\omega) = \sum_{j=1}^{n} \frac{a_j^2 k_{2jj} \omega_j^4}{\left(\omega_j^2 + \omega^2\right)^2 - 4\omega^2 \psi_j^2}. \tag{3.48}$$

Having the values of $R(s)$ or $S(\omega)$ for different s and ω, it is possible to construct, using (3.46)–(3.48), a system of linearly independent equations for the unknown numerical parameters included in them, as well as estimates β_j and $\psi_j, j = \overline{1, n}$.

At small β_j values the resonant frequencies ψ_j can be approximately determined by the peaks in the power spectral density (3.48).

Let's consider the diagnostic features determined by the probability distribution density of the mathematical model of vibrations of rolling bearings. The presence of certain defects in a bearing leads to a change in the nature of its vibrations, which, in turn, causes a change in the shape of its distribution density curve. This is confirmed by the experimental results of histogram analysis of vibrations of rolling bearings (Chap. 9 of this book).

As is known, the shape of the probability density curve is characterized by certain parameters. For example, a one-dimensional bell curve is completely determined by the first two moments (expectation and variance).

As a result of many experiments, it has been established that the mathematical expectation $M[\xi(t)] = M[\eta(1)]\sum_{j=1}^{n} a_j$ and the dispersion $R(0)$ of the research vibrations, which is determined by (3.21), are ineffective diagnostic signs. More complete information about the nature of vibrations is provided by their probability density curve, and, in particular, by the third and fourth moments, which can be used as diagnostic signs of the technical condition of bearings. However, in practice, it is more convenient to use not the third and fourth moments of vibration of bearings, but the coefficients of asymmetry k and kurtosis γ, which are related to the third and fourth moments of vibration by the relations

$$k^2 = \mu_3 \Big/ \mu_2^{3/2}, \tag{3.49}$$

$$\gamma = \beta_2 - 3, \tag{3.50}$$

where

$$\beta_2 = \mu_4 / \mu_2^2, \tag{3.51}$$

μ_2, μ_3 and μ_4 are central points, defined as

$$\mu_k = \frac{1}{n}\sum_{j=1}^{n} \left(\xi_j - \nu_1\right)^k, k = 2, 3, \ldots, \tag{3.52}$$

where $\nu_1 = \frac{1}{n}\sum_{j=1}^{n} \xi_j$ is the linear estimate for initial moments.

In vibration diagnostics of bearings, quantitative assessments $\tilde{k}$, $\tilde{\gamma}$ and $\tilde{\beta}_2$ can be obtained using a histogram analysis program, the main algorithms of which are described in [116–120].

Thus, the obtained methods for constructing estimates for the above-mentioned diagnostic parameters, which are necessary for the analysis of vibrations of bearings of electrical units, represent a theoretical and computational basis for their diagnosis using statistical methods.

When theoretically substantiating diagnostic signs using model (3.3), it was believed that the appearance of various defects in the diagnosed nodes is equivalent to a change in parameter values R_j, L_j, C_j in the proposed model (3.3), which in turn changes the initial and central moments, as well as the nature of the probability distribution function curve.

So, for diagnosing some components of electrical equipment, such as a gearbox, rolling bearings, laminated magnetic circuit of the generator stator, etc. Based on their vibrations, the following diagnostic signs can be used [121–125]:

1. attenuation coefficients β_j, $j = \overline{1, n}$;
2. resonant frequencies ω_j, $j = \overline{1, n}$;
3. magnitudes of initial and central moments;

4. the nature of the probability distribution function curve.

With a statistical approach, all vibration diagnostics are based on an appropriately selected and deeply reasoned probabilistic vibration model and are most often carried out in three stages. The first is to establish diagnostic signs that correspond to different technical conditions of the diagnosed nodes and are sensitive to their changes; the second is associated with the choice of diagnostic spaces (training stage) and involves the formation, based on experimental data, of training sets (images) that correspond to specific defects in the node being diagnosed; the third involves the construction of decisive rules for diagnosing defects in the components under study and making decisions regarding their technical condition. The last stage is implemented based on training sets, which are formed by preliminary registration of vibrations of the diagnosed EM components (according to the previously specified operating modes of the diagnosed EM).

In conclusion of this paragraph, we will dwell on the issues of determining some probabilistic parameters of the AE process and the rationale for their use as diagnostic signs when conducting AE diagnostics of electrical equipment components.

3.6.2 *Selection of Diagnostic Features Based on the Probabilistic Characteristics of AE Signals*

To determine the probabilistic characteristics of the process $\xi(t)$ at the output of the piezoelectric transducer (PET), it is necessary to select a specific type of pulse transient function $\varphi_0(\tau,\ t)$ and $\varphi(\tau,\ t)$, characterizing the object of diagnosis and PET, respectively [126]. In addition, it is necessary to take into account the properties of the generating process $\eta(t)$. In our case, AE pulses act as such a process. To justify the choice of the type of impulse transition functions $\varphi_0(\tau,\ t)$ and briefly discuss the general provisions of the method of stochastic integral representations [127–129].

The idea of the method of stochastic integral representations is that at the input of the system under study with an impulse transition function $\varphi(\tau,\ t)$, a link with an impulse transition function is mentally completed $\varphi_0(\tau,\ t)$ (Fig. 3.15).

At the diagram, presented in Fig. 3.15, a mentally completed link with a pulse transition function $\varphi_0(\tau,\ t)$ as if it forms a random process (3.3). It is not necessary to impose a condition of physical feasibility on this completed link. Function $\varphi_0(\tau,\ t)$ and the properties of the process $\zeta'(t)$ are selected in such a way that the response of the completed link would be stochastically equivalent to the impact

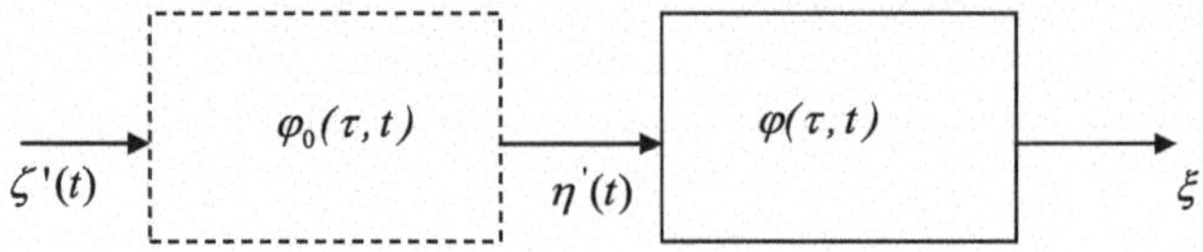

Fig. 3.15 Illustration of the method of stochastic integral representations

on the analyzed system with a pulsed transition function $\varphi(\tau, t)$. Analytically, this can be represented by the following expressions:

$$\eta(t) = \int_{-\infty}^{\infty} \phi_0(\tau, t) d\zeta(\tau), t \in (-\infty, \infty), \quad (3.53)$$

$$\xi(t) = \int_{-\infty}^{\infty} \phi(\tau, t)\eta(\tau)d\tau = \int_{-\infty}^{\infty}\int_{-\infty}^{\infty} \phi(\tau, t)\phi_0(\tau_1, \tau)d\varsigma(\tau_1)d\tau. \quad (3.54)$$

Let us introduce the notation $\phi_{01}(\tau, t) = \int_{-\infty}^{\infty} \phi(\tau_1, t)\phi_0(\tau, \tau_1)d\tau_1$, then

$$\xi(t) = \int_{-\infty}^{\infty} \phi_{01}(\tau, t)d\varsigma(\tau). \quad (3.55)$$

Physically $\phi_0(\tau, t)$ can be considered as a pulse transition function that characterizes the object being diagnosed (the medium of generation and propagation of AE pulses). $\varsigma'(\tau)$ is a generating process, which is a generalized derivative of a real random process with independent increments $\{\varsigma(\tau), \varsigma(0) = 0, \tau \in (-\infty, \infty)\}$. In our case $\varsigma'(\tau)$ precisely describes the AE process that occurs in the environment of the diagnosed object.

Then the AE process of the recorded PET on the surface of the object being diagnosed can be represented by expression (3.53), and the results of measuring this process, i.e. the AE process recorded at the output of the probe, taking into account the relation (3.55), is characterized by the expression

$$\xi(t) = \int_{-\infty}^{\infty} \eta(x)\phi(t - x)dx = \int_{-\infty}^{\infty}\int_{-\infty}^{\infty} \phi_0(\tau, x)\phi_0(t - x)dxd\varsigma(\tau). \quad (3.56)$$

To simplify the last expression, we use the previously considered property 2 for LRP [9]. Namely, based on (3.3) for a stationary LRP, we have

$$\xi(t) = \int_{-\infty}^{\infty} \phi(\tau, t)d\eta(\tau) = \int_{-\infty}^{\infty} \phi(t - \tau)d\eta(\tau), \quad t \in (-\infty, \infty). \quad (3.57)$$

Following property 2 for LRP, we‘ll get

$$\xi(t) = \int_{-\infty}^{\infty} \eta(x)\phi(t-x)dx = \int_{-\infty}^{\infty}\int_{-\infty}^{\infty} \phi_0(\tau, x)\phi(t-x)dx d\varsigma(\tau) = \int_{-\infty}^{\infty} \phi_1(\tau, t)d\varsigma(\tau), \tag{3.58}$$

where

$$\phi_1(\tau, t) = \int_{-\infty}^{\infty} \phi_0(\tau, x)\phi(t-x)dx. \tag{3.59}$$

If the process $\xi(t)$ is stationary, i.e. can be represented by expression (3.57), then expression (3.59) takes the form

$$\phi_1(\tau, t) = \phi_1(\tau - t) = \int_{-\infty}^{\infty} \phi_0(x-\tau)\phi(t-x)dx = \int_{-\infty}^{\infty} \phi_0(x)\phi(t-\tau-x)dx. \tag{3.60}$$

As noted above, determining the probabilistic characteristics of the process $\xi(t)$ at the output of the probe directly depends on the choice of the type of pulse transient functions $\phi_0(\tau, t)$ and $\phi(\tau, t)$, characterizing the diagnostic object and PET, respectively. In addition, it is necessary to take into account the properties of the generating process $\eta'(t)$. In our case, AE pulses act as such a process.

Following the results of works [9, 10], the shape of elementary pulses of the acoustic AE signal can be characterized by relaxation-type pulses, which at the point of placement of the piezoelectric transducer on the surface of the diagnosed object are described by the function

$$\phi_b(s) = \alpha \cdot e^{-\alpha s} \cdot U(s),\ s \in (-\infty, \infty),\quad s = t - \tau, \tag{3.61}$$

where $\alpha > 0$ is the attenuation coefficient. The same function can be used to describe the shape of AE pulses at the output of an amplitude detector.

According to the above results, the kernel of the integral representations of both the continuous and discrete components has the form (3.30). We denote the attenuation coefficients for the continuous and discrete components by α_c and α_b respectively. We will assume that the continuous component is produced by stationary Gaussian white noise with zero mean and variance σ^2 [129–132].

Information on the distribution of discrete AE pulse amplitudes is given, for example, in [9]. According to these sources, this distribution varies depending on the stage of deformation of the material, that is, it is determined, in particular, by the type of AE source. Thus, in the mentioned sources, at the stage of plastic deformation, the distribution is close to exponential, and the development of a crack is characterized by a high probability of the appearance of pulses of significant amplitude, so the distribution becomes more uniform.

Let us limit ourselves to studying the case of plastic deformation, when the amplitude distribution density can be described by the function

$$p(x) = b \cdot \mathrm{e}^{-bx} \cdot U(x),\ x \in (-\infty, \infty), \tag{3.62}$$

where $b > 0$ is the real constant.

As for the impulse component, then we will consider it as a Poisson flow of impulses, the distribution of amplitudes of which has the form (3.62) with intensity $\lambda \cdot t,\ t \in [0, \infty)$.

Based on (3.32) and (3.33), the one- and two-dimensional CF of such a process are respectively described as follows:

$$f_\xi(u;t) = \exp\left\{ -\frac{\sigma^2 u^2 \alpha_c}{4} + \lambda b \int_{-\infty}^{\infty} \tau \int_0^{\infty} \left(\exp\left[ixu\alpha_b \mathrm{e}^{-\alpha_d(t-\tau)} U(\tau) \right] - 1 \right) \cdot \mathrm{e}^{-bx} dx d\tau \right\};$$

$$f_\xi(u_1, u_2; t_1, t_2) = \exp\left\{ -\frac{\sigma^2}{2} \alpha_c^2 \sum_{k,j=1}^{2} u_k u_j \int_{-\infty}^{\infty} \mathrm{e}^{-\alpha_c(t_k + t_j - 2\tau)} U\left(\min\left(t_k, t_j\right) - \tau\right) d\tau \right.$$
$$\left. + \lambda b \int_{-\infty}^{\infty} \tau \int_0^{\infty} \left(\exp\left[ixu\alpha_b \sum_{k=1}^{2} u_k \mathrm{e}^{-\alpha_d(t_k - \tau)} U(t_k - \tau) \right] - 1 \right) \cdot \mathrm{e}^{-bx} dx d\tau \right\}. \tag{3.63}$$

Let us now proceed directly to determining the probabilistic characteristics of the AE signal, which is recorded at a certain point on the surface of a deformed solid. To do this, we will use the linear model (3.30), as well as theoretical and experimental results of studies of AE during the deformation of metals, a review of which is given, for example, in [9].

Expressions for moments and correlation functions are defined as follows:

$$\begin{aligned} \mathrm{M}\xi(t) &= \frac{\lambda}{b} \cdot \left(t - \frac{1}{\alpha_b} \right); \\ \mathrm{D}\xi(t) &= \frac{\sigma^2 \alpha_c}{2} + \frac{\lambda}{b^2} \cdot \left(\alpha_b t - \frac{1}{2} \right); \\ R_\xi(t_1, t_2) &= \frac{\sigma^2 \alpha_c}{2} \cdot \exp[-\alpha_c(t_1 + t_2 - 2\min(t_1, t_2))] \\ &+ \frac{\lambda}{b^2} \cdot \exp[-\alpha_b(t_1 + t_2 - 2\min(t_1, t_2))] \cdot \left(\alpha_b \min(t_1, t_2) - \frac{1}{2} \right). \end{aligned} \tag{3.64}$$

The given expressions taking into account the physical meaning of the parameters α_c,α_b, make it possible to explain the nature of the AE signal depending on the type and characteristics of the emission source. Let us emphasize some features: the average value of the AE signal is determined only by the parameters of its discrete component; the signal value correlation function is exponential; the average power of the process is determined by two components, the first of which, depending on and, is contributed by the continuous component, and the second, proportional to the emission intensity, is contributed by the discrete component. These features, within the framework of the assumptions made, are consistent with the experimental data given in [17, 18].

Options can characterize the physical and mechanical properties of the material from which the diagnosed object is made.

Regarding the characteristics of the electrical signal of a discrete AE, some authors, for example [9, 17, 18], in their studies proceeded from the fact that the shape of the elementary pulse at the output of the probe can be described by the function

$$\phi_b(\tau, t) \equiv \phi_b(s) = \gamma \cdot e^{-\beta s} \sin\theta s \cdot U(s), \; s = t - \tau, \; s \in (-\infty, \infty), \tag{3.65}$$

where $\theta > 0$, $0 < \beta < \theta$ are real constants that have a specific physical meaning, namely, oscillation frequencies and damping coefficients, respectively; $\gamma = \left(\beta^2 + \theta^2\right)/\theta$ is normalizing factor; $U(s), s \in (-\infty, \infty)$ is Heaviside function.

It should be noted that the attenuation coefficient is determined by the decay time constant of the probe and the duration of the AE acoustic pulse. The latter, in turn, depends on the properties of the material, and the type of AE source and decreases with increasing deformation. Thus, in the general case, for a particular sample, the parameter is a function of deformation. However, given that the PET attenuation constant significantly exceeds the duration of the AE acoustic pulse, this fact is usually neglected [133–136].

Let us consider the issue of determining the probabilistic characteristics of the AE signal at the output of the probe under the assumption that the AE signal pulses arriving at its input represent a simple Poisson flow with constant intensity $\lambda > 0$ and with the same amplitude. In this case, the generating process is a simple Poisson process $\{\pi(\tau), \pi(0) = 0, \tau \in (-\infty, \infty)\}$ with parameter and increment distribution function at $F_1(x) = 1$ at $x \geq 1$ and $F_1(x) = 0$ at $x < 1$. Then, taking into account (4.65), the process $\xi(t)$ at the output of the probe can be represented as follows:

$$\xi(t) = \int_{-\infty}^{\infty} \gamma e^{-\beta(t-\tau)} \sin[\theta(t-\tau)]\, U(t-\tau)\, d\pi(\tau), \tag{3.66}$$

where $\{\pi(\tau), \pi(0) = 0, t \in (-\infty, \infty)\}$ is a simple Poisson process. Then, taking into account [9] and expression (3.63), the relation for the CF process (3.66) will be written as follows:

$$f(u) = \exp\left\{\lambda \int_{-\infty}^{\infty} \left[e^{iu\gamma e^{-\beta s} \sin\theta s\, U(s)} - 1\right] ds\right\}. \tag{3.67}$$

Based on (3.64), the expressions for the mathematical expectation, dispersion, and correlation function will be written as follows:

$$\begin{aligned}
\mathrm{M}\{\xi(t)\} &= \lambda \int_{-\infty}^{\infty} \gamma e^{-\beta\tau} \sin\theta\tau\, U(\tau)\, d\tau = \lambda;\\
\mathrm{D}\{\xi(t)\} &= \lambda \int_{-\infty}^{\infty} [\gamma e^{-\beta\tau} \sin\theta\tau\, U(\tau)]^2 d\tau = \lambda \frac{\theta^2+\beta^2}{4\beta}\\
R_\xi(s) &= \lambda \frac{\theta^2+\beta^2}{4} e^{-\beta s}\left\{\frac{\cos\theta s}{\beta} + \frac{\sin\theta|s|}{\theta}\right\}, \quad s = t_2 - t_1.
\end{aligned} \tag{3.68}$$

To determine the probability distribution function of the process (3.66), you can use expressions (3.61) and (3.62). In accordance with the latter, the modulus and argument of CF (3.67) have the form

$$\begin{aligned}
R(u) &= \exp\left\{\lambda \int_0^{\infty} \left(\cos\left(u\gamma e^{-\beta\tau}\sin\theta\tau\right) - 1\right) d\tau\right\};\\
A(u) &= -\lambda \int_0^{\infty} \sin\left(u\lambda e^{-\beta\tau}\sin\theta\tau\right) d\tau.
\end{aligned} \tag{3.69}$$

The given expressions do not allow one to simply express the distribution function of the form (3.34) of the AE signal in terms of elementary or known special functions. Therefore, to obtain these dependencies, you can use numerical integration methods. In particular, in [] a similar problem was solved using Philo's quadrature formula. Figures 3.16 and 3.17 show the dependences of the distribution functions $F(x)$ AE signal (3.66) for various values λ θ(in Fig. 3.16—$\theta = 5$; in Fig. 3.17 $\theta - =$ 15). While constructing these dependencies, the value was taken. Figure 3.18 shows the dependences $F(x)$ of the AE signal distribution functions obtained for fixed values $\lambda = 2$ and $\theta = 5$, but for different values β.

Based on the analysis of the obtained curves, the following can be noted. As can be seen from Fig. 3.16, in the case of the influence of a simple Poisson process on the input of a low-Q factor (in the given example) $\{\pi(\tau), \pi(0) = 0, t \in (-\infty, \infty)\}$ the response under study (3.66) contains negative pulses with a relatively small amplitude. This can be judged by the nature of the distribution function curve $F(x)$ at negative values x. As the quality factor of the probe increases, an increase in negative values in the response (3.66) is observed. This is illustrated by the nature of the distribution functions $F(x)$ presented in Fig. 3.17. These dependencies were constructed using $\beta = 5$ and $\omega = 15$. In addition, as follows from Fig. 3.16 and Fig. 3.17, with increasing flow intensity λ nature of change in function $F(x)$ becomes slower. Analysis of Fig. 3.18 showed that an increase in

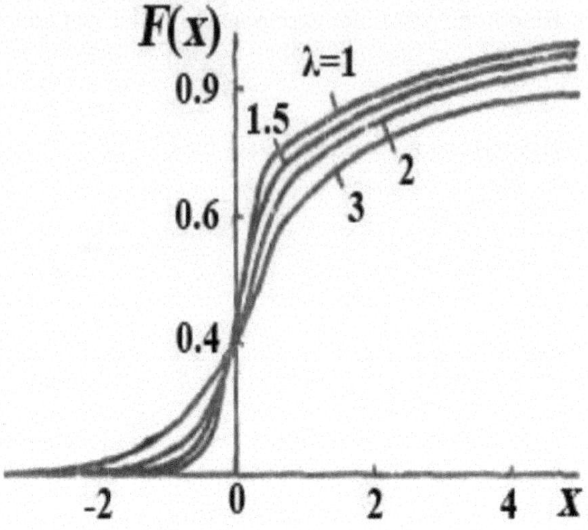

Fig. 3.16 Dependences of the probability distribution function $F(x)$ of the AE signal for various values λ and $\theta = 5$

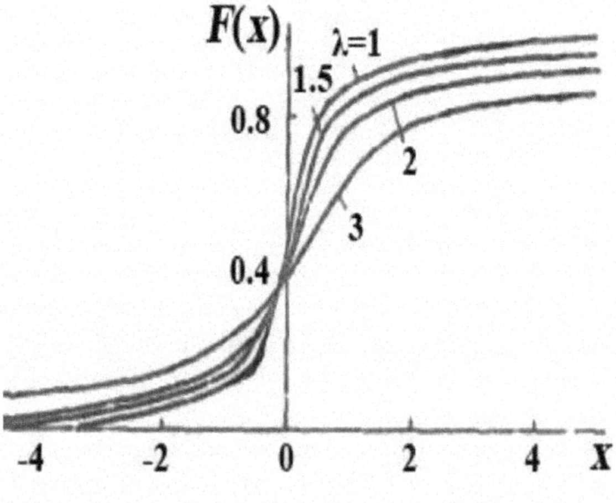

Fig. 3.17 Dependences of the probability distribution function $F(x)$ of the AE signal for various values λ and $\theta = 15$

the attenuation coefficient β at constant λ and θ leads $F(x)$ to a steeper increase in the range of its values [0.2; 0.8].

Thus, the results shown in Fig. 3.16 and Fig. 3.17 indicate an expansion of the range of probable values of the process $\xi(t)$, and in Fig. 3.18, on the contrary, a narrowing of the range of these values. The results obtained make it possible to explain the data from experimental studies of AE signals, which consist of an increase in the absolute values of these signals at maximum values of elastic deformations of metals.

Expressions (3.67)–(3.69) allow us based on the values of non-random parameters β, θ, λ to determine the probabilistic characteristics of models (3.30) and (3.66) of AE signals using the values of non-random parameters, and, therefore, to judge the stochastic properties of real AE signals. More general formulas (3.63) and (3.64)

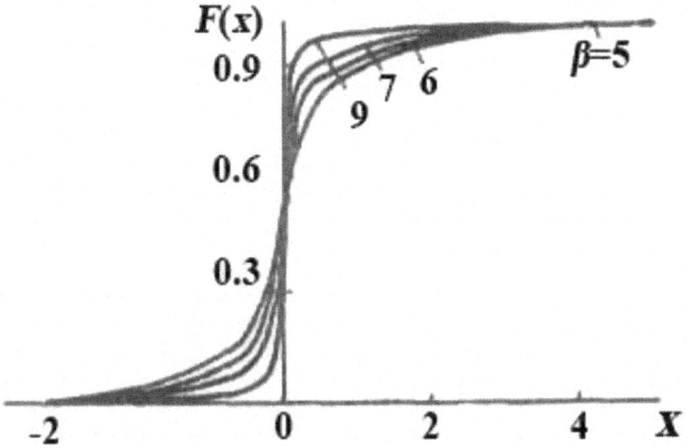

Fig. 3.18 Dependences of the probability distribution function $F(x)$ of the AE signal for $\lambda = 2$ and $\theta = 5 = 5$ and for different values β

connect the most important parameters of AE—the intensity of pulses $\lambda(t)$, their form, amplitude distribution, and distribution of intervals between them with the probabilistic characteristics observed.

In the future, we will consider the issues of simulation modeling of vibration and AE processes in electrical equipment components. In addition, some of the probabilistic parameters of these processes can be used when conducting experimental studies on vibration and AE diagnostics of the specified equipment. These issues are discussed in subsequent chapter of this book.

References

1. Rajan, D.: Probability, random variables, and stochastic processes. In: Mathematical Foundations for Signal Processing, Communications, and Networking, pp. 205–244. CRC Press (2017)
2. Grimmett, G., Stirzaker, D.: Probability and random processes. Oxford University Press (2020)
3. Florescu, I.: Probability and stochastic processes. John Wiley & Sons (2014)
4. Micheas, A.C.: Theory of stochastic objects: probability, stochastic processes and inference. Chapman and Hall/CRC (2018)
5. Miller, S., Childers, D.: Probability and Random Processes: With Applications to Signal Processing and Communications. Academic Press (2012)
6. Puchalski, A.A., Komorska, I.: Generative modelling of vibration signals in machine maintenance. Eksploatacja i Niezawodność 25(4) (2023)
7. Tama, B.A., Vania, M., Lee, S., Lim, S.: Recent advances in the application of deep learning for fault diagnosis of rotating machinery using vibration signals. Artif. Intell. Rev. **56**(5), 4667–4709 (2023)

8. Ren, H., Dai, Z., Zhou, B., Zhang, B., Yin, A., Pei, L., Cao, Q.: Vibration monitoring and semisupervised multivariate invertible deep probabilistic learning for gearbox faults identification. IEEE Sens. J. **22**(22), 22020–22029 (2022)
9. Babak, V., Zaporozhets, A., Kuts, Y., Myslovych, M., Fryz, M., Scherbak, L.: Models and characteristics of identification of noise stochastic signals of research objects. In: CEUR Workshop Proceedings, vol. 3309, pp. 349–362 (2022)
10. Babak, V., Scherbak, L., Kuts, Y., Zaporozhets, A.: Information and measurement technologies for solving problems of energy informatics. In: The 1st International Workshop on Information Technologies: Theoretical and Applied Problems 2021. CEUR Workshop Proceedings, vol. 3039, pp. 24–31 (2021)
11. Babak, V.P., Babak, S.V., Myslovych, M.V., Zaporozhets, A.O., Zvaritch, V.M. (2020). Methods and models for information data analysis. Diagnos. Syst. Energy Equipm. 23–70
12. Fageot, J., Uhlmann, V., Unser, M.: Gaussian and sparse processes are limits of generalized Poisson processes. Appl. Comput. Harmon. Anal. **48**(3), 1045–1065 (2020)
13. Jasiulis-Gołdyn, B.H., Misiewicz, J.K.: Classical definitions of the Poisson process do not coincide in the case of generalized convolutions. Lith. Math. J. **55**, 518–542 (2015)
14. Xiong, Q., Xu, Y., Peng, Y., Zhang, W., Li, Y., Tang, L.: Low-speed rolling bearing fault diagnosis based on EMD denoising and parameter estimate with alpha stable distribution. J. Mech. Sci. Technol. **31**, 1587–1601 (2017)
15. Xiong, Q., Zhang, W., Lu, T., Mei, G., Liang, S.: A fault diagnosis method for rolling bearings based on feature fusion of multifractal detrended fluctuation analysis and alpha stable distribution. Shock and Vibration, 2016 (2016)
16. Eremenko, V., Zaporozhets, A., Isaienko, V., Babikova, K.: Application of wavelet transform for determining diagnostic signs. In: Proceedings of the 15th International Conference on ICT in Education, Research and Industrial Applications. Integration, Harmonization and Knowledge Transfer, vol. 1, pp. 12–15 (2019)
17. Babak, V.P., Babak, S.V., Eremenko, V.S., Kuts, Y.V., Myslovych, M.V., Scherbak, L.M., Zaporozhets, A.O. (2021). Models and measures for the diagnosis of electric power equipment. Models Measures Measurem. Monit. 99–126
18. Babak, V., Zaporozhets, A., Kovtun, S., Myslovych, M., Kuts, Y., Scherbak, L.: Information support for identification of the technical state of electric power facilities. In: Systems, Decision and Control in Energy V, pp. 129–153. Springer Nature Switzerland, Cham (2023)
19. Li, Y., Zhang, W., Xiong, Q., Lu, T., Mei, G.: A novel fault diagnosis model for bearing of railway vehicles using vibration signals based on symmetric alpha-stable distribution feature extraction. Shock and Vibration (2016)
20. Matsiuk, A.V., Pryimak, M.V.: Mathematical model of electroretinogram in the form of linear stochastic process. In: 2005 IEEE intelligent data acquisition and advanced computing systems: technology and applications, pp. 663–666. IEEE (2005)
21. Zvaritch, V., Glazkova, E.: Application of linear AR and ARMA processes for simulation of power equipment diagnostic systems information signals. In: 2015 16th International Conference on Computational Problems of Electrical Engineering (CPEE), pp. 259–261. IEEE (2015)
22. Rosiński, J., Szulga, J.: Product random measures and double stochastic integrals. In: Martingale theory in harmonic analysis and banach spaces: proceedings of the NSF-CBMS conference held at the cleveland State University, Cleveland, Ohio, July 13–17, 1981 pp. 181–199. Springer Berlin Heidelberg, Berlin, Heidelberg (2006)
23. Devroye, L.: Simulating perpetuities. Methodol. Comput. Appl. Probab. **3**(1), 97–115 (2001)
24. Wang, Y., Xia, A., Zhang, P., Qin, G.: Probabilistic physical modeling of randomly corroded surface and its use in reliability analysis of corroded pipelines under spatiotemporal vibration. Ocean Eng. **262**, 112219 (2022)
25. En, X., Zhang, Y., Huang, X., Li, Y., Yang, G.: Time-varying reliability and global sensitivity analysis of regenerative chatter stability in turning considering tool wear. Mech. Based Des. Struct. Mach. **50**(12), 4084–4104 (2022)

26. Cong, F., Zhou, Q., Chen, L., Lin, F., Lin, X., Zhou, Y.: Hob wear state condition monitoring based on statistical distribution law. CIRP J. Manuf. Sci. Technol. **44**, 16–26 (2023)
27. Babak, V.P., Babak, S.V., Eremenko, V.S., Kuts, Y.V., Myslovych, M.V., Scherbak, L. M., Zaporozhets, A.O.: Models of measuring signals and fields. Models and Measures in Measurements and Monitoring, 33–59
28. Schreiber, T., Schmitz, A.: Improved surrogate data for nonlinearity tests. Phys. Rev. Lett. **77**(4), 635 (1996)
29. Kostenko, P.Y., Vasylyshyn, V.I.: Enhancing the efficiency of spectral analysis of signals by the Root-MUSIC method using surrogate data. Radioelectron. Commun. Syst. **57**, 31–38 (2014)
30. Babak, V.P., Babak, S.V., Myslovych, M.V., Zaporozhets, A.O., Zvaritch, V.M.: Simulation and software for diagnostic systems. Diagnostic Systems For Energy Equipments, 71–90 (2020)
31. Gemeinder, Y., Schuster, M., Radnai, B., Sauer, B., Binder, A.: Calculation and validation of a bearing impedance model for ball bearings and the influence on EDM-currents. In: 2014 International Conference on electrical machines (ICEM), pp. 1804–1810. IEEE (2014)
32. Gonda, A., Paulus, S., Graf, S., Koch, O., Götz, S., Sauer, B.: Basic experimental and numerical investigations to improve the modeling of the electrical capacitance of rolling bearings. Tribol. Int. **193**, 109354 (2024)
33. Graf, S., Werner, M., Koch, O., Götz, S., Sauer, B.: Breakdown voltages in thrust bearings: behavior and Measurement. Tribol. Trans. **66**(3), 488–496 (2023)
34. Wittek, E., Kriese, M., Tischmacher, H., Gattermann, S., Ponick, B., Poll, G.: Capacitances and lubricant film thicknesses of motor bearings under different operating conditions. In: The XIX International Conference on Electrical Machines-ICEM 2010, pp. 1–6. IEEE, (2010)
35. Medvid, V., Belyakova, I., Maruschak, O., Piscio, V.: A one-dimensional mathematical model of piezoelectric transformers for CAD system. Sci. J. TNTU **108**(4), 102–116 (2023)
36. Medvid, V., Belyakova, I., Piscio, V., Lupenko, S.: Model of transverse-transverse type piezoelectric transformer. Sci. J. TNTU **102**, 96–109 (2021)
37. Babak, V.P., Babak, S.V., Eremenko, V.S., Kuts, Y.V., Myslovych, M.V., Scherbak, L.M., Babak, V.P., Babak, S.V., Eremenko, V.S. and Kuts, Y.V., Zaporozhets, A.O.: Models and measures for measuring random angular quantities. Models Measures Measurem. Monitor. 61–97 (2021)
38. Babak, V.P., Babak, S.V., Eremenko, V.S., Kuts, Y.V., Myslovych, M.V., Scherbak, L.M., Zaporozhets, A.O.: Examples of using models and measures on the circle. Models and Measures Meas. Monit. 127–156 (2021)
39. Purushotham, V., Narayanan, S., Prasad, S.A.: Multi-fault diagnosis of rolling bearing elements using wavelet analysis and hidden Markov model based fault recognition. NDT and E Int. **38**(8), 654–664 (2005)
40. Zaporozhets, A., Eremenko, V., Babak, V., Isaienko, V., Babikova, K.: Using hilbert transform in diagnostic of composite materials by impedance method. Periodica Polytechnica Electr. Eng. Comput. Sci. **64**(4), 334–342 (2020)
41. Babak, V., Eremenko, V., Zaporozhets, A.: Research of diagnostic parameters of composite materials using Johnson distribution. Int. J. Comput. **18**(4), 483–494 (2019)
42. Yiakopoulos, C.T., Gryllias, K.C., Antoniadis, I.A.: Rolling element bearing fault detection in industrial environments based on a K-means clustering approach. Expert Syst. Appl. **38**(3), 2888–2911 (2011)
43. Kankar, P.K., Sharma, S.C., Harsha, S.P.: Rolling element bearing fault diagnosis using wavelet transform. Neurocomputing **74**(10), 1638–1645 (2011)
44. Xian, Z., Pengcheng, Z., Qingxin, Y., Lihua, Z., Xin, Z.: Numerical estimation and optimization of vibration noise due to magnetostriction and magnetic forces for laminated core structure. In: 2014 17th International Conference on Electrical Machines and Systems (ICEMS), pp. 1548–1551. IEEE (2014)
45. Zhu, L., Yang, Q., Yan, R., Li, Y., Zhang, X., Yan, W., Zhu, J.: Numerical computation for a new way to reduce vibration and noise due to magnetostriction and magnetic forces of transformer cores. J. Appl. Phys. **113**(17) (2013)

46. Gao, Y., Muramatsu, K., Hatim, M.J., Nagata, M.: The effect of laminated structure on coupled magnetic field and mechanical analyses of iron core and its homogenization technique. IEEE Trans. Magn. **47**(5), 1358–1361 (2010)
47. George, T.J., Seidt, J., Shen, M.H.H., Nicholas, T., Cross, C.J.: Development of a novel vibration-based fatigue testing methodology. Int. J. Fatigue **26**(5), 477–486 (2004)
48. Akbari, M.R., Nimafar, M., Ganji, D.D., Chalmiani, H.K.: Investigation on non-linear vibration in arched beam for bridges construction via AGM method. Appl. Math. Comput. **298**, 95–110 (2017)
49. Karnovsky, I.A., Lebed, E.: Theory of Vibration Protection, pp. 176–182. Springer International Publishing, Switzerland (2016)
50. Qing, G.H., Qiu, J.J., Hu, Y.D.: Vibration analysis of large turbo-generator stator system. In: Proceedings International Conference on Power System Technology, vol. 4, pp. 2168–2172. IEEE (2002)
51. Sewak, R., Ranjan, R., Kumar, V.: Experimental modal analysis of stator overhangs of a large turbogenerator. Energy Power Eng. **3**(3), 221–226 (2011)
52. He, Y.L., Deng, W.Q., Peng, B., Ke, M.Q., Tang, G.J., Wan, S.T., Liu, X.Y.: Stator vibration characteristic identification of turbogenerator among single and composite faults composed of static air-gap eccentricity and rotor interturn short circuit. Shock and Vibration (2016)
53. Thomson, W.: Theory of Vibration with Applications. CrC Press (2018)
54. Wirsching, P.H., Paez, T.L., Ortiz, K.: Random vibrations: theory and practice. Courier Corporation (2006)
55. Babak, V.P., Babak, S.V., Eremenko, V.S., Kuts, Y.V., Myslovych, M.V., Scherbak, L.M., Zaporozhets, A.O.: Problems and features of measurements. Models Meas. Meas. Monit. 1–31 (2021)
56. Babak, V., Zaporozhets, A., Zvaritch, V., Scherbak, L., Myslovych, M., Kuts, Y.: Models and measures in theory and practice of manufacturing processes. IFAC-PapersOnLine **55**(10), 1956–1961 (2022)
57. Babak, V.P., Babak, S.V., Myslovych, M.V., Zaporozhets, A.O., Zvaritch, V.M., Babak, V.P., Zvaritch, V.M.: Technical provision of diagnostic systems. Diagn. Syst. Energy Equipm. 91–133 (2020)
58. Reddy, D.R., Sen, A., Johnston, G.L.: Dynamics of a limit cycle oscillator under time delayed linear and nonlinear feedbacks. Physica D **144**(3–4), 335–357 (2000)
59. Babak, V.P., Babak, S.V., Myslovych, M.V., Zaporozhets, A.O., Zvaritch, V.M.: Principles of construction of systems for diagnosing the energy equipment. Diagnos. Syst. Energy Equipm. 1–22 (2020)
60. Benarab, A., Boussaada, I., Trabelsi, K., Bonnet, C.: Multiplicity–Induced–Dominancy property for second–order neutral differential equations with application in oscillation damping. Eur. J. Control. **69**, 100721 (2023)
61. Farina, L., Rinaldi, S.: Positive Linear Systems: Theory and Applications. John Wiley & Sons (2011)
62. Friedland, B.: Control system design: an introduction to state-space methods. Courier Corporat. (2012)
63. Zvaritch, V.: Some singularities of linear AR processes characterization in applied problems of power equipment and power systems diagnosis. In: Power systems research and operation: Selected Problems III, pp. 263–278. Springer Nature Switzerland, Cham (2023)
64. Lupenko, S.: The mathematical model of cyclic signals in dynamic systems as a cyclically correlated random process. Mathematics **10**(18), 3406 (2022)
65. Zvaritch V.: Some mathematical models which could be applied to develop arctic cybersecurity systems. In: Cybersecurity and Resilience in the Arctic, pp. 373–386. IOS Press, (2020)
66. Mondal, S., Mahanta, C.: Adaptive second-order sliding mode controller for a twin rotor multi-input–multi-output system. IET Control Theory Appl. **6**(14), 2157–2167 (2012)
67. Bartolini, G., Ferrara, A., Usai, E., Utkin, V.I.: On multi-input chattering-free second-order sliding mode control. IEEE Trans. Autom. Control **45**(9), 1711–1717 (2000)

68. Ahmad, S.M., Chipperfield, A.J., Tokhi, M.O.: Dynamic modelling and open-loop control of a twin rotor multi-input multi-output system. Proceedings of the institution of mechanical engineers. Part I: J. Syst. Cont. Eng. **216**(6), 477–496 (2002)
69. Yan, R., Gao, R.X.: Multi-scale enveloping spectrogram for vibration analysis in bearing defect diagnosis. Tribol. Int. **42**(2), 293–302 (2009)
70. Pham, M.T., Kim, J.M., Kim, C.H.: Accurate bearing fault diagnosis under variable shaft speed using convolutional neural networks and vibration spectrogram. Appl. Sci. **10**(18), 6385 (2020)
71. Khadersab, A., Shivakumar, S.: Vibration analysis techniques for rotating machinery and its effect on bearing faults. Proc. Manuf. **20**, 247–252 (2018)
72. Zaporozhets, A., Eremenko, V., Isaenko, V., Babikova, K.: Approach for creating reference signals for detecting defects in diagnosing of composite materials. In: Conference on Computer Science and Information Technologies, pp. 154–172. Springer International Publishing, Cham (2019)
73. Bäßler, R., Bäßler, T., Kley, M.: Classification of load and rotational speed at wire-race bearings using convolutional neural networks with vibration spectrograms. tm-Technisches Messen **89**(5), 352–362 (2022)
74. Arun, P., Madhukumar, S., Careena, P.: A Method for the investigation of bearing vibration based on spectrogram image comparison. In: IOP Conference Series: Materials Science and Engineering, vol. 396, No. 1, p. 012044. IOP Publishing (2018)
75. Zaporozhets, A.O.: Correlation analysis between the components of energy balance and pollutant emissions. Water Air Soil Pollut. **232**, 1–22 (2021)
76. Bai, T., Yang, J., Duan, L., Wang, Y.: Fault diagnosis method research of mechanical equipment based on sensor correlation analysis and deep learning. Shock. Vib. **2020**, 1–11 (2020)
77. Ming, A.B., Qin, Z.Y., Zhang, W., Chu, F.L.: Spectrum auto-correlation analysis and its application to fault diagnosis of rolling element bearings. Mech. Syst. Signal Process. **41**(1–2), 141–154 (2013)
78. Liu, H., Zhang, J., Cheng, Y., Lu, C.: Fault diagnosis of gearbox using empirical mode decomposition and multi-fractal detrended cross-correlation analysis. J. Sound Vib. **385**, 350–371 (2016)
79. Zaporozhets, A., Redko, O., Babak, V., Eremenko, V., Mokiychuk, V.: Method of indirect measurement of oxygen concentration in the air. Naukovyi Visnyk Natsionalnoho Hirnychoho Universytetu **5**, 105–114 (2018)
80. Babak, S., Babak, V., Zaporozhets, A., Sverdlova, A.: Method of statistical spline functions for solving problems of data approximation and prediction of objects state. In: Proceedings of the Second International Workshop on Computer Modeling and Intelligent Systems (CMIS-2019), pp. 15–19. Zaporizhzhia, Ukraine, April (2019)
81. Thao, N.X., Tuan, V.K., Hong, N.T.: Integral transforms of Fourier cosine and sine generalized convolution type. Int. J. Math. Mathemat. Sci. 2007 (2007)
82. Tuck, E.O.: On positivity of fourier transforms. Bull. Aust. Math. Soc. **74**(1), 133–138 (2006)
83. Nazarchuk, Z., Skalskyi, V., Serhiyenko, O.: Acoustic emission. Foundat. Eng. Mech. (2017)
84. Zhao, L., Kang, L., Yao, S.: Research and application of acoustic emission signal processing technology. IEEE Access **7**, 984–993 (2018)
85. Zaporozhets, A.A., Eremenko, V.S., Serhiienko, R.V., Ivanov, S.A.: Development of an intelligent system for diagnosing the technical condition of the heat power equipment. In: 2018 IEEE 13th International Scientific and Technical Conference on Computer Sciences and Information Technologies (CSIT), Vol. 1, pp. 48–51. IEEE (2018)
86. Zaporozhets, A., Babak, V., Sverdlova, A., Isaienko, V., Babikova, K.: Development of a system for diagnosing heat power equipment based on IEEE 802.11 s. In: Systems, Decision and Control in Energy II. Springer International Publishing, Cham, pp. 141–151 (2021)
87. Pappas, Y.Z., Kontsos, A., Loutas, T.H., Kostopoulos, V.: On the characterization of continuous fibres fracture by quantifying acoustic emission and acousto-ultrasonics waveforms. NDT E Int. **37**(5), 389–401 (2004)

88. Holford, K.M.: Acoustic emission–basic principles and future directions. Strain **36**(2), 51–54 (2000)
89. Bussiba, A., Kupiec, M., Ifergane, S., Piat, R., Böhlke, T.: Damage evolution and fracture events sequence in various composites by acoustic emission technique. Compos. Sci. Technol. **68**(5), 1144–1155 (2008)
90. Leonenko, N., Scalas, E., Trinh, M.: Limit theorems for the fractional nonhomogeneous Poisson process. J. Appl. Probab. **56**(1), 246–264 (2019)
91. Zocher, M.: Multivariate counting processes. Professoren des Inst. für Math. Stochastik (2006)
92. Djouadi, S.M., Maroulas, V., Pan, X., Xiong, J.: Consistency and asymptotics of a Poisson intensity least-squares estimator for partially observed jump–diffusion processes. Statist. Probab. Lett. **123**, 8–16 (2017)
93. Gao, X., Zhu, L.: Functional central limit theorems for stationary Hawkes processes and application to infinite-server queues. Queueing Syst. **90**, 161–206 (2018)
94. Shang, Y., Liu, Q., Niu, B., Zhang, M., Qi, W., Wu, J.: Mechanical fault diagnosis system based on acoustic feature analysis in gas insulated switchgear. In: 2017 1st International Conference on Electrical Materials and Power Equipment (ICEMPE), pp. 610–615. IEEE (2017)
95. Lai, Y., Yu, W.: Analytical solution for forced vibration of a simply-supported V-shaped folded plate roof. Thin-Walled Structures **40**(3), 215–223 (2002)
96. Kobayashi, H., Mark, B.L., Turin, W.: Probability, Random Processes, and Statistical Analysis: Applications to Communications, Signal Processing, Queueing Theory and Mathematical Finance. Cambridge University Press (2011)
97. Jorgensen, P.E.: Analysis and probability: wavelets, signals, fractals (Vol. 234). Springer Science & Business Media (2006)
98. Chen, Z., Ma, Q., Yan, X., Lou, S.: Prestress design of ellipsoid hybrid form and parameter analysis of design factors. J. Aerosp. Eng. **32**(1), 04018137 (2019)
99. Gardiner, C.: Stochastic Methods, vol. 4. Springer, Berlin (2009)
100. Debusschere, B.J., Najm, H.N., Pébay, P.P., Knio, O.M., Ghanem, R.G., Le Maitre, O.P.: Numerical challenges in the use of polynomial chaos representations for stochastic processes. SIAM J. Sci. Comput. **26**(2), 698–719 (2004)
101. Crandall, S.H., Mark, W.D.: Random Vibration in Mechanical Systems. Academic Press (2014)
102. Wong, E., Hajek, B.: Stochastic Processes in Engineering Systems. Springer Science & Business Media (2012)
103. Sverdlova, A., Zaporozhets, A.: Predicting anomaly conditions of energy equipment using neural networks. In: E3S Web of Conferences, vol. 280, p. 09005 (2021)
104. Wu, Y., Zhu, W.Q.: Stationary response of MDOF dissipated Hamiltonian systems to Poisson white noises (2008)
105. Veletić, M., Balasingham, I.: An information theory of neuro-transmission in multiple-access synaptic channels. IEEE Trans. Commun. **68**(2), 841–853 (2019)
106. Cacciapuoti, A.S., Piras, A., Caleffi, M.: Modeling the dynamic processing of the presynaptic terminals for intrabody nanonetworks. IEEE Trans. Commun. **64**(4), 1636–1645 (2016)
107. Cacciapuoti, A. S., Caleffi, M., Piras, A.: Neuronal communication: presynaptic terminals as transmitter array. In: Proceedings of the Second Annual International Conference on Nanoscale Computing and Communication (pp. 1–5) (2015)
108. Zhang, W., Li, C., Peng, G., Chen, Y., Zhang, Z.: A deep convolutional neural network with new training methods for bearing fault diagnosis under noisy environment and different working load. Mech. Syst. Signal Process. **100**, 439–453 (2018)
109. Newton, R.U., Dugan, E.: Application of strength diagnosis. Strength Condit. J. **24**(5), 50–59 (2002)
110. Kovtun, S., Kuts, Y., Malko, V., Fryz, M., Shcherbak, L., Kuts, V.: Application of Hilbert transform for power quality indicators monitoring in general purpose grids. Syst. Res. Energy **2**(77), 71–83 (2024)
111. Lu, W., Liang, B., Cheng, Y., Meng, D., Yang, J., Zhang, T.: Deep model based domain adaptation for fault diagnosis. IEEE Trans. Industr. Electron. **64**(3), 2296–2305 (2016)

112. Devroye, L., Györfi, L., Lugosi, G.: A Probabilistic Theory of Pattern Recognition, vol. 31. Springer Science & Business Media (2013)
113. Dougherty, E.R., Barrera, J.: Pattern recognition theory in nonlinear signal processing. J. Math. Imag. Vis. **16**, 181–197 (2002)
114. Theodoridis, S., Koutroumbas, K.: Pattern recognition. Elsevier (2006)
115. Lugosi, G.: Pattern classification and learning theory. In: Principles of nonparametric learning, pp. 1–56. Springer Vienna, Vienna (2002)
116. Obuchowski, J., Zimroz, R., Wyłomańska, A.: Blind equalization using combined skewness–kurtosis criterion for gearbox vibration enhancement. Measurement **88**, 34–44 (2016)
117. Eremenko, V., Babak, V., Zaporozhets, A.: Method of reference signals creating in non-destructive testing based on low-speed impact. Tekhnichna Elektrodynamika **4**, 070 (2021)
118. Babak, V., Zaporozhets, A., Kuts, Y., Scherbak, L., Eremenko, V.: Application of material measure in measurements: theoretical aspects. In: Systems, Decision and Control in Energy II, pp. 261–269. Springer International Publishing, Cham (2021)
119. Poveda-Martinez, P., Ramis-Soriano, J.: A comparison between psychoacoustic parameters and condition indicators for machinery fault diagnosis using vibration signals. Appl. Acoust. **166**, 107364 (2020)
120. Li, Y., Peter, W.T., Wang, X.: Recovery of vibration signal based on a super-exponential algorithm. J. Sound Vib. **311**(1–2), 537–553 (2008)
121. Joshi, A.: Use of acceleration spectra for determining the frequency-dependent attenuation coefficient and source parameters. Bull. Seismol. Soc. Am. **96**(6), 2165–2180 (2006)
122. Bohachev, I., Babak, V., Zaporozhets, A.: Novel small-aperture transducers based on magnetostrictive effect for diagnostic systems. Technical Electrodynamics **3**, 69–78 (2022)
123. Babak, V.P., Babak, S.V., Eremenko, V.S., Kuts, Y.V., Myslovych, M.V., Scherbak, L.M., Zaporozhets, A.O.: Models and measures for standardless measurements of the composite materials characteristics. Models Meas. Measur. Monit. 157–190 (2021)
124. Jazirehpour, M., Ebrahimi, S.S.: Effect of aspect ratio on dielectric, magnetic, percolative and microwave absorption properties of magnetite nanoparticles. J. Alloy. Compd. **638**, 188–196 (2015)
125. Lipovsky, B.P., Dunham, E.M.: Vibrational modes of hydraulic fractures: inference of fracture geometry from resonant frequencies and attenuation. J. Geophys. Res.: Solid Earth **120**(2), 1080–1107 (2015)
126. Babak, V., Bohachev, I., Zaporozhets, A., Khaidurov, V., Havrysh, V., Kalinichenko, A.: Some features of modeling ultrasound propagation in non-destructive control of metal structures based on the magnetostrictive effect. Electronics **12**(3), 477 (2023)
127. Barndorff-Nielsen, O.E., Maejima, M., Sato, K.I.: Some classes of multivariate infinitely divisible distributions admitting stochastic integral representations. Bernoulli, 1–33 (2006)
128. Di Nunno, G.: Stochastic integral representations, stochastic derivatives and minimal variance hedging. Stochast. Stochast. Rep. **73**(1–2), 181–198 (2002)
129. Ji, U.C., Obata, N.: Quantum stochastic integral representations of Fock space operators. Stochast. An Int. J. Probabil. Stochast. Proc. **81**(3–4), 367–384 (2009)
130. Morikawa, H.: An approximate method to represent the extreme values of non-stationary gaussian white noise.
131. Zhu, Z., Gong, W., Yu, Z., Wang, K.: Investigation on the EPC method in analyzing the nonlinear oscillators under both harmonic and Gaussian white noise excitations. J. Vib. Control **29**(13–14), 2935–2949 (2023)
132. Wang, K., Zhu, Z., Xu, L.: Transient probabilistic solutions of stochastic oscillator with even nonlinearities by exponential polynomial closure method. J. Vib. Control **28**(9–10), 1086–1094 (2022)
133. Sherrit, S., Mukherjee, B.K.: Characterization of piezoelectric materials for transducers. ArXiv preprint arXiv:0711.2657 (2007)
134. Arnau, A. (ed.): Piezoelectric Transducers and Applications (Vol. 2004). Springer, Berlin (2004)

135. Cappon, H., Keesman, K.J.: Numerical modeling of piezoelectric transducers using physical parameters. IEEE Trans. Ultrason. Ferroelectr. Freq. Control **59**(5), 1023–1032 (2012)
136. Roh, Y., Varadan, V.V., Varadan, V.K.: Characterization of all the elastic, dielectric, and piezoelectric constants of uniaxially oriented poled PVDF films. IEEE Trans. Ultrason. Ferroelectr. Freq. Control **49**(6), 836–847 (2002)

Chapter 4
Linear Periodic Random Processes in Constructing Models Characterizing the Operation of Electrical Equipment

Abstract This comprehensive research explores the application of the linear periodic random processes (LPRP) and the linear random processes (LRP) models' parameters in diagnosing and understanding the technical conditions and operational issues of diesel engine generators (DEG) and electric machines (EM). By integrating these models, the study aims to capture the cyclical and stochastic nature of the operational behaviors of EM and their components, particularly focusing on the vibrations and rotational dynamics of DEGs and EMs. Through the analysis of the uneven rotation of the crankshaft and the distribution of cylinder power, the research demonstrates how LPRP and LRP can be effectively used to diagnose potential malfunctions and optimize the performance of these systems. The methodology includes measuring the kinetic energy of the shaft, calculating acceleration deviations, and applying discrete Fourier transform to identify harmonics indicative of operational integrity or issues. The findings suggest that the absence or presence of specific harmonics can diagnose uneven cylinder power distribution, crucial for maintaining efficient and reliable operation. This chapter extends the application of LPRP and LRP in analyzing stochastically periodic behavior, offering significant insights for enhancing diagnostic techniques for DEGs and EMs, marking a pioneering step in applying periodic random processes to mechanical diagnostic fields.

Keywords Cyclic processes · Linear periodic random processes · Linear random processes · Rotary machines · Electric equipment · Diagnostics · Noise signals

Today, the issues of constructing diagnostic systems for fixed components of electrical machines (stator elements, laminated magnetic circuit, frame with places of attachment to the foundation, etc.) have been largely resolved [1–6]. However, the presence of rotating units (rolling bearings, rotor, brush-commutator unit, etc.) creates certain requirements for diagnostic systems of electrical machines (EM), which are used to determine their technical condition [7–13].

This chapter discusses some issues related to the description of the basic principles of the theory of cyclic processes [14–17], which are subsequently used to create and practically use diagnostic systems to determine the technical condition of rotating

V. Babak et al., *Statistical Diagnostics of Electric Power Equipment*, Studies in Systems, Decision and Control 573, https://doi.org/10.1007/978-3-031-76253-6_4

components of electric vehicles. Mathematical models have been constructed and analyzed that take into account the dynamics of these EM units, as well as the cyclicity of diagnostic signals accompanying their operation.

4.1 Models of Cyclic Processes in the Description of Diagnostic Signals

A characteristic feature of the physical processes that accompany the operation of AC EM, as well as EM containing rotating units, are, on the one hand, their cyclicity, and on the other hand, the influence of additional factors of a random nature that arise directly in the units of operating equipment [15–17]. Thus, the laminated magnetic cores of transformers are exposed to an alternating magnetic field caused by periodic currents that flow in the EM windings; rolling bearings are exposed to cyclically changing mechanical forces caused by the rotation of the shafts; compression in the cylinders of diesel-electric generators changes under the diesel operating cycle, etc. [15–17]. Moreover, the values of such processes are not repeated exactly (in the mathematical sense) in each cycle (period), but are stochastic. Thus, the mathematical model of diagnostic signals must simultaneously take into account both the cyclicity and stochasticity of the physical research processes.

The main task that arose when constructing a mathematical model of vibrations of rotating units was the development, based on the classes of linear random processes (LRP) and linear periodic random processes (LPRP), of a generalized mathematical model of vibration processes that accompany the operation of moving units of electric vehicles, for the subsequent use of this models when creating new methods and technical means for EM diagnostics. This made it possible to obtain complete probabilistic characteristics of the processes under study (for example, moments of any order) in the general case [18–22]. In turn, this made it possible to establish the most informative diagnostic signs and thus increase the accuracy, reliability, and reliability of diagnosing the technical condition of the moving components of EVs. In addition, the simultaneous use of the LRP and LPRP classes in the study of vibration diagnostic signals, measured on rotating EM components, made it possible to construct effective algorithms for their analysis and classification of EM defects.

Today there are several works [15–17], mainly of an applied nature, which talk about such processes and they play a significant role in substantiating certain cyclic physical models. First of all, let's clarify the concept of a periodic function. We will be talking about a deterministic function.

If some function $f(x)$ has a T period, then its derivative is also periodic and has a T period (in short, it is T-periodic). Its antiderivative $F(x) = \int_0^x f(t)dt$ has a T period, provided that $\int_0^T f(t)dt = 0$, otherwise the antiderivative is not periodic.

When diagnosing various components of electrical machines, the signals coming from sensors installed on these components, as a rule, are cyclic. Let's briefly look at the terminology, namely, consider the *concepts of cyclicity, rhythmicity, and stochastic periodicity*. Let us use the calculations given in [15–17].

A cyclic process is a physical concept that is defined as a physical phenomenon and refers to the case when, from a physical point of view (and not strictly mathematically), the repeatability of characteristics or physical properties of signals can be distinguished.

The concept of rhythm is used when at least two physical processes are observed, one of which has a cyclical character close to the second. Then these two processes are called rhythmic (one is considered in rhythm with the other). For example, various vibrations caused by each other or those that occur depending on another associated with it. They talk about rhythm when two or more processes or several processes are observed simultaneously.

The idea of *stochastic periodicity* belongs to Slutsky [23]. In this case, the Hilbert condition was not initially required. It should be noted that the theory of periodic random processes is a theory of measurable functions of two variables, one of which is considered as a parameter—according to it, the process is periodic. According to the second variable the process is always considered as a function of a set, which is defined on some probability space. Thanks to this second variable, a random process acquires properties that are not inherent in the deterministic functions of two variables. According to the second variable, a random process cannot have periodic properties, since it is not ordered like numbers on the number line.

Stochastic periodicity is understood as the property of the characteristics of a random process repeating in time after a known period, and implementations of the process can also be non-periodic [24–26]. The above statements are not a definition, since they do not allow us to answer the question of what is the period of a random process, so we will further clarify and specify them. If, as is often customary in energy theory, we are interested only in the characteristics of the process within the framework of the first two distribution functions, then we arrive at the so-called periodic random processes in a broad sense.

As was shown in Chap. 2, the LRP is determined by a stochastic integral of the form [27, 28]

$$\xi(t) = \int_{-\infty}^{\infty} \phi(\tau, t) d\eta(\tau), t \in T, \tag{4.1}$$

where $\varphi(\tau, t), t \in (-\infty, \infty)$—real non-random numerical function (kernel (1)), which satisfies uniformly according to the t condition $\phi(\tau, t) \in L_p(-\infty, \infty)$ at $p = 1, 2$; $\{\eta(\tau), \mathrm{P}\{\eta(0) = 0\} = 1, \tau \in (-\infty, \infty)\}$—a (generating) process with independent increments and an infinitely divisible characteristic function (CF) of the increments. It is this class of processes that allows us to construct various classes of linear processes, including the class of LPRP.

The basics of the LPRP theory are given in [29–31], where the conditions are formulated under which LPRP models can act as models of cyclic processes. By this work, *a separable random process* is called periodic in the broad sense $\xi(t)$, $t \in$

$(-\infty, \infty)$ if such a thing exists $T > 0$ that its one-dimensional and two-dimensional distribution functions satisfy the conditions

$$F(x; t) \equiv F(x; t + T), F(x_1, x_2; t_1, t_2)$$
$$= F(x_1, x_2; t_1 + T, t_2 + T), \quad t_1, t_2 \in (-\infty, \infty).$$

In many practical problems, one cannot limit oneself to considering only the first two moments of a random process. In particular, to determine the quality of the statistical assessment of the correlation function, it is necessary to know the third and fourth moments. With a more complete study of the properties of processes in EM, taking into account their periodic probabilistic characteristics above the second order, periodic processes in the understanding of E. Slutsky can be used. Although in the mentioned work [23] the author does not give a strict definition, it is clear that the periodicity of random processes is considered in terms of equivalence in distribution. By the approach of E. E. Slutsky, the definition of periodic random processes can be given in the following form.

Separable random process $\xi(t)$, $t \in (-\infty, \infty)$, is said to be *periodic* if there is a number such $T > 0$ that the finite-dimensional vectors $(\xi(t_1), \xi(t_2), \ldots, \xi(t_n))$ and $(\xi(t_1 + T), \xi(t_2 + T), \ldots, \xi(t_n + T))$, where $\{t_k\}$, $k = 1, 2, \ldots$—process separability set $\xi(t)$, for all integers are stochastically equivalent in the broad sense. The smallest number $T > 0$, for which this condition is met is called the period of the process $\xi(t)$.

By [29–31], stochastic equivalence in the broad sense is equivalence in terms of distribution functions. Thus, from the above definition it follows that all finite-dimensional distribution functions of the process $\xi(t)$ are skew-periodic, that is

$$F(x_1, ..., x_n; t_1, ..., t_n) = F(x_1, ..., x_n; t_1 + T, ..., t_n + T), \tag{4.2}$$

where $F(x_1, ..., x_n; t_1, ..., t_n)$—$n$-dimensional distribution function of the process $\xi(t)$, $n = 1, 2, \ldots$.

Thus, a periodic random process in Slutsky's understanding is periodic in the broad sense, but the opposite is not true in the general case. If for a periodic process, there are moments μ^n up to order $n \geq 1$ then they are all T periodic t, and the mixed moments are periodic in the set of arguments. A stationary random process is periodic according to Slutskii, and its period can be an arbitrary number $T > 0$.

Using the ideas of E. Slutsky, we will analyze some definitions regarding periodic random processes. Let us dwell directly on the construction of a mathematical model of the moving parts of EM.

4.2 Vibrations of Rotating Components of Electrical Machines

As shown in [10, 11], in many cases, when solving problems of diagnosing various technical objects, moments μ_n can be used as diagnostic signs. Thus, to determine quantitative estimates of the specified diagnostic signs that characterize the technical condition of certain moving parts of an EM, it is necessary to take into account the value $T > 0$. In addition, knowledge of this period provides the opportunity to reasonably solve several practical problems related to the statistical processing of diagnostic signals (selection of sample length, hospitalization, etc.). Below, as an example, we will dwell on the issues of constructing mathematical models of vibration diagnostic signals in electrical machine components, taking into account the cyclic nature of their operation.

To construct a mathematical model of vibrations, it is necessary to select the unit or element being diagnosed. The main interest is the vibrations of this particular unit, which should be considered as a diagnostic signal. Vibrations that are generated by other exciting forces must be considered as interference.

Let us assume that it is necessary to determine the state of pressing of the laminated magnetic circuit of the EM rotor. By this task, vibration measurement is carried out during the operation of the electric vehicle using an accelerometer, which is located in the end part of the electric vehicle rotor. At the measurement point, vibrations of a certain intensity are observed, which are caused by various exciting forces accompanying the EM's work. Among them, the following main factors can be identified [10, 11, 32, 33]:

- $\xi_{EM}(t)$—electromagnetic forces;
- $\xi_{RB}(t)$—rolling bearings;
- $\xi_{BC}(t)$—brush-collector unit;
- $\xi_{AD}(t)$—aerodynamic forces;
- $\xi_{MC}(t)$—laminated magnetic circuit.

Vibrational processes caused by these exciting forces originate in different places of the studied EM and arrive at the measurement point in different ways, while experiencing certain transformations (as a result of the absorption of their energy in the body of the studied EM, time delay, etc.). An accelerometer installed at the measurement point records the total influence of vibrations from all of the above factors [10, 11, 34, 35].

The considered process can be conventionally depicted by the diagram shown in Fig. 4.1, where rectangles depict multi-input linear systems that model the processes of vibrations passing from the points at which they originate to the measurement point, which is designated "B" in Fig. 4.1.

Analytically, the signal measured by the accelerometer can be represented by the relation

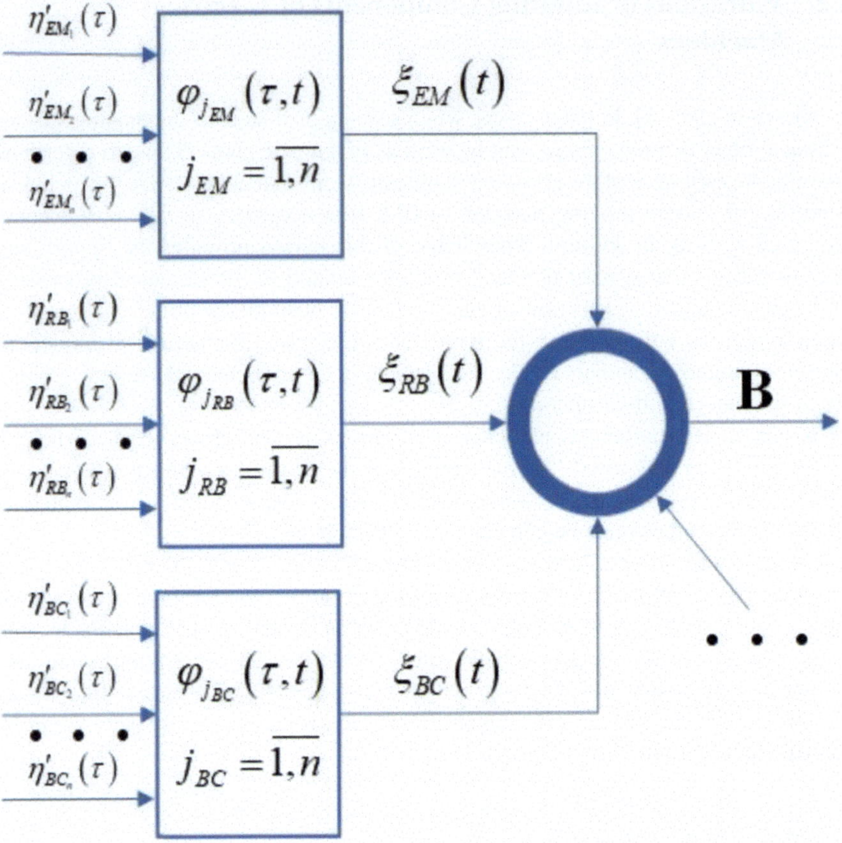

Fig. 4.1 Scheme of vibration formation at the measurement point, taking into account the influence of various exciting forces accompanying the EM's work

$$\xi(t) = \sum_{j=1}^{n} a_j \xi_{EM_j}(t) + \sum_{j=1}^{n} a_j \xi_{RB_j}(t) + \sum_{j=1}^{n} a_j \xi_{BC_j}(t) + \sum_{j=1}^{n} a_j \xi_{AD_j}(t) + \sum_{j=1}^{n} a_j \xi_{MC_j}(t), \quad (4.3)$$

where $t \in (-\infty, \infty)$, n—some positive number, which is determined by the design, technological, and operational characteristics of the unit and specifies the number of system inputs; a_j—weighting coefficients that take into account the attenuation in the corresponding channel, $a_j > 0$.

Thus, expression (4.3) can be considered as a generalized mathematical model of vibrations that accompany the operation of moving components of an electric vehicle. The proposed model takes into account both the periodicity of the vibration process and the influence on it of individual vibration sources, which have different physical natures and spatial localization. As shown by the results of experimental studies [10, 11, 36–39], the development of technical diagnostic systems for EM components based on the created model (4.3) allows for diagnostics of moving EM components with specified accuracy and reliability.

According to [10, 11], the vibrations of a moving EM unit can be represented as the response of the mechanical system of this unit to the influence of the generating pulse flow that is formed in the unit during its operation. White noise [40–43] is usually used as a model of such a generating pulse flow, and the model of the mechanical system of the EM unit can be approximately represented as a linear system of a certain order.

By its physical nature, the process that causes vibrations of nodes is a certain flow of impulses random in time and intensity. Let us assume that they are statistically independent and satisfy the condition of uniform incomplete smallness. Thus, this process can be described by a mathematical model of random processes with independent values and an infinitely divisible distribution law.

The moving part of the EM can be considered as a mechanical oscillatory system, which is affected by the generating process, which in turn causes it to resonate at certain frequencies. Resonant frequencies are determined not only by the design features of the unit under study, but also by the features of the entire electrical machine as a whole.

The use of the LRP and LPRP classes when constructing probabilistic models that are multi-resonant oscillatory allows for a complete probabilistic analysis of the proposed models that describe the operation of the diagnosed node. In turn, this makes it possible to reasonably select diagnostic features and software and hardware for technical diagnostic tools, which increases the reliability of the diagnostics performed and ultimately increases the reliability of operated EMs.

According to [10, 11, 19, 39], in a linear approximation, a mechanical oscillatory system can be represented by one or a set of linear links of the first and second order, that is, a set of oscillatory and aperiodic links. Based on this, to analyze such a mechanical oscillatory system, one can use the well-developed apparatus of the theory of linear systems.

In this case, the vibrations that are recorded at the point where the accelerometer is placed on the diagnosed EM unit represent its response to the influence of the vector-generating process.

Process $\eta(\tau)$, the probabilistic characteristics of which are largely determined by the technical condition of the node under study, can be specified using CF. In this case, the relations known from [10, 11] make it possible to obtain an expression for the n-dimensional CF of the process values $\xi(t)$, as well as its moments μ^n. Thus, it is possible to obtain analytical expressions for all probabilistic characteristics of the vibration process, which makes it possible to create effective methods for vibration diagnostics of the research EM.

As an example, let us consider the problem of determining the main statistical characteristics of vibration diagnostic signals of moving EM units.

The dynamics of the rotating component of the moving unit of the electric vehicle, for the description of which the LPRP model can be used, is represented by the equation

$$\varphi_t = \varphi_0 + \omega_0 t + \varphi_\partial(t), \tag{4.4}$$

where φ_0—the initial phase, which is recorded by a special sensor (for example, for wind-electric units—the zero mark, and diesel generators—the top dead point (TDP) mark), and therefore can subsequently be taken equal to zero, ω_0—average angular velocity, which is determined by the relation $\omega_0 = \pi n / 30$, n—shaft rpm.

The function $\varphi_\partial(t)$ represents a fluctuation in the angle of rotation of the shaft relative to its uniformly varying angle of rotation.

So, it is this value $\varphi_\partial(t)$ carries information about the uneven rotation of the EM shaft at each revolution. The latter may be associated with the appearance of defects in various units of the studied EM. Let us immediately note that $\varphi_\partial(t)$ is random and to study it is necessary to use the mathematical apparatus of the LPRP theory, and using (4.4), which includes ω_0, you can determine the period $T > 0$ the process under study. Thus, from expression (4.4) it is clear that the process of uneven rotation of the shaft is characterized by stochastic periodicity, that is, when some probabilistic characteristics of the signals behave periodically, although their implementations themselves are not necessarily periodic.

Using the LPRP class when describing such processes makes it possible to obtain their full probabilistic characteristics in terms of distribution functions (for example (4.2)) or CF.

As is known [27], the logarithm of the one-dimensional CF of the vibration process (4.1) can be represented as follows:

$$\begin{aligned} \ln f_\xi(u) = ium \sum_{j=1}^{n} a_{jn} + \int_{-\infty}^{\infty}\int_{-\infty}^{\infty} \Bigg[\exp\left(iux \sum_{j=1}^{n} a_{jn}\phi_j(\tau) \right) - \\ -1 - iux \sum_{j=1}^{n} a_{jn}\phi_j(\tau) \Bigg] \frac{dK(x)}{x^2} d\tau, \end{aligned} \tag{4.5}$$

where m and $K(x)$—parameters of infinitely divisible CF process increments $\eta(t)$ in Kolmogorov form under the assumption that $\int_{-\infty}^{\infty} \varphi(\tau, t) d\tau < \infty$.

Generative process $\eta'(\tau)$ the nature of which, as noted above, is largely determined by the technical condition of the unit under study, can be specified using a CF of the form (4.5). It is by the components included in (4.5) that we have the opportunity to study the generating process $\eta'(\tau)$.

Based on the results of the above models, it is possible to substantiate possible diagnostic signs of the technical condition of EM components, taking into account the cyclic component included in the vibration diagnostic signal $\xi(t)$. Taking this component into account is due to the size of the period $T > 0$. This quantity is introduced into analytical expressions to obtain spectral-correlation dependencies and distribution functions by analogy with expression (4.2). Otherwise, this justification is carried out in the same way as it was done in Chap. 3.

A slightly different approach is required when constructing mathematical models and subsequent substantiation of diagnostic features when solving the problem of diagnosing diesel-electric units. Let's take a closer look at this issue.

4.3 Dynamics of Changes in Cylinder Power on the Shaft of a Diesel-Electric Generator

Among autonomous energy sources, a special role belongs to diesel-electric generators (DEG), which are installed at various critical facilities where failures in the supply of electrical energy are not allowed, as an emergency power source. At the same time, increased demands are placed on the reliability of these energy sources, which, in particular, can be ensured using various monitoring and diagnostic systems [44–48].

As is known, various types of diesel engines are used as the main drive of an electric generator in a diesel generator; diagnosing the technical condition of which requires the creation of special diagnostic systems that take into account the specifics of the operation of such a diesel engine as an internal combustion engine.

According to statistical data [49–53], one of the most unreliable components in diesel engines is the cylinder-piston group (CPG). This is because this unit, which converts the reciprocating motion of the pistons into the rotational motion of the crankshaft, operates under the most severe temperature and dynamic conditions. Based on this, below we will dwell on the construction of a mathematical model of the dynamics of the rotational motion of the diesel generator shaft and its subsequent study.

While constructing a mathematical model that describes the nature of changes in cylinder power increments over time, it is necessary to take into account several factors and design features of the diesel generator under study. In this regard, to solve this problem, we will consider some laws of motion of the diesel crankshaft.

To construct a mathematical model of the rotation of the DEG shaft [54–56], we use expression (4.4).

As noted above, the function $\varphi_\partial\,(t)$ is nothing more than a fluctuation in the angle of rotation of the shaft relative to its uniform angle of rotation at a t time. It is this value $\varphi_\partial\,(t)$ will be subject to further study, since it carries information about the uneven rotation of the diesel crankshaft, and, consequently, about the non-identity of the working cycles of the diesel cylinders. Let us immediately note that $\varphi_\partial\,(t)$ is random and to study it is necessary to use the mathematical apparatus of the theory of random processes [15–17].

The use of a statistical approach when analyzing the uneven rotation of the diesel generator shaft allows:

- create a technique for dealing with interference;
- assess the reliability of the measured results;
- to develop a method for measuring the increment of cylinder power based on the uneven rotation of the diesel crankshaft.

From expression (4.4) it is clear that the process of uneven rotation of the diesel crankshaft, and, consequently, the measurement of cylinder power increments is characterized by stochastic periodicity.

To construct a mathematical model of the process of uneven rotation of the shaft $\varphi_\partial(t)$, let's use the LPRP class [27]. The logarithm of the finite-dimensional CF LPRP will be written in the form [27]

$$\begin{aligned}
&\ln f_\xi(u_1, u_2, \ldots, u_n; t_1, t_2, \ldots, t_n) \\
&= \ln f_\xi(u_1, u_2, \ldots, u_n; t_1 + T, t_2 + T, \ldots, t_n + T) \\
&= i\sum_{j=1}^{n} u_j \int_{-\infty}^{\infty} \phi(\tau, t_j) d\mu(\tau) - \frac{1}{2}\sum_{j,k=1}^{n} u_j u_k \int_{-\infty}^{\infty} \phi(\tau, t_j)\phi(\tau, t_k) dD(\tau) \\
&\quad + \int_{-\infty}^{\infty}\int_{-\infty}^{\infty} \left[\exp\left(ix\sum_{j=1}^{n} u_j\phi(\tau, t_j)\right) - 1 - \frac{ix}{1+x^2}\sum_{j=1}^{n} u_j\phi(\tau, t_j)\right]\frac{1+x^2}{x} G(d\tau, dx),
\end{aligned} \tag{4.6}$$

where $\mu(t)$—continuous function, $D(t)$—continuous and non-decreasing function, $G(dt, dx)$—spectral measure characterizing the distribution of jumps in the Poisson component of a random process with independent homogeneous increments $\eta(t)$.

It should be noted that for the existence of finite-dimensional CFs (4.6), the following conditions must be met:

$$\int_{-\infty}^{\infty} |\phi(\tau, t)| d\mu(\tau) < \infty; \quad \int_{-\infty}^{\infty} |\phi(\tau, t)| d\mu(\tau) < \infty;$$

$$\int_{-\infty}^{\infty} \left(e^{ix\phi(\tau,t)} - 1 - \frac{ix\phi(\tau, t)}{1+x}\right)\frac{1+x^2}{x} G(d\tau, dx) < \infty,$$

and additionally for each $t_j \in T$

$$\varphi(\tau, t_j) = \varphi(\tau, t_j + T).$$

Thus, using the LPRP model specified by a finite-dimensional CF of the form (4.6), it is possible to conduct a complete study of the interference that occurs during uneven rotation of the shaft of the DEG under study. When carrying out further research, we will assume that the angular velocity of rotation of the DEG shaft in a steady state is an LPRP with a period equal to the period of rotation of the shaft of the research DEG.

The adopted LPRP model makes it possible to fully, and not only within the framework of energy spectral theory, which is difficult to apply to the analysis of non-stationary processes, analyze random changes that occur during shaft rotation and the resulting errors when measuring increments of cylinder power of diesel generators.

The proposed LPRP model, which describes the non-stationary process of uneven rotation of the shaft of the DEG under study, can be successfully applied to the analysis of this process even if it has a distribution other than Gaussian [57, 58].

While carrying out further analysis, it is necessary to make various transformations of the measured signal, namely its differentiation, and integration, and calculate the difference. In this case, the chosen LPRP model turns out to be indispensable because the derivatives and integrals of the LPRP, if they exist, are also LPRP. In addition, the sum and difference of LPRP with the same periods are also LPRP with the same period. Finally, all moments of the LPRP with period T (and, consequently, semi-invariants) are periodic functions with the same period. LPRP with period T (and, consequently, semi-invariants) are periodic functions with the same period.

When using a mathematical model of the LPRP to study stochastically periodic phenomena, the main point is to determine the estimate of the LPRP period. Using this model to determine the distribution of cylinder power of a diesel generator in a steady state based on the uneven rotation of the crankshaft, there is no need to determine the period estimate. When measuring LPRP implementations at the output of a sensor (for example, an angular velocity sensor of the PDF-3 type), pulses from the output of the TDP sensor are synchronously recorded via the second channel. When conducting a more accurate analysis of the moment of passage of these points, it is possible to statistically average and estimate their dispersion over time, which will answer the question about the accuracy of the results of such measurements.

Thus, having a period value, we can measure any nested sequence of LPRP samples that interests us, which will represent the initial statistics for determining the unevenness of shaft rotation and the spread (compared to some average value) of the cylinder powers of the diesel generator.

A nested sequence, for example, about the function of measuring the angle of rotation of a shaft (angular velocity or acceleration), is understood as an arbitrary sequence of angle readings (angular velocity or acceleration), taken with a time step less than the time of one rotation of the shaft.

While researching the unevenness of shaft rotation by expression (4.4), we may be interested in the parameter ω_0 and function $\varphi_\partial(t)$ or its first two derivatives $\varphi'_\partial(t)$ $\varphi''_\partial(t)$. Parameter measurement ω_0 does not cause any difficulties. Therefore, further analysis will consist of determining $\varphi_\partial(t)$ its two derivatives. From expression (4.4), the angular speed of rotation of the shaft can be determined as follows

$$\omega_t = \phi'_t = \omega_0 + \phi'_\partial(t), \tag{4.7}$$

and acceleration

$$\omega_t' = \phi_t'' = \phi_\partial''(t). \tag{4.8}$$

It should also be noted that the deviation of the shaft rotation angle, in turn, can be represented as follows

$$\phi_\partial(t) = \phi_\partial^u(t) + \xi(t), \tag{4.9}$$

where $\varphi_\partial^u(t)$—deterministic component due to the operation of the actual cylinders of the DEG under study; $\xi(t)$—random component, which is described by the LPRP model.

Function $\varphi_\partial^u(t)$ in steady-state engine operation, is a deterministic periodic function, varying with a period equal to or a multiple of the rotation period of this engine, even in the case when deviations from the normal operating mode are observed in the cylinders (or in one of them).

In the case of proper operation of all cylinders (complete identity in their operation), this function will have a period less than the period indicated above. This remark also applies to the random component $\xi(t)$.

Further we will assume that $M\xi(t) = 0$. Otherwise, we will refer it to $\varphi_\partial^u(t)$ as an additive component.

Thus, if for $\varphi_\partial(t)$ accept the LPRP model, then, due to the properties of the LPRP stated in [27], the first two derivatives of $\varphi_\partial(t)$ there will also be LPRP, i.e. LPRP will be angular velocity ω_t and acceleration ω_t'. In addition, the angular acceleration of the shaft coincides with the second derivative of the deviation of the angle of rotation of the shaft, described by expression (4.9).

The kinetic energy of the shaft at moment t is related to the angular velocity of the shaft by the following relation

$$W_t = \frac{1}{2} J_Z \omega_t^2,$$

where J_Z—moment of inertia of the shaft relative to its axis of rotation. Then the change in the kinetic energy of the shaft over a certain time interval $[t_1, t_2]$ according to the law of change in the kinetic energy of a rotating shaft relative to a fixed axis, taking into account (4.7), will be determined as

$$\begin{aligned} W_{t_2} - W_{t_1} &= \frac{1}{2} J_Z \left(\omega_{t_2}^2 - \omega_{t_1}^2\right) \\ &= J_Z \left\{ \omega_0 \left[\phi_\partial'(t_2) - \phi_\partial'(t_1)\right] + \frac{1}{2}\left[\phi_\partial'(t_2)\right]^2 - \frac{1}{2}\left[\phi_\partial'(t_1)\right]^2 \right\}. \end{aligned} \tag{4.10}$$

Using expression (4.10), we can determine the increments of cylinder power since $\varphi_\partial'(t)$ is LPRP, then the change in kinetic energy will also be LPRP with the same period.

As a result of studies carried out using the LPRP model, relationships were established that make it possible to determine the fluctuation of the shaft rotation angle $\varphi_\partial(t)$ relative to its uniform rotation at the moment t, as well as the associated fluctuations in angular velocity $\varphi_\partial'(t)$ and acceleration $\varphi_\partial'(t)$ It is these parameters that are subsequently used as diagnostic signs characterizing the technical condition of the diesel generator during its operation.

4.4 Construction of Estimates of LPRP Describing the Dynamics of Cylinder Power on the Shaft of DEG

When using a mathematical model of the LPRP to study stochastically periodic phenomena, the main point is to determine the estimate of the LPRP. To construct statistical estimates of the LPRP period, it is necessary to specify the object of study, in other words, to select a certain type of DEG.

Based on this, we will use the proposed model to determine the distribution of cylinder power of DEG type 5D70. As a sensor for measuring LPRP implementations, we will use an angular velocity sensor of the PDF-3 type [10, 11].

While determining the distribution of cylinder power in a steady state based on the uneven rotation of the crankshaft, there is no need to determine the period estimate, because When carrying out measurements of LPRP implementations at the output of the PDF-3 sensor, which was used during the experiments, pulses from the output of the TDP sensor are synchronously recorded via its second channel. While conducting a more accurate analysis of the moment of passage of these points, it is possible to statistically average and estimate their dispersion over time, which gives us an answer to the question about the accuracy of the results of such measurements.

By expressions (4.7), (4.8), and (4.9) given above, by observation we need to determine ω_0, $\varphi_\partial(t)$, $\varphi_\partial'(t)$ and $\varphi_\partial''(t)$. To determine them, a PDF-3 sensor is used, which generates a sequence of pulses, by which the above characteristics are determined. The repetition period of these pulses varies randomly, according to the uneven rotation of the diesel generator shaft. Some technical issues associated with the generation of such pulses are outlined below.

So, a PDF-3 sensor is installed at the free end of the DEG shaft, which converts the angular velocity into the pulse repetition frequency. The output signals of the sensor are formed in the form of two series of pulses B1 and B2 along two separate channels. The number of pulses per shaft revolution in each series is 600, and the duty cycle is $2 + 10\%$.

The pulses of two series B1 and B2 are shifted relative to each other by a quarter of the pulse repetition period. When the shaft rotates clockwise, the pulses of series B2 lag behind the series of pulses B1. Each series of pulses of the output signals is presented in direct and inverse forms. Diagrams of the output signals of the PDF-3 sensor are presented in Fig. 4.2a–d. In other words, the inverse series of pulses concerning the main one is shifted by half a period.

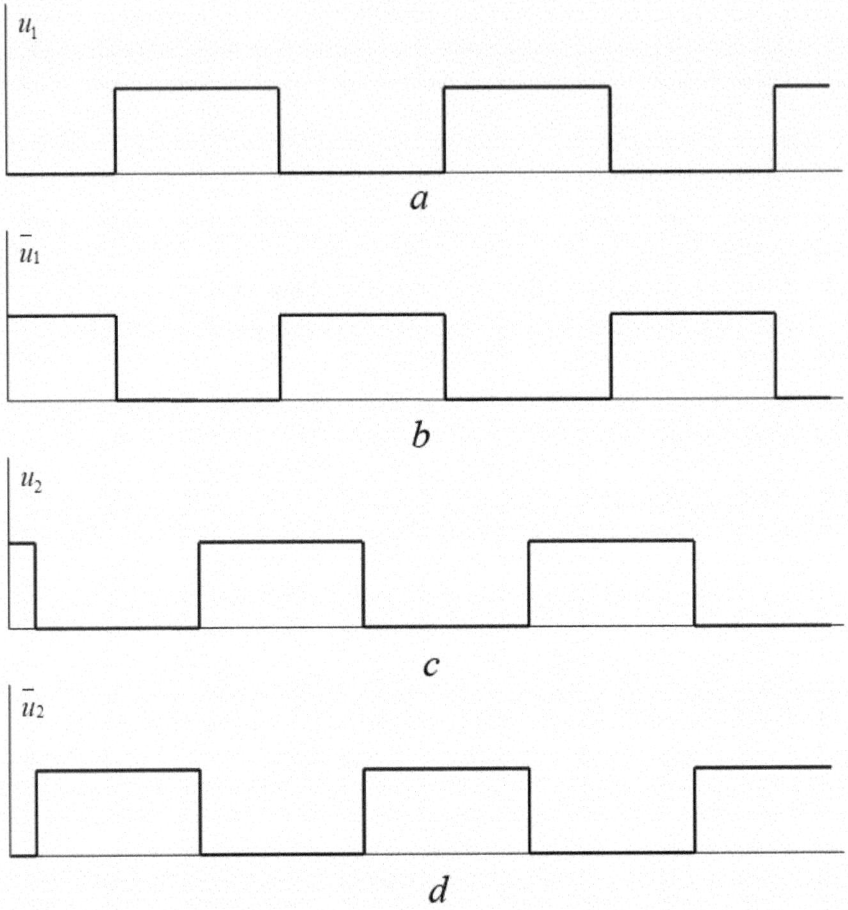

Fig. 4.2 Diagrams of PDF-3 sensor output signals

The operation of the sensor is based on the modulation of the photocurrent in the radiation receiver under the influence of rays incident on it. The main part of the sensor is a metal modulation disk mounted on the shaft and a fixed indicator plate. There are 600 grooves on the disk. There are similar grooves on the indicator plate. The slots of the first and second channels are spaced apart in such a way that they provide a phase shift of the output signals of different channels by a quarter of a period.

Let us now dwell in more detail on the description using the LPRP model of the output signals of the PDF-3 sensor.

So, as stated above, during the rotation of the engine crankshaft, a stream of pulses $\eta(t)$ is formed at the output of the PDF-3 sensor (Fig. 4.3), the shape of each of which is close to rectangular. However, due to uneven rotation of the motor shaft, due to inaccuracy in the manufacture of grooves in the modulation disk and indicator plate,

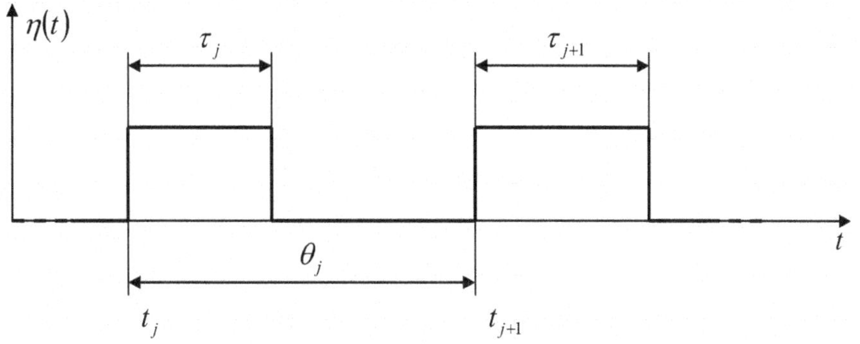

Fig. 4.3 Pulse parameters at the sensor output

as well as other types of errors, the moments at which pulses occur $\left\{t_j, j = \overline{1, n}\right\}$ and their duration $\left\{\tau_j, j = \overline{1, n}\right\}$ (Fig. 4.3) are random.

Since the height ("amplitude") of these pulses does not carry any information, we will assume that they have a strictly rectangular shape. Thus, each stream of such impulses is associated with two sequences of random variables:

$$\left\{t_j, j = \overline{1, n}\right\} \text{ and } \left\{\tau_j, j = \overline{1, n}\right\}.$$

The pulses under study do not overlap in time, because the condition is met $\Theta_j = t_{j+1} - t_j > T_j$.

The value t_j depends on the random position of the rectangular grooves on the modulation disk and the indicator plate.

When the diesel crankshaft rotates, the values t_j are significantly influenced by the uneven rotation of the shaft.

While solving the problem posed to study the uneven rotation of the DEG shaft, the main interest is the sequence $\left\{\Theta_j, j = \overline{1, n}\right\}$. As can be seen from Fig. 4.3, Θ_j which represents the time interval between samples.

Thus, a sequence of time interval values is observed (on one full cycle of diesel operation) $\left\{\Theta_j, j = \overline{1, 1200}\right\}$ between equidistant values of the angle of rotation of the rotating shaft. When conducting experimental studies, we obtain a quantized curve of the dependence of the shaft rotation time on the rotation angle with a sampling step along the angle equal to $\Delta\varphi = \frac{2\pi}{600}$. This implies

$$\frac{1}{\omega_j} = \frac{\Theta_j}{\Delta\phi} = \frac{600 \cdot \Theta_j}{2\pi}, \quad j = \overline{1, 1200}, \tag{4.11}$$

where $\Theta_j = n_j \cdot \Delta t$; Δt—time step; n_j– number of counting pulses filling the interval.

From the sequence $\frac{1}{\omega_j}$ in expression (4.11) we can go to the sequence of instantaneous angular velocities

$$\omega_j = \frac{\Delta\phi}{\Theta_j} = \frac{2\pi}{600 \cdot \Theta_j}. \tag{4.12}$$

In these expressions, the meaning ω_j represents the first difference of the corresponding rotation angle values φ_j, and the angle of rotation itself at a certain point in time is determined by the accumulated amount

$$\varphi_t = \Delta t \sum_{j=1}^{t_j \le t} \omega_j, \varphi_1 = 0. \tag{4.13}$$

We note in passing that the finite differences in ω_j in (13) are taken by equivalent rotation angle references and therefore their time grid does not have to be equidistant.

As follows from expressions (4.8) and (4.10), the deviation of kinetic energy is proportional to the angular acceleration. Therefore, it is by angular acceleration that it is convenient to study the dynamics of the cylinder powers of the DEG under study. Sequences of instantaneous values of angular acceleration are determined by the differences

$$\Delta\omega_j = \omega_{j+1} - \omega_j = \frac{1}{\Theta_{j+1}} - \frac{1}{\Theta_j} = \frac{\Theta_j - \Theta_{j+1}}{\Theta_j \cdot \Theta_{j+1}}. \tag{4.14}$$

Thus, we obtain a sequence of samples of instantaneous values of the angular acceleration of the shaft. These samples represent a discretized implementation describing the acceleration caused by the deviation of the shaft rotation angle $\varphi''_\partial(t)$.

Measurements of the kinetic energy of the shaft, taking into account expression (4.10), are determined as follows

$$W_j = \frac{1}{2} J_z \omega_j^2. \tag{4.15}$$

So, after two revolutions of the shaft, the sequence of acceleration readings is recorded $\left\{\Delta\omega_j, j = \overline{1,\ 1200}\right\}$ on one cycle (two revolutions) of rotation of the crankshaft, the duration of which will be denoted by T. N such cycles are recorded in the computer memory.

As already indicated, $\Delta\omega_j$ in (4.14) represents the counts $\phi''_\partial(t)$ is acceleration deviation, which in steady state consists of the sum of the deterministic deviation component $\frac{d^2}{dt^2}\left[\phi''(t)\right]$ and LPRP with period T. Since according to [10, 11, 27], the period $\xi''(t)$ coincides with the engine cycle period, then the readings $\phi''_\partial(t)$, taken through this period T can be represented as follows

$$\left\{\Delta\omega_{j+1200\,k}, k = \overline{0,\ N-1}\right\},$$

where N—several cycles, j—fixed number $1 \le j \le 1200$, and represents the implementation of a stationary ergodic process, from which it is possible to construct estimates of the mathematical expectations (average values) of the instantaneous accelerations of the shaft, which are consistent, i.e. become more accurate as the number N increases. Namely

$$\overline{\Delta\omega_J} = \frac{1}{N}\sum_{k=0}^{N-1}\Delta\omega_j + 1200k = \phi_{\ddot{\partial}}(j) + \frac{1}{N}\sum_{k=0}^{N-1}\xi_j + 1200k, \quad j = \overline{1,\ 1200} \tag{4.16}$$

In the last expression (4.16), the second term tends to zero in the root-mean-square sense as N increases due to the assumption that $M_{\xi}(t) = 0$ and process $\xi(t)$ has a nested ergodic sequence of samples taken at a time interval equal to cycle T [15–17].

To explain further calculations, let us dwell on the following mathematical fact.

If there is a periodic function $f(t)$ with period T, i.e.

$$f(t) = f(t+T), t \in (-\infty; \infty),$$

then a harmonic with a frequency $\omega = \frac{2\pi}{T}$ called the fundamental (main) harmonic $f(t)$. Harmonics with frequencies lower than ω are called subharmonics, and those with frequencies higher ω are higher harmonics. It is known that for a periodic function $f(t)$ with period T, the subharmonics are zero.

Let us now consider two functions $f(t)$ and $f_1(t)$, periodic according to periods T and $T_n = \frac{T}{n}$, where $n \geq 2$, n—whole number.

Let

$$f_1(t) = f(t) + \Delta f(t),$$

where function $|\Delta f(t)| << |f(t)|$, i. the difference between functions $f(t)$ and $f_1(t)$ minor.

It is easy to see that the main harmonic of the function $f(t)$ for the function $f_1(t)$ is a subharmonic. Thus, if $\Delta f(t)$ is identically equal to zero (or close to zero), then $f_1(t)$ and $f(t)$ have a T_n period and contains a harmonic with ω frequency equal to zero (or close to zero). This means that when moving from $f_1(t)$ to $f(t)$ those. from a periodic function with a T period to a periodic function with a period $T_n = \frac{T}{n}$, a fundamental harmonic function $f(t)$ with frequency ω becomes a subharmonic of the function $f_1(t)$, and, therefore, disappears. At the same time, a harmonic with a frequency ω_n from n-th goes to the main (first).

In addition, with such a transition (when $\Delta f(t) \to 0$) all harmonics whose frequency is lower than the frequency $\omega_n = \frac{2\pi n}{T}$, turn to zero and the entire spectrum of the process is this out $f_1(t)$.

All the above arguments, with some minor restrictions on the sampling grid, remain valid for the case of discrete-time, i.e. about periodic sequences of readings, which are $\overline{\Delta\omega_j}$ and to all others calculated by Θ_j. Only in this case, instead of integrals, when determining the Fourier coefficients [59–64], the formulas of the discrete Fourier transform are used.

Applying the formulas of the discrete Fourier transform to the above sequence (16) over a period equal to 1200 samples, we determine the first nine Fourier coefficients (taking into account the fact that, due to the design, the 5D70 diesel engine has 8 pairs of cylinders) and the corresponding initial amplitudes A_n and phases ϕ_n according to the formulas:

$$C_n = \frac{1}{1200}\sum_{j=1}^{1200} \Delta\overline{\omega_j} e^{-\frac{2\pi i j}{1200}};\ A_n = \sqrt{(\mathrm{Re}[C_n])^2 + (\mathrm{Im}[C_n])^2};$$

$$\phi_n = Arg[C_n];\ n = 1, 2, \ldots 9, \tag{4.17}$$

where Re—real part of a complex number, Im—imaginary part, and *Arg*—an argument.

Formulas (4.17) allow us to conclude the dynamics of cylinder powers. Let's briefly look at this issue.

As mentioned above, $\Delta\omega_j$ always has a period equal to T, expressed in discrete units, i.e. T corresponds to 1200 samples. But in the case when all cylinders are working properly, a period appears $T_p = \frac{T}{P}$, which in counts is defined as $\frac{1200}{m}$, where P is the number of identical energy quanta transmitted in one period by the cylinder-piston group. For the 5D70 diesel engine, these quanta, due to its design, are the same for pairs of cylinders (but not for each individually), i.e. $P = 8$. Thus, the absence of harmonics at frequency $\omega = \frac{2\pi}{T}$ indicates the proper operation of the cylinders and the equality of their cylinder power. This fact is the basis for the conclusion about the uniform distribution of cylinder power in the 5D70 type DEG under study.

In the case of uneven distribution of cylinder power (malfunction in the operation of the CPG, fuel system, etc.), the phase-frequency and amplitude-frequency spectrum of subharmonics is examined and from them, taking into account the learning results (training sets), the reason that caused the uneven distribution is determined cylinder capacities.

Thus, methods have been proposed for constructing estimates of some LPRP parameters that characterize the dynamics of cylinder powers on the DEG shaft. It is shown that the parameters of the amplitude-frequency and phase-frequency spectra of the angular velocity or acceleration of the shaft of the diagnosed diesel generator can be used as such estimates that make it possible to diagnose defects in the operation of the diesel cylinder head.

References

1. Stone, G.C.: Partial discharge diagnostics and electrical equipment insulation condition assessment. IEEE Trans. Dielectr. Electr. Insul. **12**(5), 891–904 (2005)

2. Jadin, M.S., Taib, S.: Recent progress in diagnosing the reliability of electrical equipment by using infrared thermography. Infrared Phys. Technol. **55**(4), 236–245 (2012)
3. Chou, Y.C., Yao, L.: Automatic diagnostic system of electrical equipment using infrared thermography. In: 2009 International Conference of Soft Computing and Pattern Recognition, pp. 155–160. IEEE (2009, December)
4. Babak, V., Eremenko, V., Zaporozhets, A.: Research of diagnostic parameters of composite materials using Johnson distribution. Int. J. Comput. **18**(4), 483–494 (2019)
5. Zaporozhets, A., Eremenko, V., Babak, V., Isaienko, V., Babikova, K.: Using Hilbert transform in diagnostic of composite materials by impedance method. Periodica polytechnica Electr. Eng. Comput. Sci. **64**(4), 334–342 (2020)
6. Babak, V.P., Babak, S.V., Eremenko, V.S., Kuts, Y.V., Myslovych, M.V., Scherbak, L.M., Zaporozhets, A.O., et al.: Models and measures for standardless measurements of the composite materials characteristics. In: Models and Measures in Measurements and Monitoring, pp. 157–190 (2021)
7. Yan, R., Gao, R.X., Chen, X.: Wavelets for fault diagnosis of rotary machines: a review with applications. Signal Process. **96**, 1–15 (2014)
8. Wang, Z.Y., Lu, C., Zhou, B.: Fault diagnosis for rotary machinery with selective ensemble neural networks. Mech. Syst. Signal Process. **113**, 112–130 (2018)
9. Zhang, S., Asakura, T., Xu, X., Xu, B.: Fault diagnosis system for rotary machine based on fuzzy neural networks. JSME Int. J. Ser. C **46**(3), 1035–1041 (2003)
10. Babak, V.P., Babak, S.V., Eremenko, V.S., Kuts, Y.V., Myslovych, M.V., Scherbak, L.M., Zaporozhets, A.O., et al.: Models and measures for the diagnosis of electric power equipment. In: Models and Measures in Measurements and Monitoring, pp. 99–126 (2021)
11. Babak, V., Zaporozhets, A., Kovtun, S., Myslovych, M., Kuts, Y., Scherbak, L.: Information support for identification of the technical state of electric power facilities. In: Systems, Decision and Control in Energy V, pp. 129–153. Springer Nature Switzerland, Cham (2023)
12. Yan, R., Gao, R.X.: Energy-based feature extraction for defect diagnosis in rotary machines. IEEE Trans. Instrum. Meas. **58**(9), 3130–3139 (2009)
13. Sohaib, M., Kim, J.M.: Fault diagnosis of rotary machine bearings under inconsistent working conditions. IEEE Trans. Instrum. Meas. **69**(6), 3334–3347 (2019)
14. Capdessus, C., Sidahmed, M., Lacoume, J.L.: Cyclostationary processes: application in gear faults early diagnosis. Mech. Syst. Signal Process. **14**(3), 371–385 (2000)
15. Babak, V., Zaporozhets, A., Kulyk, M., Kuts, Y., Scherbak, L.: Application of discrete Hilbert transform to estimate the characteristics of cyclic signals: information provision. In: Systems, Decision and Control in Energy IV: Volume I. Modern Power Systems and Clean Energy, pp. 93–115. Springer Nature Switzerland, Cham (2023)
16. Babak, V.P., Babak, S.V., Eremenko, V.S., Kuts, Y.V., Myslovych, M.V., Scherbak, L.M., Zaporozhets, A.O., et al.: Models and measures for measuring random angular quantities. In: Models and Measures in Measurements and Monitoring, pp. 61–97 (2021)
17. Babak, V.P., Babak, S.V., Eremenko, V.S., Kuts, Y.V., Myslovych, M.V., Scherbak, L.M., Zaporozhets, A.O., et al.: Examples of using models and measures on the circle. In: Models and Measures in Measurements and Monitoring, p. 127 (2021)
18. Zvaritch, V., Myslovych, M., Gyzhko, Y.: Application of linear random processes to construction of diagnostic system for power engineering equipment. In: IFIP International Conference on Advances in Production Management Systems, pp. 617–622. Springer International Publishing, Cham (2021, August)
19. Babak, V.P., Babak, S.V., Myslovych, M.V., Zaporozhets, A.O., Zvaritch, V.M.: Methods and models for information data analysis. In: Diagnostic Systems for Energy Equipments, pp. 23–70 (2020)
20. Fryz, M., Mlynko, B.: Properties of stationarity and cyclostationarity of conditional linear random processes. In: 2020 IEEE 15th International Conference on Advanced Trends in Radioelectronics, Telecommunications and Computer Engineering (TCSET), pp. 166–170. IEEE (2020, February)

21. Fryz, M., Scherbak, L., Karpinski, M.P., Mlynko, B.: Characteristic function of conditional linear random process. In: ITTAP, pp. 129–135 (2021, November).
22. Lupenko, S., Lutsyk, N., Lapusta, Y.: Cyclic linear random process as a mathematical model of cyclic signals. Acta mechanica et automatica **9**(4), 219–224 (2015)
23. Fryz, M., Mlynko, B.: Property analysis of conditional linear random process as a mathematical model of cyclostationary signal. In: ITTAP, pp. 77–82 (2022)
24. Bittanti, S.: Deterministic and stochastic linear periodic systems. In: Time Series and Linear Systems, pp. 141–182. Springer Berlin Heidelberg, Berlin, Heidelberg (2006)
25. Hurd, H.L., Miamee, A.: Periodically correlated random sequences: spectral theory and practice, vol. 355. Wiley (2007)
26. Florescu, I.: Probability and Stochastic Processes. Wiley (2014)
27. Babak, V., Zaporozhets, A., Kuts, Y., Myslovych, M., Fryz, M., Scherbak, L.: Models and characteristics of identification of noise stochastic signals of research objects. In: CEUR Workshop Proceedings (vol. 3309, pp. 349–362) (2022, December)
28. Babak, V., Scherbak, L., Kuts, Y., Zaporozhets, A.: Information and measurement technologies for solving problems of energy informatics. In: The 1st International Workshop on Information Technologies: Theoretical and Applied Problems 2021. CEUR Workshop Proceedings, vol. 3039, pp. 24–31 (2021, December)
29. Lawler, G.F.: Introduction to Stochastic Processes. Chapman and Hall/CRC (2018)
30. Jazwinski, A.H.: Stochastic Processes and Filtering Theory. Courier Corporation (2007)
31. Pavliotis, G.A.: Stochastic processes and applications. In: Texts in Applied Mathematics, p. 60 (2014)
32. Babak, V.P., Babak, S.V., Myslovych, M.V., Zaporozhets, A.O., Zvaritch, V.M., Babak, V.P., Zvaritch, V.M., et al.: Simulation and software for diagnostic systems. In: Diagnostic Systems For Energy Equipments, pp. 71–90 (2020)
33. Babak, V.P., Babak, S.V., Myslovych, M.V., Zaporozhets, A.O., Zvaritch, V.M., Babak, V.P., Zvaritch, V.M., et al.: Technical provision of diagnostic systems. In: Diagnostic Systems for Energy Equipments, pp. 91–133 (2020)
34. Raj, V.P., Natarajan, K., Girikumar, S.T.: Induction motor fault detection and diagnosis by vibration analysis using MEMS accelerometer. In: 2013 International Conference on Emerging Trends in Communication, Control, Signal Processing and Computing Applications (C2SPCA), pp. 1–6. IEEE (2013, October)
35. Varanis, M., Silva, A., Mereles, A., Pederiva, R.: MEMS accelerometers for mechanical vibrations analysis: a comprehensive review with applications. J. Braz. Soc. Mech. Sci. Eng. **40**, 1–18 (2018)
36. Babak, V., Zaporozhets, A., Kuts, Y., Scherbak, L., Eremenko, V.: Application of material measure in measurements: theoretical aspects. In: Systems, Decision and Control in Energy II, pp. 261–269. Springer International Publishing, Cham (2021)
37. Babak, V.P., Babak, S.V., Eremenko, V.: Problems and features of measurements. In: Kuts, Y.V., Myslovych, M.V., Scherbak, L.M., Zaporozhets, A.O., et al. (eds.) Models and Measures in Measurements and Monitoring, pp. 1–31 (2021)
38. Babak, V.P., Babak, S.V., Eremenko, V.S., Kuts, Y.V., Myslovych, M.V., Scherbak, L.M., Zaporozhets, A.O.: Models of measuring signals and fields. In: Models and Measures in Measurements and Monitoring, pp. 33–59 (2021)
39. Babak, V.P., Babak, S.V., Myslovych, M.V., Zaporozhets, A.O., Zvaritch, V.M.: Principles of construction of systems for diagnosing the energy equipment. In: Diagnostic Systems for Energy Equipments, pp. 1–22 (2020)
40. Zhang, L., Xiong, G., Liu, H., Zou, H., Guo, W.: Bearing fault diagnosis using multi-scale entropy and adaptive neuro-fuzzy inference. Expert Syst. Appl. **37**(8), 6077–6085 (2010)
41. Wu, S.D., Wu, C.W., Lin, S.G., Wang, C.C., Lee, K.Y.: Time series analysis using composite multiscale entropy. Entropy **15**(3), 1069–1084 (2013)
42. Lei, Y., He, Z., Zi, Y.: Application of the EEMD method to rotor fault diagnosis of rotating machinery. Mech. Syst. Signal Process. **23**(4), 1327–1338 (2009)

43. Hoang, D.T., Kang, H.J.: Rolling element bearing fault diagnosis using convolutional neural network and vibration image. Cogn. Syst. Res. **53**, 42–50 (2019)
44. Casado, A.J., Nieto, F.J., Blázquez, F., Platero, C.A.: A monitoring system for diesel engine driven generators based on electric power output oscillation assessment. IEEE Trans. Ind. Appl. **53**(3), 3182–3188 (2016)
45. Zhiravetska, A., Gasparjan, A., Terebkov, A.: Monitoring of current technical condotion of vessel diesel-generator installation. In: 2017 19th European Conference on Power Electronics and Applications (EPE'17 ECCE Europe) (pp. P-1). IEEE (2017, September)
46. Melnyk, O., Onyshchenko, S., Koskina, Y., Aleksandrovska, N., Drozhzhyn, O., Maluha, E., Bondaryuk, M., et al.: Full overlap ship security model: an integrative approach to shipboard equipment information security. In: E3S Web of Conferences, vol. 501, p. 02002. EDP Sciences (2024)
47. Melnyk, O., Onishchenko, O., Onyshchenko, S., Voloshyn, A., Ocheretna, V.: Comprehensive study and evaluation of ship energy efficiency and environmental safety management measures. In: Systems, Decision and Control in Energy V, pp. 665–679. Springer Nature Switzerland, Cham (2023)
48. Banks, J., Hines, J., Lebold, M., Campbell, R., Begg, C., Byington, C.: Failure modes and predictive diagnostics considerations for diesel engines. In: Proceedings of the 55th Meeting of the Society for Machinery Failure Prevention Technology, pp. 2–5 (2001, April)
49. Pham, B.T., Lybeck, N.J., Agarwal, V.: Online Monitoring Technical Basis and Analysis Framework for Emergency Diesel Generators-Interim Report for FY 2013 (No. INL/EXT-12–27754). Idaho National Lab.(INL), Idaho Falls, ID (United States) (2012)
50. LingAitis, L.P., Lebedevas, S., Liudvinavičius, L.: Evaluation of the operational reliability and forecasting of the operating life of the power train of the freight diesel locomotive fleet. Eksploatacja i Niezawodność **16**(1), 73–79 (2014)
51. Hou, L., Zou, J., Du, C., Zhang, J.: A fault diagnosis model of marine diesel engine cylinder based on modified genetic algorithm and multilayer perceptron. Soft. Comput. **24**, 7603–7613 (2020)
52. Onishchenko, O., Bukaros, A., Melnyk, O., Yarovenko, V., Voloshyn, A., Lohinov, O.: Ship refrigeration system operating cycle efficiency assessment and identification of ways to reduce energy consumption of maritime Transport. In: Systems, Decision and Control in Energy V, pp. 641–652. Springer Nature Switzerland, Cham (2023)
53. Artur, K., Stankevich, P., Aulin, D., Alexsandr, B.: Efficiency improvement of locomotive-type diesel engine operation due to introduction of resource-saving technologies for cleaning diesel and diesel locomotive systems. Procedia Comput. Sci. **149**, 264–273 (2019)
54. Chang, M.Y., Chen, J.K., Chang, C.Y.: A simple spinning laminated composite shaft model. Int. J. Solids Struct. **41**(3–4), 637–662 (2004)
55. Gounaris, G.D., Papadopoulos, C.A.: Crack identification in rotating shafts by coupled response measurements. Eng. Fract. Mech. **69**(3), 339–352 (2002)
56. Gao, P., Hou, L., Yang, R., Chen, Y.: Local defect modelling and nonlinear dynamic analysis for the inter-shaft bearing in a dual-rotor system. Appl. Math. Model. **68**, 29–47 (2019)
57. Babak, S., Babak, V., Zaporozhets, A., Sverdlova, A.: Method of statistical spline functions for solving problems of data approximation and prediction of objects state. In: Proceedings of the Second International Workshop on Computer Modeling and Intelligent Systems (CMIS-2019), Zaporizhzhia, Ukraine, pp. 15–19 (2019, April)
58. Babak, V., Zaporozhets, A., Zvaritch, V., Scherbak, L., Myslovych, M., Kuts, Y.: Models and measures in theory and practice of manufacturing processes. IFAC-PapersOnLine **55**(10), 1956–1961 (2022)
59. Zaporozhets, A., Redko, O., Babak, V., Eremenko, V., Mokiychuk, V.: Method of indirect measurement of oxygen concentration in the air. Naukovyi Visnyk Natsionalnoho Hirnychoho Universytetu **5**, 105–114 (2018)
60. Wang, L.H., Zhao, X.P., Wu, J.X., Xie, Y.Y., Zhang, Y.H.: Motor fault diagnosis based on short-time Fourier transform and convolutional neural network. Chin. J. Mechan. Eng. **30**, 1357–1368 (2017)

61. Eremenko, V., Zaporozhets, A., Isaienko, V., Babikova, K.: Application of wavelet transform for determining diagnostic signs. In: Proceedings of the 15th International Conference on ICT in Education, Research and Industrial Applications. Integration, Harmonization and Knowledge Transfer, vol. 1, pp. 12–15 (2019, June)
62. Eremenko, V., Babak, V., Zaporozhets, A.: Method of reference signals creating in non-destructive testing based on low-speed impact. Tekhnichna Elektrodynamika **4**, 070 (2021)
63. Argov, S., Ramesh, J., Salman, A., Sinelnikov, I., Goldstein, J., Guterman, H., Mordechai, S.: Diagnostic potential of Fourier-transform infrared microspectroscopy and advanced computational methods in colon cancer patients. J. Biomed. Opt. **7**(2), 248–254 (2002)
64. Zaporozhets, A.A., Eremenko, V.S., Serhiienko, R.V., Ivanov, S.A.: Development of an intelligent system for diagnosing the technical condition of the heat power equipment. In: 2018 IEEE 13th International Scientific and Technical Conference on Computer Sciences and Information Technologies (CSIT), vol. 1, pp. 48–51. IEEE (2018, September)

Chapter 5
Statistical Assessment of Diagnostic Parameters

Abstract This chapter presents a comprehensive exploration of statistical methods for the estimation and analysis of diagnostic signals in a variety of contexts, ranging from spectral analysis to parameter estimation and hypothesis testing. Beginning with numerical parameters of diagnostic signals, the discussion extends to spectral correlation analysis, focusing on the calculation of moments, the power spectral density, and the correlation function from signal samples. The utilization of the Goertzel algorithm for efficient spectral density estimation highlights the chapter's contribution to spectral analysis techniques. Further, the application of Pearson's system of curves for empirical data distribution approximation underscores the chapter's innovative approach to statistical analysis. The extract also delves into stationarity testing methods (F-test, t-test, and Kolmogorov–Smirnov test) critical for diagnostic reliability and defect identification, illustrating the chapter's diagnostic focus. In addressing statistical hypothesis testing, the chapter introduces a method based on the maximum likelihood principle for distinguishing hypotheses about the parameters of a multidimensional Gaussian distribution, demonstrating the chapter's depth in statistical methodology. Lastly, it explores the use of the Neyman-Pearson criterion in planning diagnostic tests, offering a practical framework for hypothesis distinction regarding system states. This exploration underscores the chapter's significant contribution to both the theoretical and practical aspects of diagnostic signal analysis, providing valuable insights and methods for researchers and practitioners in the field.

Keywords Diagnostic signals · Diagnostic parameters · Spectral correlation analysis · Probability distribution functions · Statistical hypotheses · Gaussian distribution · White noise · Stationarity

The result of the diagnostic process is to determine the actual technical condition of the diagnostic object. To do this, a diagnostic experiment is carried out, which consists of measuring diagnostic signals generated by the diagnostic object in response to certain influences—test or working—depending on the type of diagnosis. As a result of measurements, one or more implementations of diagnostic signals are obtained

V. Babak et al., *Statistical Diagnostics of Electric Power Equipment*, Studies in Systems, Decision and Control 573, https://doi.org/10.1007/978-3-031-76253-6_5

[1–5]. Based on the analysis of these implementations, it is necessary to estimate the values of diagnostic parameters characterizing the actual technical condition of the diagnostic object. Using the decision rules and these values, a decision is made about the presence of a defect at a given moment and, if necessary, a classification of the type of defect is carried out.

Let's consider the methods used in statistical diagnostics to obtain estimates of diagnostic parameters.

5.1 Numerical Parameters of Diagnostic Signals

As a result of measuring any signal using a digital information-measuring system (IMS) over a fixed period, a series of sequential time-ordered samples is obtained, which is called a *realization* or *time series*. Thus, a continuous-time signal is replaced by a sequence of values taken at separate points in time. In other words, the signal is time sampled. In order for such an operation not to lead to significant losses of information, the research signals must meet certain conditions [6–8].

First of all, from a probabilistic point of view, it is necessary to assume that the signal can be represented as a separable random process. This allows it to be described by a sequence of finite-dimensional distribution functions [8–12]. In addition, the signal must have a finite spectrum, that is, one that does not contain spectral components with frequencies above a certain maximum frequency f_{max}. Such a continuous-time signal can be accurately represented by a time series if samples are taken at a frequency at least twice the maximum frequency f_{max}.

When using an electronic computer as part of a diagnostic IMS, it should be borne in mind that the quantities, the sequence of which replaces the real signal, are quantized in level, that is, their values belong to a certain fixed set, depending on the number of bits of the analog-to-digital convertor (ADC) used. Statistical estimates of moments calculated from such sequences contain a methodological error. This error can be reduced by introducing Sheppard corrections [13–16]. But, given the fact that modern ADCs have a large number of bits, the absolute value of the error from quantization is small, and therefore Sheppard's corrections have virtually no effect on the final result of the assessment. Thus, the phenomenon of level quantization can be neglected.

Often diagnostic signals are non-stationary in nature, but the parameters that characterize them from a probabilistic point of view change rather slowly over time. This fact allows us to consider the time series obtained by measuring such signals over fairly short periods as conditionally stationary.

Additionally, it is assumed that the ergodic hypothesis is valid for the time series under study, which allows, when finding statistical estimates of the probabilistic parameters of the signal represented by the series, to replace averaging over an ensemble of realizations with time averaging over one time series [10, 17–19].

Let us now consider the expressions used to obtain estimates of the numerical characteristics of diagnostic signals.

Let X_k– time series samples, $k = \overline{0,\ N-1}$, N—integer, number of sample elements. With diagnostic signals, $\xi(t)$ they are connected by the relation

$$X_k = \xi(t_k), \tag{5.1}$$

and moments in time $t_k \in \mathbf{T}$ are most often chosen in such a way that

$$t_k = k \cdot \Delta t, k = \overline{0,\ N-1},$$

where Δt—sampling step.

Within the framework of the above assumptions, statistical estimates of the parameters of the one-dimensional distribution of the research signal have the following form [20–24].

Expected value:

$$\tilde{\nu}_1 = \frac{1}{N}\sum_{k=0}^{N-1} X_k. \tag{5.2}$$

Centered time series:

$$\overset{\circ}{X}_k = X_k - \tilde{\nu}_1, k = \overline{0,\ N-1}.$$

The initial moment of the n-th order:

$$\tilde{\nu}_n = \frac{1}{N}\sum_{k=0}^{N-1} (X_k)^n, n = 2,\ 3,\ \ldots. \tag{5.3}$$

The central moment of the n-th order:

$$\tilde{\mu}_n = \frac{1}{N}\sum_{k=0}^{N-1} \left(\overset{\circ}{X}_k\right)^n, n = 3,\ 4,\ \ldots. \tag{5.4}$$

It should be noted that the variance estimate μ_2 (central moment of the second order), calculated from expression (5.4) at $n = 2$, turns out to be biased, that is, it contains a methodological error. Therefore, instead of it, an unbiased estimate is taken, calculated using the expression

$$\tilde{\mu}_2 = \frac{1}{N-1}\sum_{k=0}^{N-1} \left(\overset{\circ}{X}_k\right)^2. \tag{5.5}$$

Before the experiment, the values X_k were not exactly known. Thus, they can be considered random variables corresponding to the values of the diagnostic signal under study at certain points in time. As a result, the result of the calculation for each of expressions (5.2)–(5.5) is also a random variable that can be characterized by mathematical expectation and variance. This fact is used to construct interval estimates—a certain interval that, with a predetermined probability, contains the actual value of the estimated parameter.

Assessment the parameters of non-stationary signals from one implementation is generally impossible since these parameters themselves depend on time. In other words, non-stationary signals are generally not ergodic. Thus, to obtain estimates of the moments of such a signal, it is necessary to perform averaging over an ensemble of realizations. But for one practically important special case—periodic linear processes (LRP)—it is possible to construct a scheme by which it is still possible to obtain estimates of the time-dependent moments of the process, using only one sufficiently long implementation. This can be achieved by forming a so-called φ-series [25–27].

Let's consider the idea of forming samples for analyzing diagnostic signals characterized by cyclicity, using the example of a physical process observed during the rotation of a rotary electric machine (EM). We will assume that the angular speed of rotation of the rotor is constant. Then we should expect that the characteristics of the research signal will be repeated at times when the rotor occupies the same angular position relative to the stator. Thus, the diagnostic signal under study can be described mathematically in the form of a periodic random process (in the strict or broad sense) with a certain period T[28–32]. Depending on the operating characteristics of the machine, the period T of a random process can be either equal to the period of rotation of the rotor, or an integer number of times less.

As follows from the definition of a periodic random process given in Chap. 2 of this book, the random variables

$$\xi(\phi), \xi(\phi + T), \xi(\phi + k \cdot T), \tag{5.6}$$

where $(\varphi + k \cdot T) \in \mathbf{T}$, $k = 0,\ 1,\ \ldots,$,—stochastically equivalent (in the corresponding sense), and for any $\varphi \in \mathbf{T}$. . Thus, sequence (5.6) can be considered as some stationary discrete random process depending on the parameter k. Such a process is called nested in relation to the process $\xi(t)$.

Obviously, any periodic random process can be decomposed into a set of nested processes for different values $\varphi \in [0,\ T)$. Since the value φ changes within one period, it can be called a "phase" of the nested process, and the folded process itself, corresponding to the phase φ,—respectively φ-series of processes $\xi(t)$.

Certain features characterize time series obtained by sampling periodic random processes. Let X_k in (5.1) be sample from some T-periodic random process, $\xi(t)$, and $z = T/\Delta t$ integer. Thus, for the period T random process $\xi(t)$ hits exactly z counts. In addition, we require that $N = z \cdot r$, where r is also whole.

Then the sequence

$$\xi_{ij} = X_{i+jz},\quad j = \overline{0,\ r-1}, \tag{5.7}$$

for every fixed $i = \overline{0,\ z-1}$ denotes elements i-th φ- series time series X_k.

Thus, for further statistical processing, the implementation of a periodic process is represented in the form of a Φ-matrix, each row of which contains all the elements of $i =$ th φ-series, and their number in each line is equal to r:

$$\Phi = \left\{\xi_{ij},\ i \in \overline{0, z-1},\ j \in \overline{0, r-1}\right\}. \tag{5.8}$$

Now we can write expressions for finding estimates of the moments of a periodic random process $\xi(t)$ one implementation at a time [33–38]. The initial moment of the n-th order:

$$\tilde{\nu}_n(i) = \frac{1}{r}\sum_{j=0}^{r-1}\left(X_{i+jz}\right)^n, n = 1,\ 2,\ \ldots,$$

the central moment of the n-th order:

$$\tilde{\mu}_n(i) = \frac{1}{r}\sum_{j=0}^{r-1}\left(X_{i+jz} - \nu_1(i)\right)^n, n = 3,\ 4,\ \ldots,$$

dispersion:

$$\tilde{\mu}_2(i) = \frac{1}{r-1}\sum_{j=0}^{r-1}\left(X_{i+jz} - \nu_1(i)\right)^2.$$

Functions $\tilde{\nu}_n(i)$, $\tilde{\mu}_n(i)$—periodic in i with z period.

Note that to construct φ-series it is necessary to know the period of the random process under study. Estimation of this period based on the obtained implementation can be carried out using several methods. The sampling step must be consistent with the period value. If the condition $z = T\big/\Delta t \in \mathbf{Z}$ is not satisfied, sequence (5.7) will no longer be stationary in the general case.

5.2 Spectral Correlation Analysis

Spectral correlation analysis is one of the main tools for studying diagnostic signals, especially vibrations of various components of the equipment being diagnosed [39–44]. This method of statistical processing allows one to analyze the behavior of individual frequency components of the signal's energy spectrum, which are often quite informative diagnostic features.

Let us consider the basic relationships used in the spectral-correlation analysis of random processes.

According to Khinchin theorem [45–49], the correlation function $R(\tau)$ of a real stationary random process can be represented as

$$R(\tau) = \frac{1}{2\pi} \int_{-\infty}^{\infty} \cos(\omega\tau) dF(\omega), \tau \in (-\infty, \infty), \quad (5.9)$$

where $F(\omega)$ is spectral function.

If the spectral function $F(\omega)$ is absolutely continuous, then there is its derivative $G(\omega)$, which is called the power spectral density and is given by the expression

$$G(\omega) = \int_{-\infty}^{\infty} R(\tau) \cos(\omega\tau) d\tau, \omega \in (-\infty, \infty). \quad (5.10)$$

Note that the $G(\omega)$ function is even, therefore, it is enough to specify it $\omega \in [0, \infty)$. Taking into account that $R(\tau)$ is also even, (5.9) and (5.10) can be rewritten in the form

$$S(\omega) = 4 \int_{0}^{\infty} R(\tau) \cos(\omega\tau) d\tau, \omega \in [0, \infty); \quad (5.11)$$

$$R(\tau) = \frac{1}{2\pi} \int_{0}^{\infty} S(\omega) \cos(\omega\tau) d\omega, \tau \in (-\infty, \infty), \quad (5.12)$$

where the power spectral density $S(\omega)$ is given on the positive semi-axis:

$$S(\omega) = \begin{cases} G(\omega), & \omega = 0, \\ 2G(\omega), & \omega > 0, \\ 0, & \omega < 0. \end{cases}$$

In general, various options for determining the power spectral density of a real random process are possible, differing in the normalizing coefficients before the integrals.

Let us consider expressions for estimating the correlation function and power spectral density from a sample of a stationary random process (5.1).

The correlation function of a random process is estimated as follows [50–53]:

$$\tilde{R}_\tau = \tilde{R}(\tau \cdot \Delta t) = \frac{1}{N - \tau - 1} \sum_{k=0}^{N-\tau-1} \overset{\circ}{X}_k \overset{\circ}{X}_{k+\tau}, \tau = \overline{0, s},$$

where s—the maximum shift.

The estimate of the normalized correlation function is found as follows [50–53]:

$$\overset{\circ}{R}_\tau = \tilde{R}_\tau / \tilde{R}_0, \tau = \overline{0, s}.$$

Value s usually chosen so that the values of the correlation function $\overset{\circ}{R}_\tau$ at $\tau > s$ were quite small.

The spectral density $S(\omega)$ can be estimated in different ways. Based on the definition of spectral density (5.11), we can construct the following estimate:

$$\tilde{S}(\omega) = 2\Delta t\left(\tilde{R}_0 + 2\sum_{k=0}^{s-1}\tilde{R}_k\cos(\omega k\Delta t) + \tilde{R}_S\cos(\omega s\Delta t)\right), \omega \in [0, \infty). \quad (5.13)$$

An estimate of this form is often used

$$S_N(\omega) = \frac{\Delta t}{N}\left(\sum_{k=0}^{N-1}\overset{\circ}{X}_k\cos(\omega k\Delta t)\right)^2,$$

which is called a periodogram. It can be shown [54–57] that the periodogram is equivalent to estimate (5.13), if instead, $\tilde{R}$ we take the biased estimate of the correlation function

$$R'_\tau = \frac{1}{N}\sum_{k=0}^{N-\tau-1}\overset{\circ}{X}_k\overset{\circ}{X}_{k+\tau}, \tau = \overline{0,\ N-1},$$

The variance of the periodogram does not depend on it N, so the estimate is untenable. To improve estimates of the spectral densities of the signal under study, the spectral window method is used.

The calculation $\tilde{S}(\omega)$ is most often carried out for discrete equally spaced frequency values

$$\omega_k = \frac{2\pi k}{M\cdot\Delta t}, k = 0,\ 1,\ \ldots,\ M/2, \quad (5.14)$$

where $M = s + 1$

To find the spectral density, you can use the discrete Fourier transform [58–62]

$$S_k = \sum_{n=0}^{M-1}\tilde{R}_n\exp\left[-2\pi i\frac{kn}{M}\right], k = 0,\ 1,\ \ldots,\ M-1, \quad (5.15)$$

and here the normalizing factor $1/\Delta t$ is included in the S_k values to simplify the notation, and in addition S_k is complex quantities.

Calculation directly from expressions (5.13) or (5.15) for s large values requires a significant amount of time. To calculate all values S_k using formula (5.15), it is necessary to perform approximately the M^2 operations of multiplication and addition of complex numbers.

In general, one such operation is equivalent to four operations of multiplication and addition of real numbers, but if we take into account that the $\tilde{R}_n$ factor is always real, then the number of operations with real numbers can be halved.

The cost of computer time when calculating using expression (5.15) can be significantly reduced if we take into account the properties of the function

$$W_M(u) = \exp\left[-2\pi i \frac{u}{M}\right].$$

In particular, it is periodic $W_M(M) = 1$ and and in front of all u and v the equality $W_M(u+v) = W_M(u)W_M(v)$.

Based on this, calculation algorithms called "Fast Fourier Transform" (FFT) were created and put into practice. The most common FFT algorithm today is the Cooley-Tukey algorithm [63–67]. Provided that $M = 2^p$ calculating the M-point FFT requires approximate multiplication and addition of complex numbers. For large M values, this is significantly less than calculations by the direct method using expression (5.15). In addition, there are FFT algorithms that focus on real input sequences, which can further reduce the number of operations by half compared to general FFT algorithms (i.e., for complex sequences).

Despite this, in some cases, FFT is not the most efficient calculation method. This is due to the fact that the FFT procedure always calculates all points M of the Fourier transform, and in practice it may be necessary to find only the first $K < M$. In this case, when using the direct method, the required number of operations with complex numbers will be $K \cdot M$.

Let us introduce the calculation acceleration factor

$$k_{\text{a.f.}}(K) = \frac{K \cdot M}{M \cdot p} = \frac{K}{p}.$$

characterizing how many times more effective is the use of FFT than the use of the direct method for calculating the required number of frequency K components.

Obviously, FFT has an advantage only when $k_{\text{a.f.}}(K) > 1$. If this condition is not met, therefore the use of the direct calculation method when $K < p$ turns out to be more effective.

Let us now return to the consideration of expression (5.13). Direct calculation shows that to calculate $\tilde{S}(\omega)$ one frequency value, you need only the M operations of multiplication and addition of real numbers. Considering the above, this is 2 times more efficient than using the direct method of calculation using expression (5.15).

The previous considerations did not take into account the operations required to evaluate $\cos(u)$ functions or $W_M(u)$. Often all the necessary values of these functions are calculated once and stored in a table for use in repeated calculations. However, this requires additional memory, which is often undesirable.

Let's consider the Goertzel algorithm [68–72] for calculating S_k values from (5.15), which has slightly higher efficiency than the direct method, and at the same time requires memorizing only one coefficient for each S_k value.

According to Goertzel's algorithm, S_k is found as a result of filtering the input numerical sequence using a second-order recurrent filter:

$$\begin{aligned} S_k &= \exp\left[-2\pi i \frac{k}{M}\right] \cdot V_{M-1} - V_{M-2}; \\ V_n &= \tilde{R}_n + 2\cos\left[2\pi \frac{k}{M}\right] \cdot V_{n-1} - V_{n-2}, \quad n = 0,\ 1,\ \ldots,\ M-1; \\ V_{-1} &= V_{-2} = 0. \end{aligned} \tag{5.16}$$

Let us modify the Goertzel algorithm to find an estimate of the spectral density $\tilde{S}(\omega)$ of a real process on a discrete set of ω_k points given in (5.14). First of all, we note that

$$\mathrm{Re} S_k = \sum_{n=0}^{M-1} \tilde{R}_n \cos\left[2\pi \frac{kn}{M}\right].$$

If in the last expression, we take $\hat{R}_n$ instead of $\tilde{R}_n$ sequence, such that

$$\hat{R}_n = \begin{cases} \frac{1}{2}\tilde{R}_0, & n = 0, \\ \tilde{R}_n, & n = 1,\ 2,\ \ldots\ M-2, \\ \frac{1}{2}\tilde{R}_{M-1}, & n = M-1, \end{cases}$$

then it is easy to see that its value coincides with (5.13) at $\omega = \omega_k$ up to a constant factor.

Thus, we write the modified Goertzel algorithm $\tilde{S}(\omega_k)$ for calculation in the following form:

$$\begin{aligned} \tilde{S}(\omega_k) &= 4\Delta t \cdot \left[\lambda_k \cdot \left(V_{M-1} - \tfrac{1}{2}\tilde{R}_{M-1}\right) - V_{M-2}\right]; \\ V_{-1} &= 0,\ V_0 = \tfrac{1}{2}\tilde{R}_0; \\ V_n &= \tilde{R}_n + 2\lambda_k \cdot V_{n-1} - V_{n-2},\ n = 1,\ 2,\ \ldots,\ M-1; \\ \lambda_k &= \cos\left[2\pi \frac{k}{M}\right]. \end{aligned} \tag{5.17}$$

To complete the consideration of the spectral analysis of stationary random processes, we write the expression (5.15) in the form

$$\tilde{S}(\omega) = \Delta t \sum_{n=0}^{M-1} \tilde{R}_n \exp[-\omega i \cdot n\Delta t]. \tag{5.18}$$

Expressions (5.13) and (5.18) are valid for any value $\omega \geq 0$, while the FFT and Goertzel algorithms can only be used for ω_k frequencies defined in (5.14). That is, using formulas (5.13) or (5.18), if necessary, it can be analyzed the behavior of the spectral density of the diagnostic signal with a higher frequency resolution than using the FFT or Goertzel algorithm.

5.3 Empirical Densities and Probability Distribution Functions

Complete information about a random process is contained in the sequence of its finite-dimensional probability distribution functions. The purpose of histogram analysis of diagnostic signals is to study the nature of the one-dimensional distribution density of these signals based on the results of measuring their implementations.

From a practical point of view, the most convenient characteristics of random processes are the numerical values of their parameters, in particular, the moments of the probability distribution. For example, in works [28, 29] the possibility of using EM estimates of the 3rd and 4th moments of their vibrations for diagnostics of rolling bearings is shown.

When defects appear in the equipment under study, not only the numerical values of the parameters can change, but also the type of distribution of the diagnostic process as a whole. The probability distribution is a functional characteristic that is less convenient from a practical point of view. In addition, in practice, the assessment of the distribution of information process values is carried out based on measured implementations. The result is an empirical histogram or empirical distribution function, rather than an analytical expression for the distribution function or its density.

The problem of finding an analytical expression for the density of the experimental distribution (the so-called histogram smoothing) can be solved using several different methods. The approach, which was founded by Bruns, Gram, Charlier, and Edgeworth, consists of expanding the density of the desired distribution into a series of derivatives of the density of the normal distribution. Another approach is to find a function of the random variable under study that, at least approximately, leads it to one of the known distributions [73–76]. Sometimes (much less often than the two previous methods) Burr distributions are also used to smooth histograms [77–81].

A fairly promising method for solving this problem can be considered the use of a system of curves introduced by K. Pearson. In this case, both the selection of an analytical expression for the distribution and the classification are simultaneously carried out, since each type of distribution is assigned a specific number. This number in some cases can act as a diagnostic parameter. When smoothing histograms, the method of moments and the Pearson curve system are used [82–85].

Let us briefly consider the basic relationships on which it is based.

The English researcher Karl Pearson at the end of the 19th and beginning of the twentieth centuries set himself the goal of finding mathematical expressions to describe the density of distributions with which he dealt in practice. Analyzing the resulting samples, Pearson noticed that in most cases they have significant skewness and kurtosis, that is, their distributions are very different from normal. The asymmetry and kurtosis of the distribution are associated with the third and fourth moments, therefore, for the theoretical curve to describe such an asymmetric experimental distribution with sufficient accuracy, it must have the same numerical values of the first four moments as the experimental curve.

Taking as a basis the difference equation satisfied by the hypergeometric distribution, and introducing some additional assumptions, Pearson obtained the differential equation

$$\frac{dp(x)}{dx} = \frac{x - a_0}{b_0 + b_1 x + b_2 x^2} p(x), \tag{5.19}$$

where $p(x)$ is distribution density.

The solutions of this equation are called Pearson curves. Options a_0, b_0, b_1 and b_2, included in Eq. (5.19) can be expressed through the first four moments of the distribution, for which an analytical expression is selected. Namely, for a centered distribution

$$a_0 = b_1, b_0 = \frac{c_0}{d}, b_1 = \frac{c_1}{d}, b_2 = \frac{c_2}{d}, \tag{5.20}$$

where

$$\left.\begin{aligned} c_0 &= -\mu_2\left(4\mu_2\mu_4 - 3\mu_3^2\right), \\ c_1 &= -\mu_3\left(\mu_4 + 3\mu_2^2\right), \\ c_2 &= -2\mu_2\mu_4 + 6\mu_2^3 + 3\mu_3^2, \\ d &= 10\mu_2\mu_4 - 18\mu_2^3 - 12\mu_3^2. \end{aligned}\right\} \tag{5.21}$$

Here μ_n, $n = 2,\ 3,\ 4$ are the central moments of distribution.

As is known, the general solution of Eq. (5.19) has the form

$$p(x) = Ce^{\phi(x)}, \phi(x) = \int \frac{x - a_0}{b_0 + b_1 x + b_2 x^2} dx. \tag{5.22}$$

The constant C included in expression (5.22) is selected from the condition satisfied by the density of any distribution, namely:

$$\int\limits_{-\infty}^{\infty} p(x)dx = 1. \tag{5.23}$$

The nature of the curve $\varphi(x)$ is determined by the roots of the equation

$$b_0 + b_1 x + b_2 x^2 = 0, \tag{5.24}$$

or rather the quantity associated with them

$$\kappa = \frac{b_1^2}{4b_0 b_2}, \tag{5.25}$$

later called the "Pearson test" or "Pearson kappa" [28, 29].

Pearson's kappa is the main criterion when choosing the type of fitting curve. In total, K. Pearson identified 12 types of curves, among which 3 are basic, and the rest are their limiting or special cases. The main types occur at the following values of Pearson's kappa: Type I at $\kappa < 0$, Type IV at $0 < \kappa < 1$, and Type VI at $\kappa > 1$. Some authors, along with the "classical" 12 Pearson types, also distinguish separately the normal distribution (considering it the 13th type) due to its enormous practical significance. K. Pearson himself in his works bypasses the normal distribution, arguing that it is "just a separate point" on his type diagram [28, 29].

The shape of each of the curves of the Pearson system (without taking into account its position relative to the origin and the axial scale) is determined by the following two parameters:

$$\beta_1 = \frac{\tilde{\mu}_3^2}{\tilde{\mu}_2^3}, \beta_2 = \frac{\tilde{\mu}_4}{\tilde{\mu}_2^2},$$

where $\tilde{\mu}_k$ is an estimation of the k central moment, $k = 2,\ 3,\ 4$.

Therefore, each specific distribution curve can be uniquely associated with a certain point on the plane $(\beta_1,\ \beta_2)$, as shown in Fig. 5.1.

The β_1 and β_2 parameters determine how asymmetrical and deformed the curve is along the vertical axis relative to the normal Gaussian curve (point G on the diagram).

Let us consider expressions for the three main types of Pearson curves. Let's introduce additional quantities:

$$l_1 = 5\beta_2 - 6\beta_1 - 9, l_2 = 3\beta_1 - 2\beta_2 + 6, S = 2\left(l_1 / l_2 + 1\right),$$
$$w = \beta_1(S+2)^2 + 16(S+1)$$

Pearson's kappa in terms of β_1 and β_2 is expressed as follows:

$$\kappa = \frac{-\beta_1(S+2)^2}{16(S+1)}.$$

Type I corresponds to the distribution

$$p(x) = p_0\left|1 + \frac{x}{a_1}\right|^{q_1}\left|1 - \frac{x}{a_2}\right|^{q_2}, -a_1 < x < a_2, \tag{5.26}$$

where $a_1 + a_2 > 0$, $q_1 > -1$, $q_2 > -1$, p_0– normalizing factor.

The parameters included in (5.26) are determined through sample moments as follows:

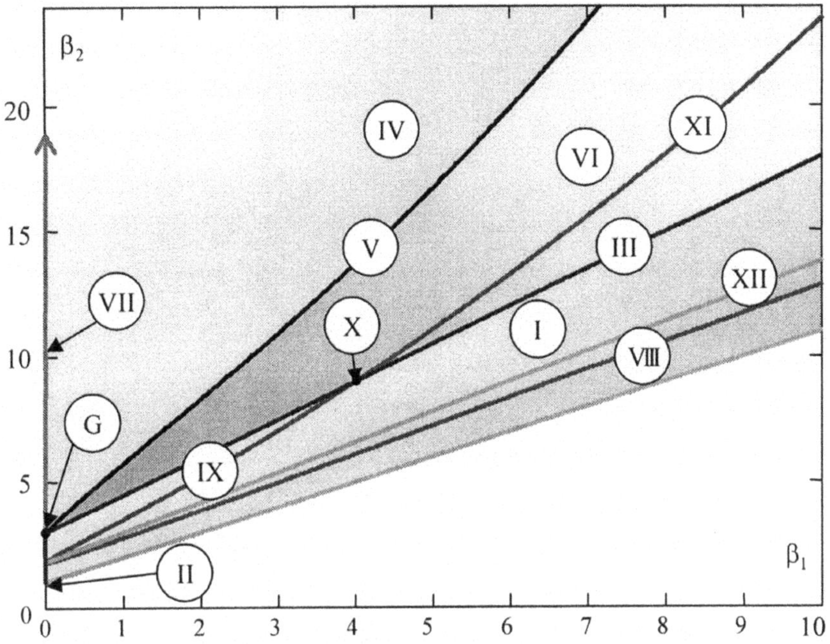

Fig. 5.1 Diagram of types of Pearson curves in the plane $(\beta_1,\ \beta_2)$

$$q_1 = \frac{1}{2}[(S-2)+\nu];\ q_2 = \frac{1}{2}[(S-2)-\nu];$$
$$a_1 = \frac{1}{2}c\sqrt{\tilde{\mu}_2 w};\ a_2 = \frac{1}{2}(1-c)\sqrt{\tilde{\mu}_2 w};$$
$$p_0 = \frac{|a_1|^{q_1}|a_2|^{q_2}\Gamma(q_1+q_2+2)}{(a_1+a_2)^{q_1+q_2+1}\Gamma(q_1+1)\Gamma(q_2+1)},$$

where

$$\nu = -sign(\tilde{\mu}_3)S(S+2)\sqrt{\beta_1/\,|w|};\ c = \frac{q_1+1}{S} - \frac{2\tilde{\nu}_1}{\sqrt{\beta_2 w}},$$

and $\tilde{\nu}_1$—an estimate of the mathematical expectation (first sample moment).

Note that with some change of variables (5.26) turns into the well-known expression for the beta distribution.

Type IV corresponds to the expression

$$p(x) = p_0\left(1+\frac{x^2}{a^2}\right)^q e^{-\nu\cdot arctg\frac{x}{a}},\ -\infty < x < \infty, \tag{5.27}$$

where $-\infty < \nu < \infty$, $a > 0$, $q < -2,5$, p_0—normalizing factor.

The parameters included in (5.27) are determined through sample moments as follows:

$$q = l_1 / l_2, a = \frac{1}{4}\sqrt{-\tilde{\mu}_2 w}, p_0 = [a \cdot F(-S,\ \nu)]^{-1},$$

where

$$\nu = -sign(\tilde{\mu}_3)S(S+2)\sqrt{\beta_1 / |w|};\ F(S,\ \nu) = e^{-\frac{\nu\pi}{2}} \int_0^{\pi} \sin^S(x) e^{\nu x} dx.$$

Type VI corresponds to the expression

$$p(x) = p_0 \left(\frac{x}{a}\right)^{q_1} \left(\frac{x}{a} - 1\right)^{q_2}, x/a > 1, \tag{5.28}$$

where $a \neq 0, q_1 < -4, q_2 > -1, p_0$– normalizing factor.

The parameters included in (5.28) are determined through sample moments as follows:

$$q_1 = \frac{1}{2}[(S-2)+\nu];\ q_2 = \frac{1}{2}[(S-2)-\nu];$$

$$a = \frac{1}{2} sign(\tilde{\mu}_3)\sqrt{\tilde{\mu}_2 w};$$

$$p_0 = \frac{\Gamma(-q_1)}{|a|\Gamma(-q_1 - q_2 - 1)\Gamma(q_2 + 1)},$$

where $\nu = -S(S+2)\sqrt{\beta_1 / |w|}$.

Distribution (5.28) is specified on the interval $-\infty < x < a$ at $\tilde{\mu}_3 < 0$ and on the interval $a < x < \infty$ at $\tilde{\mu}_3 > 0$.

The remaining types are special cases of the above expressions:

- for specific values β_1 and β_2, which corresponds to one specific point on the plane $(\beta_1,\ \beta_2)$—types X and G;
- at certain relationships between β_1 and β_2 corresponding to lines on the plane $(\beta_1,\ \beta_2)$—types II, III, V, VII, VIII, IX, XI, XII.

The diagram in Fig. 5.1 shows areas, lines, and points corresponding to specific types of Pearson curves. Detailed information about the formulas of all Pearson curves can be found in his original papers [86–88].

Any empirical distribution corresponds to a certain Pearson distribution curve, since, having a finite implementation, it is always possible to obtain finite values for the estimates of the first four moments. Using these values, a completely specific value of the Pearson criterion is calculated, and therefore a very specific type of distribution is obtained.

The curves obtained using the Pearson method do not necessarily fully correspond to the true empirical distribution, because their moments of order higher than the 4th may differ from the moments of the empirical distribution. But in most practically important cases, the approximation of the distribution of the closest curve from the Pearson system completely satisfies the requirements of the problem.

Let's consider an algorithm for selecting a specific curve from the Pearson system that optimally describes empirical data. This task is not trivial, as will be shown below.

Probably the first computer program for smoothing empirical histograms using a system of Pearson curves was written in 1977 for the MIR 1 and MIR 2 machines by Prof. B.G. Marchenko and E.S. Goncharuk [89–92]. This program used an original algorithm for selecting the most appropriate type of Pearson curve. This algorithm was further improved and in 1986, based on the results of these developments, a set of application programs was registered in the State Fund of Algorithms and Programs, developed jointly by V.B. Marchenko and E.P. Osadchim. The diagram of the Pearson curve fitting algorithm, which was formed as a result of these studies, is shown in Fig. 5.2. A feature of the algorithm is that as a result of the selection, not one type of curve is proposed, but several types at once that are closest to the empirical distribution.

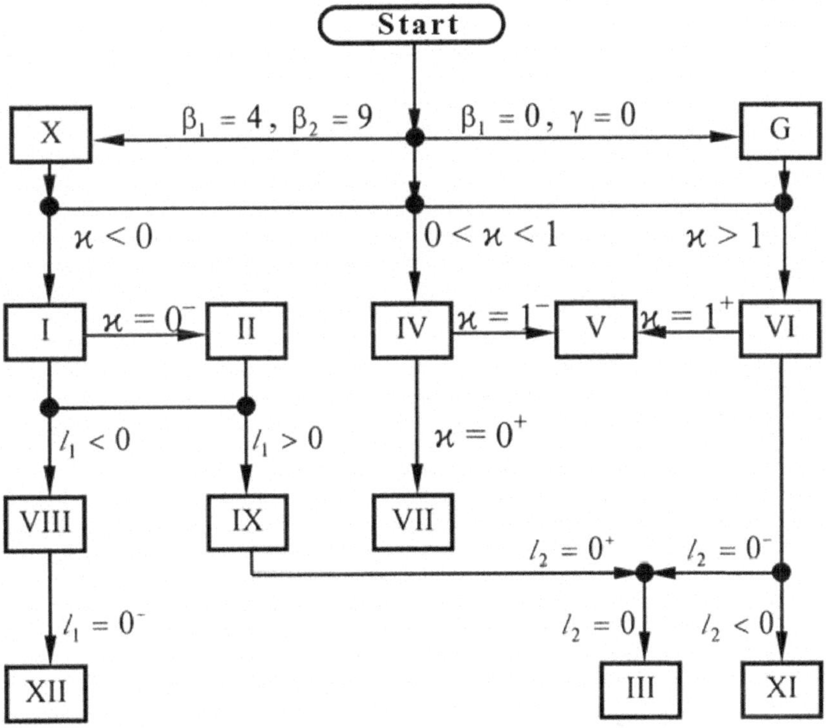

Fig. 5.2 Scheme of the Pearson curve fitting algorithm

The algorithm shown in Fig. 5.2 determines which area the point (β_1, β_2) obtained as a result of the experiment falls into. This gives the basic Pearson type: I, IV or VI. Next, it is determined which of the lines (or points) of transition types the experimental point lies close enough to. These types are offered as an alternative to basic types because their expressions contain fewer parameters than basic types. Therefore, it is easier to work with them in the future when analyzing experimental data.

This approach is justified, since the values β_1 and β_2 obtained as a result of processing the measurement results are random variables and therefore have some "scatter" relative to their "real" ("theoretical") values. The latter are unknown to us, but it is quite possible that they lie precisely on one of the lines of transitional types. At the very least, there is no reason not to take advantage of the possible simplification obtained by choosing one of the transition types instead of the main one where this is permissible.

The algorithm discussed above has one significant drawback. Here the parameters of transition types are determined for experimentally obtained values β_1 and β_2. In some cases, this leads to incorrect results, since formulas are used that do not formally apply to such parameter values. In practice, this can lead to the fact that the distribution curve takes negative values for some abscissa values, or the area limited by the curve turns out to be different from unity.

Let's consider how the Pearson curve fitting algorithm can be improved to avoid these problems.

Figure 5.3 shows a fragment of the diagram in the plane (β_1, β_2) in the region of small values β_1 and β_2, where most of the lines and points corresponding to Pearson transition types are concentrated. The point $\left(\overline{\beta}_1, \overline{\beta}_2\right)$ corresponding to the empirically obtained estimates of the moments will be denoted by "A". In the example given, it lies in the region of main type I between the lines of transition types II, III, and IX. In addition, it is "close enough" to point G (normal distribution).

With a small number of measurements, the "scatter" of parameter estimates β_1 and β_2 can be significant. The area in which, at a certain fixed level of confidence probability, there could be a point (β_1, β_2) corresponding to the "real" distribution is conventionally shown in Fig. 5.3 as part of an ellipse. As can be seen, there is a possibility that the "real" distribution of the measured quantity does not belong to the main type I, but to one of the transition types II, III, IX, or even G.

Thus, expressions of all five listed types can be used to smooth the empirical diagram. The algorithm discussed above does not answer the question of which of these expressions best describes the experimental data. In addition, as already indicated, it is incorrect to substitute empirical values $\left(\overline{\beta}_1, \overline{\beta}_2\right)$ into expressions of types II, III, IX, or G, since formally these expressions are not valid for such parameter values [93].

To find points that belong to transition types of curves, draw perpendiculars from point "A" to the corresponding lines and find the intersection points.

It is known, that a line that corresponds to type III satisfies the equation $2\beta_2 - 3\beta_1 - 6 = 0$, that is, it is direct. The line corresponding to type II is also a straight line—a fragment of the ordinate axis from point $\beta_2 = 1.8$ to point $\beta_2 = 3.0$. The

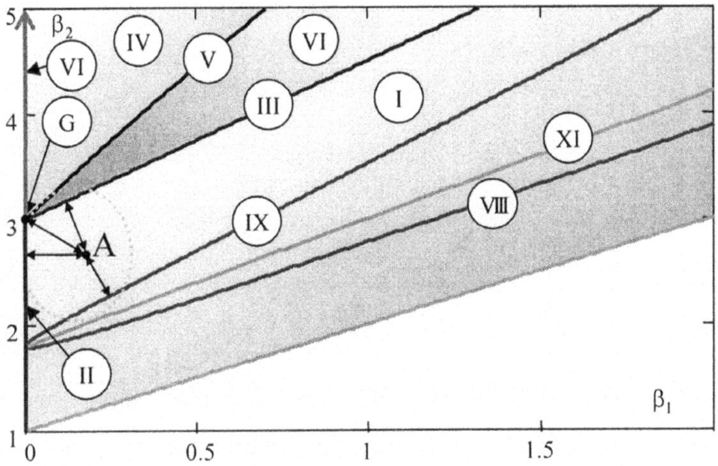

Fig. 5.3 Fragment of a diagram of Pearson curves in a plane (β_1, β_2)

distance from a point to a straight line on a plane is found according to the classical expression known from the course of analytical geometry.

A type IX line satisfies the equation

$$\beta_1(8\beta_2 - 9\beta_1 - 12)(\beta_2 + 3)^2 = (4\beta_2 - 3\beta_1)(10\beta_2 - 12\beta_1 - 18)^2.$$

That is, it is not direct. The distance from point "A" to this line can be found by minimizing the expression along the entire type IX curve $\sqrt{(\beta_1 - \overline{\beta}_1)^2 + (\beta_2 - \overline{\beta}_2)^2}$, where $(\overline{\beta}_1, \overline{\beta}_2)$ are coordinates of point "A".

The point at which the minimum is reached will be considered a point of type IX, which corresponds to the empirical distribution.

Next, according to the χ^2 criterion at a given confidence level, the statistical hypothesis is tested that the theoretical distribution corresponds to the empirical one. This procedure is performed sequentially for all candidate types: II, III, IX, and G. As a result, some types may be discarded as not corresponding to the empirical distribution with a given confidence level. The remaining types that have passed the test successfully can be sorted in ascending order of the χ^2 statistical value and thus propose the one that most closely describes the empirical distribution.

5.4 Statistical Criteria for Assessing the Stationarity of Diagnostic Signals

Since for a stationary ergodic process, when constructing statistical estimates, it is sufficient to have only one of its implementations, it becomes necessary to determine, using statistical criteria, whether the observed implementation belongs to the class of stationary processes. Note that such criteria are based on the idea of testing statistical hypotheses about the invariance of mathematical expectation, dispersion, one-dimensional distribution function, and other characteristics [28–30].

Let us assume that some random process $\xi(t)$, $t \in \mathbf{T}$, is observed and analyzed, where $\mathbf{T}$ is a continuous time interval, the duration of which can be chosen at your discretion, i.e. if necessary, it is always possible to expand it by continuing the experiment.

As a result of observations of $\xi(t)$, random sample functions appear and their realizations are fixed, which we will denote by $\xi_k(t)$ and $x_k(t)$ respectively, $t \in \mathbf{T}_k$, $k = \overline{1,\ N}$, where N is the number of sequentially observed implementations of the process $\xi(t)$ over time intervals $\mathbf{T}_k \subset \mathbf{T}$.

They can be either continuous periods (if analog recording equipment is used) or a sequence of moments in time (in the case of digital processing).

As an illustrative example, Fig. 5.4 shows five possible implementations $(N = 5)$, obtained over a certain $\mathbf{T}$ interval.

Statistical analysis of implementations should begin with a clarification of the behavior of the characteristics of the process $\xi(t)$ that interest us in time.

The stationarity of the signal for the studied implementations can be checked by analyzing the time behavior of estimates of the main statistical parameters (mathematical expectation $\mathbf{M}\xi(t)$, dispersion $\mathbf{D}\xi(t)$, correlation coefficient ρ, asymmetry k,

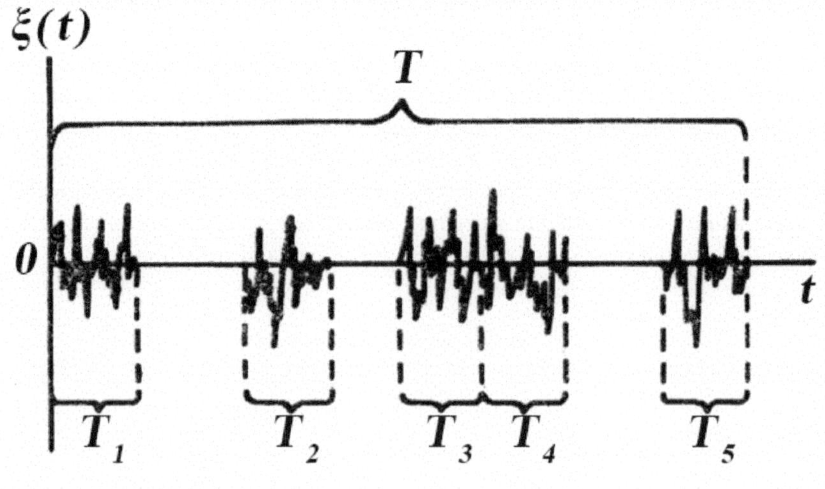

Fig. 5.4 Vibrogram segments

kurtosis γ, etc.) obtained from segments taken over sequences of non-overlapping time intervals. If the segments under study are realizations of a stationary random process and the ergodic theorem or its variations holds for $\xi(t)$, then it becomes possible to reduce the number of experimental data taken into account in the diagnostic analysis without reducing the reliability of the diagnostics performed.

Since this work is mainly aimed at using digital methods for processing diagnostic signals, we consider random processes with discrete time obtained using analog-to-digital conversion of observed and recorded signals, i.e. time series of the form

$$\{\xi(k\Delta t),\ k = 0,\ \pm 1,\ \pm 2,\ \ldots\}, \tag{5.29}$$

where Δt—sampling step. In this case, the time series is observed on the finite segments of the interval $\mathbf{T}$. As a result of the experiment, a series of numerical values x_k is obtained corresponding to $\xi(k\Delta t)$ for the $k = \overline{1,\ n}$ process (5.29)—the implementation of a sample of volume n.

The statistical hypothesis H_0 about the stationarity of the process (5.29) against the alternative hypothesis H_1 about nonstationarity is tested based on such a sample, which is equivalent to testing the original process $\xi(t)$, $t \in \mathbf{T}$ for stationarity on a discrete lattice $t_k = k\Delta t$, $k = \overline{1,\ n}$.

To check the stationarity of a signal if its distribution density has a normal law, you can use methods for testing statistical hypotheses using t and F-criteria. This allows to check the signal under study based on its mathematical expectation and dispersion, which for the normal distribution law is equivalent to checking for stationarity in the strict sense.

Hypothesis about the stationarity of a normal process represented by two independent segments of realizations ($N = 2$) volume n_1 and n_2, the variance is checked using Fisher's F test. According to this criterion, the hypothesis about the equality of variances estimated from two implementations of the process $\xi(t)$, taken at non-overlapping time intervals $\mathbf{T}_1$ and $\mathbf{T}_2$, is verified using statistics

$$W_{n_1-1,\,n_2-1} = \frac{\tilde{\mu}_2^{(1)}}{\tilde{\mu}_2^{(2)}} = \left(\frac{n_1}{n_1 - 1}\mu_2^{(1)}\right) \Big/ \left(\frac{n_2}{n_2 - 1}\mu_2^{(2)}\right), \tag{5.30}$$

where $\tilde{\mu}_2^{(1)}$, $\tilde{\mu}_2^{(2)}$ are unbiased estimates of variances over intervals $\mathbf{T}_1$ and $\mathbf{T}_2$ sampled volume samples n_1 and n_2, determined by (5.5); $\mu_2^{(1)}$ and $\mu_2^{(2)}$ are found according to (5.4) at $n = n_1$ and $n = n_2$. Statistics $W_{n_1-1,\,n_2-1}$ if the hypothesis about the constancy of dispersion over time is valid, it has an F distribution with degrees of freedom n_1-1 and n_2-1 [73, 94, 95].

In further calculations we assume $\mu_2^{(1)} > \mu_2^{(2)}$, which does not limit generality, since otherwise it is possible to renumber the observable implementations in the reverse order.

The critical region corresponding to the considered case of testing the hypothesis of stationarity in dispersion is constructed as follows. Using the table for the F-distribution for a certain level α, we find the value $F_{\alpha/2,\,n_1-1,\,n_2-1}$. These values are given only for $\alpha = 0.1$ and $\alpha = 0.02$ with n_1—1 and n_2—1 from 1 to 500 with interpolation. Forbes et al. [95] contains F distribution tables for other values of α.

If, for a given significance level α, the obtained statistic value $W_{n_1-1,\, n_2-1}$ will be less than the table $F_{\alpha/2,\, n_1-1,\, n_2-1}$, then the discrepancy between the variances can be considered insignificant. Then the hypothesis about the stationarity of the research process is accepted. Otherwise, i.e. when $W_{n_1-1,\, n_2-1} > F_{\alpha/2,\, n_1-1,\, n_2-1}$, the stationarity hypothesis is rejected.

Note that in the considered case of checking for stationarity using the F-test for dispersion, there are no assumptions regarding the mathematical expectations of the original signal. Therefore, one should begin with such a check, and only if after this there are any doubts regarding the constancy of mathematical expectations over time, then a check is carried out using the t-test.

The two-sample t-test ($N = 2$) is used to test the hypothesis about the stationarity of the studied signals according to the mathematical expectation with constant dispersion. In this case, hypothesis H_0 about the equality of mathematical expectations of two independent normal random samples [96–99] from a general population with a normal distribution function is tested using t-statistics

$$t = \frac{\nu_1^{(1)} - \nu_1^{(2)}}{\sqrt{n_1\mu_2^{(1)} + n_2\mu_2^{(2)}}}\sqrt{\frac{n_1 + n_2 - 2}{n_1 + n_2}}, \tag{5.31}$$

where $\nu_1^{(1)}$, $\nu_1^{(2)}$—estimates of mathematical expectations (5.2) of each realization; $\mu_2^{(1)}$, $\mu_2^{(2)}$—estimates of variances (5.4) for $k = 2$ of each of the realizations; n_1, n_2—sample volumes of each of the two implementations. When the sample sizes are the same, i.e. $n_1 = n_2 = n$,, expression (5.31) is transformed to the form

$$t = \frac{\nu_1^{(1)} - \nu_1^{(2)}}{\sqrt{n_1\mu_2^{(1)} + n_2\mu_2^{(2)}}}\sqrt{\frac{n_1 + n_2 - 2}{n_1 + n_2}}. \tag{5.32}$$

If the hypothesis H_0 is true, i.e., that the signal under study is a stationary process according to the mathematical expectation, determined by (5.31) or (5.32), the statistics has a Student t-distribution with $m = n_1 + n_2 - 2$ or $m = 2n - 2$ degrees of freedom.

Difference $\nu_1^{(1)} - \nu_1^{(2)}$ is considered significant for some α if $|t| > t_{m,\, 1-\alpha/2}$. In this case, the hypothesis H_0 is rejected, i.e. counts; that the signal is non-stationary. If $|t| \le t_{m,\, 1-\alpha/2}$, then the hypothesis H_0 is accepted, i.e., the signal according to the mathematical expectation is considered stationary [36].

The significance level specifies the probability of making a wrong decision. In practice, it is often expressed as a percentage and chosen to be 5%, i.e. $\alpha = 0.05$, less often—1 or 0.5%. The choice of α does not affect the validity if the sample size is fixed. To increase reliability, it is necessary to increase the sample size.

The described F- and t-criteria can be used to check the stationarity of the signals under study even if their distribution is different from normal. In this case, the signals are checked for stationarity in the broad sense.

A more effective check for stationarity of the signals under study in the case, if their distribution is different from normal, can be done using the Kolmogorov–Smirnov criterion, the applicability of which is based on the following two theorems.

Glyvenko's theorem [100–102]. Let $F(x)$ be the distribution function of the random variable ξ, and $F_n(x)$ be the empirical distribution function of the results of n independent observations on the value ξ, determined by the expression

$$F_n(x) = \frac{h_x}{n}, \tag{5.33}$$

where h_x—number of those observations (sample elements) which values are less than x. Then at $n \to \infty$

$$\mathrm{P}\left\{ \sup_{-\infty<x<\infty} |F_n(x) - F(x)| \to 0 \right\} = 1.$$

The proof of this theorem is given in [100–102]

Kolmogorov's theorem [103–105]. If the distribution function $F(x)$ of the random variable ξ is continuous, then

$$\lim_{n\to\infty} \mathrm{P}\left\{\sqrt{n}D_n < z\right\} = K(z),$$

where

$$D_n = \sup_{-\infty<x<\infty} |F_n(x) - F(x)|; \tag{5.34}$$

$$K(z) = \begin{cases} 0 & \text{at } z \le 0, \\ \sum\limits_{k=-\infty}^{\infty} (-1)^k e^{-2k^2z^2} & \text{at } z > 0, \end{cases} \tag{5.35}$$

$F(x)$ is determined according to (2.23).

In what follows, we assume that with a sufficiently large sample size, there is an approximate equality $\mathrm{P}\left\{\sqrt{n}D_n < z\right\} \approx K(z)$.

The essence of applying the Kolmogorov–Smirnov criterion [106–108] is to check that N independent samples of size n obtained at different times belong to the same random process (5.29) with the distribution function $F(x)$.

It should be noted that, taking into account the above theorem, the Kolmogorov–Smirnov criterion assumes the continuity of the function $F(x)$. When testing hypothesis H_0 relative to H_1 using this criterion, first the statistics are determined from the available samples of the same size n

$$D_{n,n}^{(i,j)} = \max \left| F_{in}(x) - F_{jn}(x) \right|, \tag{5.36}$$

where $F_{in}(x)$ and $F_{jn}(x)$ are empirical distribution functions of the i-th and j-th samples, $i, j = \overline{1, N}, i \neq j$, determined by (5.33) from the elements of the samples, under the assumption that the latter are independent.

If at $N = 2$ for a pair of samples $i = 1, j = 2$

$$D_{n,n}^{(1,2)} > K_{1-\alpha}\sqrt{2/n}, \tag{5.37}$$

then the hypothesis H_0 about the stationarity of the signal under study (5.29) is accepted. Otherwise, the competing hypothesis H_1 is accepted. The values of the quantiles of the Kolmogorov distribution $K_{1-\alpha}$ included in (5.37) are found in specific tables.

The parameter α is called the significance level of the Kolmogorov–Smirnov test and is equal to the probability of accepting a stationary process as non-stationary based on two samples, i.e. for $N = 2$. For $N > 2$, it becomes possible, when comparing increasingly longer samples, to judge whether the property of stationarity is preserved over the time interval under study.

While assessing the stationarity of the diagnostic signal for the obtained pairs of samples based on the implementation (5.29), statistics (5.36) are calculated. If the calculated statistics satisfy inequality (5.37), then with significance level α the signal under study is a stationary random process. If α is not specified, then, using the values of D_{nn}, from the specific tables it can be determined the level of significance with which the signal under study can be considered a stationary random process.

The advantage of choosing the Kolmogorov–Smirnov test for assessing the stationarity of signals is that no other criterion uses such a direct and easily interpreted measure of divergence as D_{nn}. The disadvantage is the need to have statistically independent sample elements. In conclusion, we note that the F- and t-tests are parametric, and the Kolmogorov–Smirnov test is non-parametric.

The described methods make it possible to evaluate the stationarity of diagnostic signals and thereby experimentally prove the possibility of using certain statistical diagnostic methods; carry out a preliminary selection of implementations to eliminate errors that occur during measurements, determine the stationarity interval, and judge the appearance of non-stationarity, which can significantly increase the reliability of diagnosis. The use of methods for assessing stationarity also makes it possible to detect the appearance of nonstationarity in the signals under study, which in most cases is equivalent to detecting a defect in the node being diagnosed.

5.5 Identification and Classification of Defects in Electrical Equipment

To construct decisive rules for diagnosing specific electrical equipment (EE) defects, it is necessary, first of all, to have theoretically substantiated and experimentally verified diagnostic signs, which, as can be seen from the consideration of the previous subsections, is the main condition for the selection of diagnostic spaces and the formation in them of training sets corresponding to various types of diagnosed defects. Only by having such initial data can one proceed directly to solving the problem of constructing decision rules for diagnosing and classifying specific types of EE defects, for example [109–115].

In the simplest case, the diagnostic task can be reduced to testing simple hypotheses about one or a set of parameters or a multidimensional distribution function of an observed random variable, which can be the parameters of the random process under study (diagnostic signal).

5.5.1 *Testing of Statistical Hypotheses*

Let us consider the case of testing statistical hypotheses [116–118] about parameters, for example, the frequencies of a multiresonance model, a multidimensional Gaussian quantity

$$\Xi_m = (\xi_1, \ \xi_2, \ \ldots, \ \xi_m) \tag{5.38}$$

by n volume sample, i.e. consisting of vectors

$$\Xi_{m1}, \ \Xi_{m2}, \ \ldots, \ \Xi_{mn}, \tag{5.39}$$

observed realization (resampling) of which is a rectangular matrix

$$\mathbf{X} = \left\| x_{jk} \right\|, j = \overline{1, m}, k = \overline{1, n}. \tag{5.40}$$

This means that events $\xi_{jk} = x_{jk}$ took place in front of everyone $j = \overline{1, m}$ and $k = \overline{1, n}$. Let us assume that the elements of the sequence of vectors (5.39) are statistically independent. Note that, in contrast to (5.39), the elements of the matrix (5.40) are non-random, since they are already observed values of the elements (5.39). As is known, the probability density of a Gaussian m-dimensional vector (5.38)

$$p(\mathbf{Y}) = \frac{|\mathbf{M}|^{-1/2}}{(2\pi)^{m/2}} \exp\left[-\frac{1}{2}(\mathbf{Y}-\boldsymbol{\Theta})^T\mathbf{M}^{-1}(\mathbf{Y}-\boldsymbol{\Theta})\right], \tag{5.41}$$

where **M**—correlation matrix of vector components (5.38); $\boldsymbol{\Theta}$ is matrix-column, $\boldsymbol{\Theta} = \{M\xi_j\}, j = \overline{1,\ m}$; symbol $(\mathbf{Y} - \boldsymbol{\Theta})^T$ denotes a transposed matrix—a row matrix.

In the simplest case, the multidimensional hypothesis testing problem is formulated in this way.

The implementation of the sample (5.39) is observed in the form of a matrix of non-random numbers (5.40) and two hypotheses are formulated regarding the probability density of the vector (5.38) that generated this implementation:

$$\begin{aligned} H_0 &: \text{ parameters in (41)} : \boldsymbol{\Theta} = \boldsymbol{\Theta}_0; \\ H_1 &: \text{ parameters in (41)} : \boldsymbol{\Theta} = \boldsymbol{\Theta}_1, \end{aligned} \tag{5.42}$$

with the same known matrix **M**. In this case H_0 is called the main hypothesis, and H_1 is the competing one. It is assumed that $\boldsymbol{\Theta}_0 \neq \boldsymbol{\Theta}_1$. It is necessary, based on the observed implementation (5.40), to construct a rule by which a decision is made about the validity of one of the hypotheses. This can be done in this way.

The function $p(\Xi_m)$ obtained from (5.41) by replacing the non-random argument **Y** with a random vector Ξ_m is called the likelihood function [119, 120]. It takes random values as a function of the random vector (5.38) and depends on the non-random vector parameter $\boldsymbol{\Theta}$, i.e., $p(\Xi_m,\ \boldsymbol{\Theta}) = p(\Xi_m)$. The ratio of these functions for various values $\boldsymbol{\Theta}$, defined in the formulation of hypotheses (5.42), which is called the likelihood ratio, will be denoted by $l(\Xi_m)$. Its logarithm has the form

$$\varsigma = \ln l(\Xi_m) = \frac{p(\Xi_m,\ \boldsymbol{\Theta}_1)}{p(\Xi_m,\ \boldsymbol{\Theta}_0)}, \tag{5.43}$$

where the distribution density from which the likelihood function is constructed is determined in accordance with (5.41).

If an ordered sequence of vectors is observed, that is, a sample of volume n described by expression (5.39), then for each element of such a sequence of vector observations we obtain the logarithm of the elementary likelihood ratio taking into account (5.41) in the form

$$\varsigma_k = \ln l(\Xi_{mk}) = \Xi_{mk}^T \mathbf{M}^{-1}(\Theta_1 - \Theta_0) - \frac{1}{2}(\Theta_1 + \Theta_0)^T \mathbf{M}^{-1}(\Theta_1 - \Theta_0), \quad k = \overline{1,\ n}. \tag{5.44}$$

Every $\varsigma_k,\ k = \overline{1,\ n}$ of this sequence is a scalar Gaussian random variable linearly dependent on the components of the vector (5.39), which ς_k means it can be considered as discrete white noise [30, 121] or as a scalar Gaussian random process with a discrete argument and independent values due to the independence of the elements of the sequence (5.39).

The mathematical expectation ς_k does not depend on k and, provided that the hypothesis H_0 is true, is determined taking into account (5.41) as follows:

$$\begin{aligned}\mathrm{M}\{\varsigma_k|_{H_0}\} &= \Theta_0^T\mathbf{M}^{-1}(\Theta_1-\Theta_0)-\frac{1}{2}(\Theta_1+\Theta_0)^T\mathbf{M}^{-1}(\Theta_1-\Theta_0) = \\ &= -\frac{1}{2}(\Theta_1-\Theta_0)^T\mathbf{M}^{-1}(\Theta_1-\Theta_0) = -{}^{\kappa_2}\!/_2,\end{aligned} \tag{5.45}$$

where

$$\kappa_2 = (\Theta_1-\Theta_0)^T\mathbf{M}^{-1}(\Theta_1-\Theta_0). \tag{5.46}$$

Similarly, if the hypothesis H_1 is true

$$\mathrm{M}\{\varsigma_k|_{H_1}\} = \Theta_1^T\mathbf{M}^{-1}(\Theta_1-\Theta_0)-\frac{1}{2}(\Theta_1+\Theta_0)^T\mathbf{M}^{-1}(\Theta_1-\Theta_0) = \kappa_2/2. \tag{5.47}$$

Moving on to the definition of dispersion ς_k, we note that it is the dispersion of the linear form of Gaussian random variables that have the meaning Ξ_m of components of the vector, and is written in matrix form as $\Xi_m^T\mathbf{A}$, where is an arbitrary column matrix of non-random real coefficients, where $\mathbf{A}$ is arbitrary column matrix of non-random real coefficients $\mathbf{A} = \mathbf{M}^{-1}(\Theta_1-\Theta_0)$. The variance of this form is $\mathrm{D}\left(\Xi_m^T\mathbf{A}\right) = \mathbf{A}^T\mathbf{M}\mathbf{A}$. Taking this into account, regardless of which hypothesis holds, we obtain

$$\begin{aligned}\mathrm{D}\left(\varsigma_k|_{H_0}\right) &= \mathrm{D}\left(\varsigma_k|_{H_1}\right) = \mathrm{D}\left[\Xi_m^T\mathbf{M}^{-1}(\Theta_1-\Theta_0)\right] = \\ &= \left[\mathbf{M}^{-1}(\Theta_1-\Theta_0)\right]^T\mathbf{M}\mathbf{M}^{-1}(\Theta_1-\Theta_0) = \\ &= (\Theta_1-\Theta_0)^T\mathbf{M}^{-1}(\Theta_1-\Theta_0),\end{aligned}$$

because the $\left(\mathbf{M}^{-1}\right)^T = \left(\mathbf{M}^T\right)^{-1}$. Hence

$$\mathrm{D}\left(\varsigma_k|_{H_0}\right) = \mathrm{D}\left(\varsigma_k|_{H_1}\right) = \kappa_2, \tag{5.48}$$

where κ_2 is determined according to (5.46).

Thus $\left\{\varsigma_k,\ k = \overline{1,\ n}\right\}$ is stationary Gaussian white noise [122, 123] with mathematical expectation (5.45) or (5.47) and variance κ_2 independent of k. In [30], law (5.41), which has the described property, refers to laws with a monotonic likelihood ratio. Now let us dwell on the essence of the method of testing statistical hypotheses proposed by Neumann and Pearson [28–30].

The plausibility ratio is the quantity

$$l\left(\Xi_{m1},\ \Xi_{m2,},\ \ldots,\ \Xi_{mn}\right) = \prod_{k=1}^{n}\frac{p(\Xi_{mk},\ \Theta_1)}{p(\Xi_{mk},\ \Theta_0)}, \tag{5.49}$$

whose logarithm is equal to the sum of the logarithms of elementary likelihood ratios (5.44) $\eta_n = \ln l(\Xi_{m1},\ \Xi_{m2},\ \ldots,\ \Xi_{mn}) = \sum_{k=1}^{n}\varsigma_k$, or

$$\eta_n = \left(\sum_{k=1}^{n} \Xi_{mk} \right)^T \mathbf{M}^{-1}(\Theta_1 - \Theta_0) - \frac{n}{2}(\Theta_1 + \Theta_0)^T \mathbf{M}^{-1}(\Theta_1 - \Theta_0), n = 1, 2, \ldots \tag{5.50}$$

and as a function n is a random process with a discrete argument and independent increments.

It is obvious that the quantities η_n also distributed according to the normal law with dispersion $n\kappa_2$ and mathematical expectation $n\kappa_2/2$, if the hypothesis H_1 is true, and $(-n\kappa_2/2)$ is if the hypothesis H_0 is true.

Due to the random nature of the processes under study, any decision related to the choice of one of the hypotheses is accompanied by errors of two types. An error of the first type, the probability of which is usually denoted by α, occurs when the main hypothesis H_0 is rejected when it actually occurs. A type II error, the probability of which is denoted by β, is the acceptance of a hypothesis H_0, when in fact a competing hypothesis H_1 is true.

5.5.2 Diagnostic Decisions Based on a Set of Diagnostic Parameters

In the general case, when a set of diagnostic features is used to make a decision about the technical condition of equipment, a simple decision rule for selecting hypotheses in a two-alternative situation [28–30], proposed by Neumann and Pearson, can be applied. It is based on a single-threshold analysis of the logarithm of the likelihood ratio (5.50) and comes down to the optimal choice of a certain threshold C that divides the set of admissible values of the likelihood ratio e^{η_n} into two disjoint subsets. In other words, the main point in this problem is the choice of the set of admissible values of the e^{η_n} process some threshold C for which, for a given value α, given a fixed sample size n and the smallest possible sample β, one could conclude that the H_0 hypothesis holds at $\eta_n \le \ln C$ and H_1 at $\eta_n > \ln C$. Or, taking into account (5.50), the hypothesis H_0 is accepted if

$$\tilde{\Xi}_m \mathbf{M}^{-1}(\Theta_1 - \Theta_0) \le K, \tag{5.51}$$

and the hypothesis H_1, if

$$\tilde{\Xi}_m \mathbf{M}^{-1}(\Theta_1 - \Theta_0) > K, \tag{5.52}$$

where

$$\tilde{\Xi}_m = \frac{1}{n} \sum_{k=1}^{n} \Xi_{mk}^T = \left(\frac{1}{n} \sum_{k=1}^{n} \xi_{1k}, \frac{1}{n} \sum_{k=1}^{n} \xi_{2k}, \ldots, \frac{1}{n} \sum_{k=1}^{n} \xi_{mk} \right), \tag{5.53}$$

ξ_{jk} – j-th element of the k-th vector from (5.39);

$$K = \frac{\ln C}{n} + \frac{1}{2}(\Theta_1 + \Theta_0)^T \mathbf{M}^{-1}(\Theta_1 - \Theta_0). \tag{5.54}$$

A-priory

$$\alpha = 1 - P_{\Theta_0}\{H_0\};\ \beta = P_{\Theta_1}\{H_0\}, \tag{5.55}$$

where $\mathrm{P}_\Theta\{H_0\}$ is the the probability of accepting the H_0 hypothesis when the true value of the distribution parameter is equal to Θ.

Due to the adopted decision rule and taking into account (5.55), we have

$$\begin{aligned} \mathrm{P}_{\Theta_0}\{H_0\} &= \mathrm{P}_{\Theta_0}\{\eta_n \le \ln C\} = 1 - \alpha; \\ \mathrm{P}_{\Theta_1}\{H_0\} &= \mathrm{P}_{\Theta_1}\{\eta_n \le \ln C\} = \beta. \end{aligned} \tag{5.56}$$

Moving from η_n to normalized deviations

$$\overline{\eta}_n = (\eta_n - M\,\eta_n)\,/\sqrt{D\eta_n},$$

for different values of the Θ parameter from (5.56), we obtain a system of two equations for the unknowns n and C

$$\begin{cases} \mathrm{P}_{\Theta_0}\left\{\overline{\eta}_n \le \dfrac{\ln C}{\sqrt{n\kappa_2}} + \dfrac{\sqrt{n\kappa_2}}{2}\right\} = 1 - \alpha; \\ \mathrm{P}_{\Theta_1}\left\{\overline{\eta}_n \le \dfrac{\ln C}{\sqrt{n\kappa_2}} - \dfrac{\sqrt{n\kappa_2}}{2}\right\} = \beta. \end{cases} \tag{5.57}$$

In (5.57) $\overline{\eta}_n$ is a Gaussian random variable with zero mathematical expectation and variance equal to one. Therefore, the probabilities included in the left side can be written through the probability integral

$$\Phi(x) = \frac{1}{\sqrt{2\pi}} \int\limits_{-\infty}^{x} e^{-z^2/2}\,dz,$$

as

$$\begin{cases} \alpha = 1 - \Phi\left\{\dfrac{\ln C}{\sqrt{n\kappa_2}} + \dfrac{\sqrt{n\kappa_2}}{2}\right\}; \\ \beta = \Phi\left\{\dfrac{\ln C}{\sqrt{n\kappa_2}} - \dfrac{\sqrt{n\kappa_2}}{2}\right\}. \end{cases}$$

Through the quantiles of the normal distribution u_p, which are defined as the roots of the equation $\Phi\left(u_p\right) = p$, the latter expressions can be written as

$$u_{1-\alpha} = \frac{\ln C}{\sqrt{n\kappa_2}} + \frac{\sqrt{n\kappa_2}}{2};\ u_\beta = \frac{\ln C}{\sqrt{n\kappa_2}} - \frac{\sqrt{n\kappa_2}}{2},$$

from where $u_{1-\alpha} = -u_\alpha$, taking into account we find

$$n = \frac{\left(u_\alpha + u_\beta\right)^2}{\kappa_2};\ \ln C = \frac{u_\alpha^2 - u_\beta^2}{2}. \tag{5.58}$$

Taking into account these expressions, we write the threshold value (5.54) in the form

$$K = \frac{\kappa_2\left(u_\alpha - u_\beta\right)}{2\left(u_\alpha + u_\beta\right)} + \frac{1}{2}(\Theta_1 + \Theta_0)^T\mathbf{M}^{-1}(\Theta_1 - \Theta_0). \tag{5.59}$$

Diagnostic tests are planned in this way. We set the value of the probabilities α and β, which are usually selected from a series of numbers 0.1; 0.05; 0.02; 0.01. As a result of training, they are assessed Θ_1, Θ_0 и $\mathbf{M}$, wherein one of the parameters usually corresponds to the conditionally serviceable state of the node being diagnosed. Then, using (5.46), (5.58), and (5.59), we determine the threshold value K and the number of observations required to make a decision, and rounding when using (5.58) is carried out in excess of the calculated fractional values. This completes the planning. A diagnostic experiment is carried out, as a result of which a numerical matrix (5.40) appears with the number of rows equal to the calculated value n. Taking the arithmetic mean over its columns, we obtain a row matrix

$$\tilde{X}_m = \left(\frac{1}{n}\sum_{k=1}^{n} x_{1k},\ \frac{1}{n}\sum_{k=1}^{n} x_{2k},\ \ldots,\ \frac{1}{n}\sum_{k=1}^{n} x_{mk}\right). \tag{5.60}$$

Then we check inequalities (5.51) and (5.52), but for observations of implementation (5.60). If the inequality holds $\tilde{X}_m\mathbf{M}^{-1}(\Theta_1 - \Theta_0) \le K$, then the hypothesis H_0 is accepted, otherwise is H_1.

The quantities Θ_0, Θ_1 and the matrix $\mathbf{M}$ are obtained as a result of training the IMS.

5.5.3 *Diagnostic Decisions in the One-Dimensional Gaussian Case*

Let us consider the simplest special case, when the diagnostic parameter in question is a random variable subject to a Gaussian distribution. Then (5.41) has the form

$$p(y) = \frac{1}{\sigma\sqrt{2\pi}} e^{-(y-\Theta)^2/2\sigma^2}. \tag{5.61}$$

The parameter value $\Theta = \Theta_0$ corresponds to the hypothesis H_0, and $\Theta = \Theta_1$ is to the hypothesis H_1 at constant value σ. It is assumed that $\Theta_0 \neq \Theta_1$.

In this case, the elementary likelihood ratio (5.44).

$$\varsigma_k = \frac{\Theta_1 - \Theta_0}{\sigma^2}\xi_k - \frac{\Theta_1^2 - \Theta_0^2}{2\sigma^2}.$$

Expressions (5.45), (5.47) and (5.48) do not change in form, but in them, taking into account (5.46) $\kappa_2 = (\Theta_1 - \Theta_0)^2/\sigma^2$. In this case (5.58) and (5.59) take the form

$$n = \frac{(u_\alpha + u_\beta)^2\sigma^2}{(\Theta_1 - \Theta_0)^2}, \tag{5.62}$$

and

$$K = \frac{1}{2\sigma^2}\left[(\Theta_1 - \Theta_0)^2\frac{u_\alpha - u_\beta}{u_\alpha + u_\beta}\right]. \tag{5.63}$$

Inequality (5.51) is written as

$$\frac{\Theta_1 - \Theta_0}{\sigma^2}\nu_1 \leq K,$$

where ν_1 is an estimate of the mathematical expectation of the observation results. Therefore, when $\Theta_1 > \Theta_0$ the hypothesis H_0 is accepted if

$$\frac{1}{n}\sum_{k=1}^{n} x_k \leq C, \tag{5.64}$$

where

$$C = K\frac{\sigma^2}{\Theta_1 - \Theta_0} = \frac{(\Theta_1 - \Theta_0)(u_\alpha - u_\beta)}{2(u_\alpha + u_\beta)} + \frac{1}{2}(\Theta_1 + \Theta_0); \tag{5.65}$$

$x_k, k = \overline{1, n}$ is sampling implementation ξ_k, $k = \overline{1, n}$. If inequality (5.64) is not satisfied, then the hypothesis H_1 is accepted.

Test planning in a one-dimensional case using the latest formulas is no different from the planning described above for the general (multidimensional) case.

According to Gaussian theory, measurement errors under certain conditions converge to a Gaussian random variable or vector. Then the above relations turn out to be exact. This occurs when measuring resonances, attenuations, and other parameters characterizing various diagnosed units.

As practice shows, in several cases, the method described above and the basic relationships with a certain degree of approximation can be used even when the one-dimensional distribution of the quantities under study is not Gaussian. Often it is even possible to introduce appropriate amendments to the above formulas.

References

1. Babak, V., Eremenko, V., Zaporozhets, A.: Research of diagnostic parameters of composite materials using Johnson distribution. Int. J. Comput. **18**(4), 483–494 (2019)
2. Eremenko, V., Zaporozhets, A., Isaienko, V., Babikova, K.: Application of wavelet transform for determining diagnostic signs. In: Proceedings of the 15th International Conference on ICT in Education, Research and Industrial Applications. Integration, Harmonization and Knowledge Transfer, vol. 1, pp. 12–15 (2019)
3. Zaporozhets, A.A., Eremenko, V.S., Serhiienko, R.V., Ivanov, S.A.: Development of an intelligent system for diagnosing the technical condition of the heat power equipment. In: 2018 IEEE 13th International Scientific and Technical Conference on Computer Sciences and Information Technologies (CSIT), vol. 1, pp. 48–51. IEEE (2018)
4. Babak, V.P., Babak, S.V., Myslovych, M.V., Zaporozhets, A.O., Zvaritch, V.M.: Technical provision of diagnostic systems. Diagnostic Syst. Energy Equipm. 91–133 (2020)
5. Furse, C.M., Kafal, M., Razzaghi, R., Shin, Y.J.: Fault diagnosis for electrical systems and power networks: a review. IEEE Sens. J. **21**(2), 888–906 (2020)
6. Gyzhko, Y., Myslovych, M., Sysak, R.: To use of spectral windows in analysis of vibration signals. Przeglad Electrotechnichny **89**(2A), 294–296 (2013)
7. Tytarenko, V., Tychkov, D., Bilokin, S., Bondarenko, M., Andriienko, V.: Development of a simulation model of an information-measuring system of electrical characteristics of the functional coatings of electronic devices. Math. Model. **4**(2), 68–71 (2020)
8. Tymchuk, S.O., Abramenko, I.G., Zahumenna, K.V., Miroshnyk, O.O., Shchur, T.G., Tomporowski, A., Kruszelnicka, W.: Determination of the discretion interval of the temporal series of the technological process parameter measurement in ACS TP in the noises conditions. In Journal of Physics: Conference Series, vol. 1781, No. 1, p. 012017. IOP Publishing (2021)
9. Tempelman, A.: Generalization of the Glivenko-Cantelli theorem to finite-dimensional distributions of ergodic homogeneous random fields. Statist. Probab. Lett. **195**, 109767 (2023)
10. Babak, V., Zaporozhets, A., Kuts, Y., Myslovych, M., Fryz, M., Scherbak, L.: Models and characteristics of identification of noise stochastic signals of research objects. In: CEUR Workshop Proceedings, vol. 3309, pp. 349–362 (2022)
11. Babak, V., Scherbak, L., Kuts, Y., Zaporozhets, A.: Information and measurement technologies for solving problems of energy informatics. In: The 1st International Workshop on Information Technologies: Theoretical and Applied Problems 2021. CEUR Workshop Proceedings, vol. 3039, pp. 24–31 (2021)
12. Khanıev, T.A., Unver, İ., Maden, S.: On the semi-Markovian random walk with two reflecting barriers (2001)
13. Liu, T., Zhang, B., Hu, G., Bai, Z.: Revisit of Sheppard corrections in linear regression. SCIENCE CHINA Math. **53**, 1435–1451 (2010)
14. Stuart, A., Ord, K.: Kendall's advanced theory of statistics, distribution theory, vol. 1. John Wiley & Sons (2010)
15. Lal-Jadziak, J.: Accuracy in determination of correlation functions by digital methods. Metrol. Measur. Syst. **8**(2), 153–163 (2001)

16. Murray, L.L., Bellhouse, D.R.: WF Sheppard's correspondence with Karl Pearson and the development of his tables and moment estimates. Hist. Math. **53**, 108–117 (2020)
17. La ïb, N., & Ould-Sa ÏD, E.: A robust nonparametric estimation of the autoregression function under an ergodic hypothesis. Canad. J. Statist. **28**(4), 817–828 (2000)
18. Higgins, C.W., Katul, G.G., Froidevaux, M., Simeonov, V., Parlange, M.B.: Are atmospheric surface layer flows ergodic? Geophys. Res. Lett. **40**(12), 3342–3346 (2013)
19. Loh, I. Approximation with ergodic processes and testability. J. Appl. Probabil. 1–20 (2024)
20. Zhao, H., Wu, Z.: Statistical Estimation of Uncertainty in Surface Duct Parameters Inversion. In: 2018 12th International Symposium on Antennas, Propagation and EM Theory (ISAPE), pp. 1–4. IEEE (2018)
21. Babak, S., Babak, V., Zaporozhets, A., Sverdlova, A.: Method of statistical spline functions for solving problems of data approximation and prediction of objects state. In Proceedings of the Second International Workshop on Computer Modeling and Intelligent Systems (CMIS-2019), Zaporizhzhia, Ukraine, April, pp. 15–19 (2019)
22. Babak, V., Zaporozhets, A., Kuts, Y., Scherbak, L., Eremenko, V.: Application of material measure in measurements: theoretical aspects. In: Systems, Decision and Control in Energy II, pp. 261–269. Springer International Publishing, Cham (2021)
23. Ruzgas, T., Lukauskas, M., Čepkauskas, G.: Nonparametric multivariate density estimation: case study of cauchy mixture model. Mathematics **9**(21), 2717 (2021)
24. Kaipio, J.P., Somersalo, E.: Statistical inversion theory. Statistical Comput. Inverse Probl. 49–114 (2005)
25. Lupenko, S.A., Osukhivska, H.M., Lutsyk, N.S., Stadnyk, N.B., Zozulia, A.M., Shablii, N.R.: The comparative analysis of mathematical models of cyclic signals structure and processes. Sci. J. TNTU **82**(2), 115–127 (2016)
26. Lupenko, S.: The mathematical model of cyclic signals in dynamic systems as a cyclically correlated random process. Mathematics **10**(18), 3406 (2022)
27. Motai, Y.: Data-Variant Kernel Analysis. John Wiley & Sons (2015)
28. Babak, V., Zaporozhets, A., Kovtun, S., Myslovych, M., Kuts, Y., Scherbak, L.: Information support for identification of the technical state of electric power facilities. In: Systems, Decision and Control in Energy V, pp. 129–153. Springer Nature Switzerland, Cham (2023)
29. Babak, V.P., Babak, S.V., Eremenko, V.S., Kuts, Y.V., Myslovych, M.V., Scherbak, L.M., Zaporozhets, A.O.: Models and measures for the diagnosis of electric power equipment. Models Meas. Measurem. Monit. 99–126 (2021)
30. Babak, V.P., Babak, S.V., Myslovych, M.V., Zaporozhets, A.O., Zvaritch, V.M.: Methods and models for information data analysis. Diagn. Syst. Energy Equipments 23–70 (2020)
31. Burdzik, R., Konieczny, Ł.: Diagnosing of shock-absorbers of car vehicles at changeable pressure in tires. Diagnostyka, 27–32 (2009)
32. Cholewa, W., Korbicz, J., Moczulski, W., Timofiejczuk, A.: Methods of signal analysis. Fault Diagnosis: Models, Artif. Intell. Appl. 119–153 (2004)
33. Kobayashi, H., Mark, B.L., Turin, W.: Probability, Random Processes, and Statistical Analysis: Applications to Communications, Signal Processing, Queueing Theory and Mathematical Finance. Cambridge University Press (2011)
34. Xu, Y., Hespanha, J.P.: Estimation under uncontrolled and controlled communications in networked control systems. In: Proceedings of the 44th IEEE Conference on Decision and Control, pp. 842–847. IEEE (2005)
35. Hurd, H.L., Miamee, A.: Periodically Correlated Random Sequences: Spectral Theory and Practice, vol. 355. John Wiley & Sons (2007)
36. Zaporozhets, A.O.: Correlation analysis between the components of energy balance and pollutant emissions. Water Air Soil Pollut. **232**, 1–22 (2021)
37. Lucor, D., Su, C.H., Karniadakis, G.E.: Generalized polynomial chaos and random oscillators. Int. J. Numer. Meth. Eng. **60**(3), 571–596 (2004)
38. Gray, R.M., Davisson, L.D.: An Introduction to Statistical Signal Processing. Cambridge University Press (2004)

39. Randall, R.B., Antoni, J., Chobsaard, S.: The relationship between spectral correlation and envelope analysis in the diagnostics of bearing faults and other cyclostationary machine signals. Mech. Syst. Signal Process. **15**(5), 945–962 (2001)
40. Babak, V.P., Babak, S.V., Myslovych, M.V., Zaporozhets, A.O., Zvaritch, V.M.: Simulation and software for diagnostic systems. Diagnostic Syst. Energy Equipments 71–90 (2020)
41. Zaporozhets, A., Eremenko, V., Isaenko, V., Babikova, K.: Approach for creating reference signals for detecting defects in diagnosing of composite materials. In: Conference on Computer Science and Information Technologies, pp. 154–172. Springer International Publishing, Cham (2019)
42. Isermann, R.: Fault-Diagnosis Systems: An Introduction from Fault Detection to Fault Tolerance. Springer Science & Business Media (2006)
43. Antoni, J., Randall, R.B.: The spectral kurtosis: application to the vibratory surveillance and diagnostics of rotating machines. Mech. Syst. Signal Process. **20**(2), 308–331 (2006)
44. Jardine, A.K., Lin, D., Banjevic, D.: A review on machinery diagnostics and prognostics implementing condition-based maintenance. Mech. Syst. Signal Process. **20**(7), 1483–1510 (2006)
45. Li, Y.Y., Lee, Y.W., Ho, T.S., Wei, R.T., Lai, P.Y., Jao, K.S., Huang, S.L.: Interferometry based EUV spectrometer. IEEE Photonics J. **9**(4), 1–8 (2017)
46. Marwan, N., Braun, T.: Power spectral estimate for discrete data. Chaos: an Interdisciplinary J. Nonlinear Sci. **33**(5) (2023)
47. Verma, N.K., Singh, S., Gupta, J.K., Sevakula, R.K., Dixit, S., Salour, A.: Smartphone application for fault recognition. In: 2012 Sixth International Conference on Sensing Technology (ICST), pp. 1–6. IEEE (2012)
48. Torcal-Milla, F.J., Lobera, J., Lopez, A.M., Palero, V., Andres, N., Arroyo, M.P.: Mach-Zehnder-based measurement of light emitting diodes temporal coherence. Optik **267**, 169722 (2022)
49. Tomashin, A., Leonardi, G., Wallot, S.: Four methods to distinguish between fractal dimensions in time series through recurrence quantification analysis. Entropy **24**(9), 1314 (2022)
50. Newland, D.E.: An introduction to random vibrations, spectral and wavelet analysis. Courier Corporation (2012)
51. Grimmett, G., Stirzaker, D.: Probability and Random Processes. Oxford University Press (2020)
52. Crandall, S.H., Mark, W.D.: Random vibration in mechanical systems. Academic Press (2014)
53. Bendat, J.S., Piersol, A.G.: Random Data: Analysis and Measurement Procedures. John Wiley & Sons (2011)
54. Li, T.H.: Laplace periodogram for time series analysis. J. Am. Stat. Assoc. **103**(482), 757–768 (2008)
55. Stoica, P., Li, J., He, H.: Spectral analysis of nonuniformly sampled data: a new approach versus the periodogram. IEEE Trans. Signal Process. **57**(3), 843–858 (2008)
56. Li, T.H.: A robust periodogram for high-resolution spectral analysis. Signal Process. **90**(7), 2133–2140 (2010)
57. Akin, M., Kiymik, M.K.: Application of periodogram and AR spectral analysis to EEG signals. J. Med. Syst. **24**, 247–256 (2000)
58. Candan, C., Kutay, M.A., Ozaktas, H.M.: The discrete fractional fourier transform. IEEE Trans. Signal Process. **48**(5), 1329–1337 (2000)
59. Babak, V., Zaporozhets, A., Zvaritch, V., Scherbak, L., Myslovych, M., Kuts, Y.: Models and measures in theory and practice of manufacturing processes. IFAC-PapersOnLine **55**(10), 1956–1961 (2022)
60. Zaporozhets, A., Redko, O., Babak, V., Eremenko, V., Mokiychuk, V.: Method of indirect measurement of oxygen concentration in the air. Naukovyi Visnyk Natsionalnoho Hirnychoho Universytetu **5**, 105–114 (2018)
61. CusidÓCusido, J., Romeral, L., Ortega, J.A., Rosero, J.A., Espinosa, A.G.: Fault detection in induction machines using power spectral density in wavelet decomposition. IEEE Trans. Industr. Electron. **55**(2), 633–643 (2008)

62. Polat, K., Güneş, S.: Classification of epileptiform EEG using a hybrid system based on decision tree classifier and fast fourier transform. Appl. Math. Comput. **187**(2), 1017–1026 (2007)
63. Shirbhate, R., Panse, T., Ralekar, C.: Design of parallel FFT architecture using cooley Tukey algorithm. In: 2015 International Conference on Communications and Signal Processing (ICCSP), pp. 0574–0578. IEEE (2015)
64. Ratnam, D.V., Pasha, A., Swathi, P., Rao, M.V.G.: Acquisition of GPS L1 signals using Cooley-tukey FFT algorithm. In: 2013 IEEE International Conference on Signal Processing, Computing and Control (ISPCC), pp. 1–4. IEEE (2013)
65. Voronenko, Y., Puschel, M.: Algebraic signal processing theory: cooley-Tukey type algorithms for real DFTs. IEEE Trans. Signal Process. **57**(1), 205–222 (2008)
66. Abbaci, M., & Zmirli, M.O.: Performance evaluation of PMU using modified cooley-tukey algorithm based on GDFT. In: 2018 International Conference on Applied Smart Systems (ICASS), pp. 1–6. IEEE (2018)
67. Püschel, M., Rtteler, M.: Algebraic signal processing theory: cooley-Tukey type algorithms on the 2-D hexagonal spatial lattice. Appl. Algebra Eng. Commun. Comput. **19**(3), 259–292 (2008)
68. Li, C., Du, P., Li, K., Liu, Y., Jiang, H., Quan, Z.: Accurate goertzel algorithm: error analysis, validations and applications. Mathematics **10**(11), 1788 (2022)
69. Sysel, P., Rajmic, P.: Goertzel algorithm generalized to non-integer multiples of fundamental frequency. EURASIP J. Adv. Signal Proc. **2012**, 1–8 (2012)
70. Chen, H., Chen, G., Li, J.: The generalized Goertzel algorithm and its parallel hardware implementation. Sci China Ser A Math **51**(1), 37–41 (2008)
71. Brown, S., Timoney, J., Lysaght, T.: An evaluation of the Goertzel Algorithm for low-power, embedded systems. In: 25th IET Irish Signals & Systems Conference 2014 and 2014 China-Ireland International Conference on Information and Communications Technologies (ISSC 2014/CIICT 2014), pp. 357–361. IET (2014)
72. Dabrowski, A., Marciniak, T.: Canonic Goertzel algorithm and drawbacks of various Goertzel algorithm formulations. In: 2017 Signal Processing: Algorithms, Architectures, Arrangements, and Applications (SPA), pp. 259–262. IEEE (2017)
73. Forbes, C., Evans, M., Hastings, N., Peacock, B.: Statistical distributions. John Wiley & Sons (2011)
74. Cordeiro, G.M., de Castro, M.: A new family of generalized distributions. J. Stat. Comput. Simul. **81**(7), 883–898 (2011)
75. Watanabe, T.: Convolution equivalence and distributions of random sums. Probab. Theory Relat. Fields **142**, 367–397 (2008)
76. Devroye, L.: Nonuniform random variate generation. Handbooks Oper. Res. Manage. Sci. **13**, 83–121 (2006)
77. Yari, G., Tondpour, Z.: The new Burr distribution and its application. Math. Sci. **11**(1), 47–54 (2017)
78. Chakraborty, T., Das, S., Chattopadhyay, S.: A new method for generalizing burr and related distributions. Math. Slovaca **72**(1), 241–264 (2022)
79. Pawlas, P., Szynal, D.: Recurrence relations for single and product moments of generalized order statistics from Pareto, generalized Pareto, and Burr distributions. Commun. Stat.-Theo. Methods **30**(4), 739–746 (2001)
80. Jamal, F., Abuzaid, A.H., Tahir, M.H., Nasir, M.A., Khan, S., Mashwani, W.K.: New modified burr iii distribution, properties and applications. Math. Comput. Appl. **26**(4), 82 (2021)
81. Nadarajah, S., Pogány, T.K., Saxena, R.K.: On the characteristic function for Burr distributions. Statistics **46**(3), 419–428 (2012)
82. Borroni, C.G.: Understanding Karl Pearson's influence on Italian statistics in the early 20th century. Int. Stat. Rev. **77**(1), 81–95 (2009)
83. Babak, V.P., Babak, S.V., Eremenko, V.S., Kuts, Y.V., Myslovych, M.V., Scherbak, L.M., Zaporozhets, A.O.: Problems and features of measurements. Models Meas. Meas. Monit. 1–31 (2021)

84. Babak, V.P., Babak, S.V., Myslovych, M.V., Zaporozhets, A.O., Zvaritch, V.M.: Principles of construction of systems for diagnosing the energy equipment. Diagnost. Syst. Energy Equip. 1–22 (2020)
85. Xiong, Q., Du, P., Deng, J., Huang, D., Song, G., Qian, L., Luo, Y.: Global sensitivity analysis for nuclear reactor LBLOCA with time-dependent outputs. Reliab. Eng. Syst. Saf. **221**, 108337 (2022)
86. Pearson, K.: Mathematical contributions to the theory of evolution on a form of spurious correlation which may arise when indices are used in the measurement of organs. Proc. Royal Soc. london, **60**(359–367), 489–498 (1897)
87. Pearson, K.: Contributions to the mathematical theory of evolution. Philosoph. Trans. Royal Soc. London. A, **185**, 71–110 (1894)
88. Pearson, K.: I. Mathematical contributions to the theory of evolution.—VII. On the correlation of characters not quantitatively measurable. Philosophical Trans. Royal Soc. London. Series A, Containing Papers of a Mathematical or Physical Character, **195**(262–273), 1–47 (1900)
89. Zvarich, V.N., Marchenko, B.G.: Linear autoregressive processes with periodic structures as models of information signals. Radioelectron. Commun. Syst. **54**(7), 367–372 (2011)
90. Marchenko, B.G., Tkachuk, R.A., Palamar, M.I.: Harmonic analysis using in human visual system diagnosnics by means of elektroretinograms identification. In: Pattern recognition and information processing (1997)
91. Guzii, V.I., Marchenko, B.G., Protsenko, L.D.: Transformations providing for detection invariance with changes in the spectral parameters of low-frequency noise. Radioehlektronika **28**, 3–10 (1985)
92. Marchenko, B., Shcherbak, L.: 1975, p. 144. Naukova Dumka, Linear random processes and their applications. Kyiv (1975)
93. Babak, V.P., Babak, S.V., Eremenko, V.S., Kuts, Y.V., Myslovych, M.V., Scherbak, L.M., Zaporozhets, A.O.: Models of measuring signals and fields. Models Meas. Measurements Monit. 33–59 (2021)
94. Smyth, G.K.: Linear models and empirical bayes methods for assessing differential expression in microarray experiments. Stat. Appl. Genet. Molecular Biol. **3**(1) (2004)
95. Dobson, A.J., Barnett, A.G.: An Introduction to Generalized Linear Models. Chapman and Hall/CRC (2018)
96. Hommel, G.: Adaptive modifications of hypotheses after an interim analysis. Biomet. J.: J. Mathemat. Methods Biosci. **43**(5), 581–589 (2001)
97. Székely, G.J., Rizzo, M.L.: Testing for Equal Distributions in High Dimension. InterStat, **5**(16.10), 1249–1272 (2004)
98. Manoukian, E.B.: Modern Concepts and Theorems of Mathematical Statistics. Springer Science & Business Media (2012)
99. Dodge, Y.: The Concise Encyclopedia of Statistics. Springer Science & Business Media (2008)
100. Haeusler, E.H.: How many times do we need an assumption to prove a tautology in minimal logic? examples on the compression power of classical reasoning. Electron. Notes Theoret. Comput. Sci. **315**, 31–46 (2015)
101. Kesner, D., Viana, P.: Proceedings seventh workshop on logical and semantic frameworks, with applications (2013). ArXiv preprint arXiv:1303.7136.
102. Drago, A.: The birth of the non-Euclidean geometries as the more significant crisis in the foundations of modern mathematics. Logic Philos Sci **9**(1), 103–110 (2011)
103. Schmidt-Hieber, J.: The Kolmogorov-Arnold representation theorem revisited. Neural Netw. **137**, 119–126 (2021)
104. Köppen, M.: On the training of a kolmogorov network. In: Artificial Neural Networks—ICANN 2002: International Conference Madrid, Spain, August 28–30, 2002 Proceedings 12, pp. 474–479. Springer Berlin Heidelberg (2002)
105. Arnold, V.I.: Proof of a theorem of AN Kolmogorov on the invariance of quasi-periodic motions under small perturbations of the Hamiltonian. Collected Works: Representations of Functions, Celestial Mechanics and KAM Theory **1957–1965**, 267–294 (2009)

106. Rassokhin, D.N., Agrafiotis, D.K.: Kolmogorov-Smirnov statistic and its application in library design. J. Mol. Graph. Model. **18**(4–5), 368–382 (2000)
107. Greenwell, R.N., Finch, S.J.: Randomized rejection procedure for the two-sample Kolmogorov-Smirnov statistic. Comput. Stat. Data Anal. **46**(2), 257–267 (2004)
108. Nguyen, H.D.: A Two-sample Kolmogorov-Smirnov-like test for big data. In: Data Mining: 15th Australasian Conference, AusDM 2017, Melbourne, VIC, Australia, August 19–20, 2017, Revised Selected Papers 15, pp. 89–106. Springer Singapore (2018)
109. Sverdlova, A., Zaporozhets, A.: Predicting anomaly conditions of energy equipment using neural networks. In: E3S Web of Conferences, vol. 280, p. 09005 (2021)
110. Zaporozhets, A., Eremenko, V., Babak, V., Isaienko, V., Babikova, K.: Using Hilbert transform in diagnostic of composite materials by impedance method. Period. Polytechnica Electr. Eng. Comput. Sci. **64**(4), 334–342 (2020)
111. Eremenko, V., Babak, V., Zaporozhets, A.: Method of reference signals creating in non-destructive testing based on low-speed impact. Tekhnichna ElektroDynamika **4**, 070 (2021)
112. Babak, V., Bohachev, I., Zaporozhets, A., Khaidurov, V., Havrysh, V., Kalinichenko, A.: Some features of modeling ultrasound propagation in non-destructive control of metal structures based on the magnetostrictive effect. Electronics **12**(3), 477 (2023)
113. Zaporozhets, A., Babak, V., Sverdlova, A., Isaienko, V., Babikova, K.: Development of a system for diagnosing heat power equipment based on IEEE 802.11 s. In: Systems, Decision and Control in Energy II, pp. 141–151. Springer International Publishing, Cham (2021)
114. Babak, V.P., Babak, S.V., Eremenko, V.S., Kuts, Y.V., Myslovych, M.V., Scherbak, L.M., Zaporozhets, A.O.: Models and measures for standardless measurements of the composite materials characteristics. Models Meas. Meas. Monit. 157–190 (2021)
115. Bohachev, I., Babak, V., Zaporozhets, A.: Novel small-aperture transducers based on magnetostrictive effect for diagnostic systems. Tech. Electrodynamics **3**, 69–78 (2022)
116. Santhakumaran, A.: Fundamentals of Testing Statistical Hypotheses. Atlantic Publishers & Dist (2001)
117. Holický, M.: Testing of statistical hypotheses. Introduction to Probability and Statistics for Engineers, pp. 125–138 (2013)
118. Kachiashvili, G.K., Kachiashvili, K.J., Mueed, A.: Specific features of regions of acceptance of hypotheses in conditional Bayesian problems of statistical hypotheses testing. Sankhya A **74**, 112–125 (2012)
119. He, J., Jones, J.W., Graham, W.D., Dukes, M.D.: Influence of likelihood function choice for estimating crop model parameters using the generalized likelihood uncertainty estimation method. Agric. Syst. **103**(5), 256–264 (2010)
120. Rohde, C.A.: Introductory Statistical Inference with the Likelihood Function. Springer, New York (2014)
121. Li, F., Luan, P.: ARMA model for predicting the number of new outbreaks of newcastle disease during the month. In: 2011 IEEE International Conference on Computer Science and Automation Engineering, vol. 4, pp. 660–663. IEEE (2011)
122. Huang, D., Yang, J., Zhou, D., Litak, G., Liu, H.: Influence of Poisson white noise on the response statistics of nonlinear system and its applications to bearing fault diagnosis. J. Comput. Nonlinear Dyn. **14**(3), 031010 (2019)
123. Guo, S.L., Yang, Y.G., Sun, Y.H.: Stochastic response of an energy harvesting system with viscoelastic element under Gaussian white noise excitation. Chaos Solitons Fractals **151**, 111231 (2021)

Chapter 6
Simulation of Diagnostic Signals of Electric Equipment

Abstract This chapter explores the advancement in diagnostic signal simulations for electrical equipment, emphasizing acoustic emission (AE) signals, vibration during shock diagnostics, and the detection of rod fastening defects using vibration analysis. The first section introduces a comprehensive approach for simulating diagnostic signals utilizing linear random processes (LRPs), providing a foundation for the efficient analysis and verification of diagnostic features without extensive physical experimentation. It elaborates on the development and application of a simulation model for AE signals in electrical equipment, showcasing how this model facilitates the understanding of signal propagation. Further, the chapter delves into the simulation of AE signals and vibration in electrical equipment under shock diagnostics, highlighting a developed program that generates AE signals mimicking those from a piezoelectric transducer. The program's ability to modify various parameters allows for a detailed analysis of their influence on AE and vibration signals, offering insights consistent with theoretical models and real measurements. Moreover, the study proposes the use of the parameter β, derived from the power spectrum, to detect defects in rod fastening through vibration analysis, marking a significant step towards predictive maintenance and fault diagnosis in mechanical systems. By comparing the β values of suspended and fixed rods, the chapter validates the effectiveness of numerical models in replicating real-world dynamics. The investigation extends to the simulation of failure modes in diesel-electric generator cylinders, suggesting a novel approach to diagnosing specific cylinder failures through amplitude-frequency and phase-frequency characteristics analysis. Collectively, this research presents a valuable contribution to the field of diagnostic signal simulation, offering novel methods and tools for the analysis, verification, and simulation of diagnostic signals in electrical and mechanical systems. The findings promise to enhance the precision of fault diagnostics, reduce dependency on physical experiments, and pave the way for more effective predictive maintenance strategies.

Keywords Linear random processes · Spectral-correlation analysis · Statistical diagnostics acoustic emission · Noise signals · Information-measuring system · Electric equipment

V. Babak et al., *Statistical Diagnostics of Electric Power Equipment*, Studies in Systems, Decision and Control 573, https://doi.org/10.1007/978-3-031-76253-6_6

An important step in solving problems of statistical diagnostics is simulation modeling aimed at testing the method being developed. The use of modern computers for this purpose makes it possible to significantly reduce the number of experiments on real equipment, as a result of which a reduction in the cost of money and time for the design of diagnostic information-measuring system (IMS) is achieved [1–5].

The main advantage of simulation modeling compared to other modeling methods is the ability to solve problems of exceptional complexity: the system under study may contain linear or nonlinear links, elements of continuous or discrete action, be influenced by numerous random factors of a complex nature, described by cumbersome relationships, etc.

The task of statistical simulation modeling is to obtain discrete implementations of random processes with given parameters. All methods and techniques for modeling such implementations are based on the use of random numbers that have a uniform distribution on the interval [0;1). A random variable with such a distribution is called basic [6–10].

Success in computer simulation tasks is determined by two main factors:

- quality of random number source;
- selection of a rational calculation algorithm.

The issue of choosing a method for obtaining random numbers is a priority since the success of solving the simulation problem as a whole largely depends on its successful solution. The calculation algorithm is specific to each task and must be considered separately in each specific case. Therefore, let us dwell on the consideration of the first question.

6.1 Formation of Random Sequences with Specified Parameters

There are several known ways to obtain random numbers. Each of them has its advantages and disadvantages. Let's consider some methods from the point of view of their use in simulation modeling using a computer.

While computing using a computer to obtain random numbers, special electronic devices were developed, which are called physical random variable generators. To build the generator, natural primary sources of random noise were used (radioactive decay, noise of electronic and semiconductor devices, etc.) [11–14]. Physical generators were manufactured as special peripheral devices for computers and as separate devices—so-called noise generators. Although the main purpose of noise generators is analog modeling, analysis of amplifier characteristics, and the like, the signals received from such generators can be "digitized" using an analog-to-digital converter (ADC), recorded in computer memory, and then used as the initial sequence for simulation modeling [15–18].

Physical generators have several disadvantages, including, firstly, the need to constantly check the quality of the device's operation, due to the instability of the characteristics of the generated signals, and secondly, the numbers generated by such devices cannot be reproduced repeatedly, which practically eliminates verification of the received results for control of calculations [19–21].

The suitability of random numbers used for simulation modeling is determined not by how they are obtained, but by whether they satisfy the stated objectives. Basically, this is checked using several generally accepted statistical tests. Therefore, instead of the methods described above, you can use a certain formula to obtain random numbers, provided that the basic statistical properties of such numbers coincide with the theoretical properties of a random sequence of values with a uniform distribution. The numbers obtained in this way are deterministic, but they have properties close to the properties of implementing a random sequence. That's why they are called pseudorandom. Random numbers produced by physical generators are sometimes called "true" random to emphasize their difference from pseudo-random numbers [22–24].

Most of the algorithms that are used to obtain pseudorandom numbers are quite economical and fast-acting, that is, their operation does not require either significant amounts of computer memory or large amounts of computer time. Since each sequence of pseudorandom numbers can always be obtained again, there are no problems in verifying the results of calculations, as can be the case when using physical generators. In addition, having carried out the necessary statistical tests for such a sequence once, you can use it as many times as necessary the next time you solve a given class of problems. In this regard, when simulating statistical modeling using a computer, it is most advisable to use pseudo-random sequences as a source of random numbers.

Currently, to obtain pseudo-random sequences, recurrence relations of the form are most often used:

$$a_{k+1} = F(a_k), k = 0, 1, \ldots, \tag{6.1}$$

where a_0—given real number from a half-interval $[0, 1)$.

This is due to the simplicity and high efficiency of the algorithms that correspond to them. The most common now is the so-called congruent method (sometimes also called the method of comparisons or the method of deductions), developed by D. Lehmer. It consists of the recurrent calculation of a sequence, the elements of which are obtained according to the formula

$$x_k = Mx_{k-1} + b(\mathrm{mod}P), k = 1, 2, \ldots, \tag{6.2}$$

where M and b are natural numbers, and M is large; x_0 is given natural number; the notation $x(\mathrm{mod}P)$ denotes the modulo operation P of x number (in other words, $x(\mathrm{mod}P)$ is equal to the remainder of dividing celestially by).

From expression (6.2) it is directly seen that all x_k, $k = 1, 2, \ldots$, are the natural numbers not exceeding P. Evenly distributed numbers $a_k \in [0, 1)$ are obtained from numbers x_k by the formula

$$a_k = \frac{x_k}{P}, k = 1, 2, \ldots. \quad (6.3)$$

The simplest form of (6.2) has at $b = 0$. In this case, the algorithm for obtaining a pseudorandom sequence is called the multiplicative congruent method [25–27]. The name "mixed method" is used when $b \neq 0$.

If M and P in (6.2) are mutually prime numbers, and it is chosen large enough, the sequence (6.3) is approximately uncorrelated. When computing on a computer, it is convenient to choose $P = 2^m$, where m is an integer, then calculations by formula (6.3) are faster.

Note that each number in the computer memory is represented by some finite number of bits (binary digits). Such real numbers reflect on the computer the set of all real numbers with limited accuracy. As a consequence, all sequences of the form (6.3) appear to be periodic. This is easily understood from the following considerations.

Since only a finite number of numbers can be represented by a finite number of bits in the computer memory, there will always be a natural number L such that an element a_L of the sequence coincides with one of the previous values a_l. Then, by (6.1), the following equality will be fulfilled

$$a_{k+L} = a_{k+l} \text{ for all integers } k \geq 0. \quad (6.4)$$

Let L be the smallest integer at which expression (6.4) is valid for some $l < L$. The set of numbers is called $a_1, a_2, \ldots, a_L$ the aperiodicity segment of the sequence (6.1), the number L is called the length of the aperiodicity segment, and $T = L - l$ is the length of the period. All numbers that make up the aperiodicity segment are different.

For the case when $P = 2^m$, the maximum length of the period of the sequence (6.3) is equal to 2^m, does not depend on. The entry denotes equality modulo, that is, all numbers whose remainder from dividing by is equal to. Given this choice, and the sequence (6.3) consists of all numbers from 0 to 2^{m-1}.

Let us now consider methods of obtaining pseudorandom values with a given distribution other than uniform.

The main method of obtaining random sequences with a continuous distribution law $F(x)$ is the nonlinear transformation method [28–31]. For its application, there must exist an inverse function $F^{-1}(x)$. A random variable ξ, distributed according to the law $F(x)$, is obtained from a uniformly distributed a using transformation

$$\xi = F^{-1}(a). \quad (6.5)$$

The advantage of this method of modeling random numbers [22–24] is the absence of methodological errors inherent in approximate methods, and hence the high accuracy of the reproduction of distribution laws and simplicity of preparatory work, because the transformation of the uniform law into the necessary law is given in the form of a ready analytical dependence. Its main disadvantage is its relatively low speed, because the computer calculation of nonlinear transformations often requires a significant number of elementary operations.

One of the most important distributions from the point of view of practical application is the Gaussian distribution [32–35]. Random variables and processes distributed according to the Gaussian law are often used in modeling physical processes and phenomena. For example, the Gaussian law describes thermal noise in radio-technical devices, measurement errors of continuous physical quantities, etc.

The probability distribution density of a Gaussian random variable is described by the expression:

$$p(x) = \frac{1}{\sigma\sqrt{2\pi}} e^{-\frac{(x-m)^2}{2\sigma^2}}, \sigma > 0,$$

where m—mathematical expectation of the random variable; σ—its standard deviation.

To model a Gaussian random variable, it is difficult to use the method of nonlinear transformation (6.5), because the integral function of the Gaussian distribution is not expressed in elementary functions $F(x)$ and it is impossible to find the inverse function to it explicitly. But there is another way, which uses a property known in probability theory: the distribution of the product of two independent random variables, one of which has a Rayleigh distribution with parameters $\sigma = 1$, and the other is distributed according to the Arcsinus law with parameters $m = 0$, $b = 1$, is a Gaussian law with zero mathematical expectation and unit variance.

The density of the Rayleigh law distribution [36–38] is given by the expression

$$p(x) = \frac{x}{\sigma^2} e^{-\frac{x^2}{2\sigma^2}}, x \geq 0.$$

The integral distribution function of the Rayleigh law has the form

$$F(x) = 1 - e^{-\frac{x^2}{2\sigma^2}}, x \geq 0,$$

and the nonlinear transformation to obtain a Rayleigh random variable from a uniformly distributed variable a is given by the expression

$$y = \sigma\sqrt{-2 \cdot \ln(1 - a)}.$$

Further, since the value obeys the same distribution as $(1 - a)$, we can apply the expression

$$y = \sigma\sqrt{-2 \cdot \ln a}.$$

The distribution density of the arcsine laws [39–42]

$$p(x) = \frac{1}{\pi b \sqrt{1 - \frac{(z-m)^2}{b^2}}},$$

integral distribution function

$$F(x) = \frac{1}{\pi} \arcsin \frac{z-m}{b} + \frac{1}{2},$$

and the transformation of a uniformly distributed quantity a

$$z = b \sin\left[\pi\left(a - \frac{1}{2}\right)\right] + m.$$

Using the symmetry properties of trigonometric functions $\pi(a - 1/2)$, the argument can be replaced by $2\pi a$.

Finally, to form random variables distributed according to the Gaussian law, we use the obtained values y and z. Finally, taking into account that $\sigma = 1$, $m = 0$, $b = 1$, we obtain the expression

$$\xi = \sqrt{-2 \cdot \ln a_1} \sin(2\pi a_2), \tag{6.6}$$

where ξ—Gaussian random variable with parameters $m = 0$ and $\sigma = 1$; a_1, a_2–independent variables [0, 1) uniformly distributed on the half-interval.

Along with (6.6) the following expression is used

$$\eta = \sqrt{-2 \cdot \ln a_1} \cos(2\pi a_2), \tag{6.7}$$

where η—Gaussian random variable with parameters $m = 0$ and $\sigma = 1$; a_1, a_2—the same as in (6.6).

The values ξ and η appear to be independent. Thus, having 2 values of a uniformly distributed random variable, we obtain 2 values of a variable with Gaussian distribution, which is important when it is necessary to model long sequences.

In problems that do not impose high requirements to the quality of random numbers, it is recommended to use more economical approximate methods to reduce the number of elementary operations. One such method is based on the central limit theorem [43–47]. A Gaussian random variable is formed as follows:

$$\xi = \sum_{j=1}^{12} a_j - 6. \tag{6.8}$$

The disadvantages of this method are as follows: at $x > 6$ the density distribution $p(x)$ of the random variable (6.8) is identical to zero, and in addition, to obtain one number with a Gaussian distribution needs 12 uniformly distributed numbers. The latter can be a significant obstacle if you want to form a long sequence of

Gaussian numbers using pseudo-random uniformly distributed numbers obtained by the congruent method, because the periodicity of the base sequence may appear.

To model a Gaussian sequence with given parameters and use the obvious transformation

$$\varsigma = \sigma \cdot \xi + m, \tag{6.9}$$

where ξ– Gaussian random variable with parameters $m = 0$, $\sigma = 1$; ς—Gaussian random variable with the given parameters m and σ.

6.2 Modeling Realizations of Linear Random Fields and Processes

As was said above, only random processes with discrete time can be processed using the computer. Therefore, any analog signal measured by the diagnostic system is first of all subjected to time discretization with some step Δt, and then to analog-to-digital conversion with some quantization step Δh.

It should be noted that quantization by level, generally speaking, leads to distortion of the true process distribution since the initial process at the corresponding moment can have any value from the admissible range, while the quantized value can take only a finite number of different values. But modern ADCs are characterized by a large number of digits and therefore this distortion can be neglected. Thus, we will not take into account the effects of quantization in the simulation modeling of diagnostic signals.

Time discretization is, in fact, the replacement of the original process with continuous time by a sequence of values of this process at fixed moments. (This corresponds to the principle of time variation model "Δt" [48–50].) Given that realizations of finite length T are most often measured, this is equivalent to replacing the process with N a set of random variables, which is defined as the integer part of the ratio $T/\Delta t$. In most cases, these random variables turn out to be mutually dependent.

In a more general case, when modeling a random field is considered, it is also necessary to obtain a finite number of values at fixed points of some multidimensional space.

Thus, the problem can be reduced to the modeling of a set of random variables (N-dimensional random vector). The set of random variables N is fully described by their N-dimensional probability distribution, which can be considered as a given, because it can be found based on a priori information about the random process or random field to be modeled. The modeling process in the most general case consists in using conditional distributions to obtain each subsequent element of the vector. Thus, it is necessary to know the sequence of finite-dimensional distributions up to N order inclusive. But in practically important cases, the total dimensionality of the vector to be modeled can be very large, which makes the use of this method impossible. In this case, a rather simple and effective way out of the situation is the use of integral images of linear random processes (LRPs) and fields.

Let us first consider a way of modeling a random linear field, which is based on using its image in the form of a stochastic integral

$$\xi(\boldsymbol{x}, t) = \int_{\boldsymbol{R}^n} \int_{-\infty}^{\infty} \phi(\tau,\theta;\boldsymbol{x}, t) d\eta(\tau,\theta), \tag{6.10}$$

where $\eta(\tau,\theta)$—spatio-temporal random field with independent increments, with probability 1 equal to zero if at least one of the coordinates is zero; $\phi(\boldsymbol{\tau}, \theta; \boldsymbol{x}, t)$—non-random function integrable with square over $\boldsymbol{\tau}$ and θ at any $\boldsymbol{x} \in \boldsymbol{R}^3$, and $t \in \boldsymbol{R}^1$.

Expression (6.10) can be approximated by an infinite integral sum. But, taking into account the conditions imposed on the function $\phi(\boldsymbol{\tau}, \theta)$, infinite sums can be replaced by finite sums with sufficient accuracy. Thus, we obtain the image of a discrete linear random field in the form of

$$\xi_{j,k,l,m} = h_0^3 \Delta t \sum_{n=-N}^{N} \sum_{p=-P}^{P} \sum_{q=-Q}^{Q} \sum_{s=-S}^{S} \phi_{n,p,q,s,j,k,l,m} \cdot \varsigma_{n,p,q,s}, \tag{6.11}$$

where h_0—discretization step by variables x_1, x_2 and x_3; Δt—discretization step by time;

$$\phi_{n,p,q,s,j,k,l,m} = \phi(nh_0, ph_0, qh_0, s\Delta t; jh_0, kh_0, lh_0, m\Delta t);$$

$\varsigma_{n,p,q,s}$—samples of discrete white noise; j, k, l, m, n, p, q, s—integers.

Using expression (6.11), one can obtain realizations of random fields whose characteristics depend on the function $\phi(\boldsymbol{\tau}, \theta)$ and parameters of the generating field $\varsigma_{j,k,l,m}$. In the partial case, when the generating field has a Gaussian distribution, the resulting linear field $\varsigma_{j,k,l,m}$ will also be Gaussian, but its points will no longer be independent: the correlation between them uniquely depends on the type of function $\phi(\tau, \theta)$.

LRPs are modeled similarly. A discrete linear process [51–54] $\{\xi_k, t_k = k \cdot \Delta t, k = 0, \pm 1, \ldots\}$ can be represented as

$$\xi_k = \sum_{j=-\infty}^{\infty} \phi_{j,k} \zeta_j, k = 0, \pm 1, \pm 2 \ldots, \tag{6.12}$$

where $\{\zeta_k, t_k = k \cdot \Delta t, k = 0, \pm 1, \ldots\}$ that produces white noise; $\phi_{j,k}$, $j, k = 0, \pm 1, \ldots$ are function values $\phi(\tau; t)$ at the points $\tau_j = j \cdot \Delta t$, $t_k = k \cdot \Delta t$; Δt—time sampling step.

In practice, in the expression (6.12) the limits of the sum are chosen finite, for example, from 0 to N, where the value N is chosen based on the properties of the function $\phi(\tau; t)$ and the required accuracy.

Given the properties of linear random processes, expression (6.12) makes it possible to use similar methods to operate with any distribution laws from the class of infinitely divisible.

6.3 Simulation of Diagnostic Signals in Electrical Equipment

In diagnostics simulation models allow by substituting specific numerical values of parameters to trace the tendency of changing quantitative estimates of diagnostic features depending on the possible technical state of objects. Thus, they allow to substantiate and carry out the verification of diagnostic features. The use of simulation models significantly reduces the time of the experiment, as the preparation and conduct of research on physical models or real objects would require much more time and material costs.

Let us consider examples of simulation model applications to verify the adequacy of the developed mathematical models and the informativeness of the selected diagnostic features.

6.3.1 Acoustic Emission Signals

Let us consider the methodology and program of acoustic emission (AE) processes modelling accompanying the electrical equipment (EE) operation.

The mathematical model of AE signals was developed in [55–57] based on a LRP model. Taking into account [55–57], the AE process is the result of linear filtering of a white noise type process. The channel of AE signal origin and propagation in the research object is schematically represented in Fig. 6.1.

Based on the LRP properties [58–60], the AE $\xi(t)$ signal measured at the output of a piezoelectric transducer (PET) [2, 61–65] is written as follows:

$$\xi(t) = \int_{-\infty}^{\infty} \phi(\tau, t)\eta'(\tau)d\tau = \int_{-\infty}^{\infty}\int_{-\infty}^{\infty} \phi(\tau, t)\phi_0(s, \tau)d\varsigma(s)d\tau. \quad (6.13)$$

Let's denote $\phi_1(\tau, t) = \int_{-\infty}^{\infty} \phi(s, t)\phi_0(\tau, s)ds$. Then (13) will take the form

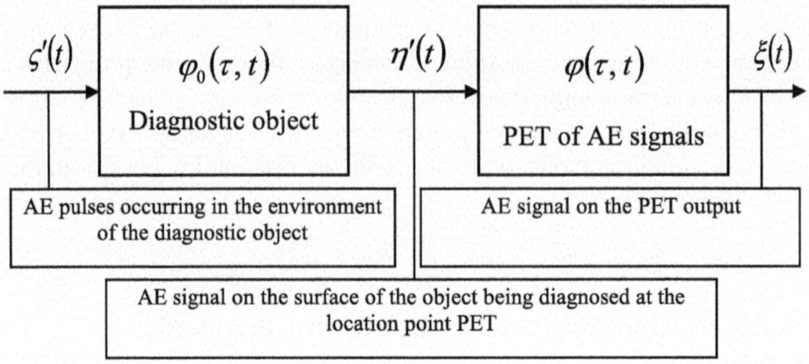

Fig. 6.1 Model of AE occurrence and propagation

$$\xi(t) = \int_{-\infty}^{\infty} \phi_1(\tau, t) d\varsigma(\tau). \tag{6.14}$$

In this case, the probabilistic characteristics of the $\xi(t)$ process are completely determined by the choice of a particular type of impulse transition functions $\phi_0(\tau, t)$ and $\phi(\tau, t)$, characterizing the object of diagnosis and the PET, respectively. Besides, it is necessary to take into account the properties of the generating process $\varsigma\prime(t)$. It is on these points that the scheme of the AE process modelling is based.

The AE pulses appearing in the environment of the diagnosctic object can be represented by the expression

$$\phi_0(\tau, t) \equiv \phi_0(t - \tau) \equiv \phi_0(s) = \alpha \cdot \mathrm{e}^{-\alpha s} \cdot U(s), s \in (-\infty, \infty), \tag{6.15}$$

where $\alpha > 0$ is the attenuation coefficient; $U(s)$ is the Heaviside function.

The impulse transition function of the PET has the following form

$$\phi(\tau, t) \equiv \phi(t - \tau) \equiv \phi(s) = \gamma \cdot \mathrm{e}^{-\beta s} \sin\theta s \cdot U(s), s \in (-\infty, \infty), \tag{6.16}$$

where $\theta > 0$, $0 < \beta < \theta$ are actual constants, which have a specific physical meaning, namely oscillation frequency and attenuation coefficient, respectively; $\gamma = (\beta^2 + \theta^2)/\theta$ is normalizing factor; $\theta = \sqrt{\omega^2 - \beta^2}$.

Proceeding from (13)

$$\phi_1(\tau, t) = \phi_1(\tau - t) = \int_{-\infty}^{\infty} \phi_0(x - \tau)\phi(t - x)dx = \int_{-\infty}^{\infty} \phi_0(x)\phi(t - \tau - x)dx. \tag{6.17}$$

Considering (6.15) and (6.16), we obtain

$$\phi_1(\tau, t) = \phi_1(t-\tau) = \alpha\,\gamma\, e^{-\beta(t-\tau)} \int\limits_0^\infty e^{-(\alpha-\beta)x} \sin[-\theta x + \theta(t-\tau)]dx. \quad (6.18)$$

The integral in the last expression is reduced to the tabular integral. As a result of transformations we obtain

$$\phi_1(t-\tau) = \frac{\alpha\gamma e^{-\beta(t-\tau)}}{(\alpha-\beta)^2+\theta^2}\{\theta\cos[\theta(t-\tau)] + (\alpha-\beta)\sin[\theta(t-\tau)]\}. \quad (6.19)$$

Having obtained explicitly the convolution (6.19), the AE process at the PET output, taking into account (6.13), can be written as follows:

$$\begin{aligned}\xi(t) = \frac{\alpha\cdot\gamma}{(\alpha-\beta)^2+\theta^2}\Bigg[\theta \int\limits_0^t e^{-\beta(t-\tau)} \cos[\theta(t-\tau)]d\tau \\ +(\alpha-\beta)\int\limits_0^t e^{-\beta(t-\tau)} \sin[\theta(t-\tau)]d\tau\Bigg].\end{aligned} \quad (6.20)$$

Expression (6.19) is a joint impulse transition function characterizing the AE signal propagation channel in a system consisting of two series-connected links of the first and second order. These links describe the system: diagnostics object—PET.

In simulation, this system can be represented by a series connection of RC and RLC circuits [66–68] with impulse transition functions (6.15) and (6.16), respectively. As is usually done while solving such problems, we will assume that the pulse signals of AE measured by the PET at the point of its placement on the surface of the research object do not significantly affect the intrinsic characteristics of the PET. Besides, while solving the modelling problem for PET with the pulse transition function (6.16), the most interesting from the point of view of practice is the oscillatory mode of PET operation. That is, when in expression (6.16) $\theta > 0$, or, based on the relation $\theta = \sqrt{\omega^2 - \beta^2}$, $\omega > \beta$. Therefore, we will limit ourselves to consider only such a case.

Based on the above expressions, a program has been developed that allows modelling and studying pulse signals of AE. The program consists of several modules combined to perform a common task.

The main module of the program provides its startup and main window output. The interface module contains subroutines that provide:

- processing of system events;
- processing of user commands (pressing buttons, selecting menu items, etc.) and data input.

The modelling module performs the following operations:

- calculation of input signal realization in the form of a single pulse of relaxation type (decaying exponentially);
- calculation of input signal realization as a Poisson sequence of input pulses of relaxation type;
- calculation of the PET impulse response, which is further used in the digital filter (filtering using a finite impulse response is realized);
- calculation of the PET response to the input signal.

Graphs of the input signal, PET impulse response, and PET response can be displayed in the lower part of the main window by selecting the corresponding tab.

The following parameters can be set on the main window of the created program:

- General parameters (upper part of the window) include: the sampling step $\Delta\tau$, s. The developed simulation program provides a range of changes in sampling steps. In the same part of the window, there is a button "Simulate", clicking on which starts the modelling process. The results of calculations are displayed in the "Graphs" area at the bottom of the window.
- Parameters of the oscillatory system include: duration of realization (in seconds), parameters R (in ohms), L (in mH), and C (in μF), which characterize the PET. In addition, the left part of the window displays the results of calculations of quantitative estimates of parameters β (1/s), Ψ (rad/s), ω (rad/s), f_{real} (1/s). Clicking on the "Graph" button provides the calculation and display of the graph of the impulse response of the system at the given parameters *R*, *L*, and *C*.
- Input signal parameters depend on the selected type of input signal: a single pulse or a sequence of pulses. For a single pulse: amplitude (in conventional units), start delay (in seconds). For a series of pulses: intensity (1/s), average amplitude (in conventional units), standard deviation of amplitudes of individual pulses (in conventional units). In addition, for both types of input signals are set: duration of realization (s) and attenuation coefficient (1/s). The graph of the input influence is displayed by pressing the "Graph" button, which is located on the right.

Let us consider some results of the simulation of the AE signal using the created program. Figure 6.2 shows a graph of one of the obtained realizations of the pulse sequence at the PET output.

The main objectives of the study were as follows:

1. Analysis of the response of the PET depends on its internal parameters (values of *R*, *L*, and *C*—components included in the PET) when a single relaxation pulse with a characteristic of the form (6.15) is applied to its input.
2. Analysis of the signal at the PET output when a Poisson pulse sequence of a given intensity $\lambda = const$ is applied to its input and the changing parameters *R*, *L*, and *C* characterizing the PET.
3. Analysis of the response at the PET output when its input is affected by a Poisson pulse sequence with different intensity values λ. In this case, the parameters *R*, *L*, and *C* characterizing the PET remain unchanged.

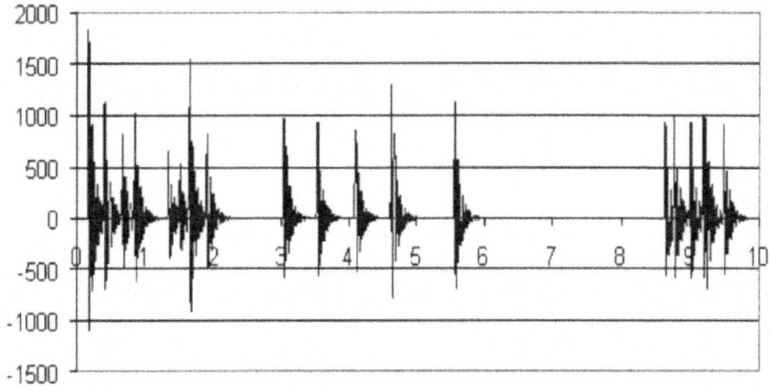

Fig. 6.2 Realization of the simulated AE signal

Varying the input parameters of the program, a series of realizations of the AE process was obtained. The analysis of these realizations showed that the growth of the coefficient values α (which is achieved by a successive increase of the R-value) leads to a decrease in the response length and a decrease in the number of its oscillations. The obtained results of modeling confirm the necessity of taking into account the intrinsic characteristics of the PET when measuring AE signals.

The second task, which was investigated using the developed program of AE processes modelling, is the formation and determination of the character of the pulse sequence at the PET output when its R, L, and C parameters change, but at constant intensity λ of the input process.

It was shown that when the PET has sufficiently high goodness, the signal at its output is a sequence in which the trailing edge of each pulse has a pronounced oscillatory character. When the goodness of the PET decreases, the oscillatory character of the trailing edge of pulses becomes less pronounced.

And the third task of the study—formation and determination of the nature of change in the pulse sequence at the output of the PET at changing the intensity of the input process λ and with unchanged parameters R, L, C. It was shown that increasing the value of the parameter λ leads to an increase in the number of pulses at a fixed time interval.

Thus, the results of AE modeling signals are in good agreement with the conclusions that can be drawn by analyzing their mathematical model.

6.3.2 Vibrations in Massive Assemblies of EE During Shock Diagnostics

Chapter 3 of this book considers the vibration model of the EE node in the form of a LRP:

$$\xi(t) = \sum_{j=1}^{n} a_j \xi_j(t), t \in (-\infty, \infty), \tag{6.21}$$

where a_j are weighting coefficients,

$$\xi_j(t) = \int_{-\infty}^{\infty} \phi_j(t-\tau) d\eta_j(\tau), j = \overline{1,n}, t \in (-\infty, \infty); \tag{6.22}$$

$$\phi_j(\tau) = \frac{\omega_j^2}{\psi_j} e^{-\beta_j \tau} \sin(\psi_j \tau) U(\tau), j = \overline{1,n}. \tag{6.23}$$

Here $\beta_j > 0$—damping coefficients; ω_j—own frequencies; $\psi_j = \sqrt{\omega_j^2 - \beta_j^2}$; n—several resonances; $\{\eta_j(\tau), j = \overline{1,n}\}$—vector of processes with independent increments; $U(\tau)$—normalized Heaviside function.

In Chap. 5, the relations for spectral-correlation analysis [69–73] of the vibration process are given, provided that it can be considered stationary. In the case of impact diagnostics, the assumption of stationarity of the process is violated, since the vibrations are damped after the impact. Strictly speaking, in this case, the previously obtained expressions for spectral analysis become inapplicable. But at diagnostics, it is possible to consider fragments of realizations of vibration processes, short enough to vibrations that did not manage to attenuate significantly during the measurement time. Such segments can be conditionally considered stationary. However, with this approach, there are methodological errors that are quite difficult to quantify using only analytical methods.

To determine how the spectral parameter estimates proposed in Chap. 5 behave in the case of the analysis of vibration processes excited by the shock method, it is possible to use simulation modeling. For this purpose, it is necessary to be able to obtain realizations of vibrations arising in the nodes of electrical equipment as a result of shock excitation. Let us consider in detail the method of obtaining such realizations.

We will use the vibration model in the form (6.21) with components (6.22), and the kernel of each component is given by expression (6.23).

Let us consider in detail the excitation properties of the system $\zeta(\tau) = \eta'(\tau)$. At impact there is a short-term process of transferring the energy of the impact hammer (or other shock application device) to the body of the node under study. The duration of impact, and hence the shape of the spectrum at shock excitation, is determined by the mass and stiffness of both the impact hammer and the material of the structure. The impact force spectrum is continuous, with maximum amplitude at 0 Hz and a subsequent decrease with increasing frequency. The duration of the impact is usually much shorter than the duration of the realizations registered in the time domain. In the initial approximation, the duration of the shock can be considered infinitesimal, and the applied force can be modeled by a delta impulse, which is known to have an equal-dimensional spectrum. But for a more adequate model, it is necessary to take into account that in real conditions the impact spectrum is limited.

In radio engineering, several "standard" pulse shapes are used: rectangular, triangular, trapezoidal, exponential, bell-shaped. In terms of shape and spectral characteristics, the bell-shaped pulse can be considered the most suitable for modeling the shock impact, which is approximated by the expression:

$$a(t) = Ae^{-2.77t^2 / t_i^2},$$

where t_i—so-called active pulse duration; A is pulse amplitude.

The main energy of the bell-shaped pulse is concentrated in the frequency region from 0 to $1/t_i$ Hz.

In practice, the "ideal" pulse shape is distorted due to various noises. For example, it can be parasitic mechanical vibrations of the impact hammer itself, various vibration and seismic waves caused by the operation of other equipment located near the unit to be diagnosed, and the like. Such noises can be taken into account in the model by additive superposition of the white noise type process.

Based on the proposed vibration model, a simulation program was developed to simulate the response of multi-resonant vibrating systems to shock. The program generates a shock in the form of a bell-shaped pulse with specified parameters, on which a discrete Gaussian white noise [74–78] is additively superimposed. The response of the system is modeled by expression (6.12), that is, by digital filtering of such an impact using the final impulse response calculated by expressions (6.23) and (6.21). The thus obtained realization of a discrete linear random process is a model of experimentally measured shock oscillations.

As an example, Fig. 6.3 shows the estimation of the power spectrum of the modelled realization. The values of parameters set during modelling are given in Table 6.1. In this case, the values of all weight a_j coefficients in formula (6.21) were equal to 1.

For comparison, we present graphs of power spectra estimates obtained as a result of measuring and analyzing several responses of real objects to shock. Measurements were carried out with the help of a laboratory sample of the IMS, consisting of a personal computer, a device for converting analogue signals into a digital code, a block for pre-processing analogue signals, and accelerometers for converting vibration signals into electrical ones. The input influences were applied using an impact hammer in the form of shock excitation. In detail, the structure and use of the laboratory sample of IMS of shock vibrodiagnostics, as well as some experimental data obtained with its help, are considered in Chaps. 7 and 8 of this book.

The recorded realizations of vibration signals were processed using specially developed software, resulting in plots of correlation function estimates and power spectra estimates of the measured vibration signals.

Some of them are shown in Figs. 6.4 and 6.5.

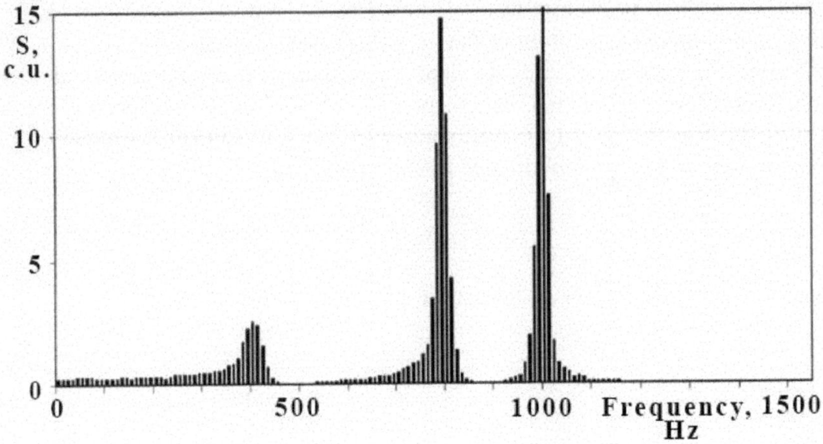

Fig. 6.3 Estimation of the power spectrum of modelled shock oscillations

Table 6.1 Parameter values set during modeling

Number of resonances	3		
Parameters of oscillating systems	1	2	3
β_j, 1/c	125	62.5	41.67
ω_j, rad/s	2582	5000	6262.24
ψ_j, rad/s	2579	4999.6	6262.10
Active pulse duration, ms	0.5		
Pulse amplitude, c.u	1		
RMS deviation of noise, u.s.e	0.02		
Sampling step, ms	0.1		
Realization length, s	0.1		

From the analysis of the graphs of estimates of correlation functions, we can see that the oscillations fade with time, and it can be assumed that the decrease in amplitude occurs according to a law close to exponential.

As can be seen in the graphs of spectral power estimates, the oscillations have a well-defined resonance character, and the number of resonances is different for different nodes. For example, in the vibration spectrum of the reinforced concrete support of the air switch type VV-220 (Fig. 6.4) is observed one clear resonance peak, while in the vibration spectrum of the tank mounting beam of the same switch (Fig. 6.5)—two clear peaks. It should be noted that among the studied realizations there were cases when up to ten clear resonance peaks could be distinguished in the spectra.

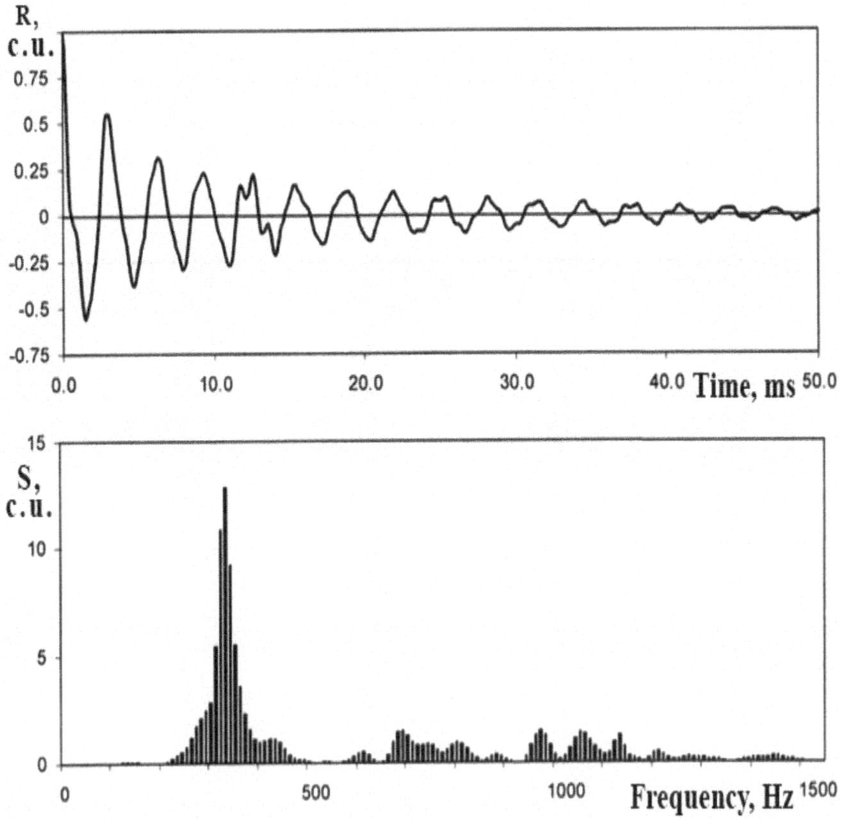

Fig. 6.4 Normalized estimation of correlation function (top) and estimation of the power spectrum (bottom) of shock oscillations of reinforced concrete support of air switch type VV-220

Visually comparing the above graphs, we can conclude that the character of the power spectrum of the modeled process is qualitatively close to the character of the spectra obtained as a result of measurements on real objects.

To compare how numerical results obtained by simulation modeling, correspond to parameter estimates derived from the analysis of real signals, the adequacy of the proposed mathematical model was also tested on physical models—test samples. Several steel parts of different geometric shapes were used as research objects. Because they have different shapes and sizes, the parameters of their vibrations are also different.

A steel rod with a diameter of about 1.5 cm and a length of about 1 m was chosen as the first test sample. The experiments were conducted under the following conditions:

(1) the rod was suspended on a metal cable;
(2) the rod was fixed on a soft support.

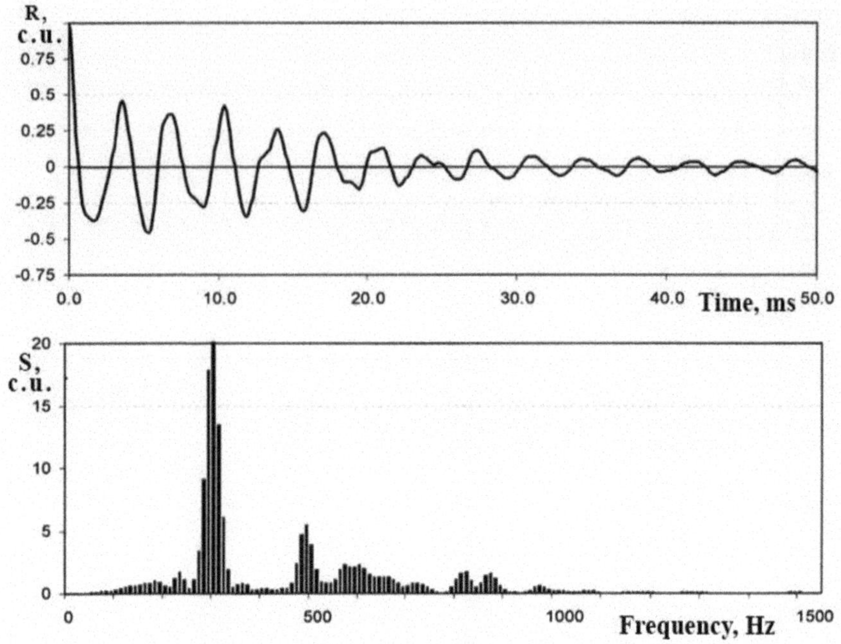

Fig. 6.5 Normalized estimation of the correlation function (top) and power spectrum (bottom) of shock vibrations of the beam of the tank attachment of the air circuit breaker type BB-220

This was done to determine the resonance parameters of the rod at different damping vibrations, which should model different technical states of the equipment being diagnosed. For example, the rod can act as a physical model of a part of the structure of the diagnostic object, which is securely fastened in a serviceable condition. As a result of a defect, the fastening is broken, and this part of the structure is released.

Let's set the following task: to find a diagnostic parameter to detect this type of defect.

In the course of the experiment, several measurements were made, and the tests were carried out with different impact forces. A typical view of the normalized evaluation of the power spectrum of rod vibration is shown in Figs. 6.6 and 6.7.

Special software was created to obtain numerical values of the estimates of resonant vibration parameters from the estimates of their power spectrum. The algorithm of its operation consists of searching for frequency maxima over a given frequency range and calculating the estimates $\tilde{\beta}_i$ and $\tilde{\psi}_i$. The relations underlying this program are discussed in detail in paragraph 3.6.

The results of the program are given in Tables 6.2 and 6.3. Comparing the corresponding lines from these two tables, it is easy to see that the differences in resonance frequencies between the two cases under consideration are very small. At the same time, the estimates for $\tilde{\beta}_i$ differ significantly. This means that in the considered case

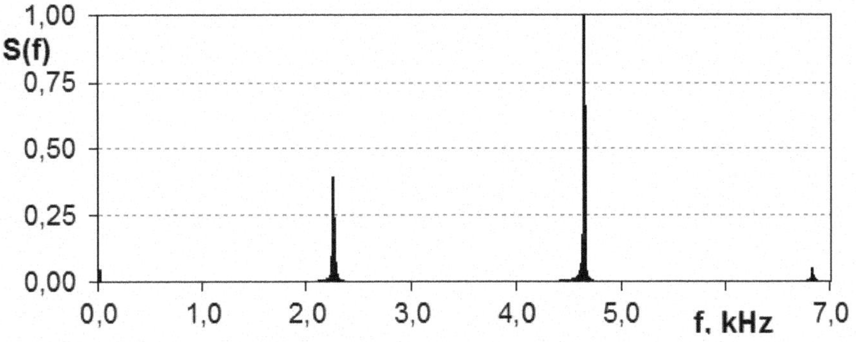

Fig. 6.6 Estimation of the vibration power spectrum of the suspended rod

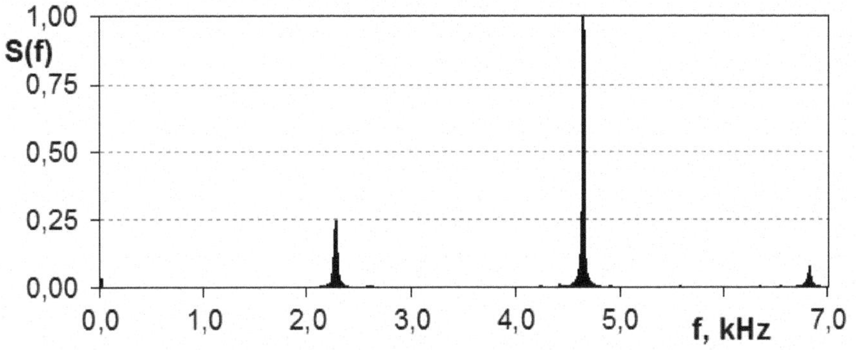

Fig. 6.7 Estimation of the vibration power spectrum of the rod, fixed on a soft support

the values $\tilde{\beta}_i$ can be used as a diagnostic parameter determining the condition of the rod fastening.

Table 6.2 Resonance parameters of the suspended rod (experimental results)

i	$\tilde{\omega}_{\max}$	$\tilde{\omega}_{h-}$	$\tilde{\omega}_{h+}$	$\tilde{\beta}_i$	$\tilde{\psi}_i$
1	14,141.39	14,095.68	14,184.84	44.58	14,141.46
2	29,183.98	29,147.73	29,212.69	32.47	29,184.00
3	42,922.70	42,846.74	43,004.78	79.03	42,922.77

Table 6.3 Parameters of resonances of the fixed rod (experimental results)

I	$\tilde{\omega}_{\max}$	$\tilde{\omega}_{h-}$	$\tilde{\omega}_{h+}$	$\tilde{\beta}_i$	$\tilde{\psi}_i$
1	14,246.85	14,130.40	14,339.38	104.40	14,247.23
2	29,183.98	29,135.50	29,228.25	46.37	29,184.02
3	42,893.94	42,775.08	43,018.30	121.62	42,894.11

Using numerical values of resonance frequencies and software, simulation modeling of vibrations in the fixed rod was carried out. Qualitatively, the graph of the power spectrum of the simulated vibration process does not differ from the graph shown in Fig. 6.7. The numerical values of the estimates of the resonance parameters of the simulated vibrations are given in Table 6.4.

The results obtained by modeling are quite close to the measurement results. This confirms that the used model adequately describes the investigated processes in real equipment. The obtained differences in the parameter estimates can be explained by the fact that the estimates are random variables and therefore cannot coincide exactly.

A steel part in the form of a hollow cylinder with several holes was chosen as the second test sample. The outer diameter of the cylinder was about 25 cm and its height was about 10 cm.

The graph of the vibration power spectrum estimation of the suspended cylinder is shown in Fig. 6.8. It contains only one well-defined resonance peak. This fact allows us to use the relations to estimate the resonance parameters in the time domain (i.e., by estimating the correlation function).

Figure 6.9 shows a fragment of the graph of the estimation of the normalized correlation function of the cylinder vibrations. The point estimates $\tilde{s}_{0,k}$ (zeros of the correlation function) and $\tilde{s}_{m,k}$ (maxima of the correlation function), which are included in the expressions for the estimates of the resonance parameters $\tilde{\beta}_i$ and $\tilde{\psi}_i$ are also marked. Note that the total length of the obtained estimate is more than 82 ms. The function oscillates with a relatively large frequency and decays relatively

Table 6.4 Parameters of resonances of the fixed rod (modeling results)

I	$\tilde{\omega}_{max}$	$\tilde{\omega}_{h-}$	$\tilde{\omega}_{h+}$	$\tilde{\beta}_i$	$\tilde{\psi}_i$
1	14,237.26	14,142.03	14,329.47	93.71	14,237.57
2	29,193.57	29,138.76	29,229.29	45.25	29,193.61
3	42,913.11	42,773.67	42,981.31	103.73	42,913.24

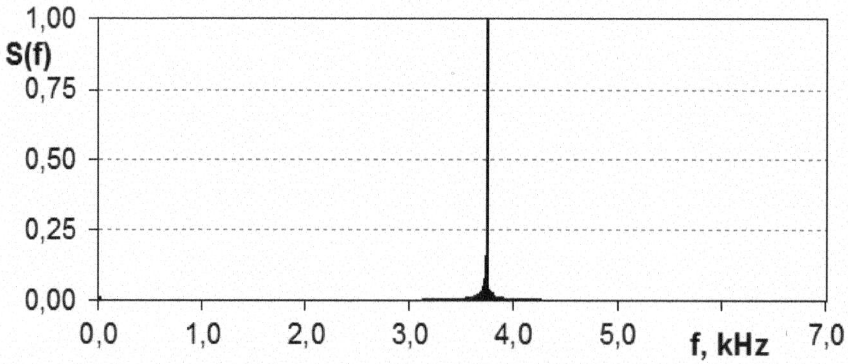

Fig. 6.8 Estimation of the vibration power spectrum of the suspended cylinder

slowly. To make the illustration more informative, only a few first oscillations are shown in Fig. 6.9.

The estimates of the parameters obtained with the help of the program are given in Table 6.5, with the estimates obtained both in the time domain and the frequency domain. As can be seen, different methods of obtaining estimates give quite close results.

Thus, it can be seen from the above that the vibration wave in the massive nodes of the EE caused by the shock can be represented as a linear random process with a kernel, which is a linear combination of impulse reactions of second-order oscillatory systems, and an input excitation in the form of a noisy bell-shaped pulse. Simulation modeling using the proposed model gives results that are close enough to those obtained experimentally, which confirms the validity of the assumptions made and the adequacy of the chosen model. The obtained results are important for substantiation of diagnostic parameters and selection of statistical methods of processing measured realizations at the creation of IMS of statistical shock diagnostics of EE.

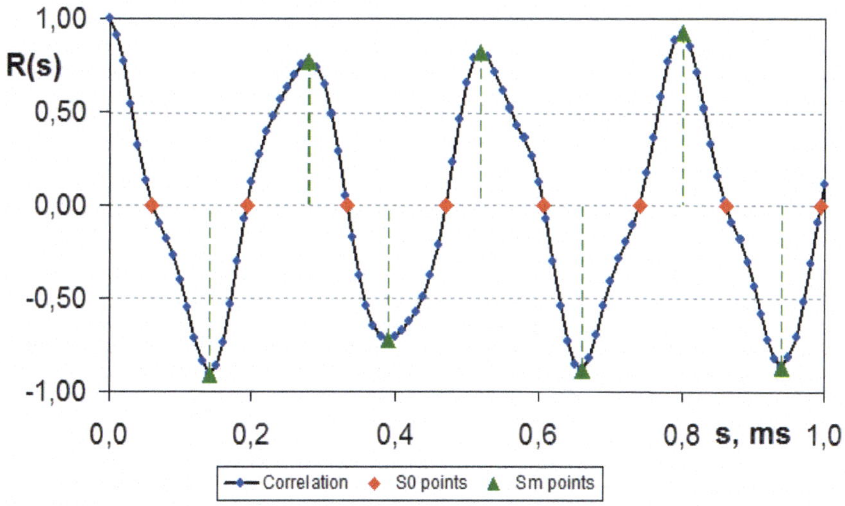

Fig. 6.9 Correlation function of cylinder vibrations (dots—estimation of the correlation function; rhombuses—estimations of points $s_{0,k}$; triangles—estimates of points $s_{m,k}$)

Table 6.5 Estimates of resonance parameters of the suspended cylinder obtained using different approaches

Method of obtaining estimates	$\tilde{\beta}_1$	$\tilde{\psi}_1$
Time domain, estimates of $s_{m,k}$ maximum points of the correlation function	36.21	23,569.91
Time domain, estimates on the points $s_{0,k}$ of the zeros of the correlation function	–	23,575.13
Frequency domain	37.06	23,584.98

6.3.3 Dynamics of Cylinder Powers on the Shaft of the Diesel-Electric Generator

While modeling the dynamics of cylinder power on the shaft of the diesel-electric generator (DEG), it is necessary to specify the type of generator under study. Since the experimental verification of the developed methodology of diagnostics of cylinder-piston group (CPG) was carried out on the DEG of 5D70 type (AD2000C), all issues related to the modeling of this process, will be considered about this type of generator [79–81].

This unit is designed for operation at important national economic facilities as an autonomous automated power source of three-phase alternating current with a frequency of 50 Hz. Since the IMS and methods considered in this work are focused on the diagnosis of the CPG of DEG, the design features of this DEG are of fundamental importance. So, the investigated DEG is a 16-cylinder four-stroke with a V-shaped arrangement of cylinders. Consequently, taking into account the known order of cylinder operation, one full cycle of each cylinder is carried out for two revolutions of the diesel crankshaft.

The pie diagram of the diesel CPG operation has been discussed above and illustrated in Fig. 6.10.

To simulate the sequence of pulses characterizing the energy pulses during the operating cycle in the diesel engine 5D70, we can use a function of the form $xe^{-\alpha x}$, where α characterizes the degree of attenuation of the pulse.

Taking into account this function, the sequence of energy pulses for one operating cycle of the diesel engine can be represented by the expression

$$\xi_c(t) = \sum_{k=1}^{16} [t - (k-1)h] \cdot e^{-\alpha[t-(k-1)h]} \cdot U[t - (k-1)h], \tag{6.24}$$

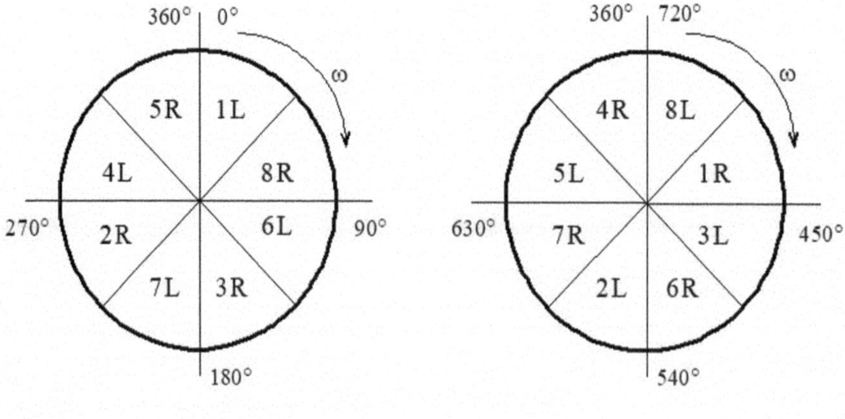

Fig. 6.10 Circular diagram of DEG's CPG work

where k—cylinder number of the research DEG, h—duration of the working stroke in the corresponding cylinder, $U(t)$—unit Heaviside function, which is defined as follows:

$$U(t) = \begin{cases} 0 \; at \; t < 0, \\ 1 \text{ at } t \geq 0. \end{cases}$$

Simulation of the failure of the corresponding cylinder is carried out by zeroing at a certain time interval the function characterizing the impulse of energy coming from the work of the cylinder under study. This can be illustrated as follows.

So, if the first cylinder failed, it can be modeled by zeroing the first summand included in (6.24), i.e. at $k = 1$. Similarly, the failure of the remaining cylinders is modeled. Analytically, it can be represented as follows

$$t \cdot e^{-\alpha t} U(t) = 0 \text{ at } k = 1;$$
$$(t - h)e^{-\alpha(t-h)} U(t - h) = 0 \text{ at } k = 2;$$
$$\ldots\ldots\ldots\ldots\ldots\ldots\ldots\ldots\ldots\ldots\ldots\ldots$$
$$(t - 15h)e^{-\alpha(t-15h)} U(t - 15h) = 0 \text{ at } k = 16.$$

Modeling of the operation of the 5D70 type DEG was carried out with the help of a special program based on the expression (6.24). Figure 6.11 shows the pulse sequence modeled by expression (6.24) using the computer at $\alpha = 2$ and $h = 1$. This sequence corresponds to one operating cycle of the diesel engine, i.e. two full revolutions of the crankshaft. The duration of the presented pulse sequence as T_s, however, and the duration of the T_i energy pulse in each cylinder, depends on the angular speed of rotation of the shaft of the studied DEG.

An estimate of the frequency region in which the main frequency component is caused by the operation of each cylinder can be determined as follows [27, 82]. Based on the designations of periods T_s and T_i, presented in Fig. 6.11, their corresponding frequencies are determined by simple expressions

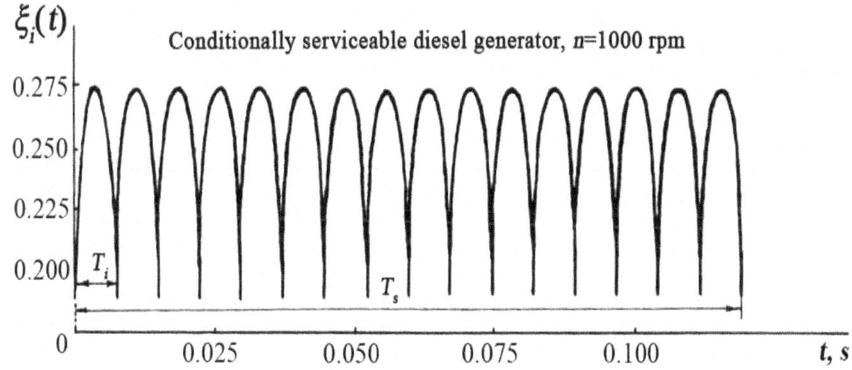

Fig. 6.11 Sequence of energy pulses for a conditionally serviceable DEG for one operating cycle

$$f_s = \frac{1}{T_s}, T_i = \frac{T_s}{16}, f_i = \frac{1}{T_i} = \frac{16}{T_s}.$$

If we assume that the nominal number of revolutions of the crankshaft of type 5D70 DEG is $n = 1000$ rpm and this diesel engine is a four-stroke, i.e. the complete working cycle in all 16 cylinders occurs for two full revolutions of the diesel crankshaft, then for the nominal speed there are the following ratios

$$n_1 = \frac{n}{2} = 500\,\text{rpm}; \quad f_s = \frac{n_1}{60} = 8.3\,\text{Hz};$$
$$f_i = 16 f_s = 132.8\,\text{Hz}.$$

Amplitude-frequency (AFC) and phase-frequency (PFC) characteristics, obtained using the Fourier transform [83–87] of the sequence of Fig. 6.11, are shown in Fig. 6.12a, b, respectively. They give a visual representation of the distribution of the main frequency components, as well as their phases. The dependences plotted in Fig. 6.12 are obtained by modeling the conditionally serviceable state of all cylinders of the investigated DEG.

One of the most important moments in solving the diagnostics problem is the determination of diagnostic signs, i.e., the parameters that allow us to unambiguously determine the change in the state of the diagnosed unit by the change like the processes

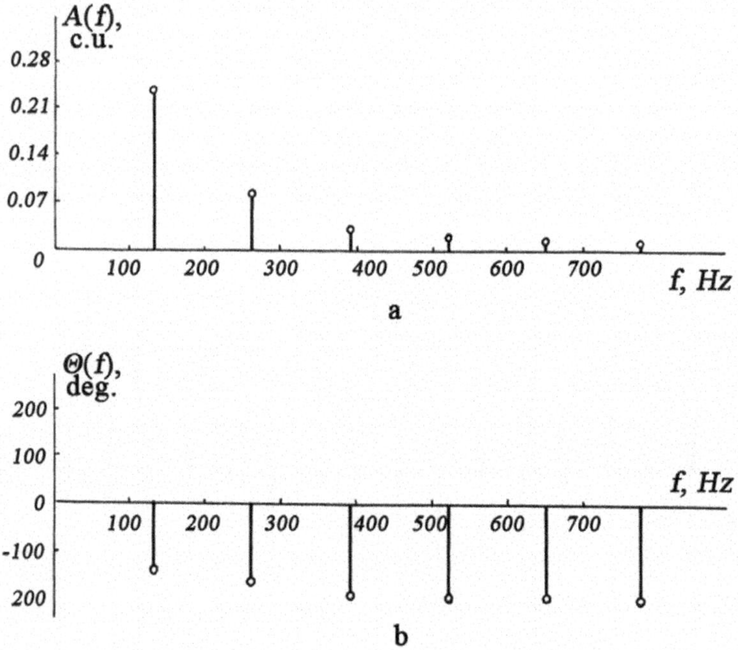

Fig. 6.12 AFC (**a**) and PFC (**b**) of a sequence of energy pulses for a conditionally serviceable DEG

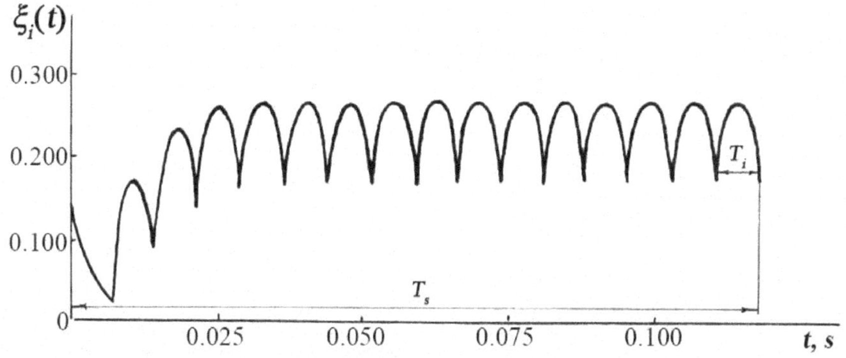

Fig. 6.13 Sequence of energy pulses at simulation of failure of the first left cylinder of the DEG

occurring in this unit, let us study on the proposed model (6.24) the behavior of its AFC and PFC in case of failure of one or a group of cylinders.

Figure 6.13 shows the pulse sequence while simulating the failure of the first left cylinder, and Fig. 6.14a, b—respectively, the AFC and FFC of this sequence. As can be seen from the presented dependences, the zeroing of one of the pulses leads to a change in the obtained characteristics. In the AFC it is manifested in the appearance of subharmonics in the low-frequency part of the spectrum and in some decrease in the amplitude of the main harmonic. It should be immediately noted that this characteristic remains unchanged when simulating the failure of any cylinder number.

Thus, by the AFC of the sequence (6.24) we can judge only the failure of any of the cylinders, without specifying which cylinder.

To answer the question, of which cylinder has failed, we can use the PFC of the research sequence (6.24). As can be seen from the comparison of Figs. 14b and 15a, ..., *n*, where the PFC of the sequence (6.24) at the failure of different numbers of cylinders is presented, simulation of failure of a certain cylinder leads to a corresponding change in the PFC, i.e. a certain number of the failed cylinder corresponds to a quite definite PFC.

Thus, taking into account the results of modeling as diagnostic signs in diagnostics of the DEG's CPG can be used:

- to determine the failure of any one cylinder—the presence of subharmonics $d_j, j = \overline{1, 15}$ in the AFC of the angular speed of shaft rotation;
- to determine the number of cylinders that have failed—phase angles $\alpha_j^0, j = \overline{1, 15}$ of subharmonics in the PFC of the angular velocity of shaft rotation.

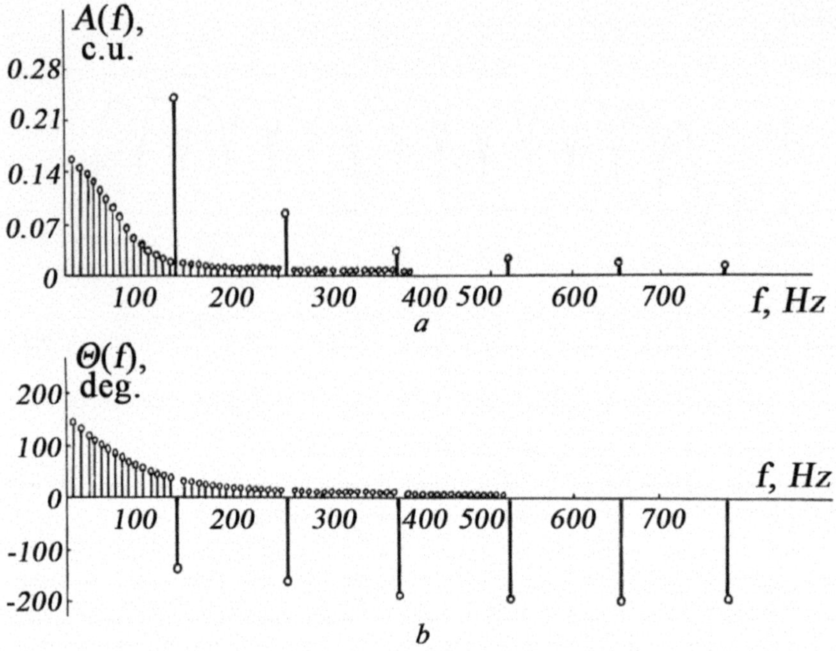

Fig. 6.14 AFC (**a**) and PFC (**b**) of the energy pulse sequence while simulating the failure of the first left cylinder of the DEG

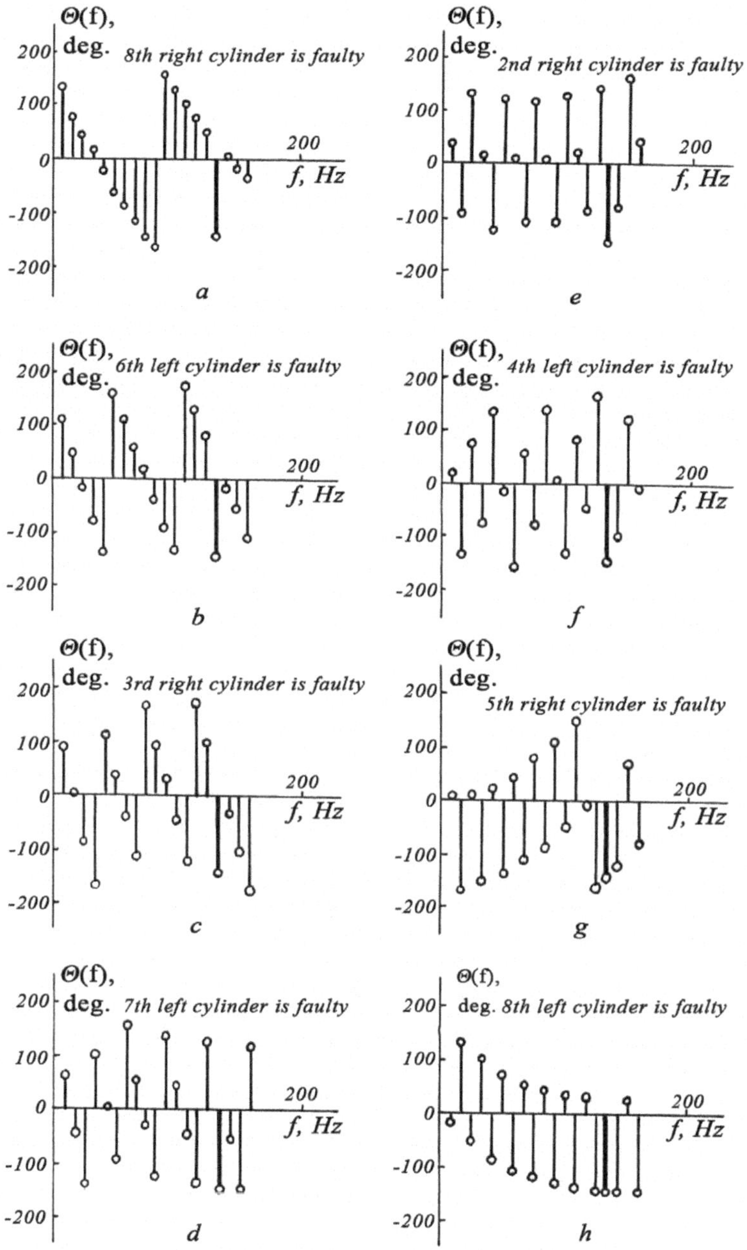

Fig. 6.15 PFC of the sequence of energy pulses while simulating the failure of one of the DEG cylinders

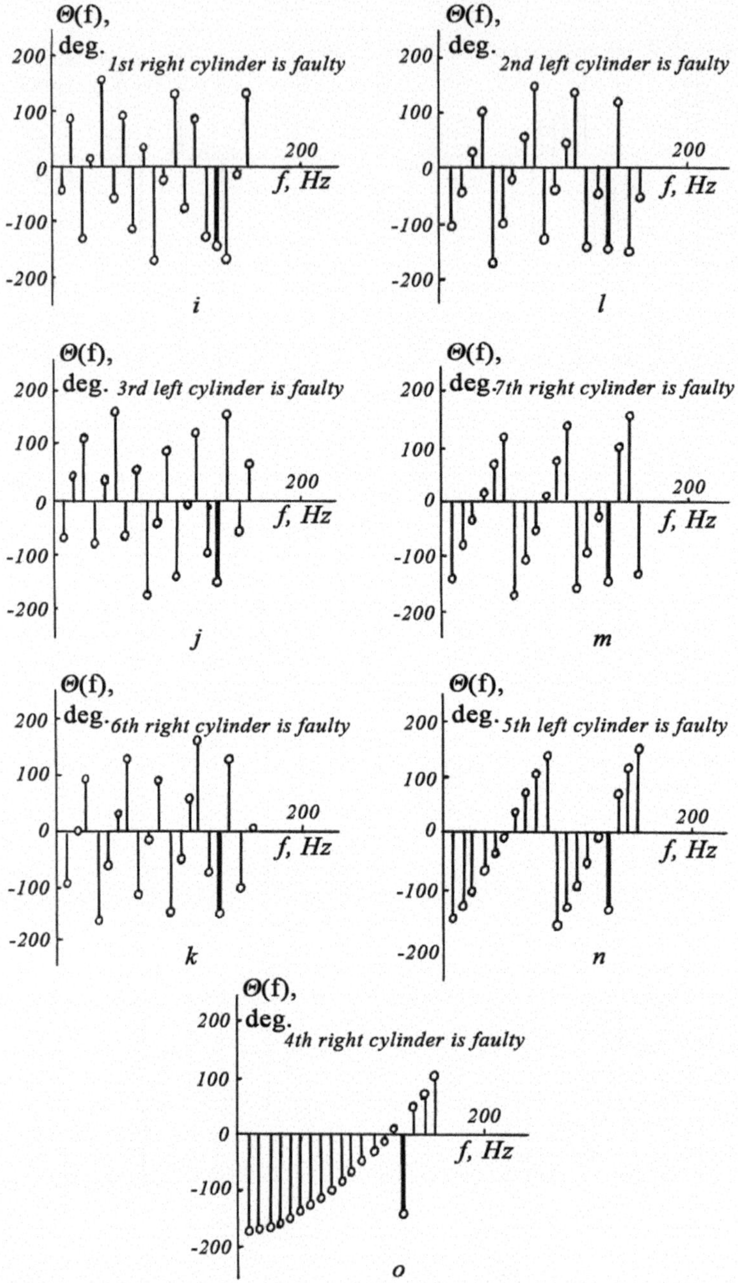

Fig. 6.15 (continued)

References

1. Mellit, A., Tina, G.M., Kalogirou, S.A.: Fault detection and diagnosis methods for photovoltaic systems: a review. Renew. Sustain. Energy Rev. **91**, 1–17 (2018)
2. Babak, V.P., Babak, S.V., Myslovych, M.V., Zaporozhets, A.O., Zvaritch, V.M.: Technical provision of diagnostic systems. In: Diagnostic Systems for Energy Equipments, pp. 91–133 (2020)
3. Babak, V.P., Babak, S.V., Eremenko, V.S., Kuts, Y.V., Myslovych, M.V., Scherbak, L.M., Zaporozhets, A.O.: Models and measures for standardless measurements of the composite materials characteristics. In: Models and Measures in Measurements and Monitoring, pp. 157–190 (2021)
4. Zaporozhets, A., Babak, V., Sverdlova, A., Isaienko, V., Babikova, K.: Development of a system for diagnosing heat power equipment based on IEEE 802.11s. In: Systems, Decision and Control in Energy II, pp. 141–151. Springer International Publishing, Cham (2021)
5. Chen, H., Jiang, B., Ding, S.X., Huang, B.: Data-driven fault diagnosis for traction systems in high-speed trains: a survey, challenges, and perspectives. IEEE Trans. Intell. Transp. Syst. **23**(3), 1700–1716 (2020)
6. Coakley, D., Raftery, P., Keane, M.: A review of methods to match building energy simulation models to measured data. Renew. Sustain. Energy Rev. **37**, 123–141 (2014)
7. Babak, V.P., Babak, S.V., Myslovych, M.V., Zaporozhets, A.O., Zvaritch, V.M.: Simulation and software for diagnostic systems. In: Diagnostic Systems for Energy Equipments, pp. 71–90 (2020)
8. Babak, V.P., Babak, S.V., Myslovych, M.V., Zaporozhets, A.O., Zvaritch, V.M.: Methods and models for information data analysis. In: Diagnostic Systems for Energy Equipments, pp. 23–70 (2020)
9. Vogl, G.W., Weiss, B.A., Helu, M.: A review of diagnostic and prognostic capabilities and best practices for manufacturing. J. Intell. Manuf. **30**, 79–95 (2019)
10. Triki-Lahiani, A., Abdelghani, A.B.B., Slama-Belkhodja, I.: Fault detection and monitoring systems for photovoltaic installations: a review. Renew. Sustain. Energy Rev. **82**, 2680–2692 (2018)
11. Babak, V., Zaporozhets, A., Kuts, Y., Myslovych, M., Fryz, M., Scherbak, L.: Models and characteristics of identification of noise stochastic signals of research objects. In: CEUR Workshop Proceedings, vol. 3309, pp. 349–362 (2022, December)
12. Babak, V., Scherbak, L., Kuts, Y., Zaporozhets, A.: Information and measurement technologies for solving problems of energy informatics. In: The 1st International Workshop on Information Technologies: Theoretical and Applied Problems 2021. CEUR Workshop Proceedings, vol. 3039, pp. 24–31 (2021, December)
13. Theunissen, F.E., Elie, J.E.: Neural processing of natural sounds. Nat. Rev. Neurosci. **15**(6), 355–366 (2014)
14. Harne, R.L., Wang, K.W.: A review of the recent research on vibration energy harvesting via bistable systems. Smart Mater. Struct. **22**(2), 023001 (2013)
15. Yuan, F.: Design techniques for time-mode noise-shaping analog-to-digital converters: a state-of-the-art review. Analog Integr. Circ. Sig. Process **79**, 191–206 (2014)
16. Babak, V.P., Babak, S.V., Eremenko, V.S., Kuts, Y.V., Myslovych, M.V., Scherbak, L.M., Zaporozhets, A.O.: Models and measures for the diagnosis of electric power equipment. In: Models and Measures in Measurements and Monitoring, pp. 99–126 (2021)
17. Zheng, Y., Zhao, Y., Zhou, N., Wang, H., Jiang, D.: A short review of some analog-to-digital converters resolution enhancement methods. Measurement **180**, 109554 (2021)
18. Li, D., Zhao, X., Liu, S., Liu, M., Ding, R., Liang, Y., Zhu, Z.: Radio frequency analog-to-digital converters: Systems and circuits review. Microelectron. J. **119**, 105331 (2022)
19. Gao, Z., Liu, X.: An overview on fault diagnosis, prognosis and resilient control for wind turbine systems. Processes **9**(2), 300 (2021)

20. Babak, V.P., Babak, S.V., Eremenko, V.S., Kuts, Y.V., Myslovych, M.V., Scherbak, L.M., Zaporozhets, A.O.: Problems and features of measurements. In: Models and Measures in Measurements and Monitoring, pp. 1–31 (2021)
21. Hare, J., Shi, X., Gupta, S., Bazzi, A.: Fault diagnostics in smart micro-grids: a survey. Renew. Sustain. Energy Rev. **60**, 1114–1124 (2016)
22. Herrero-Collantes, M., Garcia-Escartin, J.C.: Quantum random number generators. Rev. Mod. Phys. **89**(1), 015004 (2017)
23. James, F., Moneta, L.: Review of high-quality random number generators. Comput. Softw. Big Sci. **4**, 1–12 (2020)
24. Yu, F., Li, L., Tang, Q., Cai, S., Song, Y., Xu, Q.: A survey on true random number generators based on chaos. Discret. Dyn. Nat. Soc. **2019**, 1–10 (2019)
25. Kaur, S., Gupta, D.: A review of image encryption schemes based on the chaotic map. Int. J. Comput. Technol. Appl. **5**(1), 144 (2014)
26. Sundararajan, A.D.D., Rajashree, R.: A comprehensive survey on lightweight asymmetric key cryptographic algorithm for resource constrained devices. ECS Trans. **107**(1), 7457 (2022)
27. Babak, V.P., Babak, S.V., Eremenko, V.S., Kuts, Y.V., Myslovych, M.V., Scherbak, L.M., Zaporozhets, A.O.: Models and measures for measuring random angular quantities. In: Models and Measures in Measurements and Monitoring, pp. 61–97 (2021)
28. Babak, V.P., Babak, S.V., Eremenko, V.S., Kuts, Y.V., Myslovych, M.V., Scherbak, L.M., Zaporozhets, A.O.: Models of measuring signals and fields. In: Models and Measures in Measurements and Monitoring, pp. 33–59 (2021)
29. Salles, R., Belloze, K., Porto, F., Gonzalez, P.H., Ogasawara, E.: Nonstationary time series transformation methods: an experimental review. Knowl.-Based Syst. **164**, 274–291 (2019)
30. Turitsyn, S.K., Prilepsky, J.E., Le, S.T., Wahls, S., Frumin, L.L., Kamalian, M., Derevyanko, S.A.: Nonlinear Fourier transform for optical data processing and transmission: advances and perspectives. Optica **4**(3), 307–322 (2017)
31. Schneider, M.: A review of nonlinear FFT-based computational homogenization methods. Acta Mech. **232**(6), 2051–2100 (2021)
32. Zaporozhets, A.O.: Correlation analysis between the components of energy balance and pollutant emissions. Water Air Soil Pollut. **232**, 1–22 (2021)
33. Babak, V.P., Babak, S.V., Myslovych, M.V., Zaporozhets, A.O., Zvaritch, V.M.: Principles of construction of systems for diagnosing the energy equipment. In: Diagnostic Systems for Energy Equipments, pp. 1–22 (2020)
34. Ding, S.X.: Data-Driven Design of Fault Diagnosis and Fault-Tolerant Control Systems, pp. 117–199. Springer London, London (2014)
35. Chen, Z., Ding, S.X., Peng, T., Yang, C., Gui, W.: Fault detection for non-Gaussian processes using generalized canonical correlation analysis and randomized algorithms. IEEE Trans. Industr. Electron. **65**(2), 1559–1567 (2017)
36. Dey, S., Dey, T., Kundu, D.: Two-parameter Rayleigh distribution: different methods of estimation. Am. J. Math. Manag. Sci. **33**(1), 55–74 (2014)
37. Bidaoui, H., El Abbassi, I., El Bouardi, A., Darcherif, A.: Wind speed data analysis using Weibull and Rayleigh distribution functions, case study: five cities northern Morocco. Procedia Manuf. **32**, 786–793 (2019)
38. Gomes, A.E., da-Silva, C.Q., Cordeiro, G.M., Ortega, E.M.: A new lifetime model: the Kumaraswamy generalized Rayleigh distribution. J. Stat. Comput. Simul. **84**(2), 290–309 (2014)
39. Eamaz, A., Yeganegi, F., Soltanalian, M.: Modified arcsine law for one-bit sampled stationary signals with time-varying thresholds. In: ICASSP 2021–2021 IEEE International Conference on Acoustics, Speech and Signal Processing (ICASSP), pp. 5459–5463. IEEE (2021, June)
40. Eamaz, A., Yeganegi, F., Soltanalian, M.: Covariance recovery for one-bit sampled stationary signals with time-varying sampling thresholds. Signal Process. **206**, 108899 (2023)
41. Eamaz, A., Yeganegi, F., Soltanalian, M.: Covariance recovery for one-bit sampled non-stationary signals with time-varying sampling thresholds. IEEE Trans. Signal Process. **70**, 5222–5236 (2022)

42. Majumder, S.: One-bit spectrum sensing using Gustafson-Kessel fuzzy clustering for cognitive radio network. Phys. Commun. **62**, 102239 (2024)
43. Kwak, S.G., Kim, J.H.: Central limit theorem: the cornerstone of modern statistics. Korean J. Anesthesiol. **70**(2), 144 (2017)
44. Li, X., Ding, P.: General forms of finite population central limit theorems with applications to causal inference. J. Am. Stat. Assoc. **112**(520), 1759–1769 (2017)
45. Heyde, C.C.: Central limit theorem. In: Wiley StatsRef: Statistics Reference Online (2014)
46. Levin, K., Athreya, A., Tang, M., Lyzinski, V., Priebe, C.E.: A central limit theorem for an omnibus embedding of multiple random dot product graphs. In: 2017 IEEE international conference on data mining workshops (ICDMW), pp. 964–967. IEEE (2017, November)
47. Liu, H., Ren, J., Yang, Y.: Randomization-based joint central limit theorem and efficient covariate adjustment in randomized block 2 k factorial experiments. J. Am. Stat. Assoc. **119**(545), 136–150 (2024)
48. Fritzson, P.: Principles of Object-Oriented Modeling and Simulation with Modelica 3.3: A Cyber-Physical Approach. Wiley (2014)
49. Kline, R.B.: Principles and Practice of Structural Equation Modeling. Guilford Publications (2023)
50. Wang, W., Den Brinker, A.C., Stuijk, S., De Haan, G.: Algorithmic principles of remote PPG. IEEE Trans. Biomed. Eng. **64**(7), 1479–1491 (2016)
51. Najafzadeh, N., Fotouhi, M., Kasaei, S.: Object tracking using Kalman filter with adaptive sampled histogram. In: 2015 23rd Iranian Conference on Electrical Engineering, pp. 781–786. IEEE (2015, May)
52. Jawaid, A., Seewig, J.: Discrete filter and non-gaussian noise for fast roughness simulations with gaussian processes. In: International Research Training Group Conference on Physical Modeling for Virtual Manufacturing Systems and Processes, pp. 1–15. Springer International Publishing, Cham (2023, June)
53. Akashi, F., Liu, Y., Taniguchi, M.: An Empirical Likelihood Approach for Symmetric α-Stable Processes (2015)
54. Choi, B., Bergés, M., Bou-Zeid, E., Pozzi, M.: Short-term probabilistic forecasting of meso-scale near-surface urban temperature fields. Environ Model Softw. **145**, 105189 (2021)
55. Sause, M.G., Richler, S.: Finite element modelling of cracks as acoustic emission sources. J. Nondestr. Eval. **34**, 1–13 (2015)
56. Zhao, L., Kang, L., Yao, S.: Research and application of acoustic emission signal processing technology. IEEE Access **7**, 984–993 (2018)
57. Nazarchuk, Z., Skalskyi, V., Serhiyenko, O.: Acoustic emission. In: Foundations of Engineering Mechanics (2017)
58. Crandall, S.H., Mark, W.D.: Random Vibration in Mechanical Systems. Academic Press (2014)
59. Babak, V., Zaporozhets, A., Kovtun, S., Myslovych, M., Kuts, Y., Scherbak, L.: Information support for identification of the technical state of electric power facilities. In: Systems, Decision and Control in Energy V, pp. 129–153. Springer Nature Switzerland, Cham (2023)
60. Babak, V., Zaporozhets, A., Kuts, Y., Scherbak, L., Eremenko, V.: Application of material measure in measurements: theoretical aspects. In: Systems, Decision and Control in Energy II, pp. 261–269. Springer International Publishing, Cham (2021)
61. Lee, H.J., Zhang, S., Bar-Cohen, Y., Sherrit, S.: High temperature, high power piezoelectric composite transducers. Sensors **14**(8), 14526–14552 (2014)
62. Wang, P., Wang, Z., Youn, B.D., Lee, S.: Reliability-based robust design of smart sensing systems for failure diagnostics using piezoelectric materials. Comput. Struct. **156**, 110–121 (2015)
63. Babak, V., Bohachev, I., Zaporozhets, A., Khaidurov, V., Havrysh, V., Kalinichenko, A.: Some features of modeling ultrasound propagation in non-destructive control of metal structures based on the magnetostrictive effect. Electronics **12**(3), 477 (2023)
64. An, Y.K., Kim, M., Sohn, H.: Piezoelectric transducers for assessing and monitoring civil infrastructures. In: Sensor technologies for civil infrastructures, pp. 86–120. Woodhead Publishing (2014)

65. Jiang, X., Zhang, X., Tang, T., Zhang, Y.: Electromechanical impedance based self-diagnosis of piezoelectric smart structure using principal component analysis and LibSVM. Sci. Rep. **11**(1), 11345 (2021)
66. Gómez-Aguilar, J.F., Atangana, A., Morales-Delgado, V.F.: Electrical circuits RC, LC, and RL described by Atangana-Baleanu fractional derivatives. Int. J. Circuit Theory Appl. **45**(11), 1514–1533 (2017)
67. Sarafraz, M.S., Tavazoei, M.S.: Passive realization of fractional-order impedances by a fractional element and RLC components: Conditions and procedure. IEEE Trans. Circuits Syst. I Regul. Pap. **64**(3), 585–595 (2016)
68. Sarafraz, M.S., Tavazoei, M.S.: Realizability of fractional-order impedances by passive electrical networks composed of a fractional capacitor and RLC components. IEEE Trans. Circuits Syst. I Regul. Pap. **62**(12), 2829–2835 (2015)
69. Ugwiri, M.A., Carratú, M., Paciello, V., Liguori, C.: Benefits of enhanced techniques combining negentropy, spectral correlation and kurtogram for bearing fault diagnosis. Measurement **185**, 110013 (2021)
70. Babak, V., Zaporozhets, A., Kuts, Y., Scherbak, L.: Some features of Hilbert transform and their use in energy informatics. Probl. Gener. Energy 1–2 (2022)
71. Tang, G., Tian, T.: Compound fault diagnosis of rolling bearing based on singular negentropy difference spectrum and integrated fast spectral correlation. Entropy **22**(3), 367 (2020)
72. Mauricio, A., Gryllias, K.: Bearing diagnostics using kurtosis spectral correlation based on cyclic modulation spectrum estimation. In: Turbo Expo: Power for Land, Sea, and Air, vol. 85987, p. V002T05A025. American Society of Mechanical Engineers (2022, June)
73. Babak, V., Zaporozhets, A., Zvaritch, V., Scherbak, L., Myslovych, M., Kuts, Y.: Models and measures in theory and practice of manufacturing processes. IFAC-PapersOnLine **55**(10), 1956–1961 (2022)
74. Jondral, F.K.: White gaussian noise–models for engineers. Frequenz **72**(5–6), 293–299 (2018)
75. Han, Y., Jiao, J., Mukherjee, R.: On estimation of L r-norms in Gaussian white noise models. Probab. Theory Relat. Fields **177**(3), 1243–1294 (2020)
76. Waubke, H., Kasess, C.H.: Gaussian closure technique applied to the hysteretic Bouc model with non-zero mean white noise excitation. J. Sound Vib. **382**, 258–273 (2016)
77. Sørbye, S.H., Rue, H.: Fractional Gaussian noise: Prior specification and model comparison. Environmetrics **29**(5–6), e2457 (2018)
78. Abreu, L.D.: Local maxima of white noise spectrograms and Gaussian Entire Functions. J. Fourier Anal. Appl. **28**(6), 88 (2022)
79. Hvozdeva, I., Myrhorod, V., Budashko, V., Shevchenko, V.: Problems of improving the diagnostic systems of marine diesel generator sets. In: 2020 IEEE 15th International Conference on Advanced Trends in Radioelectronics, Telecommunications and Computer Engineering (TCSET), pp. 350–354. IEEE (2020, February)
80. Onishchenko, O.A., Melnyk, O.M., Yarovenko, V.A., Aleksandrovska, N.I., Kurdiuk, S.V., Parmenova, D.G., Storchak, O.O.: Study of efficiency and advancement of marine engine oil purification and filtration technologies. J. Chem. Technol. **31**(4), 762–774 (2023)
81. Zacharewicz, M., Kniaziewicz, T.: Method of evaluation of the technical condition of the Diesel-electric unit. Diagnostyka **20**(2), 113–119 (2019)
82. Babak, V.P., Babak, S.V., Eremenko, V.S., Kuts, Y.V., Myslovych, M.V., Scherbak, L.M., Zaporozhets, A.O.: Examples of using models and measures on the circle. In: Models and Measures in Measurements and Monitoring, pp. 127–156 (2021)
83. Burriel-Valencia, J., Puche-Panadero, R., Martinez-Roman, J., Sapena-Bano, A., Pineda-Sanchez, M.: Short-frequency Fourier transform for fault diagnosis of induction machines working in transient regime. IEEE Trans. Instrum. Meas. **66**(3), 432–440 (2017)
84. Zaporozhets, A., Redko, O., Babak, V., Eremenko, V., Mokiychuk, V.: Method of indirect measurement of oxygen concentration in the air. Naukovyi Visnyk Natsionalnoho Hirnychoho Universytetu **5**, 105–114 (2018)
85. Zaporozhets, A.O., Khaidurov, V.V.: Mathematical models of inverse problems for finding the main characteristics of air pollution sources. Water Air Soil Pollut. **231**(12), 563 (2020)

86. Lin, H.C., Ye, Y.C., Huang, B.J., Su, J.L.: Bearing vibration detection and analysis using enhanced fast Fourier transform algorithm. Adv. Mech. Eng. **8**(10), 1687814016675080 (2016)
87. Khadersab, A., Shivakumar, S.: Vibration analysis techniques for rotating machinery and its effect on bearing faults. Procedia Manuf. **20**, 247–252 (2018)

Chapter 7
Information and Measuring Systems for Statistical Diagnostics of Electrical Equipment

Abstract This chapter explores the implementation and effectiveness of statistical diagnostics systems for monitoring and diagnosing electrical equipment (EE) components using information-measuring systems (IMSs). The chapter delves into the limitations of traditional diagnostic methods and how IMSs, through advanced data analysis and automation, overcome these limitations. By proposing a generalized structure for IMSs – incorporating sensors, processing units, and software components, the chapter showcases specific applications such as diagnosing rolling bearings and various other EE units including internal combustion engines and electrical machines. It emphasizes the role of statistical hypothesis testing, linear random process models, and the critical use of training sets in constructing accurate diagnostic decision rules. Experimental examples, including shock and vibration diagnostics, illustrate the system's capabilities in real-world applications. Through detailed analysis of errors including methodological, instrumental, discretization, and quantization errors, the chapter presents calculation relations for metrological control, thereby enhancing diagnostic accuracy. This comprehensive overview confirms the potential of IMSs to significantly improve the diagnostic processes of EE, highlighting the importance of statistical methods, advanced data processing, and the strategic selection of diagnostic signals in reducing errors and improving reliability.

Keywords Information-measuring system · Diagnostic system · Statistical analysis · Acoustic emission · Vibration diagnostics · Electric machine · Data processing · Software

7.1 Peculiarities of Construction of Statistical Diagnostic Systems

A relatively large number of devices and systems for the measurement and processing of random diagnostic signals accompanying the operation of various machines and mechanisms, as a rule, are oriented to the measurement of peak values of these

V. Babak et al., *Statistical Diagnostics of Electric Power Equipment*, Studies in Systems, Decision and Control 573, https://doi.org/10.1007/978-3-031-76253-6_7

signals. Thus, for example, for measurement of peak values of vibration accelerations [1–5] devices IVU-1P, VM-67 can be used, and for determination of several parameters of vibrations themselves: amplitude, average, RMS values of vibration displacement, vibration velocity and vibration acceleration—for example, VIP-2, PSV-10 V. However, their application for diagnosing the technical condition of machine units and mechanisms is unacceptable for several reasons, the main of which are the following: the use of these devices and systems assumes, mainly, the non-random nature of measured vibration signals; amplitude, average and RMS values of displacement, velocity and acceleration of vibrations only in limited cases can act as diagnostic signs; the impossibility of providing automation of some operations of diagnostics; and the inability to use these devices and systems as a diagnostic tool.

The necessity to analyze random diagnostic signals when diagnosing various components of EE, to determine the number of their statistical parameters and characteristics used as diagnostic attributes, and, finally, to ensure the learning and decision-making processes put before researchers the task of creating special diagnostic devices and systems that would combine the fulfilment of all these requirements [6–10].

Information-measuring systems (IMSs) have great possibilities for diagnosing various units of EE. Due to the presence of a computing device or computer, they implement the operations of analyzing the diagnostic signal as a random process and also provide automation of the learning process and diagnostic decision-making [11–16].

In comparison with the above-mentioned devices and other known systems, the IMS of statistical diagnostics of EE units considered in this chapter has wider opportunities for classification of the type of defect in the diagnosed unit. This is achieved due to the use of a statistical approach in the diagnostics of the investigated objects, as well as the simultaneous use of several statistical parameters of the analyzed diagnostic signals as diagnostic features [17–20]. Several examples of the classification of specific defects in some units of EE (charge magnetic conductors, rolling bearings, samples of electrical steel, etc.) are given in the Chap. 8 of this book.

Let us consider the peculiarities of the construction of statistical diagnostics systems for some units of EE. Common for such systems is their purpose, which implies:

- measurement, registration, analogue filtering, and amplification of diagnostic signals;
- conversion of analogue signals into a digital code and input of the latter into a computer;
- determination of quantitative estimations of the limits of diagnostic signs corresponding to conditionally serviceable units and the same units having different types of defects;
- training and diagnostics of the units under study.

Common for statistical diagnostic systems for different diagnostic signals is also the structural diagram (Fig. 7.1), which should contain the following [21–25]:

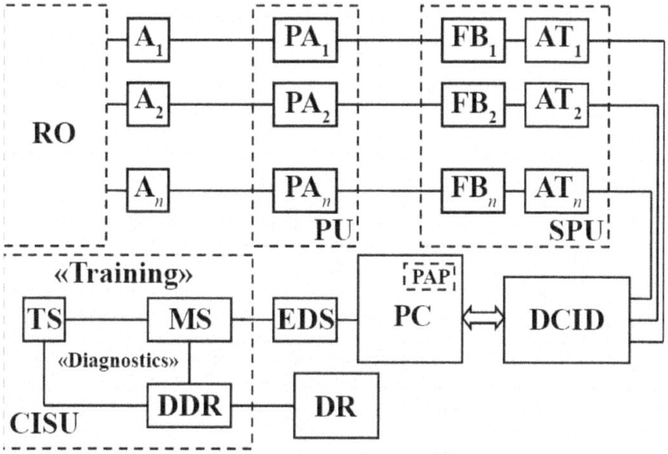

Fig. 7.1 Generalized block diagram of the IMS for statistical diagnostics

- measuring sensors $A_1, \ldots, A_n$;
- preamplifier unit (PU);
- secondary processing unit with a set of filters (SPU);
- device for data conversion and input to the computer (DCID);
- computer;
- package of application programs (PAP);
- control and information and support unit (CISU).

Various EE nodes can act as the object of the RO study. The functioning of these nodes is accompanied by various physical processes, some of which can act as diagnostic signals. Exactly this factor is decisive for concretization of the structure of the IMS of statistical diagnostics, oriented on diagnostics of a certain node of EE. Besides, the choice of this or that diagnostics node entails the selection (or development) of the corresponding software, which is carried out taking into account the results of the analysis of the mathematical model of the selected diagnostic signal.

A necessary condition for the functioning of the statistical diagnostics IMS is the availability of information and reference support (IRS), which allows to diagnose and classify possible types of defects in RO. This IRS includes training sets (TS), corresponding to different technical states of the diagnosed OM; diagnostic decision rules (DDR) for diagnosing and classifying possible types of defects in OM; block of evaluation of diagnostic signs (EDS), which can be used either for formation of TS, or for diagnosing the technical state of OM [26–32].

The transition from the training mode to the diagnostic mode of the RO is performed by means of the mode switch (MS).

It is necessary to make a brief mention (these issues are discussed in detail in paragraph 2 of this chapter) of the statistical diagnostics IMS software.

As a rule, this software includes three main parts:

- programs providing functioning of IMS of statistical diagnostics;
- programs intended for the statistical processing of random diagnostic signals;
- programs that implement algorithms for building algorithms of decisive rules for diagnostics and classification of defects in EE nodes.

The first and the third parts of the software are part of the CISU and DCID block, and the second part forms a PAP for statistical processing of diagnostic signals. It should be noted that the configuration and components of the software for solving the problem of the functioning of the statistical diagnostics IMS can vary depending on the purpose of the IMS and the composition of its electronic components. The programs realizing algorithms of construction of decisive rules for diagnostics and classification of the type of defect in the investigated unit of EE directly depend on the accepted rule of statistical hypothesis testing, as well as on the quantitative estimates of diagnostic features, based on which the TS on the diagnosed object are formed.

The programs for statistical processing of random diagnostic signals from the application package provide estimates of the following characteristics:

- auto- and mutual-correlation functions;
- spectral density;
- histograms with Pearson curve smoothing.

The developed application package also includes programs that allow to carry out:

- check the signals under study for stationarity;
- digital filtering of signals.

Processing of diagnostic signals can be carried out either by using all of the above programs included in the application package, or by selecting some of them. The choice of one or another program from the PAP for determining quantitative estimates of diagnostic attributes is determined, first of all, by the informativeness of the attributes, i.e. their ability to determine the technical state of the diagnosed object.

Below we will stop on consideration of questions connected with the application of the above-mentioned typical structure for the construction of IMS of statistical diagnostics of separate units of EE.

7.2 IMS of Vibration Diagnostics Rolling Bearings of Electric Machines

Among the known systems and devices for vibrodiagnostics of rolling bearings, it is necessary to note devices of SPM company (SPM-32, SPM-43A, BEA-52, BAS-10), and also the device of ISP-1 type, the principle of operation of which is based on the method of shock pulses [33–37]. Shock pulses are defined as short-term pressure waves arising in the bearing rotation zone. The magnitude of individual impulses, as well as the ratio between weak and strong impulses serve as information characterizing the bearing condition. To obtain this information, the basic bearing data (dimensions, type, and number of shaft revolutions) are set with the help of these devices, and then the signal value obtained during the measurement is compared with the value of shock pulses characteristic for the corresponding bearing. The BEA-52 and BAS-10 can be used to provide information on the thickness of the oil film or the amount of damage to the bearing, as described in the BEA-52 and BAS-10.

The experimental sample of the rolling bearing vibrodiagnostics IMS considered below has wider possibilities of classifying the type of defect in a bearing in comparison with the above-mentioned devices. The description of one of the early modifications of the experimental sample of vibrodiagnostic IMS of rolling bearings EM, as well as the results of its practical use are given in [20, 33, 38–42].

Let us consider some peculiarities of this system construction. The developed IMS provides operation in the modes of training and diagnostics; synchronous measurement and registration of vibrations coming from both (from the drive side and current guide side) bearing units of the EM being diagnosed; measurement and registration of vibrations of the bearing tested on a special installation; the possibility of analogue filtering of the vibration diagnostic signal; determination of quantitative estimates of the boundaries of diagnostic features corresponding to a conditionally serviceable bearing and a bearing with defects such as misalignment, lack of lubrication, pitting of the inner or outer bearing rings (in the learning mode); diagnosis of the above-mentioned bearing defects using the available training sets.

The structural scheme of IMS of vibration diagnostics of rolling bearings EM is presented in Fig. 7.2. This system can be conditionally divided into two parts: analogue, which provides measurement and registration of vibration signal, and digital, the main purpose of which is its transformation and processing on the computer. The analogue part includes accelerometers A_1, A_2, and A_3; PU; a SPU with a set of filters. The digital part contains a device for data conversion and input into the computer of DCID; the computer itself. Let's consider these components in detail.

Accelerometers of ABC-017 type were used as transducers of elastic vibrations into electric signals.

For the pre-processing of the investigated vibration signals, a vibration pre-processing device was developed, manufactured, and debugged. Structurally the device is made in the form of two blocks—PU and SPU. The PU has three identical channels consisting of:

- voltage amplifiers VA_1, VA_2, and VA_3 with high input resistance $R_{in} = 1$ GOhm and voltage gain equal to 10 dB;
- refrigeration filters RF_1, RF_2, and RF_3 filters with 50 Hz and 100 Hz cutoff frequencies for each channel.

The PU power supply unit is designed as a voltage converter and converts the DC voltage of 27 V into an AC voltage with a frequency of 20 kHz, which is then

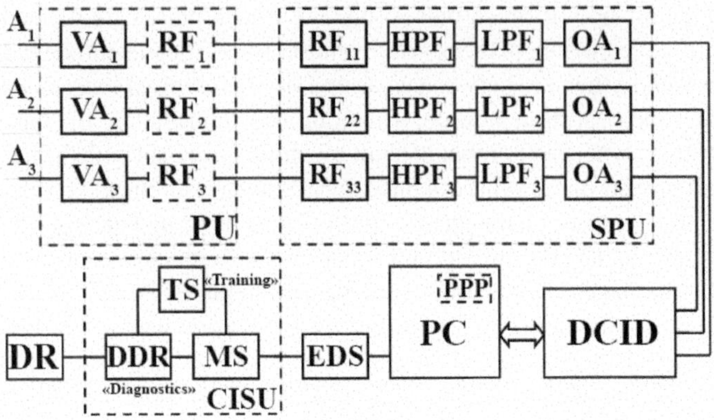

Fig. 7.2 Block diagram of the IMS for vibration diagnostics of rolling bearings

rectified and stabilized. As a result, a stabilized voltage of 5 V and -15 V is obtained to supply the PU. The use of such a converter provides galvanic isolation of the amplifier through the power supply circuit.

The secondary processing unit SPU consists of:

- filters RF_{11}, RF_{22}, and RF_{33}, tuned to the frequencies of 50 Hz and 100 Hz with a depth of rectification of about 40 dB;
- high-pass filters HPF_1, HPF_2 and HPF_3 with frequencies 0.5; 1; 5 and 10 kHz. The attenuation of the HPF in the fence band is about 45 dB/oct. and the attenuation in the transition band is 36 dB/oct;
- low-pass filters LPF_1, LPF_2, and LPF_3 with upper boundary frequencies of 2, 4, 5, and 10 kHz;
- output amplifiers OA_1, OA_2 and OA_3, allowing to amplify the signal by 0; 10; 20; 30; 40 dB.

The use of switchable systems of LPF and HPF allows to formation of the necessary frequency band for the study of vibration signals.

As can be seen from Fig. 7.2, the vibration signal in this IMS can be synchronously measured and processed by three identical channels. This feature allows using vector or vector-phase receivers. These receivers were first considered in the work of S.N. Rzhevkin and L.N. Zakharov. Subsequently, they became widespread in acoustics and vibrometry. The main advantage of their application in vibro-diagnostics is that when accelerometers are located in three mutually orthogonal directions, it is possible to study the vibration vector, which makes it possible to carry out deeper diagnostics. Figure 7.3 shows one of the variants of the vector receiver, which was used in some experiments on the study of EM vibrations. It should be noted that the choice of installation location, mutual arrangement, and fastening of accelerometers on the object depends on the purpose of the research and selected methods of diagnostics.

It should be noted that one of the main characteristics, according to which the equipment included in the analog part of the system is selected, is the range of

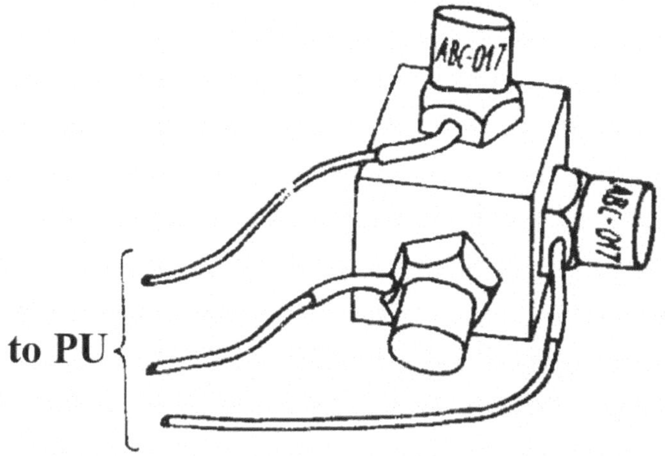

Fig. 7.3 Vector vibration receiver

vibration frequencies of the diagnosed bearings. As it is known, these vibration frequencies lie in a wide enough range, the upper boundary of which reaches 35 kHz (for roller bearings). However, as the results of the analysis of experimental data shown, often at diagnosing the technical condition of rolling bearings EM can be limited to the study of the registered vibrograms in the band up to 10 kHz. Therefore, many experiments on diagnostics of defects in bearings, considered in this book (unless it is specially specified), were conducted with the help of the system in the frequency band of 20 Hz…10 kHz.

Let us consider the digital part of the system.

Conversion of analog signal samples into a digital code is a necessary condition for their subsequent processing on a computer. The device for conversion and input of initial data is designed to work in conjunction with a personal computer, which has a system bus ISA.

The DCID consists of:

- input stage;
- 12-bit analogue-to-digital converter (ADC), which provides conversion of the input analogue signal into digital code;
- data input stage, which provides matching the ADC output with the system bus of the PC and input of the digitized signal into the PC.

Structurally, the DCID is designed as a separate board connected to the PC directly through the system bus connector and has the following characteristics:

- number of differential channels—16;
- digit capacity—12 bits;
- conversion time—2 μs;
- differential nonlinearity—0.02%;

- input resistance—1 MOhm;
- input signal range—± 5.12 V; ± 2.56 V; ± 1.28 V;
- maximum conversion frequency—up to 250 kHz.

The DCID contains a fast-acting sampling-storage device and provides the following modes of operation:

program start;

start from the built-in timer;

start from an external signal;

multi-channel data input;

programmable delay between channels during multichannel data input.

In addition to the described technical means to ensure the functioning of the system specially developed software and IRS.

IRS of the system includes:

- TS corresponding to different technical conditions of diagnosed bearings;
- diagnostics decision rules (DDR) for diagnostics and classification of some types of bearing defects;
- EDS.

The last one can be used either for the formation of the TS or for diagnostics of the bearing state. The transition from the teaching mode to the diagnostics mode is carried out by means of the MS.

The issues of diagnostic measurements with the help of the considered experimental sample of vibrodiagnostic IMS both for single rolling bearings tested on the experimental stand and as a part of EM bearing units are considered in Chap. 8 of this book.

7.2.1 IMS of Shock Diagnostics of Charge Cores

The peculiarity of the shock diagnostics method (one of the most widespread methods of test diagnostics), in comparison with other vibration methods, is that at shock excitation it is necessary to study relatively short-duration random vibration signals. The analysis of such vibration signals, determination of their statistical parameters and characteristics used as diagnostic features, as well as the training and decision-making processes set the task of creating special IMS of shock vibration diagnostics, which would allow to determine, for example, the state of pressings of the charge cores with the specified accuracy and reliability.

Purpose of the experimental sample of IMS of shock vibrodiagnostics. The developed sample of IMS of shock vibrodiagnostics provides:

- excitation of vibration waves in the investigated object of diagnostics;
- measurement, registration, analog filtering, and amplification of a random vibration signal;
- conversion of the analog signal into a digital code and input of the latter PC;

- analysis of the random vibration signal and determination of quantitative estimations of the boundaries of diagnostic features corresponding to a conditionally serviceable object and an object with various types of defects;
- training and diagnostics of the press condition of the investigated object.

Composition of IMS of shock vibration diagnostics. As can be seen from the structural diagram (Fig. 7.4), the laboratory sample of the IMS provides for the use of up to four measuring vibration converters (accelerometers), of which one (A_1) is part of the impact hammer (IH) and serves to fix the moment of impact, and the others (A_2, A_3, A_4) are fixed on the object of diagnosis DO and are used to convert into electrical signals of mechanical vibrations occurring at the points of their installation [20, 22, 30, 43–46].

From the output of accelerometers A_1, A_2, A_3, and A_4 signals are fed to preamplifiers PA_2 … PA_4, after which they are further processed in the analog filtering blocks AFB_1 … AFB_3. Further, through the channel selector (CS), the research signals are fed to the SPU.

The SPU consists of:

- 50 and 100 Hz (with a depth of about 40 dB), designed to eliminate interference in the form of frequency components of the first and second harmonics of the industrial frequency from the diagnostic signal;
- HPF with frequencies of 0.5, 1, 5, and 10 kHz;
- LPF with upper limiting frequencies of 2; 4; 5 and 10 kHz.

The set of HPF and LPF allows to formation of the necessary frequency band for vibration wave research.

As an impact hammer, IH in the created sample of IMS of impact diagnostics was used IH of PCB Piezotronics Modally Tuned Impulse Hammer type 083C03 [47].

The DCID is designed to convert analog signals into a digital code with the subsequent input of this code into the PC. The main conversion device in the DCID is used

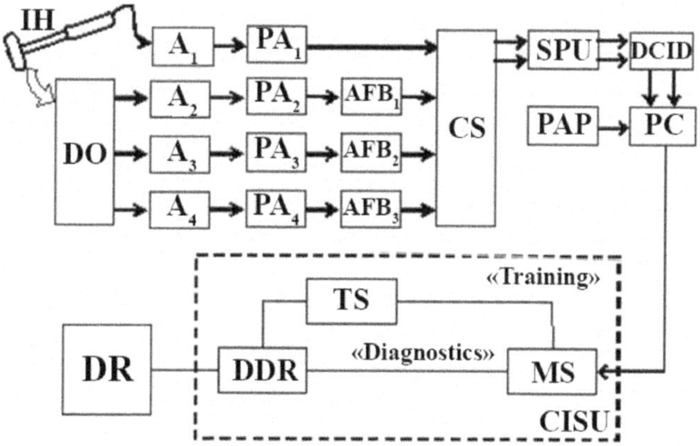

Fig. 7.4 Structural diagram of IMS of shock diagnostics of EE

ADC unit—National Instruments multifunction data acquisition device type USB-6218. This device allows to realization of multichannel conversions of measured signals with frequencies up to 250 kHz (samples per second). The DCID is powered by the USB bus to which it is connected.

The PC performs statistical processing with the help of the created PAP. The PAP includes programs of digital statistical processing, which allow for obtaining quantitative estimates of diagnostic features.

As diagnostic object (DO) can be, for example, investigated: the charge magnetic conductor (stator, transformer, etc.), the frame of the electric motor (especially in the places of its attachment to the foundation), elements of high-voltage insulator, etc.

While conducting experiments with the help of this IMS of shock diagnostics we used a three-component accelerometer (A_2, A_3, A_4 are combined in one case) of Brüel & Kjær miniature triaxial piezoelectric charge accelerometer type 4326A for vibration and shock measurements on low-mass structures, which is a miniature triaxial accelerometer with charge output for vibration and shock measurements on small-sized structures [48–50]. This accelerometer has three independent, individually calibrated outputs for simultaneous measurements in three mutually perpendicular directions.

For the experiments described in Chap. 8, only one output, labeled "Z" on the accelerometer body, was used. The frequency range of measurements with the accelerometer 4326A is from 1 to 8000 Hz for each channel.

A Brüel & Kjær Type 2634 two-stage charge amplifier (Brüel & Kjær Type 2634 two-stage charge amplifier) was used as a preamplifier (PU1…PU4) in the scheme of the shock diagnostics IMS under consideration. This preamplifier converts the signal from the output of a piezo-electric vibration sensor before converting the diagnostic signal into a digital code using an ATDC. It is designed for measurements in an industrial environment [51, 52].

The lab sample can be operated by any computer with a USB port and the necessary software.

During the diagnostics of EE units using the developed experimental model of IMS, the impact was applied using the above-described IH, which is equipped with a set of tips with different masses and hardness. By changing the tips, it is possible to change the width of the shock pulse spectrum. For example, in the experiments described in the next chapter of this book, the hardest tip was used because it provides the widest possible shock spectrum.

The accelerometer is attached to the part under study using special wax or special accelerometer fastening studs manufactured by Brüel & Kjær.

The described sample of IMS of shock diagnostics allowed us to carry out several experiments on diagnostics of the degree of pressing of the charge mag- -nanowire, the results of which are given in Chap. 8 of this book.

7.2.2 IMS of Vibration Diagnostics of Moving Parts of Electric Machines

To date, the systems of diagnostics of EE nodes, oriented on the measurement of information signals coming from sensors placed on stationary objects, are well developed and used in practice. At the same time, the solution to the problem of measurement and, especially, the transmission of information signals registered on moving nodes of EM is of considerable interest. This, first of all, concerns separate units (winding, magnetic conductor, rolling bearings, etc.) placed on the EM rotor. The modern information technologies developed today allow us to realize technically and programmatically the measurements and transfer of diagnostic signals measured directly on the moving units of EM [53–57]. Let us dwell on these issues in more detail.

The methods and technical realization of diagnostics of EM moving units proposed in [20, 22] are realized using a special diagnostics IMS. Below we will dwell on the issues of construction of such IMS, emphasizing the measurement and conversion of the vibration diagnostic signal registered on the rotating shaft of the EM. IMS of diagnostics has blocks providing measurement, conversion, and transmission of information (technical means) and blocks realizing program control of IMS, statistical processing of measured signals, and construction of decisive rules to determine the technical condition of the diagnosed equipment (software means).

Let's dwell more on the structure of the main functional parts of the IMS and consider the measured signal processing channel.

The technical part of the developed IMS consists of two main components:

- sensor block, located directly on the object of diagnostics (Fig. 7.5);
- block of information reception and processing, as the main core of which can be any PC (Fig. 7.6).

Bluetooth transceiver is connected to the used PC with the help of a USB (universal serial bus—an external bus that supports the installation of devices that are configured in the operating system). The main component of this unit is the software that performs all the necessary functions to control the operation of the IMS itself, processing the diagnostic signal, and constructing decisive rules for diagnosing the technical condition of EE units.

The sensor unit (Fig. 7.5) performs the function of measurement, processing, and transmission of measured information signals at individual nodes of the EM.

The number of sensor blocks depends on the required number of information signals received from different nodes of the EM and is limited to 256 devices, which is more than enough to measure the vibration signal of a given range from individual nodes.

Using an accelerometer of the ADXL202 type as sensor. The analog signal measured at the rotating node of the EM and formed at the sensor output has the form of a DCM signal (duty cycle modulation) [58–60], the approximate form of which is shown in Fig. 7.7.

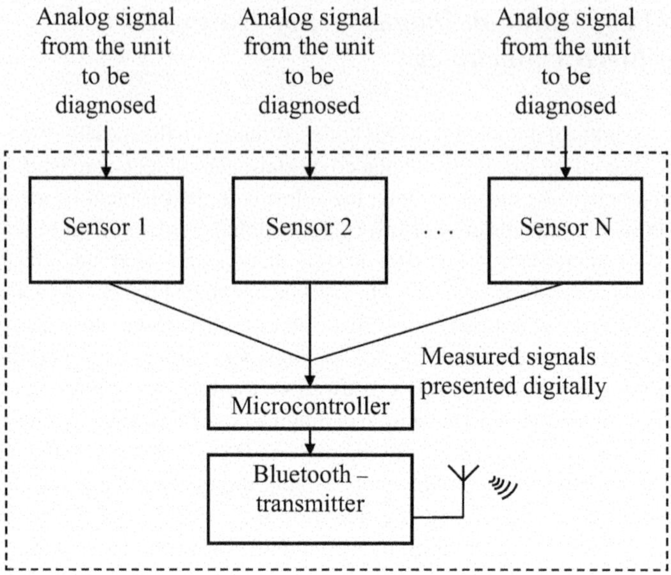

Fig. 7.5 Sensor unit

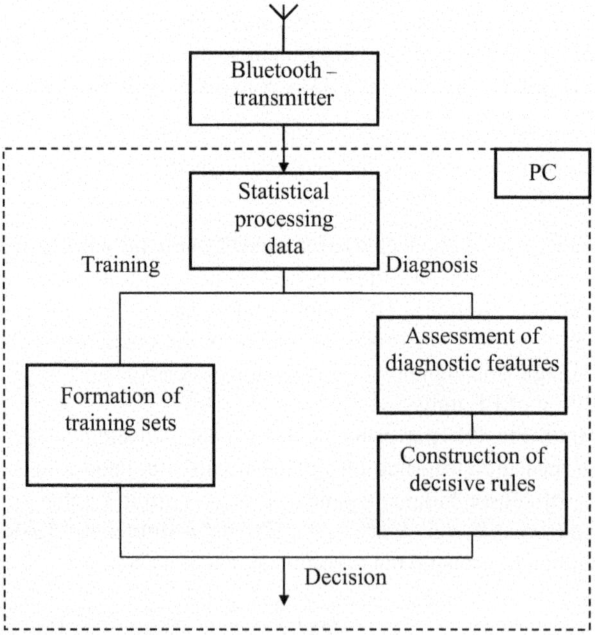

Fig. 7.6 Information receiving and processing unit

Fig. 7.7 DCM signal

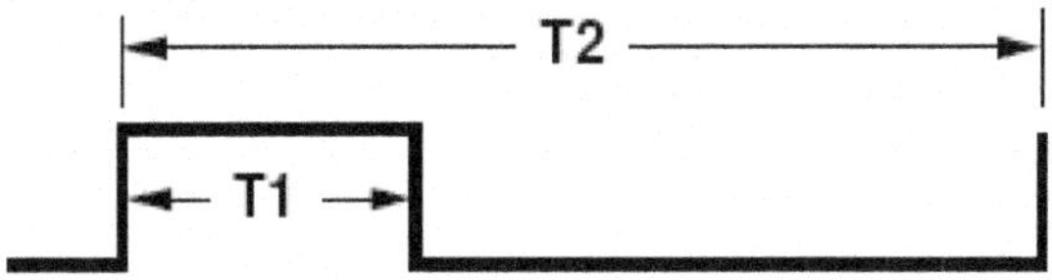

The frequency of the measured signal can be set in the range from 0.01 Hz to 6 kHz, which is sufficient for diagnosing the technical condition of windings or rolling bearings installed on the EM rotor, as well as possible thermal imbalance arising from the short circuit of individual turns of the rotor winding. It is assumed that the rotor speed of the rotor of the diagnosed EM does not exceed 1000 rpm.

Accelerometer ADXL202 [61–63] is a biaxial acceleration measurement system, which is presented on a single monolithic integral circuit. The structural–functional scheme of this accelerometer is shown in Fig. 7.8. The acceleration measurement unit in analog form is based on the principle of changing the capacitance of the measuring element due to the physical movement of a part of the measuring element, which occurs under the action of the measured acceleration. Further, after passing through the demodulation block and resistive filter, the measured signal enters the block of analog signal conversion into the DCM signal.

The received DCM signal is transmitted via a radio channel to the information receiving and processing unit (Fig. 7.6).

The measured signal can be obtained by measuring the pulse lengths T1 and T2 formed at the accelerometer output (Fig. 7.7). The amplitude of the measured signal is directly proportional to the ratio T1/T2, with zero acceleration corresponding to the case T1 = 0.5·T2.

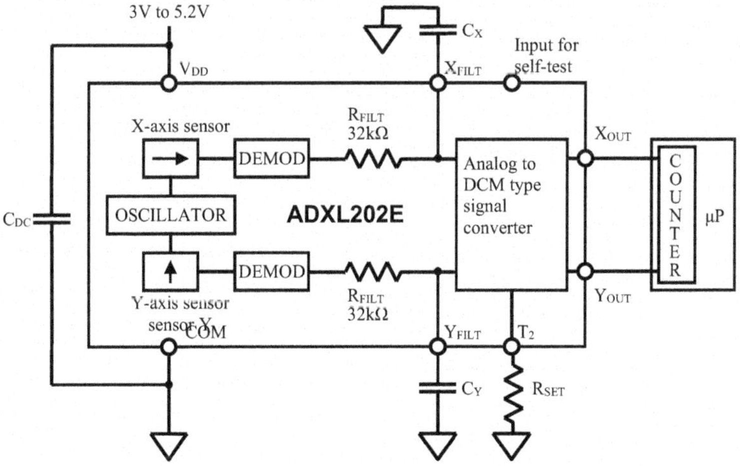

Fig. 7.8 Structural and functional diagram of accelerometer ADXL202

At T1 < 0.5·T2 the "negative" acceleration is measured relative to the selected axis and in the direction of positive acceleration. At T1 > 0.5·T2 we have a measure of positive acceleration relative to the selected axis. The scaling factor is 12.5% per 1 g (manufacturer's data).

The duration of the period T2 is set by setting the resistor R_{SET} (Fig. 7.8). The frequency bandwidth (up to 6 kHz for ADXL202) is set employing capacitors C_x and C_y (Fig. 7.8), connected to the outputs X_{FILT} and $YFILT$.

Physically, the sensor unit, which is used in the created vibration diagnostics IMS experimental sample, consists of the following main elements:

- two ADXL202 accelerometers;
- microcontroller PIC16LF873-041 manufactured by Microchip Technology;
- Bluetooth module EYMF2CAMM-XX manufactured by TAIYO YUDEN;
- related radio-electronic elements.

The convenience of using sensors, from the outputs of which DCM signals are taken, is that such sensors can be directly connected to the input of the microprocessor counter without the use of analog-to-digital conversion or linking logic chips, which simplifies the process of processing the measured signal and increases the autonomous operation time of the devices.

Use of accelerometer type ADXL001 as sensor. The vibration signal measured on the moving node of the EM being diagnosed is presented in analog form at the sensor output. The frequency of the vibration signal that is measured by the ADXL001 accelerometer can be set from 0.01 Hz to 22 kHz, this allows the investigation of higher frequency signals compared to the ADXL202 accelerometer.

Accelerometer ADXL001 is a uniaxial acceleration measurement system presented on a single monolithic integrated circuit. The structure-functional diagram of this accelerometer is presented in Fig. 7.9.

The acceleration measurement unit in analog form is built on the principle of changing the capacitance of the measured element, which arises due to the physical movement of a part of the measured element under the action of the measured acceleration. Then the measured signal, after passing through the modulator, differential sensor, and demodulator, enters the analog signal amplification unit. The differential sensor module is used to compare readings of two values of vibration acceleration, which are measured by separate parts of one sensor to achieve greater accuracy of measurement. In this case, special algorithms are used, for which the signal is first converted into another form using a modulator, and after obtaining the final value of the measured vibration acceleration, the signal enters the demodulator and output amplifier unit, at the output of which we get the analog value X_{OUT} of the measured vibration signal.

The received analog X_{OUT} signal goes to the input of the ten-bit ADC of the microcontroller of the sensor unit, where it is converted into a digital signal for further transmission via radio channel to the information receiving and processing unit.

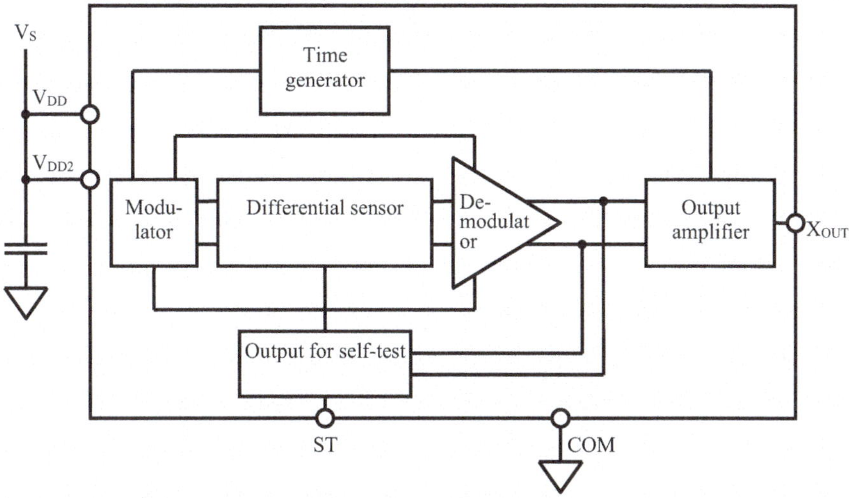

Fig. 7.9 Structure-functional diagram of accelerometer ADXL001

Fig. 7.10 General view of the sensor unit

The use of ADXL001 sensors allows for an increase in the frequency range of the measured signal, which, in turn, provides more opportunities for diagnostics of moving components of EM.

Taking into account the modular construction of the sensor unit and the presence of ADC in the control controller, it is possible to adapt the developed vibration diagnostics IMS for solving a wider range of tasks without significant changes in the technical part of the IMS itself (e.g., control of the state of temperature fields in the rotor array of EM, control of the rotor speed, etc.).

In cases of the use of different types of sensors on the output of the microcontroller, the digital signal containing the measured information is received. This signal by command from the block of reception and processing of information or at certain time intervals is transmitted to the computer, where statistical processing of information signals received from the block of sensors is carried out, according to the results of which the technical condition of rotating units of EM is diagnosed.

7.2.3 IMS of Acoustic Emission Diagnostics

One of the most promising and currently still insufficiently developed is the method of acoustic emission (AE) diagnostics. This method is especially effective for diagnostics of parts and units of EE of large geometric dimensions, complex configuration, and with a large number of welds, which are subjected to significant dynamic effects (reactor walls, pressure vessels, pipelines, turbine and generator housings, bearing units of high-power EM, ceramic insulators, etc.) [64–69]. The specified method can be most successfully implemented using specialized AE-diagnostics IMS. Below we will consider one of the variants of realization of such IMS.

The system of statistical AE diagnostics is focused on the measurement and processing of elastic oscillations occurring in the EE units during its operation.

The necessity of analyzing the AE signal, which is a random process, to solve the problem of diagnostics of various EE units is due to the need to determine the stochastic parameters and characteristics used as diagnostic features. In addition, the experimental sample of the AE IMS should provide the formation of training sets and decision-making for the detection and classification of defects in EE units.

IMS of AE-diagnostic considered in this paragraph has wider possibilities in comparison with the known systems. This is due to the fact that in the created system the model based on the use of statistical parameters of the AE-process as diagnostic features is realized.

In this paragraph let us consider the structure and peculiarities of AE IMS construction.

Structure of the experimental sample of IMS of AE diagnostics of EE. Based on the generalized structure of IMS (Fig. 7..1 of the Chap. 1) depending on the type of diagnostic tasks to be solved, a specific type of IMS of statistical diagnostics is created. So, along with stationary IMS located in immediate proximity to power objects, mobile IMS of statistical diagnostics can be used for operative measurement of information signals on objects to which measuring channels are not connected. The general view of the laboratory sample of such IMS is shown in Fig. 7.11.

The developed laboratory sample of IMS of AE-diagnostics allows to fulfill the following main functions:

- measurement of AE and vibration signals;
- accumulation, storage of measured data;

Fig. 7.11 General view of the laboratory sample of the IMS of AE-diagnostics

- statistical analysis of measured data to obtain quantitative estimates of diagnostic signs;
- formation of TS and construction of decisive rules for diagnostics and classification of defects of EE units.

The development of the equipment was carried out using portable computers, which allows to application of known and specialized software and mathematical methods of processing the signals of sensors installed on the object under study. The structural diagram of the laboratory sample of the IMS of AE-diagnostics, which was used for experimental studies of the AE process [70–74], is presented in Fig. 7.12. Structurally, the sample consists of:

- primary sensors $S_1, \ldots, S_4$, installed directly on the DO;
- module of data conversion and input into the PC (controller);
- PC;
- a package of programs for processing diagnostic signals;

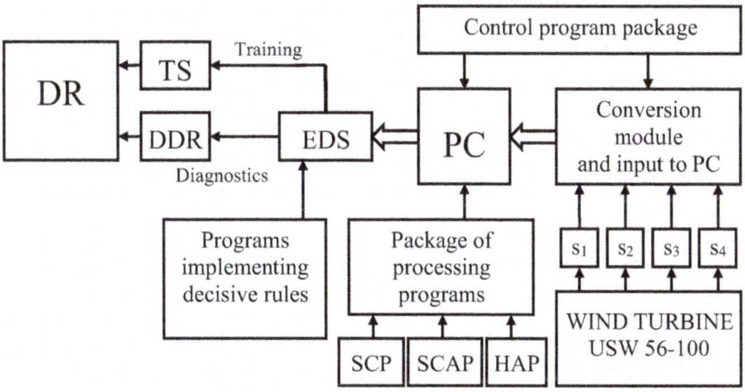

Fig. 7.12 Structural diagram of the laboratory sample of AE-diagnostics IMS

- a package of programs implementing the solving rules;
- a package of programs ensuring the operation of the AE-diagnostics IMS;
- decision-making unit, which produces the diagnosis result.

In the scheme presented in Fig. 7.12, the following designations are accepted: SCAP—spectral-correlation analysis program, HAP—histogram analysis program, SCP—stationarity check program.

Let us briefly describe each of the blocks included in the created laboratory model of the IMS.

Primary sensors (S_1, S_2, S_3, S_4) for measuring diagnostic signals were installed on the most critical nodes of the diagnosed object. During experiments, wind turbines (wind turbine USW 56–100) operating as part of wind power plants were chosen as the object of research, as well as a sample of electrical steel, which was tested on a special tensile machine. The developed laboratory sample of the IMS allows to carry out vibrodiagnostics of wind turbine units (bearings of the main shaft of the wind wheel, gearbox, generator, etc.) as well as AE-diagnostics of wind turbine units experiencing significant mechanical loads (for example, the place of articulation of the wind turbine nacelle and the mast-platform). A more detailed description of the experiments on the measurement and processing of AE signals is given in Chap. 8 of this book.

The conversion module is a microcontroller, which matches the parameters of sensor signals with the inputs of the conversion board and converts analog signals into digital code. The structural diagram and the principle of operation of individual components of the microcontroller are described in more detail below.

As a computational core of the created sample IMS of AE-diagnostics, we used components of an industrial computer mounted in a special case (Fig. 7.11). The choice of components was made based on the class of tasks and characteristics of diagnostic signals, which a particular diagnostic IMS is aimed at analyzing.

The final stage of diagnosing the EE nodes using the created IMS sample is realized with the help of the decision-making block that gives the result of diagnosis (Fig. 7.12). This block is realized programmatically. The developed IMS sample can work in two modes: TRAINING and DIAGNOSTICS. Operation of the IMS in one or another of these modes is realized with the help of control software.

In the TRAINING mode, according to quantitative EDS, obtained from the results of statistical processing of AE signals, in pre-selected diagnostic spaces, the formation of training sets of OS, corresponding to different technical states of the diagnosed unit is carried out. In the DIAGNOSTICS mode, the technical state of the diagnosed node (diagnosis result) is determined using the program for building solving rules based on the obtained EDS and performed TS.

The issues related to the operation of the decision-making unit included in the created IMS are discussed in more detail in Chap. 5 of this book.

One of the main devices included in the AE-diagnostics IMS sample is a special controller that provides the conversion of analog AE signals into a digital code. Let us briefly dwell on the consideration of this device. Since the AE-diagnostics IMS is designed to work with rather high-frequency diagnostic signals (up to the megahertz range), let us dwell in more detail on the description of the device for converting analog AE signals into a digital code.

The principle of operation of the device for converting analog signals into digital code, which is a part of the laboratory sample of IMS of AE-diagnostics. The operation of the conversion and data input equipment included in the developed sample of the AE-diagnostics IMS (Fig. 7.12) is carried out under the influence of control signals coming from the parallel port of the computer and formed by special software.

The structural diagram of the module of data conversion and input is presented in Fig. 7..13. Let us describe the principle of operation of this module.

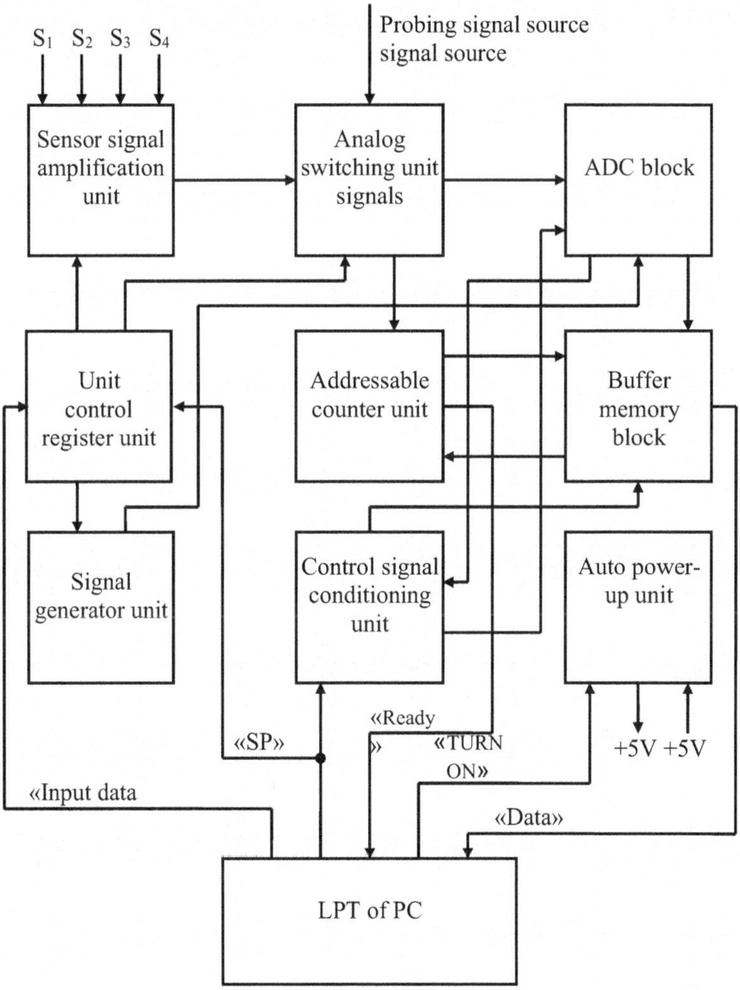

Fig. 7.13 Structural diagram of the module of data conversion and input of AE-diagnostics IMS

Initial data, accompanied by synchronous pulses "SP", are entered into the control register block to set the required mode of operation of the equipment. This data contains:

- gain coefficients for 4 sensor channels;
- information about random switching of sensor amplifiers;
- sign of sensors' alternate switching;
- selection of the frequency of signal conversion into digital form;
- sign of power supply voltage value determination.

The block of amplification of sensor signals consists of 4 independent amplifiers with variable amplification factors set by the control register block.

The analog signal switching unit connects the amplifier output of a given sensor to the ADC input in any order, or provides an alternate connection of all amplifiers during one run of the equipment.

ADC block converts amplified analog signals into digital form for recording into the microcontroller buffer memory at specified frequencies: 5, 2, 1 MHz, 500, 250, 125, and 62.5 kHz. It is implemented on the AD1672 chip, which has a conversion frequency limit of 5 MHz and a digit capacity of 12 bits.

The buffer memory block serves to match the high conversion rate with the speed of data input to the computer through the LPT port. The total capacity of the buffer memory is 128 kB.

The address counter block is designed to address the buffer memory when recording and replaying the sensor signals after starting the equipment.

The control signal generator block generates signals that control the operation of microcontroller blocks in a certain time sequence.

A signal generator block is a quartz relaxation generator that produces rectangular pulses with a given frequency specified in the ADC block.

The block of automatic power supply switches on and off the microcontroller power supply from the computer side to save power consumption of the autonomous power supply.

The described blocks form a specialized controller designed to record in the computer memory signals coming in digital form from 4 sensors in the form of a numerical sequence for their further software processing to determine quantitative estimates of diagnostic signs of the technical condition of the RO. The controller is made in the form of a separate board placed in a plastic case.

Technical characteristics of the developed controller [71] are presented in Table 7.1.

The functioning of separate nodes and blocks of the created experimental samples of IMS of statistical diagnostics is realized with the help of special software, which is considered in paragraph 2.

Table 7.1 Technical characteristics data conversion and input devices

№	Parameter	Value
1	Number of input channels	4
2	Number of amplifiers	4
3	Permissible gain values	2, 6, 18, 54, 78
4	Starting the conversion	Internal, external
5	Analog-to-digital conversion frequency	5; 2,5; 1 MHz; 500; 250; 125; 62,5 kHz
6	ADC capacity	12 bit
7	Buffer memory capacity	128 kB
8	Turning the power on and off	Automatic
9	Interface	LPT
10	Controller dimensions	150 × 150 × 30 mm
11	Controller weight	0.3 kg
12	Power supply	4 elements by 1.2 V
13	Maximum current consumption	60 mA

7.2.4 IMS of Diagnostics of Cylinder-Piston Group DEGs

Let us consider the issues related to the measurement and registration of experimental data on the non-uniformity of the diesel-electric generator (DEG) shaft rotation speed, based on the analysis of the mathematical model describing the non-uniformity of the DEG shaft rotation, and taking into account the results of modeling the dynamics of cylinder power on the DEG shaft [75–80].

Physical interpretation of the model parameters, describing the cylinder power distribution of DEG. As a concrete example of the application of the model describing the distribution of cylinder power of DEGs, consider a 5D70 type DEG, the main technical characteristics of which are presented in Chap. 5 of this book. In this case, the peculiarities of the crankshaft design, cylinder-piston group (CPG), cylinder operation sequence, etc. are of decisive importance.

When testing a four-stroke (i.e., for two revolutions of the crankshaft to perform work on all 16 cylinders) diesel engine 5D70 one cycle is put by 1200 impulses in the sequence formed at the output of the PDF-3 sensor. In other words, for one cycle of operation of the 5D70 diesel engine, we have the following sequence $\boldsymbol{\Theta} = \left\{\Theta_j,\ j = \overline{1,\ 1200}\right\}$.

As mentioned in Chap. 4, to measure the nested post-sequence of the linear periodic random process (LPRP) counts, it is necessary to know the values of its period. This parameter is measured using the top dead point (TDP) sensor [81–84]. The sequence of operation of the 5D70 diesel engine cylinder pairs, starting from the location of the TDP sensor, is given in Table 7.2.

Using Table 7.2, each operating cylinder during one cycle can be matched to a certain group of pulses from the sequence $\boldsymbol{\Theta}$.

Table 7.2 Sequence of operation of pairs of cylinders of diesel engine 5D70

Cylinder row	Cylinder numbers															
Right		8		3		2		5		1		6		7		4
Left	1		6		7		4		8		3		2		5	

Let us now dwell on the description of the parameters of the pulses forming the sequence **Θ**. Determination of the parameters of these pulses was carried out in real conditions on the experimental stand DEG 5D70.

From the output of the PDF-3 sensor, installed on the DEG, the signal was input to the receiver of the information transmission channel located in the vicinity of the PDF-3 sensor. The transmitter contains two identical channels (one redundant). From the output of the transmitter information signal through a connecting cable (length 20 m) went to the input of the receiver. The receiver contains three identical channels. The signal generated at the receiver output was fed to the input of oscilloscope C1 107. The shape of pulses, as well as the deviation of the frequency deviation of rectangular pulses from the output of the PDF-3 sensor at different operating modes of DEG 5D70, namely, at a different number of rotational speeds of DEG [85–89], as well as at "build-up" and "reset" of power, was studied.

1. DEG operation mode: 500 rpm. Oscillogram of signals of PDF-3 sensor in this case is presented in Fig. 7.14a.

Duration of pulses in this mode = 100 μs. Deviation of the trailing edge = 12 μs, i.e. more than 10%.

2. DEG operation mode: 1000 rpm, load 1000 kW. oscillogram of the signal at the receiver output is presented in Fig. 7.14b.

The pulse duration in this case is = 50 μs. The deviation of the trailing edge of the pulses is = 6 μs, i.e. > 10%. The duration of the fronts remained unchanged at 400 rpm and 1000 rpm.

3. DEG operation mode: 1000 rpm, power reset from 1000 kW to 0. The oscillogram of the signal at the receiver output is shown in Fig. 14c.

At power reset there is a short-term (within 2 s) decrease in the duration of pulses at the receiver output by 10 μs. The deviation of the leading edge practically did not change.

4. DEG operation mode: 1000 rpm, power ramp from 0 to 1000 kW. Oscillogram of the signal at the receiver output is presented in Fig. 7.14d.

In this case, a short-term (within 2…3 s) increase in the duration of pulses at the receiver output by 10 μs. The deviation of the trailing edge of the output pulses did not change.

The pulse sequence **Θ** with the above-described parameters of pulses was transmitted through the information transmission channel and with the help of the conversion device input data were entered into the computer for further processing, using a package of programs developed based on mathematical calculations given in Chap. 6 of this book.

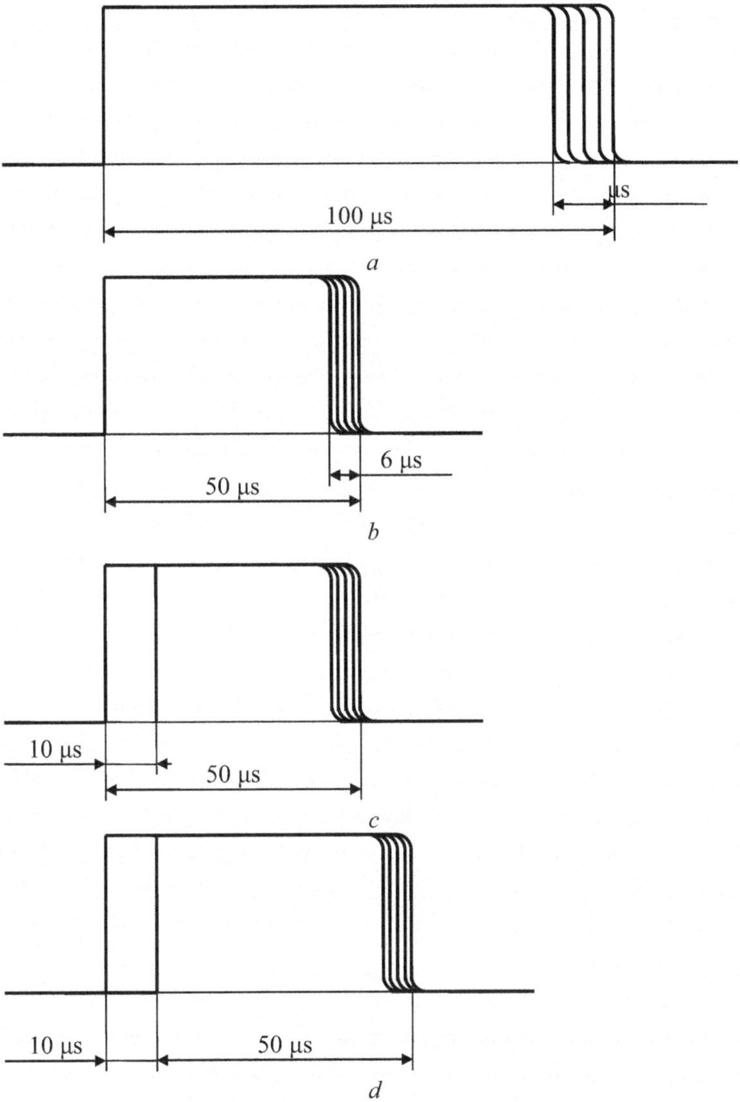

Fig. 7.14 Oscillograms of signals of PDF-3 sensor at different operating modes of the DEG 5D70

Below is a brief description of the information transmission channel from the PDF-3 sensor to the PC, as well as the experimental device for conversion and input of initial data into the PC.

Formation of experimental data on diagnostics of the DEG's CPG. This chapter gives a brief description of the experimental sample of IMS for CPG-diagnostics. The structural diagram of this IMS and the principle of operation of the device for converting raw data into a digital code and input of these data into the PC are considered.

Measurement channel and data transfer to the computer. The developed IMS as a primary sensor (sensor) to measure the initial data characterizing the non-uniformity of the rotation speed of crankshaft DEG, used photoelectric discrete displacement sensor type PDF-3.

The sensor is designed to convert the displacement (angle of rotation) of the working bodies of industrial mechanisms in the number of pulses and angular velocity in the pulse repetition rate. As applied to the DEG, at the mechanical connection of the PDF-3 sensor to the free end of the crankshaft of the tested generator, at the output of the sensor is formed a sequence of rectangular pulses, the number and duration of which can be used to determine the instantaneous values of the angular speed of rotation of the crankshaft of the tested generator.

Basic technical data of the PDF-3 sensor:

Nominal supply voltage—24 V;

voltage of output signals:

signal "1"—24 + 2,4–4,8 V,

0" signal—0.5 V;

maximum input shaft rotation speed—3000 rpm;

number of pulses per shaft revolution—600;

current consumption, not more than 300 mA;

overall dimensions, mm—150 × 150x172;

mass not more than 4 kg.

Sensors are manufactured in climatic versions U2 and T2 according to GOST 15150–69.

To transmit the flow of pulses generated at the output of the PDF-3 sensor through a coaxial cable and coordination with the input of the device for converting data into a digital code and input of these data into a computer were specially designed and manufactured experimental devices that realize the functions of a receiver and transmitter. This receiver and transmitter together with a coaxial cable form an analog channel for transmission of the original data. Its structural diagram is shown in Fig. 7.15.

Transmitter 1 is designed to convert the output signals of the PDF-3 sensor with the maximum level of + 24 V. It performs matching of the output resistance of the PDF-3 sensor with the 75 Ohm coaxial cable. The scheme of transmitter 1 allows the use of cable up to 50 m. Receiver 2 performs the matching of signals transmitted via coaxial cable with microcircuits, which are used in the conversion device.

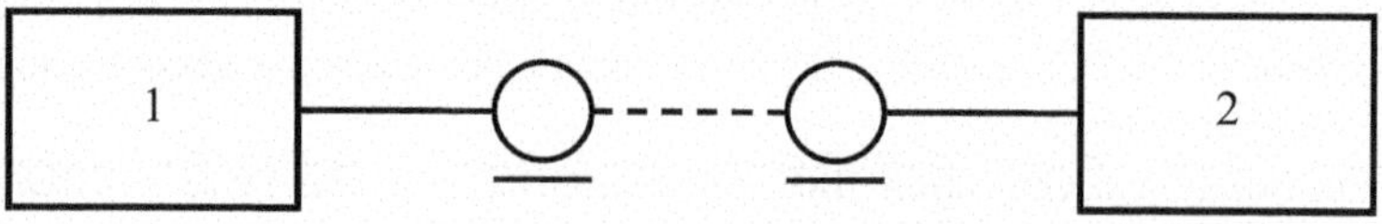

Fig. 7.15 Structural diagram of the data transmission channel from the PDF-3 sensor to the computer: 1—transmitter of signals from the PDF-3 sensor; 2—receiver of signals PDF-3 sensor

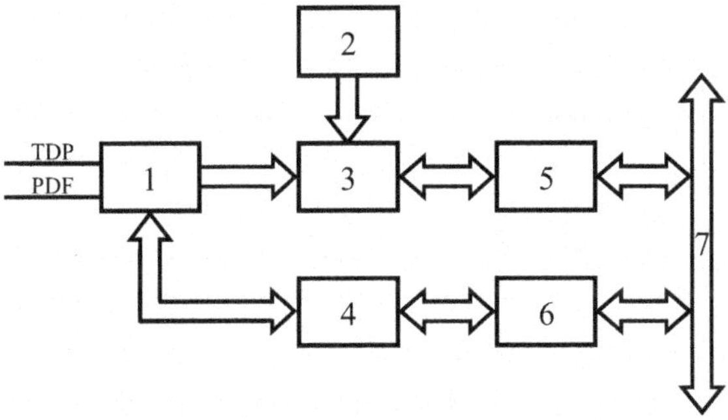

Fig. 7.16 Schematic diagram of the device for data conversion and input into PC: 1—measuring interval shaper; 2—quartz generator; 3—program-controlled timer; 4—direct access block; 5 and 6—channel interfaces; 7—communication channel with the computer; TDP—signal from the top dead point; PDF—signal from the PDF-3 sensor

Device for converting experimental data into a digital code and inputting this code into PC. Conversion of the pulse stream formed at the output of the PDF-3 sensor into a digital code is a necessary condition for their subsequent processing on PC. The developed and manufactured experimental sample of the device for data conversion and input into the PC has the following characteristics:

- number of information transmission channels: 1;
- period of repetition of measuring pulses: from $3.2 \bullet 10^{-3}$ to $0.3 \bullet 10^{-5}$ s;
- number of readings per one rotation of the DEG shaft: up to 600;
- the volume of numerical arrays registered in the computer: up to 10,240.

Transformation of signals coming from the PDF-3 sensor is realized by the formation of measuring intervals and their subsequent measurement. The structural diagram of the device for conversion and input of experimental data into the computer is shown in Fig. 7.16.

The signal formed at the output of PDF-3 sensor (Fig. 17a) is related to the speed of rotation of the DEG shaft.

To measure the pulse repetition period, the measuring interval is formed with the help of the shaper 1, as shown in the timing diagram in Fig. 17b. Then with the help of generator 2 stabilized by quartz, the measuring interval is filled with counting pulses (Fig. 7..17c).

The counting of impulses is carried out by timer 4. Using the direct access device 3 the results of counting are transferred to the PC memory. Processing of the measured pulse sequences is carried out according to the developed software.

The functioning of separate nodes and blocks of the created experimental samples of IMS of statistical diagnostics is realized with the help of different types of software. Let us briefly dwell on the consideration of this issue.

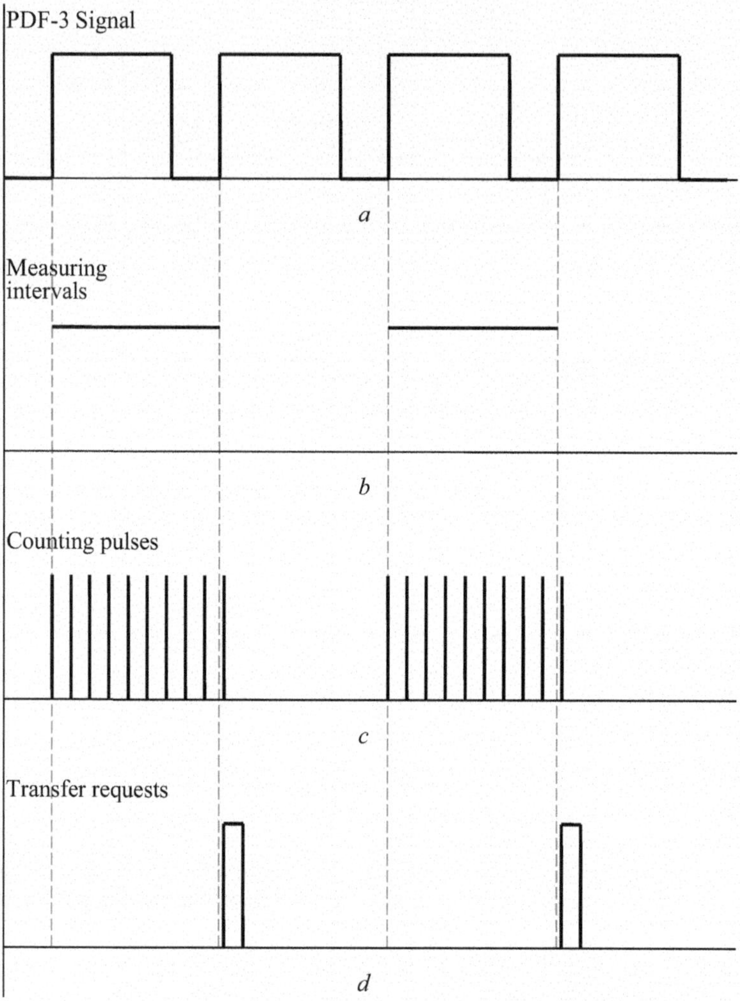

Fig. 7.17 Principle of converting the signal generated at the output of the PDF-3 sensor into a digital code

7.3 IMS Software of Statistical Diagnostics of EE

The software included in the created experimental samples of IMS of statistical diagnostics consists of three main parts:

- programs providing functioning of the created IMS;
- programs designed for the statistical processing of diagnostic signals;
- programs, realizing algorithms of construction of decisive rules for diagnostics and classification of defects in EE units.

Let us consider the description of these products.

7.3.1 Software for Controlling the Operation of the IMS

The main functions of the developed IMS are the collection, processing, and storage of diagnostic information. The peculiarity of any experimental sample of IMS of statistical diagnostics is that the software should be adapted to work on different types of computers:

- desktop computer systems;
- laptops;
- industrial computers.

It should be noted that industrial computers and laptops have relatively low performance.

The following goals were set during the development:

- operator and developer interface convenience;
- development efficiency, i.e. possible time gain and resource-saving achieved during the development process;
- reliability of the software functioning during solving monitoring, control, and diagnostics tasks;
- dynamic adaptation to user requests;
- optimization of the final code for the possibility of work on low-production computers.

The programs, providing functioning of IMS of statistical diagnostics [90–95], consist of four modules:

- basic;
- storage and processing of information;
- display of information;
- providing the interface of IMS for statistical diagnostics.

Figure 7.18 shows the structural scheme of the software. In this scheme, the main module of the program connects the other modules and provides their control [96–98].

The data interface module provides data storage to the storage medium and data loading for further processing. Storage and processing of information is one of the most critical elements in the created system. Therefore, strict requirements are imposed on them, both to technical and hardware means, and to the software, realizing them:

- long-term—information should be stored for an unlimited period;
- reliability—to exclude possible data losses as a result of hardware or software failures;

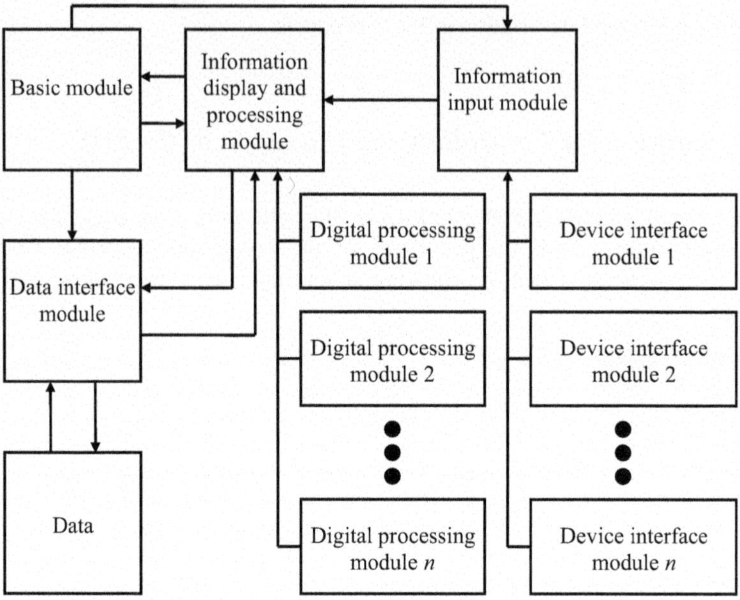

Fig. 7.18 Structural diagram of the software for controlling the operation of IMS diagnostics

- data integrity—to exclude possible violations of data structure and its connections due to incorrect actions of users;
- scalability—simplicity and ease of increasing the capacity to store and access data, due to the growth of data volumes, the number of monitoring objects, and system users;
- data size—to envisage, after preliminary analysis, the volume of stored information taking into account potential growth;
- backup duplication and data transfer—to provide automatic and manual copying of information to different sources, e.g. other computers, and optical disks.

The main window of the developed program, which is displayed on the monitor, is shown in Fig. 7.19.

Based on the requirements, a tree structure was chosen for data storage. At the lower level, the data are represented in the form of files. The file structure contains the following blocks: Signature; Test number; Date and time of test start; Date and time of test completion; Name of the investigated object; Additional information on the object; Quantization frequency; and Data. This file structure allows for scalability, fast loading, and saving of data. Data in the block can be stored in compressed form, the type of compression algorithm can vary depending on the type of data.

During working with data, the program provides a "quick view" with previously set parameters, which allows you to quickly find the necessary information. For each level, there is a "card", which specifies the conditions and characteristics of the measurements made (location of the object, date of the experiment, etc.).

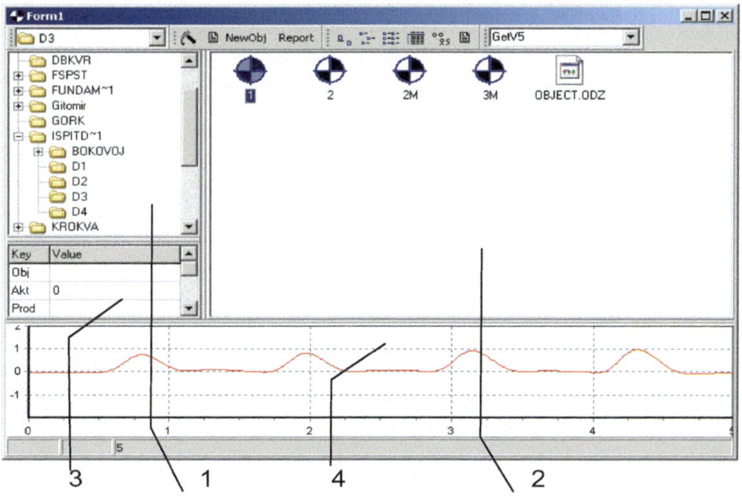

Fig. 7.19 Main program window: 1—structure; 2—list of tests; 3—test «card»; 4—preview

During viewing existing data and recording new data, the "Display and processing" window is displayed (Fig. 7.20).

The interface is based on the "Dock Managements" technology, which allows to make the interface as flexible as possible. The main and invariable element of this window is the diagram, which displays the processed signal. Other elements are connected as necessary depending on the type of processed data.

Processing modules are dynamically connected libraries (plug-ins), which contain inside the processing procedure itself and a visual element for setting parameters. The program connects plug-ins automatically. To do this, the plug-in file must be located in the same directory as the main module of the program.

It should be noted that during signal processing the initial data are not changed, the set of used algorithms and their parameters are saved in the test file.

If necessary, the data can be converted into the widespread format "XY", which allows using various mathematical software packages for data processing.

Providing the interface with the experimental sample of IMS of statistical diagnostics is also implemented as a "plug-in", which allows, without changing the main program code, to implement of the connection of various interface devices.

The following functions are realized in the developed module:

- quantization frequency setting;
- selection of the working channel;
- setting the amplifier coefficient for each channel;
- display of the device status;
- visualization of the information exchange process.

The "plug-in" window (Fig. 7.21) is displayed together with the "Display and processing" window and is a part of it.

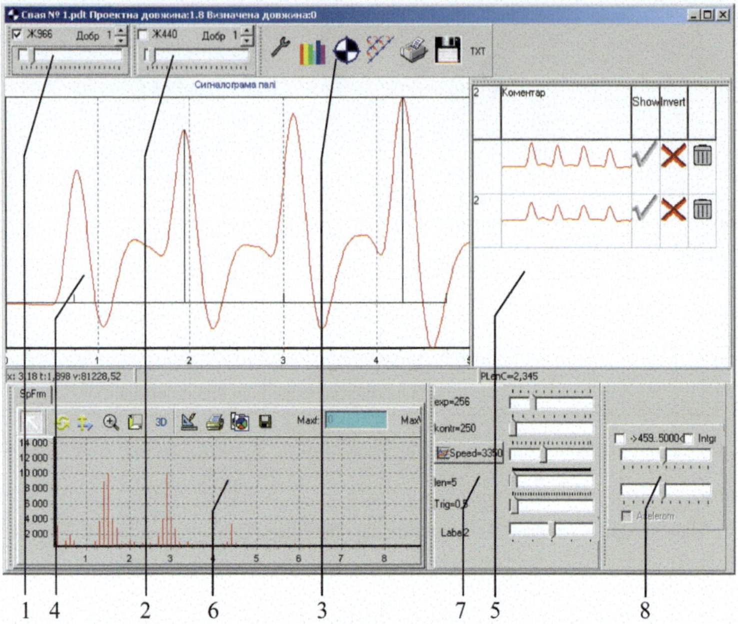

Fig. 7.20 Example of the "Display and processing" window: 1,2—parameters of filters; 3—toolbar; 4—processed signal (realization); 5—list of realizations; 6—results of processing (histogram or spectrogram) of the diagnostic signal; 7—parameters of information display; 8—bandpass filter parameters

Fig. 7.21 View of the "plug-in" dialog window

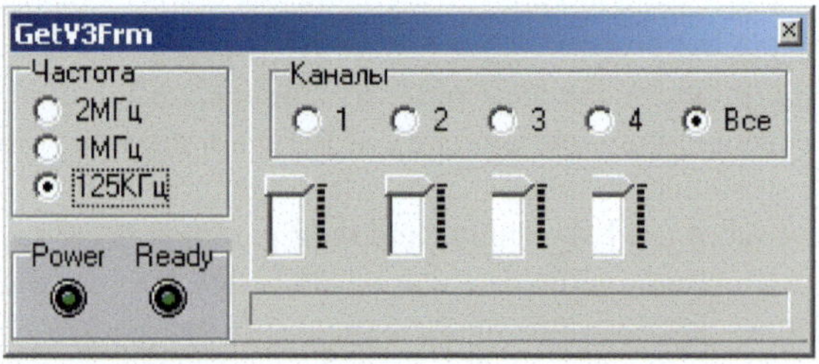

7.3.2 Software Statistical Processing of Diagnostic Signals

Programs for statistical processing of diagnostic signals, included in the created experimental sample of IMS statistical diagnostics work in Microsoft Windows environment and include the main processing programs:

- construction of histograms and their smoothing by the system of Pearson curves PEARSON;
- spectral-correlation analysis SPECTR.

The developed application package also includes auxiliary programs allowing to carry out:

- checking of the studied signals for stationarity;
- digital filtering of signals.

Recall that the basic mathematical relations and algorithms underlying these programs are discussed in detail in Chap. 6 of this book. Proceeding from this, below we shall dwell only on some general questions of practical use of the specified software for processing diagnostic signals as a part of IMS of statistical diagnostics [99–103].

As mentioned above, the processing of diagnostic signals can be carried out either by using all the above programs included in the PAP or by selecting some of them. The choice of one or another program from the PAP to determine quantitative estimates of the selected diagnostic attributes is determined, first of all, by their informativeness, i.e. their ability to determine the technical condition of the object being diagnosed.

The PEARSON program is intended for obtaining quantitative estimates of diagnostic attributes based on the analysis of the distribution of the diagnostic signals under study. This program allows us to determine estimates of initial and central moments up to the fourth inclusive, asymmetry, and excess coefficients, variation interval, build histograms, and smooth them using the system of Pearson curves.

The measurement of a diagnostic signal results in a series of consecutive time-ordered samples (time series). The ergodicity condition allows for replacement averaging over the ensemble of such realizations represented by the time series by averaging over time when finding statistical estimates of the time series.

The program has a dialog mode of communication with the user. During obtaining estimates of moments in this program, the method of averaging over time is used. When smoothing histograms, the method of moments and the system of Pearson curves are used. The relations underlying it, as well as the corresponding algorithm, based on which the PEARSON program was developed, were discussed in detail in Chap. 6 of this book.

The main window of the PEARSON program is shown in Fig. 7.22. It is divided into several semantic parts. At the top there is a menu area of standard functions (Fig. 7.23), with the help of which the following actions are performed:

- selecting a data file for processing;
- printing the graph;
- export graphs and tables to Microsoft Word program;
- exit the program.

With selecting the menu item "Open", the standard for Microsoft Windows environment dialog box of document opening appears. You can also select a file for analysis using the button, which is located on the left side immediately below the main menu area of the program. The name of the selected file to be analyzed will be displayed in the field to the right of this button. The input data for the PEARSON program is formatted as a text file (in ASCII codes).

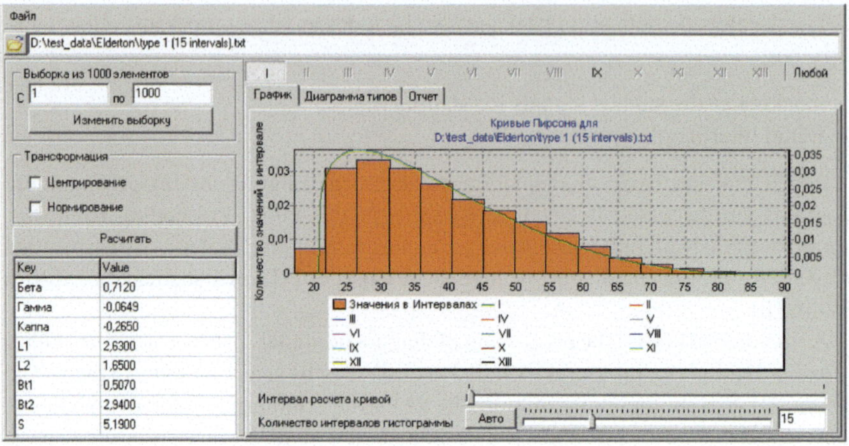

Fig. 7.22 Main window of the PEARSON program

Fig. 7.23 PEARSON standard functions menu

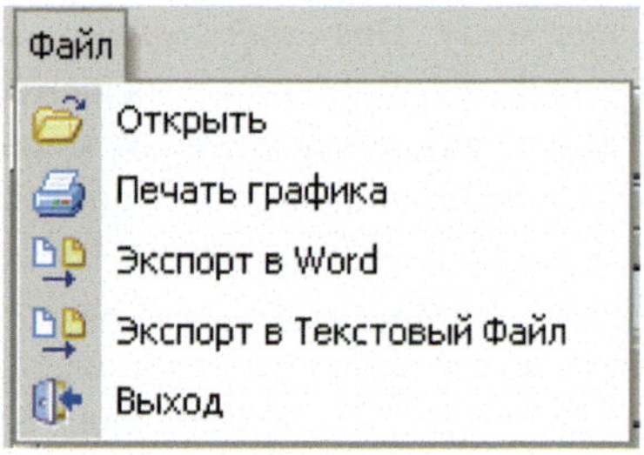

The left part of the window contains controls for selecting some part of the sample under study: "From (…) to (…)", centering and normalization of the time series, as well as the button for starting the calculation "Calculate". After pressing it, the lower left corner of the program window displays the numerical values of the calculated parameters, and the right part of the window displays graphical or tabular results, the choice between which is made using the tabs.

It is possible to display the following results:

- histogram with the graph of the selected smoothing curve superimposed on it (Fig. 7.22);
- diagram of smoothing curve types (Fig. 7.24);
- report on the calculation results in text form (Fig. 7.25), containing the values of calculated parameters, as well as tables that can be used to build graphs in the environment of other software.

The report contains histogram type (centered, normalized or not); sample size, number of intervals in the histogram; estimates of initial moments; estimates of dispersion, standard deviation, histogram interval, smallest and largest observations;

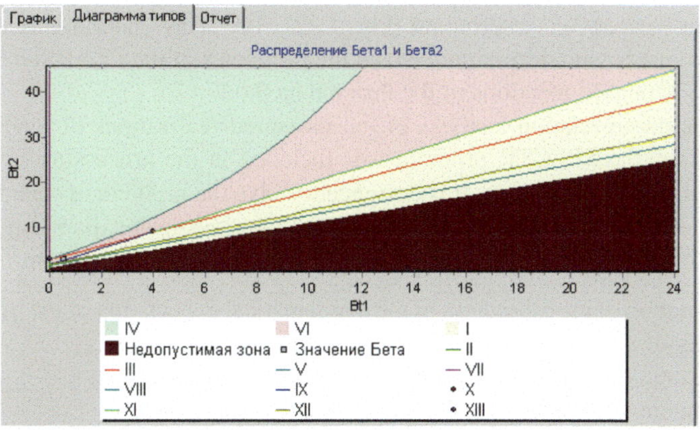

Fig. 7.24 Diagram of smoothing curve types

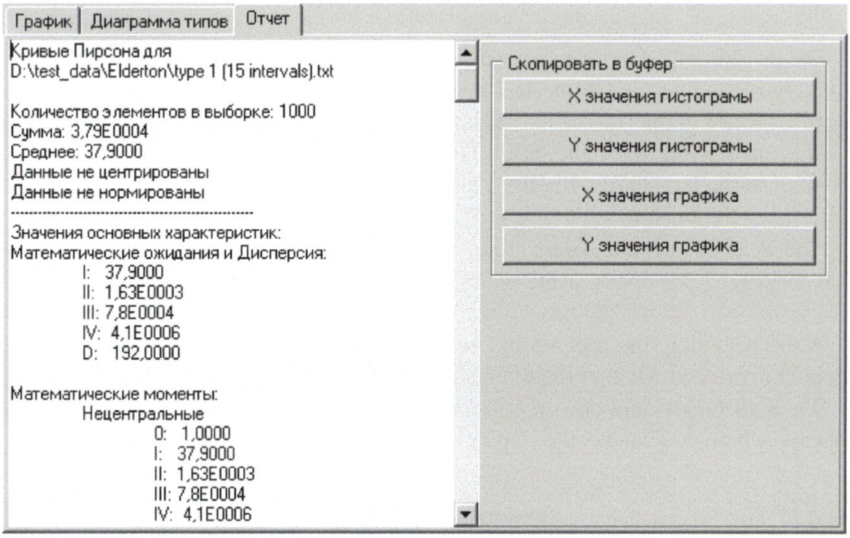

Fig. 7.25 Report on the results of PEARSON program calculations

estimates of central moments, skewness and kurtosis coefficients, Pearson's kappa value; smoothing curve type and its parameters (if possible – more than one curve type); tables of smoothing curve and histogram values, their graphs.

The tab that displays the report of the calculation results in text form has additional buttons that allow you to selectively write some tabular results to a text file.

If the simultaneous selection of several types of curves for histogram smoothing is allowed, it is possible to select only those of them that are of interest to the researcher, while hiding the graphs of those curves that are not of interest to him. For this purpose, the buttons from "I" to "XIII", also "Any" are at the top of the right part of the main window of the program (Fig. 7.22). The "Any" button is intended for experienced users who deeply understand the principles of the PEARSON program and the theoretical foundations of the Pearson method.

For spectral-correlation analysis of the measured realizations of random diagnostic signals, the SPECTR program was included in the processing software of IMS for statistical diagnostics. It is designed to solve the problem of calculating the autocorrelation function and power spectrum of the investigated diagnostic signal.

The program works according to the following generalized algorithm:

- determination of the estimation of the mathematical expectation of the time series;
- centering of the time series;
- determination of the estimation of the autocorrelation function;
- determination of the normalized autocorrelation function;
- for the estimates of normalized correlation functions of Gaussian time series the hypothesis of "zero-correlation" is tested;
- Fourier transform is used to estimate the spectral density of the process under study.

The investigated diagnostic signal (time series) is read by the SPECTR program from the input data file. Numerical results of the program execution are written to the file, spectra graphs are displayed on the monitor screen.

The main window of the SPECTR program is empty immediately after the program starts. It contains only the main program menu, which consists of the items "File" and "Parameters".

After selecting the menu item "File → Open", the window of data file selection for analysis (Fig. 7.26) opens, using which the user should set the name of the existing file.

In the next step, the user should set the file format parameters: data separator, decimal point, and whether the file has column headers (Fig. 7.27).

When all parameters are set correctly, click "OK", after which a new window appears, where you can set the step or sampling rate of the source signal, as well as

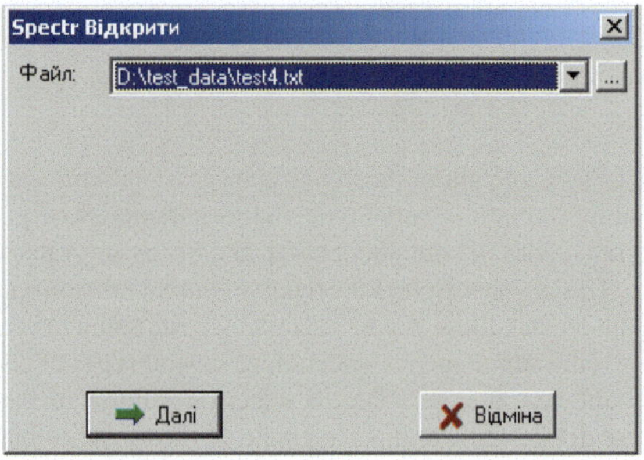

Fig. 7.26 Data file selection window

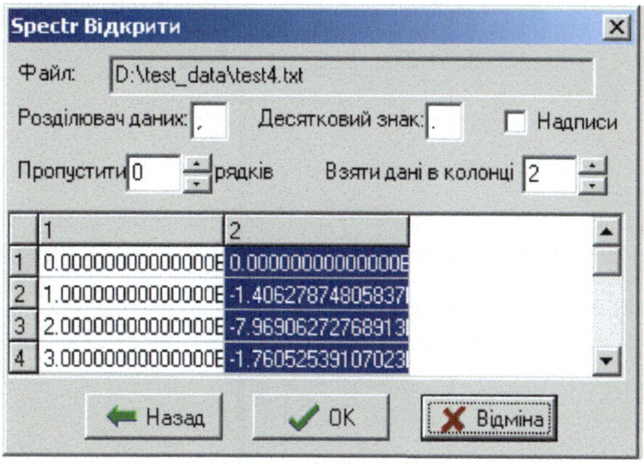

Fig. 7.27 Data file formatting selection window

the way of the graphical display of spectrum estimation (namely – what scale on the axes should be used: linear or logarithmic) (Fig. 7.28).

After pressing the "OK" button, the graph of power spectrum estimation of the investigated signal is displayed (Fig. 7.29), and text files with data tables on correlation function and power spectrum estimation are automatically created.

Because the processing programs PEARSON and SPECTR used in the IMS of statistical diagnostics are intended for processing the realizations of stationary random processes, before the subsequent statistical processing of these realizations they should be checked for stationarity. This check is performed by the program, the basic relations for which and the algorithms implementing them are given in Chap. 3. Here we will not dwell specifically on the description of the use of this program,

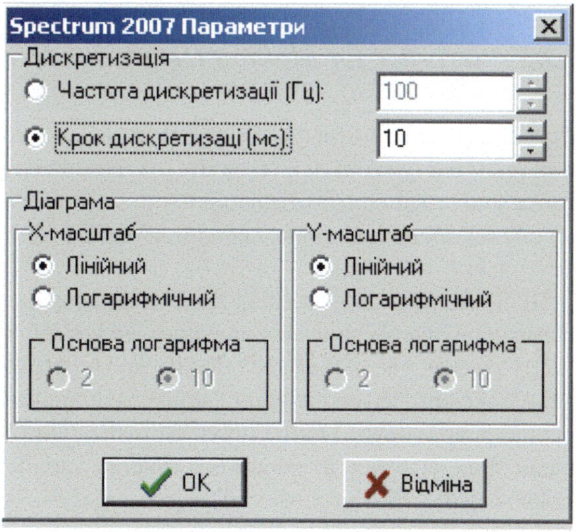

Fig. 7.28 Spectrum display parameters selection window

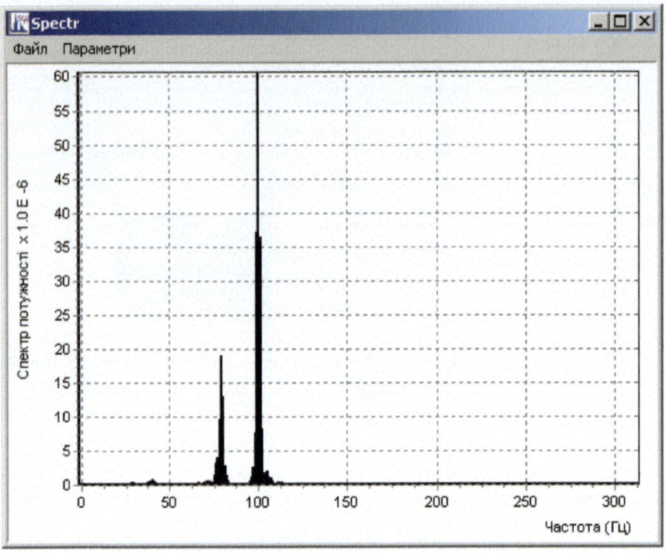

Fig. 7.29 Example of power spectrum graph

because the check of stationarity of realizations of random diagnostic signals is not very difficult.

Let us proceed to the consideration of the third type of software included in the IMS of AE-diagnostics, namely the software realizing the solving rules for diagnostics and classification of defects in EE units.

7.3.3 Software for Making Diagnostic Decisions

The program for constructing DDR is designed to determine the technical condition or classify the types of defects in the diagnosed components of EE. This program is prepared based on analytical expressions given in Chap. 5. It provides testing of hypotheses about the technical condition of EE units based on the Neyman-Pearson criterion [19, 104–107].

DDR construction is the final stage of diagnostics preparation. Various methods of statistical hypothesis testing are known and widely used in practice practice [108–112]. However, one of the most effective and relatively simple practical applications is the Neyman-Pearson method of statistical hypothesis testing. The essence of this method is as follows. On the observed realization of the process under study, we obtain some statistics X, which is a random variable and is characterized by the probability distribution density of the form $p\{x; \theta\}$. Two hypotheses are formulated concerning this probability distribution density:

$$H_0 - \text{ density parameters} \, p\{x; \theta\} : \theta = \theta_0;$$
$$H_1 - \text{ density parameters} \, p\{x; \theta\} : \theta = \theta_1,$$

where $\theta_0 \neq \theta_1$.

Next, it is necessary to obtain a criterion with the help of which it is possible to choose between the main hypothesis H_0 and the alternative H_1 hypothesis according to the observed value x of the random variable X. The construction of such a criterion is carried out according to the following principle. Some set of values of the random variable X is selected and, if the random variable takes a value from this set, the hypothesis H_0 is accepted and the hypothesis H_1 is rejected. Otherwise, the alternative hypothesis is accepted H_1.

If the hypothesis H_0 is true, and the results of the test reject it, then thereby an error is made, which is usually called an error of the first kind, and its probability is denoted by α. Hypotheses H_1 are usually chosen in such a way that the first-order error should occur as rarely as possible. In practice,α the probability is chosen to be 0.05 or 0.01. The second kind of error β occurs when the hypothesis H_0, although the hypothesis H_1 is true, is accepted according to the results of the test.

The Neumann-Pearson proposed decisive rule is reduced to the choice of some threshold C, dividing the set of admissible values of the likelihood ratio $p\{x, \theta_0\}/p\{x, \theta_1\}$ into two non-intersecting subsets. The relations determining the number of trials n and the threshold C at given values α and β, are given in Chap. 6.

A simplified scheme of the algorithm implementing the Neyman-Pearson statistical hypothesis testing rules (NEIMAN program) is shown in Fig. 7.30.

7.4 Statistical Diagnostics IMS Errors

Measurement and subsequent processing of diagnostic signals of different physical nature with the use of IMS of statistical diagnostics, and, consequently, the results of diagnostics contain errors, which are caused by: the random nature of the research processes; the impact of side processes generated in the working EM various factors; the presence of PC; the use of analog instruments and devices in the path of measurement, conversion and registration of signals under study; conversion of analog signals into digital codes for the aftermath of the diagnostic process. According to the known classification, the first three types of errors can be attributed to methodical errors, and the other two—instrumental. First of all, let us briefly dwell on the consideration of methodical errors.

Since the signals appearing in the diagnosed units of the EE are random processes, it is practically impossible to determine the technical condition of these units by a single measurement of these processes. Statistical parameters of the measured signals depend on the type of EE, operating modes of this equipment, and several other factors. Therefore, when carrying out diagnostics using statistical methods it is necessary to process at least several segments of realizations of the signals under study,

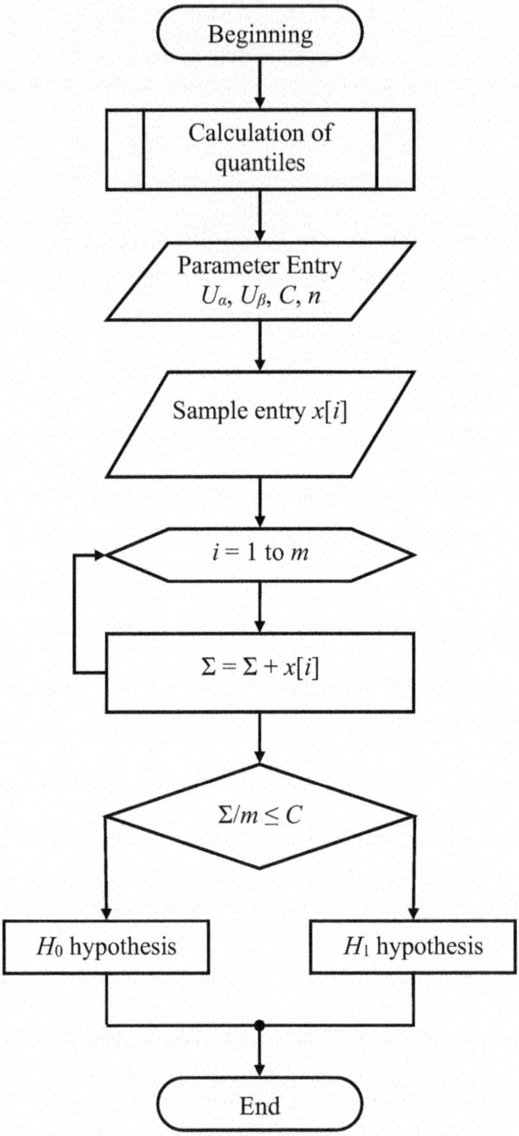

Fig. 7.30 Scheme of the NEIMAN program algorithm

characterizing the operation of both conditionally serviceable equipment and equipment having nodes with various types of defects. The stationarity of the registered diagnostic signals has a significant influence on the reliability of EE diagnostics. For estimation of stationarity of the analyzed processes a specially developed program was used, which is based on the relations given in Chap. 5 of this book. Confirmation (in the sense of the selected criteria) of stationarity makes it possible, on the one hand, to increase the accuracy of diagnostics, and on the other hand, to reduce the number of experimental data considered in the diagnostic analysis, without reducing the accuracy of diagnostics.

One of the key points when carrying out measurements on the diagnosed nodes is the choice of the signals to be investigated. In this case, all other processes occurring in the EM are considered as interference. Let's explain it in the example of vibration diagnostics of EM-bearing units. In this case, vibration displacement, vibration velocity, or vibration acceleration of vibrations are usually measured [113–117]. The informative signals are vibrations generated by the bearings themselves, and as interference are vibrations generated by the following causes:

- mechanical unbalance of the rotors;
- electromagnetic forces;
- brush-collector unit;
- aerodynamic noise.

Each of the listed causes excites characteristic vibrations in a specific frequency band, which can be determined either by eliminating different vibration sources one by one or by attenuating their effect as much as possible. Analog or digital narrow-band filtering of the measured vibrograms was used for the maximum attenuation of disturbances.

The presence of computer as a part of IMS of statistical diagnostics causes methodical errors caused by the calculation of statistical estimations of diagnostic signs on large volumes of samples. In this case, in the behavior of calculations in the computer the rounding of intermediate results is made. As a result, due to rounding errors that tend to accumulate, we get a result that has some error.

Besides the considered methodical errors, the study of measured signals is accompanied by instrumental errors caused by analog and digital devices included in the used IMS of statistical diagnostics [118–121]. Let us dwell on some questions connected with the determination of these errors.

7.4.1 Instrumental Errors of the IMS, Due to the Presence of Analog and Digital Components

The error introduced by the analog equipment is determined mainly by the instrumental error of the measuring transducers, PU, and SPU units (which include a system of analog filters) and is caused by several factors: non-uniformity of the frequency response, error of the transmission coefficient, etc.

The main instrumental error arising in digital components is introduced by the DCID. One of the main functions of this device is the conversion of analog signals into digital code. The errors arising at such conversion are connected with the presence of ADC, which in each of the three channels provides discretization of input signals in time and quantization of their level. As it is known, in this case, the errors are caused mainly by two components:

- the error associated with the time discretization of the input signal;
- the error of representation of each sample.

Let us analyze each of them.

In ADC, when processing the realizations of continuous vibration signals received from the output of accelerometers, these signals are discretized, i.e. the transition from the process with continuous time to the process with discrete time takes place. In this case, in the ideal case, we have

$$x(k\,\Delta t) = \int_{-\infty}^{\infty} x(t)\,d_t U\,(t - k\,\Delta t), \quad \Delta t > 0, \quad k = 0,\ \pm 1,\ \pm 2,\ \ldots \tag{7.1}$$

Here $x(t)$—initial continuous realization, which in the case of finite observation time $t \in [a,\ b]$ is assumed to be identically equal to zero outside this interval; $U\,(t)$—Heaviside function; Δt—discretization step.

The transition to a process with discrete time leads in the general case to irretrievable loss of information due to "thrown" values of realizations $x(t)$, and in general, it is meaningless to speak about errors arising at discretization. However, if some a priori properties of realizations are known $x(t)$, then it is possible to give the discretization errors a certain meaning, for example, understanding under it the values depending on the differences between the realizations $x(t)$ and some interpolating function dependent on the values of $x(t)$. Usually, as such a priori properties are analytical properties of realizations $x(k\,\Delta t)$, continuity, differentiability, and spectral properties.

The simplest way to reconstruct a continuous realization $x(t)$ from a discrete one is to construct a piecewise constant signal $x(k\,\Delta t)$ with continuous time from the samples. Such a process can be easily expressed through the original process as follows $x(t)$. Let us denote by $[t]$ the integer part, i.e., the other function taking the smaller t of the two nearest integer values. Then we obtain a piecewise constant process

$$y(t) = \int x(\tau)\,d\left[\frac{\tau}{\Delta t}\right], \quad t \in (-\infty,\ \infty), \tag{7.2}$$

values of which coincide with $x(k\,\Delta t)$, determined by (7.1). The graph of functions $[t]$ is shown in Fig. 7.31.

From (7.2) and the mean theorem [122, 123] it follows that for a continuous realization, $x(t)$ values $y(t)$ tend to $x(t)$. When it speaks about the discretization error, it gives a different meaning to its definition. Thus, in the case (7.2) such an error will be the difference $x(t) - y(t)$. In the general case, this error depends on the method of reconstructing the discrete process, since only two processes with continuous time can be compared.

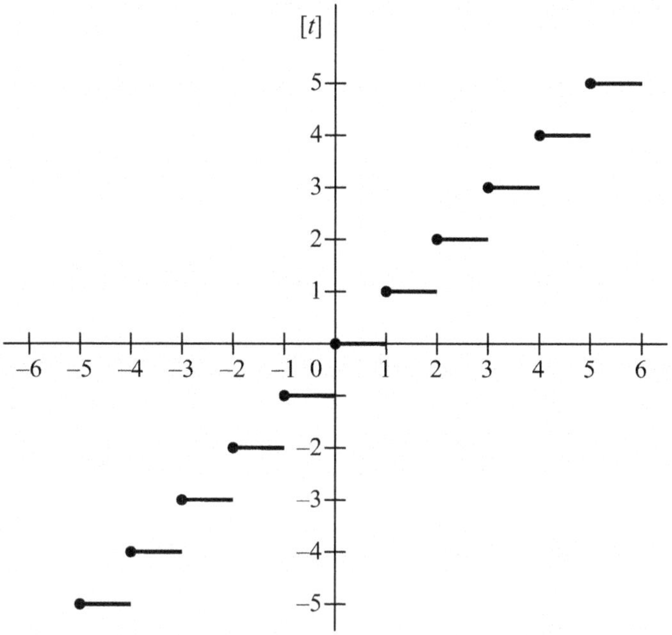

Fig. 7.31 Graph of the $[t]$ function

The sampling step Δt has a certain influence on the specified recovery error, which is often implied only as potentially possible, but not realized by the device. One way of such choice is made in the case of processes with finite energy spectra. They are completely reconstructed without errors if equidistant samples with sampling steps are chosen:

$$\Delta t = \pi / \omega_u = 1 / 2f_u, \tag{7.3}$$

on an infinite time interval, since the process with finite spectrum can take values different from zero on the whole axis $t \in (-\infty, \infty)$. Here f_u is the upper boundary frequency present in the analyzed signal.

It is impossible to realize such calculations practically, although theoretically in this case we have an interpolation formula according to Kotelnikov's theorem of samples [124–126]

$$x(t) = \sum x(k\Delta t) \frac{\sin 2\pi f_u (t - k\Delta t)}{2\pi f_u (t - k\Delta t)},$$

which, using the values given on the lattice $\{x(k\Delta t), k = 0, \pm 1, \pm 2, \ldots\}$, allows to obtain values $x(t)$ at all $t \in (-\infty, \infty)$. In practice, this series is truncated, which leads to errors, so the choice of the sampling frequency by the counting theorem does not allow us to get rid of errors in the counting results even with a finite spectrum.

The second difficulty arising when using Kotelnikov's theorem is the impossibility of fulfilling the main requirement—the finiteness of the spectrum. Thus, if a given spectrum is set over a limited interval of observations, it cannot have a finite spectral density.

To some extent, it is possible to take these errors more fully into account if we use interpolation formulas, but under the assumption that for the process under study, $\xi(t)$ any realization $x(t)$ has the required number of derivatives. In this way, the sampling step can be estimated.

The choice of the sampling interval is directly related to the question of interpolation of the signal $N \to \infty$ by its discretizations, i.e. finding the value $x(t)$ of the signal at any moment $t \in [a, b]$, at which the function is $x(t)$. Consequently, the original signal can be represented by the interpolation formula

$$x(t) = L(t) + \frac{x^{(N+1)}(\xi)}{(N+1)!}\,\omega(t), \tag{7.4}$$

where $\omega(t) = \prod_{j=0}^{N}\left(t - t_j\right)$, $\xi \in (a, b)$; $L(t)$—interpolation polynomial; $L(t) = \sum_{j=1}^{N} \frac{x(t_j)\,\omega(t)}{\omega(t_j)\,(t - t_j)}$; $\{t_1, t_2, \ldots, t_N\}$;—equidistant lattice.

The second summand in the right-hand side of (7.4) is called the residual term and determines the estimate of the error of approximation of the realization $x(t)$ by the polynomial term $L(t)$. At $N \to \infty$ the residual term with probability equal to 1, tends to zero,

$$\frac{x^{(N+1)}(\xi)}{(N+1)!}\,\omega(t) \to 0, \tag{7.5}$$

i.e. in this case the possibility of realization of the limit transition can be considered $L(t) \to x(t)$.

When analyzing a random process $\xi(t)$, the error of approximation of its realizations $x_j(t), j = \overline{1, K}$, is estimated by the mathematical expectation of the residual term (7.5). The relationship between the values of this mathematical expectation, the sample size N, and the discretization interval of the realization $x(t)$ of the random process $\xi(t)$ is established by the relations given, for example, in [7, 26, 36]. As follows from (7.5), as N increases, the approximation error decreases. Since in several cases there are no systems realizing interpolation with a given accuracy, and the durations of realizations are truncated, there is a need to increase the number of samples per period compared to the recommended by Kotelnikov. Sometimes the sampling interval is chosen based on the relationship

$$\Delta t = \frac{1}{(5 \ldots 10) f_u}. \tag{7.6}$$

However, in this case, the error should be evaluated only taking into account the problem to be solved.

The error due to the discretization of the signal without taking into account random factors in the presence of interpolation is shown in Fig. 7.32. The replacement of the real and in this case non-random periodic signal $x(t)$ by a piecewise constant function $x_D(t)$ leads to the discretization error ε^2, which can be defined as

$$\varepsilon^2 = \int [x(t) - x_D(t)]^2 dt,$$

where $x(t)$—continuous periodic signal; $x_D(t)$—discretized signal. The number of samples falling on the period of signal change, taking into account the subsequent interpolation is based on the fact that the researcher is given a certain amount of error ε^2.

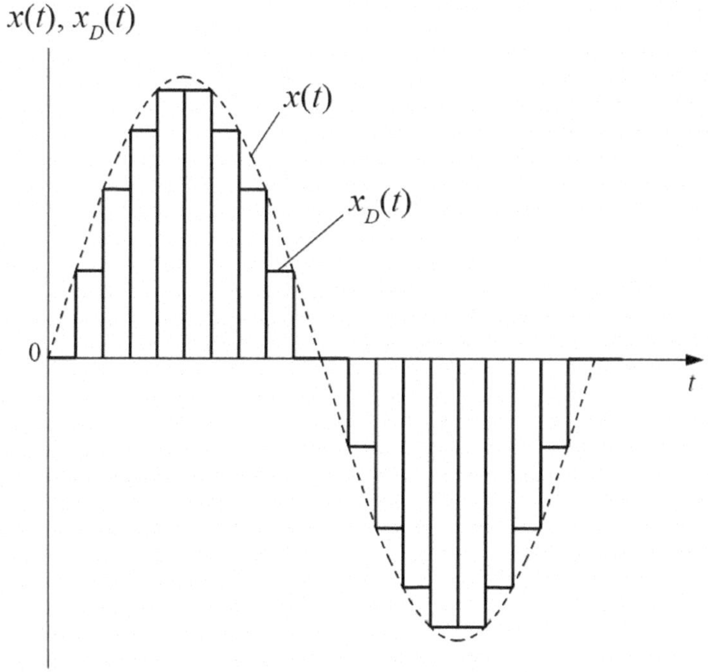

Fig. 7.32 Error due to signal discretization without taking into account random factors and using interpolation

Let's proceed to consideration of the second component of the error arising at conversion of analog signals into digital code—the error of reproduction of a sample or quantization by level.

The error of quantization by level for an ideal quantizer arises as a result of the replacement of an infinite number of possible levels of signal values by a finite number, i.e. it is caused by rounding. There are several ways of rounding; let's consider two of them.

In the following statement for the designation of the voltage coming to the input of the quantizing device, instead $x(t)$ we will use $U_x(t)$.

At rounding to the average level $[-U_m, U_m]$ the range of input voltage variation U_x is subdivided into grouping intervals of width $h = U_m / N_p$ where N_p—number of allowed levels on positive and negative half-strands, $N_p = 2^m - 1$; m–number of digits without sign except for the extreme left and right, the value of which is equal to $h/2$.

The specified transformation is carried out as follows. All values U_x, satisfying the condition $kh - h/2 < U_x \leq kh + h/2$, where $k = 0, \pm 1, \pm 2, \pm \ldots \pm N$, refer to one code value of quantization level equal to the middle of the interval kh. Each of the quantization levels is coded using binary symbols 0 and 1. The levels are denoted by corresponding binary codes.

Figure 33a shows the characteristic of quantization by the method described above, establishing the connection between the input voltage and the output code N_x (the high digit of the code is sign), assuming that $k = 3$. The largest absolute error in this case is equal to $h/2$.

When rounding to the nearest level not exceeding U_x, the input voltage range $[-Um, Um]$ is divided into grouping intervals of equal width $h = U_m/N$ as

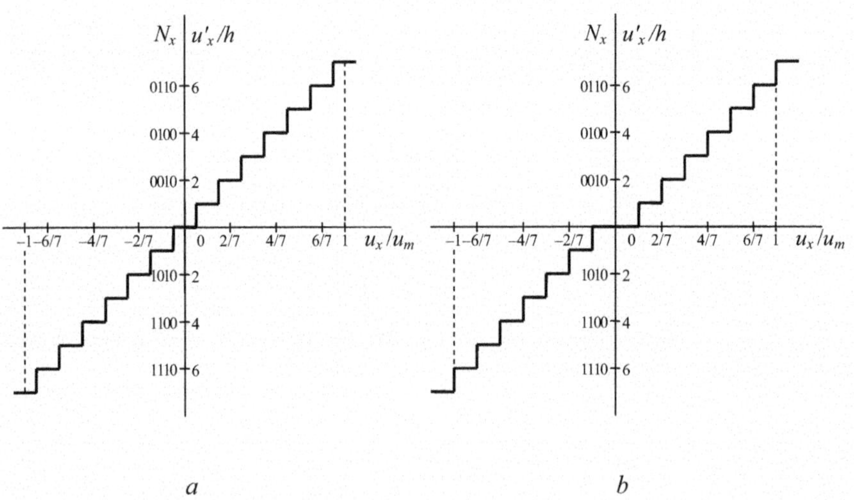

Fig. 7.33 Characteristics of quantization of different types

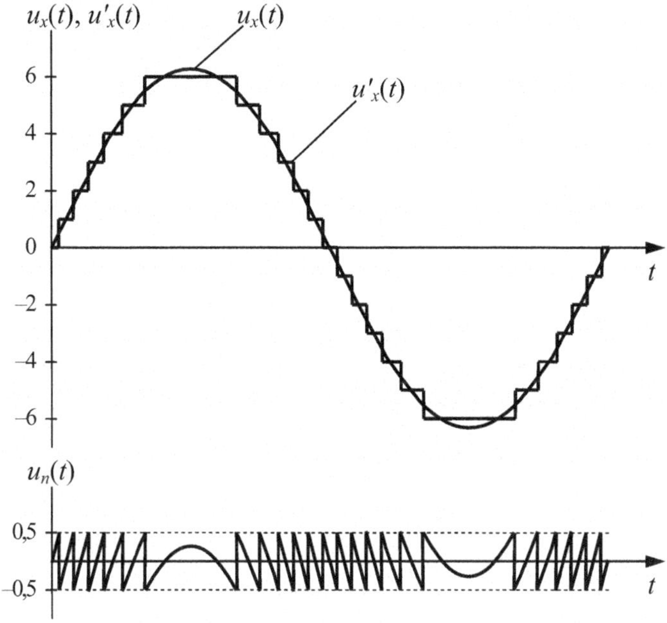

Fig. 7.34 Error of quantization of a signal of sinusoidal form

follows. All values U_x satisfying the condition, $kh < |U_x| \le (k+1)h$, $k = 0, 1, \ldots, N-1$, are referred to as the nearest boundary value $(k+1)\,h$.

Figure 33b shows the characteristics of quantization by this method. In this case, the largest absolute error is equal to h.

The difference between input and quantized signals is quantization error

$$U_n(t) = U_x(t) - U'_x(t), \quad t \in (-\infty, \infty). \tag{7.7}$$

The dependence $U_n(t)$ on $U_x(t)$ and $U'_x(t)$ for a signal of sinusoidal form is shown in Fig. 7.34. All of the above refers to the quantization error of deterministic functions, i.e. the description of the error at any moment of time for deterministic signals is determined exactly by (7.7).

Let us dwell on some peculiarities arising from the transformation of realizations of random processes into digital code with the help of ADC.

7.4.2 *Features of Errors of Analog-to-Digital Conversion of Realizations of Random Processes*

When ADC works with realizations of a random process, the meaning of the expression (7.7) changes, and the functions included in it are considered random. In this case, quantization error $U_n(t)$ is called quantization noise and it uniquely depends on the input signal. Let's stop on questions of the definition of quantization errors for realizations of random processes.

At the use of quantized signals for the calculation of estimations of statistical parameters of the initial process $\xi(t)$ because of rounding the estimation variance increases and also the bias can occur [7, 26, 36]. This can be shown as follows. Let us represent (7.7) in the form $U'_x(t) = U_x(t) - U_n(t)$. If $M\{U_x(t)\} = 0$, then the mathematical expectation of the voltage at the output of the quantizing device is

$$M\left\{U'_x(t)\right\} = M\{U_n(t)\}.$$

The correlation function $R'_x(\tau)$, $\tau \in (-\infty, \infty)$, of the voltage at the output of the quantizing device is defined as

$$R'_x(\tau) = M\left\{\overline{U}_x(t)\,U'_x(t+\tau)\right\} = R_x(\tau) - R_{xn}(\tau) - R_{nx}(\tau) + R_n(\tau),$$

where

$$R_x(\tau) = M\left\{\overline{U}_x(t)\,\overline{U}_x(t+\tau)\right\};\, R_{xn}(\tau) = M\left\{\overline{U}_x(t)\,\overline{U}_n(t+\tau)\right\};$$
$$R_{nx}(\tau) = M\left\{\overline{U}_n(t)\,\overline{U}_x(t+\tau)\right\};\, R_n(\tau) = M\left\{\overline{U}_n(t)\,\overline{U}_n(t+\tau)\right\};$$

and the line on top means the transition to the centered process. Consequently, as a result of quantization the estimates of mathematical expectation and correlation function of the input voltage get some shift. Similarly, we can show an increase in the variance of these estimates.

Thus, the signal at the ADC output can always be considered random, even when a deterministic influence is applied to its input. This is caused by the very specifics of the operation of such converters – instability of response levels, the influence of random factors in rounding, the departure of operating parameters in time, etc. Quantization and discretization errors arising during ADC operation worsen the properties of estimations of statistical parameters of input signals. Naturally, such errors should be evaluated using mathematical statistics methods. In real converters, which are electronic devices, besides the described errors, there are also errors due to the imperfection of circuit elements and parameter variations depending on time, temperature, supply voltage, etc.

The above-mentioned errors, characteristic of the statistical diagnostics IMS, serve as a basis for estimating the accuracy with which this IMS diagnoses the state of individual components of the EE.

During analyzing the realizations of random processes passing through the ADC, it is possible to use the method of statistical linearization developed by I.E. Kazakov for systems described by stochastic differential equations. Applied to ADC the method of statistical linearization can be formulated as follows. The quantized random process $x_k(t)$ at the output of ADC, obtained as a result of the nonlinear transformation of the realization of the input random process $x'(t)$, is approximated by such a linear function from the input random process, which would have the same expectation and variance as the output quantized process $x_k(t)$. The approximating dependence can be represented in the following form

$$x_k(t) = bm + b_1 x^0(t),$$

where $m = M\{x(t)\}, x^0(t) = x(t) - m$.

Let us denote $m' = M\{x_k(t)\}$ – the mathematical expectation of the quantized process $x_k(t)$ at the ADC output, $D\{x_k(t)\}$ – its variance and require the following relations:

$$bm = m', b_1^2 D\{x(t)\} = D\{x_k(t)\}.$$

Hence we obtain expressions for the coefficients b and b_1:

$$b = \frac{m'}{m}, b_1 = \sqrt{\frac{D\{x_k(t)\}}{D\{x(t)\}}}.$$

The sign of the coefficient b_1, due to the non-decreasing nature of the converter quantization characteristic, is chosen positively.

The coefficient b can be called the transmission coefficient of the device by mathematical expectation, and b_1 – the transmission coefficient of the device by standard deviation. Experimental determination of the coefficients can be carried out using a mathematical model of linear regression in the following way. Let η and ξ— two stochastically related random variables. The linear regression of the variable η relatively ξ is the function

$$M\{\eta/\xi\} = a + b\xi. \tag{7.8}$$

The task of linear regression is to construct estimates of the parameters a and b, included in expression (7.8), based on independent observations $y_1, y_2, \ldots, y_n$ of the random variable η. It is assumed that for an arbitrary value x of the random variable ξ, the random variable η has a normal distribution with mathematical expectation

$$M\eta = a + bx,$$

and dispersion σ^2, independent of x.

As applied to ADC represents the parameters of input signals, a $y_1, y_2, \ldots, y_n$—the parameters of the output digital signal, the parameter a has the meaning of the zero offset of the converter characteristic, b represents the transfer coefficient by mathematical expectation.

Expression (7.8) can be accepted as a mathematical model of the conversion device with linearized characteristics.

Because the output digital signals in real conditions are distorted by random errors, it is possible to obtain from the measurement results only an estimate of the coefficients included in expression (7.8), i.e. to find the relationship

$$\overline{y} = \overline{a} + \overline{b}x, \tag{7.9}$$

where $\overline{a}$ and $\overline{b}$ are statistical estimates of the unknown parameters a and b.

The issues of construction of these estimates are described in detail in [7, 26, 36]. When considering the linear model (7.9) analyzed in [19], the dependence of the conversion steepness on frequency was not taken into account. Due to the imperfection of frequency properties of elements and nodes of the conversion device, the steepness of conversion, determined by the parameter of the model (7.9), has some dependence on the upper boundary frequency of the signals under study. This dependence in general case cannot be calculated analytically and is determined experimentally by supplying signals with specified parameter values to the inputs of the conversion device and measuring these parameter values for the output digital signals.

Since the developed IMS is designed to work with realizations of random processes, the most suitable for the conditions of operation signals of white noise type. It is convenient to carry out the above-described test by feeding to the ADC input of the IMS conversion device a signal of white noise type with a given intensity and upper boundary frequency not exceeding half of the sampling frequency. Then using the program of statistical spectral analysis SPECTR, which was described above, on several realizations with the subsequent averaging estimated spectral density of the output digital signal as a function of frequency, resulting in the dependence of the steepness of pre-transformation on the frequency of each of the ADCs. However, such experimental verification is not always possible because the researcher often does not have generators of random signals with specified characteristics, and the characteristics of the generators themselves are very different from each other, which reduces the accuracy of this method of verification.

To study the frequency properties of converters can also be used signals consisting of a constant voltage and additively superimposed on it deterministic voltage of sinusoidal or triangular form. The amplitude of the superimposed signal is usually chosen from the condition $A = (2 \ \ldots \ 3)h$, where h – quantization step.

The controlled parameter in this case is the time average of input analog and output digital signals. However, when they are used, there are specific errors in determining such averages. Let's consider these errors when using a constant voltage and superimposed on its voltage sinusoidal form, analyzing only the sinusoidal component $U = A \sin 2\pi f t$ arriving at the input of the converter.

The memory device receives N samples of sinusoidal voltage, taken in time increments Δt ($N\,\Delta t$– time of registration). It is assumed that the first reading of the initial period occurs at the moment ϕ_0, $0 \leq \phi_0 < \Delta t$, then there are n complete oscillations in time $2\pi n$ and there is still a part of the incomplete period, which in sum with the initial incomplete period does not form a complete period, and in the worst case this sum is equal to the unipolar half-wave of the sinusoid. Thus, the constant component obtains a deviation consisting of two components. The first arises from the asymmetry of the sum of the counts on the positive and negative half-waves of the whole period caused by discretization, and the second arises from the residual of the incomplete period caused by discretization and the ratio of duration $N\,\Delta t$ and 2π.

The first component can be defined as

$$\Delta_1 = \delta_1/N, \tag{7.10}$$

where

$$\delta_1 = \frac{A}{\sin \pi f/f_D} \sin \frac{\pi\{1 - F[nf/f_D]\}f}{f_D} \sin\left\{\phi_0 - \frac{\pi f F[nf/f_D]}{f_D}\right\},$$

A – signal amplitude; f and f_D – signal frequency and sampling frequency; $F[\cdot]$ –function, fractional part.

At $f_D/f > 2$ (since $\phi_0 < \Delta t < \pi$) we have $|\delta_1| < A \sin(\pi f/f_D)$ i.e. δ_1 under the above assumptions does not exceed the signal amplitude. Therefore $|\Delta_1| < \frac{A}{N \sin \pi f/f_D}$ and practically it can always be neglected.

The second component will reach the highest value if, after excluding an integer number of periods, the remainder is exactly a unipolar half-wave of a sinusoid of frequency f. It can be shown that

$$\Delta_2 = \delta_2/N, \tag{7.11}$$

where $\delta_2 \leq A\, cosec\,(\pi f/f_D)$.

Thus, the total error $\Delta = (\delta_1 + \delta_2)/N$ is estimated as

$$|\Delta| = \frac{A}{N}\left[\sin(\pi f/f_D) + cosec(\pi f/f_D)\right].$$

Denoting the integer part $[f_D/f] = N_0$, we have

$$\begin{aligned} |\Delta| &\leq \frac{A(N_0 + 1)}{\pi N[2 - \pi/(N_0 + 1)]}; \\ |\Delta| &\leq \frac{2A}{N}\left[\frac{\pi}{2N_0} + \frac{N_0 + 1}{\pi(2 - \pi(N_0 + 1))}\right] \leq \frac{A\pi}{NN_0} + \frac{2A}{\pi}. \end{aligned} \tag{7.12}$$

It follows from the above analysis that when using periodic signals for experimental testing of ADC, it is necessary to match the frequency of the test signal and the sampling frequency, i.e., to obtain an integer number of periods in the registered realization.

The issues of checking the accuracy characteristics of the system are described in [7, 20, 22–24] in sufficient detail. Let us dwell on some issues of error estimation of the experimental sample of the IMS of statistical diagnostics.

7.4.3 Estimation of Errors of IMS of Statistical Diagnostics

Physical signals arising during the operation of diagnosed EE units have a fairly wide frequency (from a few hertz to several tens of kilohertz) and dynamic (about 100 dV) ranges. During diagnosing EE components using IMS of statistical diagnostic, it is necessary to ensure the conversion of analog signals into digital code, without, if possible, reducing the dynamic or narrowing the frequency ranges of the signals.

In real converters, which are electronic devices, except for the errors considered in paragraph 3.1 of this chapter, there are errors due to the imperfection of circuit elements and the departure of parameters depending on time, temperature, supply voltage, and some other factors. All these factors are mainly random.

Taking into account also the random character of measured diagnostic signals, below we will stop on some questions of estimation of errors arising at the practical use of IMS of statistical diagnostics.

As an example, we will determine the main types of errors for the experimental sample of the IMS of statistical vibrodiagnostics. Since this IMS sample is designed for processing of random diagnostic vibration signals, the errors arising in it are due to both the errors occurring in the processing of signals represented by one realization and the average value of these errors when using an ensemble of realizations. In general, the estimation of the end-to-end error of the whole vibration diagnostics IMS sample is a complex task, which can be the subject of an independent study. However, in the present work, when estimating the errors of the considered vibrodiagnostic IMS sample, some a priori assumptions were made concerning the causes and place of error occurrence. On this basis, when analyzing the errors, the whole investigated IMS sample was conditionally divided into several parts.

As it was mentioned in paragraph 1 of this chapter, the experimental sample of vibration diagnostics IMS (Fig. 7.1) consists of analog and digital parts. Let us briefly dwell on the analysis of errors arising in these parts of the layout. Unless it is specifically stated, we will assume that the errors considered below are inherent to each realization with probability 1. On this basis, we will focus first of all on the errors arising in the specified IMS sample when processing a single realization, and then on the issues related to the probabilistic nature of these errors (i.e., on the determination of probabilistic characteristics of errors inherent in the ensemble of realizations).

The error introduced by the analog part of the IMS is determined mainly by the instrumental error of vibration transducers and preamplifiers and is caused by several factors (non-uniformity of the frequency response, error of the transmission coefficient, etc.).

The non-uniformity of the frequency response of vibration transducers ABC-017, which were used as part of the IMS sample, in the frequency band 20 Hz...30 kHz does not exceed 3%. It means that the measurement of amplitudes of sinusoidal signals at any frequency from the specified range with the help of vibro-converters ABC-017 is made with relative errors < 3%.

The errors of the developed preamplifier, which was used in a set with vibro-converter ABC-017, are mainly caused by the non-uniformity of their frequency response. At correct matching with the signal coming from the vibration transducer, the relative error of measuring the amplitudes of sinusoidal signals using the developed preamplifier is $\delta_{pr} < 1.5\%$.

Since the above vibration transducers and preamplifiers can be conditionally considered linear links and assuming that the phase shifts in these links are negligibly small at each fixed moment, the total relative error of the analog part of the vibrodiagnostic IMS sample is between the largest and the smallest of the relative errors of the links included in the analog part of the IMS.

In this case, the relative error of the analog part of the IMS layout for each of the channels was determined by the relative errors of the vibration transducer and preamplifier and was as follows

$$\delta_{pr} \le \delta_1 \le \delta_V, 1.5\% \le \delta_1 \le 3\%.$$

The main instrumental error arising in the digital part of the developed sample of the vibration diagnostics IMS is introduced by the DCID. Below we will dwell on the estimation of errors caused by the presence of this conversion device.

As it was noted in Chap. 3, the errors arising at the conversion of analog signals into digital code, are caused by the presence of ADC, which provides discretization of input signals by time and quantization of their level. Let's use the expression for the average value of the relative error of ADC in the whole range of input voltage values ($U_{\min}$, U_m), which (assuming a uniform distribution law $p(U) = 1/(U_m - U_{\min})$ of the measured quantity) is as follows

$$\delta_{oADC} = \int \delta_o(U)p(U)dU = \delta_m \frac{U_m - U_{\min}}{U_m} + \delta_{ad} \ln \frac{U_m}{U_{\min}}, \quad (7.13)$$

where $\delta_o(U) = \delta_m + \delta_{ad} U_m / U_{\min}$; δ_m – multiplicative component of the relative error; δ_{ad} – its additive component.

The values δ_m and δ_{ad} can be determined based on the accuracy class of the ADC used. Thus, for the ADC K1108 PV2 used in the DCID, which has an accuracy class of 0.3/0.2, the values δ_{0m} and δ_{0ad} are respectively equal to $\delta_{0m} = 0.05\%$; $\delta_{0ad} = 0.2\%$. Then based on (13) and taking into account that for ADC K1108 PV2 $U_m = 1023$ mV, $U_{\min} = 2$ mV, the range average $\{U_{\min}, U_m\}$ value of the relative error will be $\delta_{0ADC} = 2.1\%$.

Thus, the total relative instrumental error of the analog and digital parts of the developed IMS sample for each of the 3 channels, provided that in the analog part of the IMS signals from the output of the preamplifier go directly to the input of the DCID, will be $1.5\% \leq \delta_{0tot} \leq 3\%$.

Now let us consider the methodological errors that arise when processing the measured realization on the computer, which is a part of the IMS vibrodiagnostic layout. As a result of processing of each realization on the computer with the help of the software described above, quantitative estimates of diagnostic features are determined. When determining these estimates, errors inevitably occur, which are caused primarily by rounding errors, as well as by the methods underlying the system of working programs.

When performing calculations with large sample sizes, rounding errors tend to accumulate. To reduce these errors, all calculations in the computer were performed with numbers rounded to the m-th decimal place, which was determined by Chebotarev law [127–129] from the expression

$$\Delta_s = \sqrt{3N} \cdot 0.5 \cdot 10^{-m}, \tag{7.14}$$

where Δ_s—absolute error of calculations, which is required to provide the obtained results; N—number of samples in the registered sample, which was chosen equal to $N = 2048$.

To obtain estimates of diagnostic signs with four correct signs (i.e. $\Delta = 0.0001$) by the expression (7.14) all calculations in the computer must be made with numbers rounded to $m \geq -\lg\left(2.6 \cdot 10^{-6}\right) \approx 6$ decimal places. The computer with an Intel processor used as a part of the developed sample IMS of vibration diagnostics provided calculations with the specified number of decimal places. As will be shown in Chap. 8, one of the most informative diagnostic signs used in vibrodiagnostics of rolling bearings is the coefficient of excess of measured vibration signals γ (or a related parameter β_2 from the Pearson curve system) and the parameter ϕ. Below we will dwell on the estimation of the latter.

The estimation of the accuracy of the parameter ϕ, determined according to expression (8.5), is obtained using the method of statistical estimation. The use of this method allows us to determine the accuracy of the estimates using the ensemble of realizations, i.e., taking into account the random nature of the bearing vibrations obtained as a result of the experiment. Let us dwell on this issue in more detail, namely, let us construct a 95% confidence interval for estimating the true mean M_ϕ of the random variable ϕ. As initial data, we will use the estimates ϕ, obtained as a result of processing 60 realizations of vibrations of the serviceable bearing 309 ESH2, installed on the collector side in the DC EM. Quantitative estimates of the parameter were obtained based on the experimental data given and also discussed in Chap. 8 of this book.

The obtained estimates ϕ are approximated by the normal distribution law with a sufficient degree of accuracy. The confidence interval for $M\,\phi$ is determined by the inequality

$$\left[\tilde{\phi} - \frac{\sqrt{D\phi}\,u_{1-\alpha}}{\sqrt{N}}\right] \leq M\,\phi < \left[\tilde{\phi} + \frac{\sqrt{D\phi}\,u_{1-\alpha}}{\sqrt{N}}\right],$$

where $\tilde{\phi}$ and $D\phi$—sample mean and variance, which for the considered series of estimates ϕ, respectively, are equal to $\tilde{\phi} = 56.73$ and $D\phi = 0.099$; $u_{1-\alpha}$—quantile of normal distribution, which at $\alpha = 0.05$ is equal to $u_{1-\alpha} = 1.645$; N—sample size. In this example, $N = 60$.

Thus, taking into account the specified initial data, the 95% confidence interval for the mean value $M\,\phi$ is as follows

$$56.66 \le M\,\phi < 56.80.$$

As will be shown in Chap. 8 of this book, obtaining the average values of the parameter ϕ with the specified accuracy provides stable diagnostics and classification of defects in bearings.

During carrying out the checks of the developed conversion device, discussed above, as well as when examining the experimental data, the program of autocorrelation analysis included in the developed software was used. When outputting the results of counting by the specified program for each ordinate of the correlation function, for each ordinate of the correlation function the estimates of the biased and unbiased correlation function, the estimate of the variance of the estimate of the correlation function, as well as the confidence probability with which the obtained estimates of the correlation function are in a certain confidence interval are printed. The length of the specified confidence interval is determined using Chebyshev's inequality [130–132]

$$P\left\{ \left|\tilde{B}(s) - B(s)\right| > QD^{1/2}\left[\tilde{B}(s)\right] \right\} \le \frac{1}{Q^2},$$

where $B(s)$—true value of the correlation function; $\tilde{B}(s)$—estimate of the correlation function; Q – confidence coefficient; $D\left[\tilde{B}(s)\right]$—estimate of the variance of the correlation function estimate.

During performing the DCID checks, sinus-modal signals were used, and the sampling frequency f_D and the frequency of the test signal f were matched to ensure that an integer number of periods was obtained in the registered realization. In addition, $f_D/f > 2$ was chosen in all tests. Consequently, the error Δ_1, determined according to (7.10), can be neglected.

The second component of the error can be estimated by (7.12). This error will reach the highest value if after excluding a whole number of periods in the registered sample, the remainder will be a unipolar sine wave of frequency f. In this case, at the maximum signal amplitude equal to 1 V, the second component of the error Δ_2, according to (7.12), can reach

$$|\Delta| \le \frac{1 \cdot 3.14}{2048 \cdot 10} + \frac{2 \cdot 1}{3.14} = 0.637,$$

or percentage

$$\delta_2 = \frac{637}{1023} \cdot 100\% \approx 62\%.$$

According to the results of this chapter the following conclusions can be made:

- on the example of an experimental sample of IMS of vibrodiagnostics of rolling bearings the nature of occurrence of main errors of IMS of statistical diagnostics was analyzed and investigated;
- basic calculation relations for metrological control of IMS of statistical diagnostics were obtained;
- using the method of confidence intervals the accuracy of the diagnostic attributes estimations obtained using a computer is established.

Having obtained the basic relations for determining the metrological characteristics of IMS of statistical diagnostics, let us proceed to the consideration of one of the most important components of such IMS, namely, the component containing pre-formed training sets corresponding to different technical states of the object of diagnostics.

7.5 Training of OMS of Statistical Diagnostic

Diagnostics using the considered IMS is based on the general statistical theory of detection and estimation, as its task is to detect defects and assess the state of the object. Statistical diagnostics is a part of diagnostics that uses diagnostic information the realizations of random processes measured at different nodes of EE.

Statistical estimation of separate parameters and characteristics of the observed random diagnostic signal, though it allows to carrying out of diagnostics with the required accuracy and reliability, however, is only the first stage of this process, as it is not supported by the estimates of statistical reliability and usually ends with obtaining the quantitative estimation of the parameters of the observed diagnostic signals.

The issues of diagnostics of EE units considered below are based on classical methods of statistical hypothesis testing using the model of linear random processes. The main attention is paid to the study of training populations formed by the most informative diagnostic features in the corresponding diagnostic spaces, as well as to the solving rules that allow to establish the presence of a quite definite defect in the diagnosed EE units with predetermined probabilities of errors of the first and second kind.

First of all, let us dwell on the basic elements of the simplest statistical diagnosis problem, which is always based on some probabilistic model of the observed process.

The first element is the source of diagnostic information, which creates some process depending on diagnostic parameters, each of which can be a scalar or a vector. In the simplest case, only one such parameter is considered, which can take two different values. Mutually exclusive hypotheses are associated with the occurrence of each value: main H_0 and alternative H_1 hypotheses. In a more complex situation, the parameter under study can take a finite or countable number of values. In this case, one vector is usually considered as one value of the vector parameter. In such a case, a multi-alternative situation takes place, and H_0 is always considered the main hypothesis, and the others are considered alternative hypotheses. It is not known a priori which hypothesis takes place for the object being diagnosed.

The second element is preliminary statistical processing, which is done as if it is known which hypothesis is true. Based on this knowledge, a well-defined point in the observation space is obtained. It is assumed that at this stage some additive random component independent of hypotheses is added, which can occur both at the output of the source of diagnostic information and in the process of its statistical processing. It distorts the value of the diagnostic parameter, so each hypothetical value of the parameter is associated with a probability law of distribution that unambiguously corresponds to each of the above-mentioned hypotheses.

The observation space is *the third element.* As a result of interference in the observation space sets of points corresponding to the hypotheses, intersect, and sometimes completely coincide. Therefore, it becomes necessary to divide the entire observation space into non-intersecting subsets - images of each hypothesis.

The fourth element is the decision rule, which is constructed taking into account the training sets and represents the optimal partitioning of some transformed observation space into non-intersecting subsets. In a two-alternative situation, the set of points on the numerical axis most often serves as such a transformed space.

Training sets - a set of parameter values that correspond to the conditionally serviceable state of the diagnosed unit or its states associated with the presence of one of the diagnosed defects. Without having such sets, it is impossible in principle to build a decisive rule, and therefore it is impossible to carry out diagnostics in the way described above. Training sets may overlap, which is most often the practice case. If the decisive rule is constructed, then the decision itself is made based on some diagnostic sample, the volume of which is calculated by the decisive rule, which completes the choice of hypothesis, i.e. the solution to the diagnosis problem is obtained.

Figure 7.35 shows the main elements of the diagnosis task: diagnostic information source (DIS); statistical preliminary processing (SPP); observation space (OS); TS; and DDR.

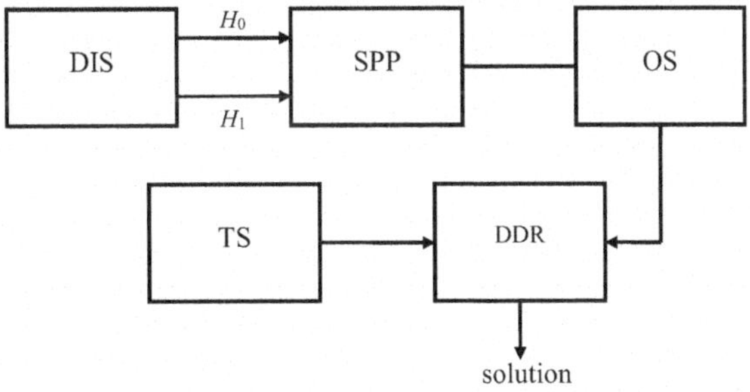

Fig. 7.35 Main elements of the diagnostics task

The process of experimental construction of training sets is usually called training in diagnostics. Its results significantly determine the accuracy and reliability of diagnostics, as it is the basis for the calculation and planning of the whole diagnostics experiment. It is in the process of training that images of technical states of the investigated units are formed, and also estimates of probabilistic laws are constructed, with the help of which hypothesis formulation and preliminary statistical processing of diagnostic information are carried out.

Since the main elements of the diagnostics task in a general form have been considered above, further it remains to dwell on the issues related to training and formation of images corresponding to possible technical states of various nodes of the EE.

The main essence of the learning process, selection of diagnostic spaces, and formation of images corresponding to different technical states of the diagnosed units will be presented on the examples of the study of signals accompanying the operation of the bearing units, the machined magnetic core of EM, and the CPG of DEG, described in the next chapter.

References

1. Eremenko, V., Zaporozhets, A., Isaienko, V., Babikova, K.: Application of wavelet transform for determining diagnostic signs. In: Proceedings of the 15th International Conference on ICT in Education, Research and Industrial Applications. Integration, Harmonization and Knowledge Transfer, Vol. 1, pp. 12–15 (2019)
2. Babak, V., Eremenko, V., Zaporozhets, A.: Research of diagnostic parameters of composite materials using Johnson distribution. Int. J. Comput. **18**(4), 483–494 (2019)
3. Zaporozhets, A., Eremenko, V., Isaenko, V., Babikova, K.: Approach for creating reference signals for detecting defects in diagnosing of composite materials. In: Conference on Computer Science and Information Technologies, pp. 154–172. Springer International Publishing, Cham (2019)
4. Lei, Y., Kong, D., Lin, J., Zuo, M.J.: Fault detection of planetary gearboxes using new diagnostic parameters. Meas. Sci. Technol. **23**(5), 055605 (2012)
5. Song, L., Wang, H., Chen, P.: Vibration-based intelligent fault diagnosis for roller bearings in low-speed rotating machinery. IEEE Trans. Instrum. Meas. **67**(8), 1887–1899 (2018)
6. Sverdlova, A., Zaporozhets, A. Predicting anomaly conditions of energy equipment using neural networks. In: E3S Web of Conferences (Vol. 280, p. 09005) (2021)
7. Babak, V., Scherbak, L., Kuts, Y., Zaporozhets, A.: Information and measurement technologies for solving problems of energy informatics. In: The 1st International Workshop on Information Technologies: Theoretical and Applied Problems 2021. CEUR Workshop Proceedings, vol. 3039, pp. 24–31 (2021)
8. Das, U.K., Tey, K.S., Seyedmahmoudian, M., Mekhilef, S., Idris, M.Y.I., Van Deventer, W., Stojcevski, A.: Forecasting of photovoltaic power generation and model optimization: a review. Renew. Sustain. Energy Rev. **81**, 912–928 (2018)
9. Jiang, C., Zhang, H., Ren, Y., Han, Z., Chen, K.C., Hanzo, L.: Machine learning paradigms for next-generation wireless networks. IEEE Wirel. Commun. **24**(2), 98–105 (2016)
10. Conejo, A.J., Carrión, M., Morales, J.M.: Decision making under uncertainty in electricity markets, vol. 1, pp. 376–384. Springer, New York (2010)

11. Malik, H., Iqbal, A., Yadav, A.K.: Soft computing in condition monitoring and diagnostics of electrical and mechanical systems, vol. 1096, p. 499. Springer, Berlin/Heidelberg, Germany (2020)
12. Henao, H., Capolino, G.A., Fernandez-Cabanas, M., Filippetti, F., Bruzzese, C., Strangas, E., Hedayati-Kia, S.: Trends in fault diagnosis for electrical machines: a review of diagnostic techniques. IEEE Ind. Electron. Mag. **8**(2), 31–42 (2014)
13. Witczak, M.: Fault diagnosis and fault-tolerant control strategies for non-linear systems. vol. 266. Springer International Publishing, Heidelberg, Germany, pp. 375–392 (2014)
14. Eremenko, V., Babak, V., Zaporozhets, A.: Method of reference signals creating in non-destructive testing based on low-speed impact. Tekhnichna Elektrodynamika **4**, 070 (2021)
15. Zaporozhets, A., Babak, V., Sverdlova, A., Isaienko, V., Babikova, K.: Development of a system for diagnosing heat power equipment based on IEEE 802.11 s. In: Systems, Decision and Control in Energy II, pp. 141–151. Springer International Publishing, Cham (2021)
16. Zaporozhets, A.: Analysis of control system of fuel combustion in boilers with oxygen sensor. Period. Polytech. Mech. Eng. **63**(4), 241–248 (2018)
17. Karlaftis, M.G., Vlahogianni, E.I.: Statistical methods versus neural networks in transportation research: differences, similarities and some insights. Transp. Res. Part C: Emerg. Technol. **19**(3), 387–399 (2011)
18. Yin, S., Ding, S.X., Xie, X., Luo, H.: A review on basic data-driven approaches for industrial process monitoring. IEEE Trans. Industr. Electron. **61**(11), 6418–6428 (2014)
19. Zaporozhets, A.O.: Correlation analysis between the components of energy balance and pollutant emissions. Water Air Soil Pollut. **232**, 1–22 (2021)
20. Babak, V. P., Babak, S. V., Eremenko, V. S., Kuts, Y. V., Myslovych, M. V., Scherbak, L. M., Zaporozhets, A.O.: Models and measures for the diagnosis of electric power equipment. Models Measures in Measurements Monitoring, pp. 99–126 (2021)
21. Zaporozhets, A., Eremenko, V., Serhiienko, R., Ivanov, S.: Methods and hardware for diagnosing thermal power equipment based on smart grid technology. In: Advances in Intelligent Systems and Computing III: Selected Papers from the International Conference on Computer Science and Information Technologies, CSIT 2018, September 11–14, Lviv, Ukraine, pp. 476–489. Springer International Publishing (2019)
22. Babak, V.P., Babak, S.V., Myslovych, M.V., Zaporozhets, A.O., Zvaritch, V.M.: Principles of construction of systems for diagnosing the energy equipment. Diagnostic Systems for Energy Equipments, pp. 1–22 (2020)
23. Babak, V.P., Babak, S.V., Eremenko, V.S., Kuts, Y.V., Myslovych, M.V., Scherbak, L.M., Zaporozhets, A.O.: Models and measures for standardless measurements of the composite materials characteristics. Models and Measures in Measurements Monitor. 157–190 (2021)
24. Babak, V., Zaporozhets, A., Zvaritch, V., Scherbak, L., Myslovych, M., Kuts, Y.: Models and measures in theory and practice of manufacturing processes. IFAC-PapersOnLine **55**(10), 1956–1961 (2022)
25. Babak, V., Zaporozhets, A., Kovtun, S., Kuts, Y., Fryz, M., Scherbak, L.: Information provision for monitoring the current state of electric power facilities. In: Advanced System Development Technologies I, pp. 283–309. Springer Nature Switzerland, Cham (2023)
26. Babak, V., Zaporozhets, A., Kuts, Y., Myslovych, M., Fryz, M., Scherbak, L.: Models and characteristics of identification of noise stochastic signals of research objects. In: CEUR Workshop Proceedings, vol. 3309, pp. 349–362 (2022)
27. Babak, V., Zaporozhets, A., Kuts, Y., Scherbak, L.: Some features of Hilbert transform and their use in energy informatics. The Probl General Energy 1–2 (2022)
28. Babak, V.P., Babak, S.V., Myslovych, M.V., Zaporozhets, A.O., Zvaritch, V.M.: Methods and models for information data analysis. Diagnost. Syst. Energy Equipments 23–70 (2020)

29. Zvaritch, V., Myslovych, M., Gyzhko, Y.: Application of linear random processes to construction of diagnostic system for power engineering equipment. In: IFIP International Conference on Advances in Production Management Systems. Springer International Publishing, Cham, pp. 617–622 (2021)
30. Myslovych, M., Hyzhko, Y., Hutorova, M., Ostapchuk, L.: On peculiarities of development of information support for technical diagnostics multilevel systems of electrical equipment. Comput. Probl. Electr. Eng. **9**(2), 34–41 (2019)
31. Babak, S., Myslovych, M., Sysak, R.: Computerized systems for remote monitoring and diagnostics of electric power facilities. Comput. Probl. Electr. Eng. **5**(1), 1–4. (2015)
32. Zaitsev, I., Levytskyi, A., Bereznychenko, V.:. Hybrid diagnostics systems for power generators faults: systems design principle and shaft run-out sensors. Power Systems Research and Operation: Selected Problems, 71–98 (2022)
33. Šiniković, G., Gubeljak, N., Veg, E., Regodić, M., Jagarinec, D., Medojević, I., Jeli, Z.: Rolling bearing fault detection in the range of ultrasound. Tehnički vjesnik **30**(3), 830–836 (2023)
34. Mazal, P., Nohal, L., Komenda, L.: Diagnostics of damage development of bearings and gears by acoustic emission method. Int. J. Microstruct. Mater. Prop. **9**(3–5), 433–447 (2014)
35. Golonka, E., Pająk, M., Kolar, D.: Selected diagnostic methods used in Hydroelectric Power Plants. In: MATEC Web of Conferences, vol. 332, p. 01011. EDP Sciences (2021)
36. Babak, V., Zaporozhets, A., Kovtun, S., Myslovych, M., Kuts, Y., Scherbak, L.: Information support for identification of the technical state of electric power facilities. In: Systems, Decision and Control in Energy V, pp. 129–153. Springer Nature Switzerland, Cham (2023)
37. Kucera, L., Gajdosik, T., Bucala, J.: The Vibrodiagnostics of damaged gears of planetary gearboxes. Commun.-Sci. Lett. Univ. Zilina **16**(3A), 67–73 (2014)
38. Dybała, J.: Diagnosing of rolling-element bearings using amplitude level-based decomposition of machine vibration signal. Measurement **126**, 143–155 (2018)
39. Mitrovic, R.M., Miskovic, Z.Z., Djukic, M.B., Bakic, G.M.: Statistical correlation between vibration characteristics, surface temperatures and service life of rolling bearings–artificially contaminated by open pit coal mine debris particles. Proced. Struct. Integrity **2**, 2338–2346 (2016)
40. Atanasovska, I.D., Soldat, N.D., Patil, S.S., Mitrovic, R.M., Tomovic, R.Z.: Damage factor calculation for condition monitoring of rolling bearings. Arab. J. Sci. Eng. **48**(3), 3181–3194 (2023)
41. Barszcz, T., Barszcz, T.: Standard vibration analysis methods. Vibration-Based Cond. Monitor. Wind Turbines, 33–86 (2019)
42. Arhun, S., Migal, V., Hnatov, A., Ponikarovska, S., Hnatova, A., Novichonok, S.: Determining the quality of electric motors by vibro-diagnostic characteristics. EAI Endorsed Trans. Energy Web **7**(29), e6–e6 (2020)
43. Mansurov, T., Aliyev, C.: Development of an information-measuring system for measuring the dynamic parameters of moving objects. In: 2021 IEEE 8th International Conference on Problems of Infocommunications, Science and Technology (PIC S&T), pp. 23–27. IEEE (2021)
44. Skliarov, V.: Nondestructive evaluation and monitoring of vibration characteristics of equipment by MEMS sensors. In: Nondestructive Characterization and Monitoring of Advanced Materials, Aerospace, Civil Infrastructure, and Transportation XIV, vol. 11380, pp. 215–226 SPIE (2020)
45. Zaitsev, I., Bereznychenko, V.: Condition monitoring and fault diagnosis systems of power generators with non-contact shaft runout electrocapacitive transducer. In: 2023 IEEE 4th KhPI Week on Advanced Technology (KhPIWeek) (pp. 1–6). IEEE (2023)
46. Glowacz, A., Tadeusiewicz, R., Legutko, S., Caesarendra, W., Irfan, M., Liu, H., Xiang, J.: Fault diagnosis of angle grinders and electric impact drills using acoustic signals. Appl. Acoust. **179**, 108070 (2021)
47. Zhang, L., Shang, H.C., Zhao, Z.Q., Zhang, D.F.: Evaluation of the dynamic characteristics of AZ91D crankcase covers and boxes. Strength Mater. **47**, 128–135 (2015)

48. Addabbo, T., Fort, A., Landi, E., Moretti, R., Mugnaini, M., Parri, L., Vignoli, V.: A Characterization system for bearing condition monitoring sensors, a case study with a low power wireless Triaxial MEMS based sensor. In: 2020 IEEE International Workshop on Metrology for Industry 4.0 & IoT, pp. 11–15. IEEE (2020)
49. Soloiu, V., Weaver, A., Parker, L., Brant, A., Smith, R., III., Ilie, M., Carapia, C.: Constant volume combustion chamber (CVCC) investigations of aerospace F-24 and Jet-A in low-temperature heat release and negative temperature coefficient regions. Energy Convers. Manage. **263**, 115687 (2022)
50. Landi, E., Spinelli, F., Intravaia, M., Mugnaini, M., Fort, A., Bianchini, M., Corradini, B.T., Scarselli, F., Tanfoni, M. (2023) A mobilenet neural network model for fault diagnosis in roller bearings. In: 2023 IEEE International Instrumentation and Measurement Technology Conference (I2MTC), pp. 01–06. IEEE.
51. Ftoutou, E., Chouchane, M.: Diesel engine injection faults' detection and classification utilizing unsupervised fuzzy clustering techniques. Proc. Inst. Mech. Eng. C J. Mech. Eng. Sci. **233**(16), 5622–5636 (2019)
52. Kovenburg, R., Slezak, A., George, C., Gale, R., Aksak, B.: The orientation dependence of the fingerprint effect for slip speed estimation and control. IEEE Sens. J. **23**(5), 5437–5447 (2023)
53. Babak, V.P., Babak, S.V., Eremenko, V.S., Kuts, Y.V., Myslovych, M.V., Scherbak, L.M., Zaporozhets, A.O.: Problems and features of measurements. Models Measures Measurements Monitor. pp. 1–31 (2021)
54. Babak, V.P., Babak, S.V., Eremenko, V.S., Kuts, Y.V., Myslovych, M.V., Scherbak, L.M., Zaporozhets, A.O.: Models of measuring signals and fields. Models Measures in Measurements and Monit. 33–59 (2021)
55. Guo, L., Lei, Y., Xing, S., Yan, T., Li, N.: Deep convolutional transfer learning network: a new method for intelligent fault diagnosis of machines with unlabeled data. IEEE Trans. Industr. Electron. **66**(9), 7316–7325 (2018)
56. Lei, Y., Yang, B., Jiang, X., Jia, F., Li, N., Nandi, A.K.: Applications of machine learning to machine fault diagnosis: a review and roadmap. Mech. Syst. Signal Process. **138**, 106587 (2020)
57. Stone, G.C., Culbert, I., Boulter, E.A., & Dhirani, H.: Electrical insulation for rotating machines: design, evaluation, aging, testing, and repair, vol. 83. John Wiley & Sons (2014)
58. Nguefack, L.T., Pauné, F., Mbihi, J.: Experimental study of ECG signal transmission system via a coaxial cable line using duty-cycle modulation. Algerian J. Eng. Technol. **4**, 1–6 (2021)
59. Jarutus, N., Kumsuwan, Y.: A unified duty-cycle modulation algorithm for a three-level NPC inverter. AEU-Int. J. Electron. Commun. **172**, 154931 (2023)
60. Niu, F., Huang, X., Ge, L., Zhang, J., Wu, L., Wang, Y., Fang, Y.: A simple and practical duty cycle modulated direct torque control for permanent magnet synchronous motors. IEEE Trans. Power Electron. **34**(2), 1572–1579 (2018)
61. Agoston, K.: Accelerometer characteristics, errors and signal conditioning. In: The International Conference Interdisciplinarity in Engineering INTER-ENG, p. 276. Elsevier Limited (2012)
62. Gómez Pau, Á., Sanahuja Moliner, R., Balado Suárez, L.M., Figueras Pàmies, J.: Built-In test of MEMS capacitive accelerometers for field failures and aging degradation. In: Proceedings of XXVIIth conference on design of circuits and integrated systems, pp. 223–228 (2012)
63. Hernandez, W., de Vicente, J., Sergiyenko, O., Tyrsa, V.: Estimation of the acceleration of a car under performance tests by using an optimal observer. In: IECON 2010–36th Annual Conference on IEEE Industrial Electronics Society, pp. 2834–2838. IEEE (2010)
64. Dornfeld, D.: Application of acoustic emission techniques in manufacturing. NDT and E Int. **25**(6), 259–269 (1992)
65. Babak, V., Bohachev, I., Zaporozhets, A., Khaidurov, V., Havrysh, V., Kalinichenko, A.: Some features of modeling ultrasound propagation in non-destructive control of metal structures based on the magnetostrictive effect. Electronics **12**(3), 477 (2023)

66. Shcherban, A., Eremenko, V., Mokiichuk, V., Zaporozhets, A.: Model of information signals formation in the diagnostics of composite products. In: Advanced Information-Measuring Technologies and Systems I, pp. 209–223. Springer Nature Switzerland, Cham (2023)
67. AlShorman, O., Alkahatni, F., Masadeh, M., Irfan, M., Glowacz, A., Althobiani, F., Glowacz, W.: Sounds and acoustic emission-based early fault diagnosis of induction motor: a review study. Adv. Mech. Eng. **13**(2), 1687814021996915 (2021)
68. Shiroishi, J.Y.S.T., Li, Y., Liang, S., Kurfess, T., Danyluk, S.: Bearing condition diagnostics via vibration and acoustic emission measurements. Mech. Syst. Signal Process. **11**(5), 693–705 (1997)
69. Nazarchuk, Z., Skalskyi, V., Serhiyenko, O.: Acoustic emission. Foundat. Eng. Mech. (2017)
70. Zaporozhets, A., Eremenko, V., Babak, V., Isaienko, V., Babikova, K.: Using Hilbert transform in diagnostic of composite materials by impedance method. Period. Polytechnica Elect. Eng. Comput. Sci. **64**(4), 334–342 (2020)
71. Babak, V.P., Babak, S.V., Myslovych, M.V., Zaporozhets, A.O., Zvaritch, V.M. (2020). Technical provision of diagnostic systems. Diagnost. Syst. Energy Equipm. 91–133
72. Lytvynenko, V., Lure, I., Boskin, O., Okipnyi, I.: Automation of acoustic-emission diagnostic systems control processes. Sci. J. TNTU **97**(1), 88–96 (2020)
73. Stakhova, A., Kvasnikov, V.: Structure construction of acoustic emission system for diagnostic of friction clusters. In: 2018 IEEE 13th international scientific and technical conference on computer sciences and information technologies (CSIT), vol. 1, pp. 64–67. IEEE (2018)
74. Marasanov, V., Stepanchikov, D., Sharko, A., Sharko, A.: Operator of the dynamic process of the appearance of acoustic emission signals during deforming the structure of materials. In: 2020 IEEE 40th International Conference on Electronics and Nanotechnology (ELNANO), pp. 646–650. IEEE (2020)
75. Zaporozhets, A.O.: Experimental research of a computer system for the control of the fuel combustion process. Cont. Fuel Combust. Boilers, pp. 89–123 (2020)
76. Zaporozhets, A.A., Eremenko, V.S., Serhiienko, R.V., Ivanov, S.A.: Development of an intelligent system for diagnosing the technical condition of the heat power equipment. In: 2018 IEEE 13th International Scientific and Technical Conference on Computer Sciences and Information Technologies (CSIT), vol. 1, pp. 48–51. IEEE (2018)
77. Gasparjans, A., Terebkovs, A., Ziravecka, A.: Application of mathematical methods for current diagnostics of cylinder-piston group elements of diesel engines. In: Engineering for Rural Development. Proceedings of the International Scientific Conference (Latvia) (No. 19). Latvia University of Life Sciences and Technologies (2020)
78. Hrynkiv, A., Aulin, V., Lysenko, S., Rogovskii, I., Titova, L., Zagurskiy, O., Kolosok, I.: Development of a system for determining the informativeness of the diagnosing parameters for a cylinder-piston group in the diesel engine during operation. Eastern-Eur J Enterprise Technol. **3**(5–105), 19–29 (2020)
79. Kapłan, M., Klimek, K., Maj, G., Zhuravel, D., Bondar, A., Lemeshchenko-Lagoda, V., Wałowski, G.: Method of evaluation of materials wear of cylinder-piston group of diesel engines in the biodiesel fuel environment. Energies **15**(9), 3416 (2022)
80. Wang, J., Li, H., Grinkevych, K., Huang, R., Xu, J., Tsybanov, G., Tkachenko, I.: Cyclic indentation method applied to evaluating surface degradation of cylinder-piston group parts. Strength Mater. 1–9 (2021)
81. Gasparjans, A., Terebkovs, A., Zhiravetska, A.: Voltage spectral structure as a parameter of system technical diagnostics of ship diesel engine-synchronous generators. Electr. Control. Commun. Eng. **8**(1), 37–42 (2015)
82. Gasparjan, A., Terebkov, A., Zhiravetska, A.: Monitoring of electro-mechanical system diesel-synchronous generator. In: 2015 IEEE 5th international conference on power engineering, energy and electrical drives (POWERENG), pp. 103–108. IEEE (2015)
83. Morelos, M.A.M., Marín, J.A.: Fuzzy control strategy for stoichiometric air-fuel mixture in automotive systems. In: World Automation Congress 2012, pp. 1–6. IEEE (2012)

84. Zhiravetska, A., Gasparjan, A., Terebkov, A.: Monitoring of current technical condotion of vessel diesel-generator installation. In: 2017 19th European Conference on Power Electronics and Applications (EPE'17 ECCE Europe), pp. P-1. IEEE (2017)
85. Xu, L., Wei, B., Yu, Y., Guerrero, J.M., Vasquez, J.: Coordinated control of diesel generators and batteries in dc hybrid electric shipboard power system. Energies **14**(19), 6246 (2021)
86. Zaporozhets, A.O.: Methods and means for the control of the fuel combustion process. Cont. Fuel Combust. Boilers, 1–33 (2020)
87. Zaporozhets, A.O.: Research of the process of fuel combustion in boilers. Cont. Fuel Combust. Boilers, 35–60 (2020)
88. Mohammed, A., Pasupuleti, J., Khatib, T., Elmenreich, W.: A review of process and operational system control of hybrid photovoltaic/diesel generator systems. Renew. Sustain. Energy Rev. **44**, 436–446 (2015)
89. Mobarra, M., Rezkallah, M., Ilinca, A.: Variable speed diesel generators: performance and characteristic comparison. Energies **15**(2), 592 (2022)
90. Babak, S., Babak, V., Zaporozhets, A., Sverdlova, A.: Method of statistical spline functions for solving problems of data approximation and prediction of objects state. In: Proceedings of the Second International Workshop on Computer Modeling and Intelligent Systems (CMIS-2019), Zaporizhzhia, Ukraine, April, pp. 15–19 (2019)
91. Babak, V., Zaporozhets, A., Kuts, Y., Scherbak, L., Eremenko, V.: Application of material measure in measurements: theoretical aspects. In: Systems, Decision and Control in Energy II, pp. 261–269. Springer International Publishing, Cham (2021)
92. Babak, V., Zaporozhets, A., Kulyk, M., Kuts, Y., Scherbak, L.: Application of discrete Hilbert transform to estimate the characteristics of cyclic signals: information provision. In: Systems, Decision and Control in Energy IV: Volume I. Modern Power Systems and Clean Energy, pp. 93–115. Springer Nature Switzerland, Cham (2023)
93. Gertler, J.: Fault detection and diagnosis in engineering systems. CRC Press (2017)
94. Templin, J., Henson, R.A.: Diagnostic measurement: theory, methods, and applications. Guilford press (2010)
95. Patton, R.J., Frank, P.M., Clark, R.N. (eds.): Issues of Fault Diagnosis for Dynamic Systems. Springer Science & Business Media (2013)
96. Babak, V.P., Babak, S.V., Myslovych, M.V., Zaporozhets, A.O., Zvaritch, V.M.: Simulation and software for diagnostic systems. Diagnost. Syst. Energy Equipments 71–90 (2020)
97. Afzal, W., Torkar, R., Feldt, R.: A systematic review of search-based testing for non-functional system properties. Inf. Softw. Technol. **51**(6), 957–976 (2009)
98. Breivold, H.P., Crnkovic, I., Larsson, M.: A systematic review of software architecture evolution research. Inf. Softw. Technol. **54**(1), 16–40 (2012)
99. Rai, A., Upadhyay, S.H.: A review on signal processing techniques utilized in the fault diagnosis of rolling element bearings. Tribol. Int. **96**, 289–306 (2016)
100. Dai, X., Gao, Z.: From model, signal to knowledge: a data-driven perspective of fault detection and diagnosis. IEEE Trans. Industr. Inf. **9**(4), 2226–2238 (2013)
101. Gangsar, P., Tiwari, R.: Signal based condition monitoring techniques for fault detection and diagnosis of induction motors: a state-of-the-art review. Mech. Syst. Signal Process. **144**, 106908 (2020)
102. Yan, R., Gao, R.X., Chen, X.: Wavelets for fault diagnosis of rotary machines: a review with applications. Signal Process. **96**, 1–15 (2014)
103. Wang, Y., Xiang, J., Markert, R., Liang, M.: Spectral kurtosis for fault detection, diagnosis and prognostics of rotating machines: a review with applications. Mech. Syst. Signal Process. **66**, 679–698 (2016)
104. Tong, X., Feng, Y., Zhao, A.: A survey on Neyman-Pearson classification and suggestions for future research. Wiley Interdisciplinary Rev.: Comput. Statist. **8**(2), 64–81 (2016)

105. Tang, L., Tian, H., Huang, H., Shi, S., Ji, Q.: A survey of mechanical fault diagnosis based on audio signal analysis. Measurement 113294 (2023)
106. Kelter, R.: Bayesian and frequentist testing for differences between two groups with parametric and nonparametric two-sample tests. Wiley Interdiscipl. Rev. Comput. Stat. **13**(6), e1523 (2021)
107. Shaikh, K.B., Jawarkar, N.P., Ahmed, V.: Machine diagnosis using acoustic analysis: A review. In: 2021 IEEE Conference on Norbert Wiener in the 21st Century (21CW), pp. 1–6). IEEE (2021)
108. Kim, W., Katipamula, S.: A review of fault detection and diagnostics methods for building systems. Sci. Technol. Built Environ. **24**(1), 3–21 (2018)
109. Tidriri, K., Chatti, N., Verron, S., Tiplica, T.: Bridging data-driven and model-based approaches for process fault diagnosis and health monitoring: A review of researches and future challenges. Annu. Rev. Control. **42**, 63–81 (2016)
110. Bollen, K.A., Biemer, P.P., Karr, A.F., Tueller, S., Berzofsky, M.E.: Are survey weights needed? A review of diagnostic tests in regression analysis. Ann. Rev. Statist. Appl. **3**, 375–392 (2016)
111. Chen, H., Jiang, B.: A review of fault detection and diagnosis for the traction system in high-speed trains. IEEE Trans. Intell. Transp. Syst. **21**(2), 450–465 (2019)
112. Dormann, C.F., Schymanski, S.J., Cabral, J., Chuine, I., Graham, C., Hartig, F., Singer, A.: Correlation and process in species distribution models: bridging a dichotomy. J. Biogeogr. **39**(12), 2119–2131 (2012)
113. Hong, Y.H., Kim, H.K., Lee, H.S.: Reconstruction of dynamic displacement and velocity from measured accelerations using the variational statement of an inverse problem. J. Sound Vib. **329**(23), 4980–5003 (2010)
114. Vedmitskyi, Y.G., Kukharchuk, V.V., Hraniak, V.F., Wójcik, W., Zhassandykyzy, M., Yesmakhanova, L.: New non-system physical quantities for vibration monitoring of transient processes at hydropower facilities, integral vibratory accelerations. Przeglad Elektrotechniczny **93**(3), 69–72 (2017)
115. Xia, H., Cao, Y.M., De Roeck, G.: Theoretical modeling and characteristic analysis of moving-train induced ground vibrations. J. Sound Vib. **329**(7), 819–832 (2010)
116. Bohachev, I., Babak, V., Zaporozhets, A.: Novel small-aperture transducers based on magnetostrictive effect for diagnostic systems. Technical Electrodynamics **3**, 69–78 (2022)
117. Myklestad, N.O.: Fundamentals of vibration analysis. Courier Dover Publications (2018)
118. Malaric, R.: Instrumentation and measurement in electrical engineering. Universal-Publishers (2011)
119. Muhlheim, M.D., Wood, R.: Technical basis for evaluating software-related common-cause failures (No. ORNL/SR-2016/130). Oak Ridge National Lab.(ORNL), Oak Ridge, TN (United States) (2016)
120. Torbert, R.B., Russell, C.T., Magnes, W., Ergun, R.E., Lindqvist, P.A., LeContel, O., Lappalainen, K.: The FIELDS instrument suite on MMS: scientific objectives, measurements, and data products. Space Sci. Rev. **199**, 105–135 (2016)
121. Liptak, B.G., Piovoso, M.J., Shinskey, F.G., Eren, H., Totherow, G.K., Jamison, J.E., Morgan, D., Hertanu, H.I., Marszal, E.M., Berge, J., McQuillan, D.: Instrument Engineers' Handbook, Volume Two: Process Control and Optimization. CRC press (2018)
122. Grabowski, A.: Cauchy mean theorem. Formalized Math **22**(2), 157–166 (2014)
123. Cheng, J.: On multivariate fractional taylor's and cauchy'mean value theorem. J. Math. Study **52**(1), 38–52 (2019)
124. Pertusa, J.F., Morante-Redolat, J.M.: Main steps in image processing and quantification: the analysis workflow. Comput. Opt. Microsc. Methods Protocols, 3–21 (2019)
125. Manna, A.K., Moulick, H.N., Patra, J., Bera, K.: Image processing with sampling and noise filtration in image reconigation process. Image **3**(7), 35–47 (2014)
126. Manríquez, R.: Metric relations in the fuzzy right triangle. Mathematics **11**(19), 4056 (2023)
127. Ueki, J.: Chebotarev links are stably generic. Bull. Lond. Math. Soc. **53**(1), 82–91 (2021)

128. Niibo, H., Ueki, J.: Idèlic class field theory for 3-manifolds and very admissible links. Trans. Am. Math. Soc. **371**(12), 8467–8488 (2019)
129. Niibo, H., Ueki, J.: A Hilbert reciprocity law on 3-manifolds. Res. Math. Sci. **10**(1), 3 (2023)
130. Stolarsky, K.B.: From wythoff's nim to chebyshev's inequality. Am. Math. Mon. **98**(10), 889–900 (1991)
131. Cohen, J.E.: Markov's inequality and Chebyshev's inequality for tail probabilities: a sharper image. Am. Stat. **69**(1), 5–7 (2015)
132. Navarro, J.: A very simple proof of the multivariate Chebyshev's inequality. Commun. Statist.-Theory Methods **45**(12), 3458–3463 (2016)

Chapter 8
Experimental Studies of Statistical Diagnostics Information and Measuring Systems

Abstract This comprehensive research chapter delves into the application of innovative diagnostic methods for electrical equipment (EE), focusing on vibration and acoustic analyses to detect and diagnose faults in rolling bearings of electric motors (EM) and the technical condition of diesel-electric generators (DEGs). Through extensive experimental setups, the chapter validates probabilistic models for diagnostic signals by utilizing real-world EE units and bespoke test benches. It introduces a specialized installation for analyzing rolling bearing vibrations, demonstrating the significance of frequency characteristics and the impact hammer test in pinpointing natural frequencies crucial for diagnostics. Furthermore, the study explores the use of shock diagnostics and acoustic emission (AE) analysis for evaluating the mechanical loading of EE components, highlighting the effectiveness of vibration maxima, and AE signal coefficients as reliable diagnostic indicators. Employing statistical hypothesis testing, particularly the Neyman-Pearson procedure, the research identifies specific faults in DEG cylinder-piston groups, showcasing the method's potential in early fault detection and preventive maintenance to enhance engine reliability. The findings contribute to the advancement of diagnostic methodologies for EE, promising improved maintenance strategies and operational efficiency.

Keywords Vibration analysis · Acoustic emission · Electrical equipment · Diesel-electric generator · Statistical hypothesis testing · Neyman-Pearson procedure · Diagnostic methods · Preventive maintenance

Probabilistic models of diagnostic signals, diagnostic signs based on the results of their analysis, real electrical objects and experimental test benches, as well as laboratory samples of statistical diagnostics of various units of electrical equipment (EE) are the necessary components that allow us to directly proceed to experiments on the diagnosis of such objects. During their implementation, considerable attention was paid to identifying the mutual influence of processes of various physical natures (vibrations, electromagnetic processes and fields, temperature, aerodynamic noise, etc.). An experimental study of the proposed diagnostic features is carried out using the example of vibration diagnostics of electric machine (EM) bearing units, the state

V. Babak et al., *Statistical Diagnostics of Electric Power Equipment*, Studies in Systems, Decision and Control 573, https://doi.org/10.1007/978-3-031-76253-6_8

of pressing of an iron package of a laminated magnetic circuit, and diagnostics of cylinder capacities during the operation of a diesel-electric generator (DEG).

8.1 Installation for Studying Vibrations of Rolling Bearings

For further research, it is expedient to analyze the vibrations of the rotating bearing itself, "decoupling" its vibrations from the vibrations of the electric drive. To realization of this idea, it was designed and manufactured an experimental installation for vibration tests of single rolling bearings, the main purpose of which is as follows:

- verification of the mathematical model of bearing vibrations discussed in Chap. 3;
- verification of diagnostic signs proposed based on this model;
- simulation of defects such as misalignment, and lack of lubrication; study of pitting of the outer or inner rings of the bearing;
- study of the correlation of vibrations of the operating drive and the bearing unit itself.

However, before starting to study the vibrations of rolling bearings using a special unit, it is necessary to determine the vibration characteristics of its components and the complete unit in static mode.

Let's proceed to consideration of these questions.

The special installation for investigation of bearing vibrations structurally consists of three main units: electric drive, massive shaft, mounting unit, and vibration measurement of the tested bearing [1–6]. The general view of the installation is shown in Fig. 8.1. Besides, there are several auxiliary units, the description and purpose of which will be described below.

The rotation of the research bearing, installed in the unit of fixing and vibration measurement 3, is provided by an electric motor DC type P-51 through solid shaft 2.

Fig. 8.1 General view of the experimental installation for vibration tests of single rolling bearings: 1—electric drive, 2—massive shaft, 3—unit of fixing and measuring vibrations of the tested bearing

This 11-kW motor provides rotation of the test bearing with any speed in the range from 10 to 1500 rpm. The use of coupling 4 with rubber fingers makes it possible to maximally reduce vibrations caused by the electric drive 1. Besides, the reduction of vibrations of shaft 2 of the experimental unit is promoted by its fixing in supports 5 with sliding bearings 6 made of fluoroplastic, and also by placing shaft 2 and the unit of fixing and measuring the vibrations of the test bearing 3 on the massive plate 7. The latter in turn is fixed to the main foundation with the help of six shock-absorbers 8 of AKSS-80 type and is acoustically separated from the separate base on which the electric drive is installed. In Fig. 8.1 only three shock absorbers are visible, the other three are located on the opposite side of plate 7, symmetrically to the first three.

The bearing mounting and vibration measurement unit is shown in more detail in Fig. 8.2. It consists of a rotary plate 1, in which the bearing under test is mounted, accelerometer 2, transom 3 for mounting the rotary plate, fixing bolts 4, bearing skew angle device 5 with arrow indicator 6 and base 7. The rotary plate 1 together with the shaft 8 is designed for testing bearings of 309 size, as one of the most widespread in EM of small and medium power. For measurement of vibration accelerations of the investigated bearing an accelerometer 2 of ABC-017 type is used, which allows to measure of bearing vibrations in the frequency band 20 Hz … 30 kHz. 30 kHz. It is installed in a radial direction about the research bearing.

The rotary plate 1 is designed to create a misalignment in the bearing up to 0.5°, which is provided by two squeeze screws 9 installed on both sides of the rotary plate (in the figure, the second squeeze screw is not visible because it is located symmetrically concerning the axis of the rotary plate, in which the bearing cage is fixed). The amount of misalignment in the bearing tested in this unit is found as the tangent of the angle formed at misalignment by the plane of the section passing radially through the middle of the inner ring of the bearing with the plane of the similar section of the outer ring (schematically shown in Fig. 8.3).

In experiments, the correlation between the vibrations of the bearing under test and the vibrations of the electric drive of the unit is of considerable interest. To make it possible to study this correlation, places for fixing accelerometers of ABC 017 type are provided on the bearing shield on the electric motor drive side.

It should be noted that similar installations for the study of rolling bearing vibrations have been used by other authors as well. For example, various versions of experimental setups for the study of bearing vibration are given in [1, 7–10]. When creating each of them, the authors pursued the goal of not only simply measuring vibrations (displacement, velocity, or acceleration) of the bearing under test, but also provided for the artificial creation of certain modes of its operation to solve the problems of vibrodiagnostics, prediction of residual life, etc. In [11–13], the authors used a modified Timken unit, with the help of which tapered roller bearings were tested for durability. Its undoubted advantage was the possibility of creating with the help of a hydraulic piston different load values of the tested bearings, which maximizes the proximity of bearing tests to the conditions of real operation.

During the operation of the experimental unit, the vibrations of the bearing under test are influenced by vibrations caused by the natural frequencies of the units in which it is fixed (rotary plate, shaft of the unit, etc.). To determine the natural frequencies

Fig. 8.2 Bearing mounting and vibration measurement unit

of various units of the considered installation in static mode, experimental studies were carried out. These studies were carried out with the help of an impact hammer and specially designed devices. The results of these experiments are described in [14–17].

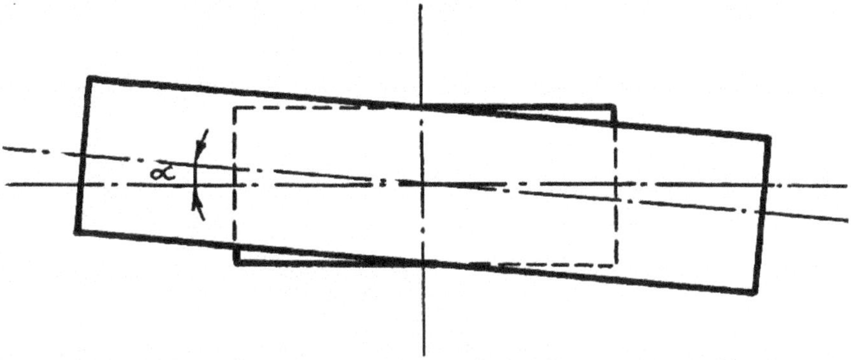

Fig. 8.3 Bearing misalignment

The experiments conducted in the static mode of the unit allowed us to determine the frequency characteristics of vibrations arising in the bearing mounting unit during hammer blows at various points of the unit. Besides, the experiments clearly showed the character of the change of natural frequencies of vibrations of separate units, in particular, the rotary plate, when they are mounted in the unit. The obtained information about the frequencies measured at the bearing mounting unit is taken into account during the experiments on vibration diagnostics of bearings, i.e. during the operation of the installation in dynamic mode. Namely, the information about these frequencies was used in the selection of filtering parameters of diagnostic vibration signals. This approach allowed to form of a certain frequency band in which the main contribution to vibrations is made by the diagnosed unit, and thus, to leave for further study the vibrations generated by this particular unit. Vibrations generated by other nodes were attenuated to the maximum extent possible using filtering. As a result, this approach allowed to increase the accuracy and reliability of the diagnostics [18–20].

Let us briefly review the results of the study of the frequency characteristics of the plant in static mode.

The study of vibration characteristics of individual units of the plant in static mode required the use and creation of several special devices and equipment. First, let us dwell on some of them. To determine the natural frequencies, the vibrodiagnostic IMS was used in a complex with a special impact hammer (IH) PCB Piezotronics and a launching device, which was used in the IMS launching scheme.

The PCB Piezotronics 086C01 IH (Fig. 8.4) contains an impact force sensor integrated into the hammerhead. This sensor provides the ability to measure the amplitude and frequency composition of the excitation energy, which is transmitted to the object of diagnosis. The hammer is supplied with a set of tips (strikers) of different mass and hardness, which provides the ability to form the necessary energy and spectrum of shock excitation depending on the characteristics of the structure under investigation. During the experiments, a hard metal tip was used, which excludes the multiple impact when the hammer touches the object under test. Such a tip

Fig. 8.4 Impact hammer

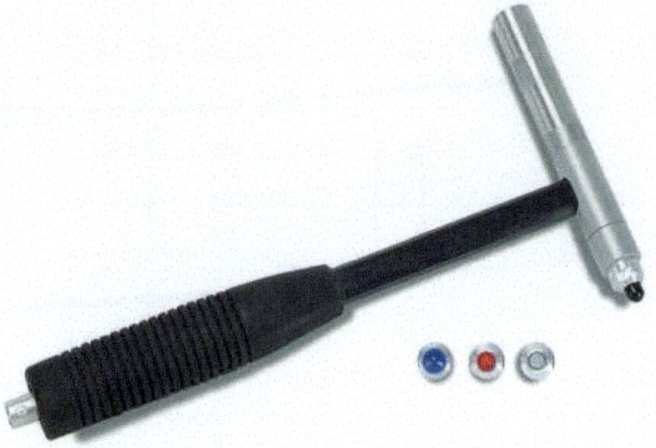

provides a short-duration impact with a significant amplitude on the tested node (approximation of δ-impulse). The response of this node to such an impact measured by another accelerometer makes it possible to determine its intrinsic characteristics after appropriate processing [21–23].

The sensor mounted in the hammer head was also used to start the electronic circuits of the IMS, which provide measurement and conversion of the vibration signal propagating in the body of various units of the plant when the hammer strikes them.

Let us briefly describe the operation of the circuit in Fig. 8.5, designed to study the vibration characteristics of the plant in the static mode. The hammer blows are carried out in points 1…7, marked by arrows on the schematic representation of the installation. A shock vibration wave is excited in the investigated units, measured by an accelerometer placed on the bearing mounting unit. Then the signal from this accelerometer is fed to the preliminary amplifier (PA) of the secondary processing unit (SPU). Simultaneously with this, a vibration wave is excited in the hammer body, which is recorded by a sensor mounted on the hammer. The signal from this sensor through the PA goes to the starting device (SD), which produces a signal for starting the data conversion and input device (DCID). At the same time, the DCID realizes the conversion into a digital code and inputs into the computer the initial data coming from the accelerometer located on the experimental unit. Further processing of experimental data is performed on the computer using any of the programs included in the package. The results of this processing are reflected on the display or printed [5, 6].

The use of PCB Piezotronics 086C01 IH and vibro-diagnostics IMS allowed to obtain some characteristics of vibrations of individual units and the whole plant. The use of the IH in the experiments to determine the intrinsic vibrations of the plant nodes made it possible to study the frequency composition and propagation channels of vibrations.

Among the characteristics of vibrations of the units of the installation, the most interesting are spectral characteristics, which give an idea of the natural frequencies of vibrations of different units. Therefore, in the following, they were given the main

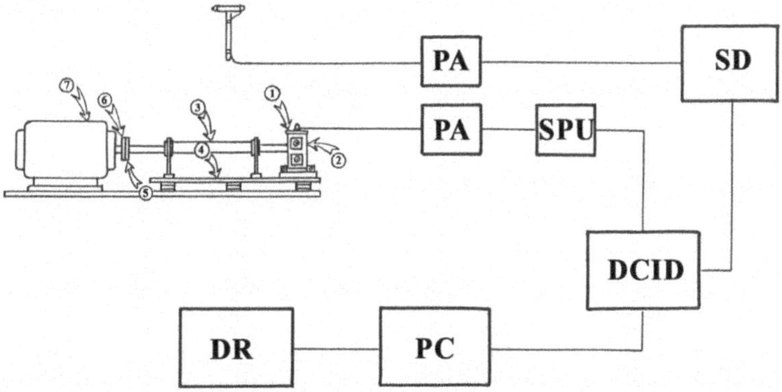

Fig. 8.5 Schematic diagram of vibration characteristics study of the installation in static mode

attention. Below we will briefly dwell only on some basic results characterizing frequency properties of a special unit for vibration tests of rolling bearings [24–27].

To trace the mutual influence exerted on each other by vibrations of different units of the installation, a series of experiments was carried out, which included:

1. investigation of natural frequencies of vibrations of separate units of the installation being in the disassembled state, namely, investigation of frequency properties of vibrations of the rotary plate (Fig. 8.2) of the special installation with and without the bearing fixed in it;
2. Determination of frequency properties of vibrations of different units within the installation.

In the first series of experiments, the main attention was paid to vibration shock tests of the rotary plate (Fig. 8.2). According to their results, it was found that the investigated plate without a bearing is a high-voltage vibrating system. From the spectral analysis of these vibrations, it was found that the vibrations excited in the plate have the main frequency in the region of 5 kHz.

While the bearing is installed in the rotary plate (Fig. 8.2), the vibrations arising in it during hammering are significantly changed. The installation of the bearing in the plate causes a rather rapid attenuation, i.e. the tested assembly has become less high-quietness. This can be explained by the appearance of a boundary between the plate housing and the outer ring of the rolling bearing, which in turn causes multiple reflections of the vibration waves excited by the impact hammer when this boundary is reached. In addition, the impact tests showed that the installation of the bearing in the rotary plate results in some broadening of the fundamental frequency bandwidth in the region of 5 kHz.

The second series of experiments involved determining the vibration characteristics of the experimental assembly. These characteristics were also determined for the static mode of the unit.

Excitation of vibrations and finding their spectral composition in the rotary plate 2, which is the main element of the bearing mounting unit, showed that in addition to the main frequency of 5 kHz, it also has frequency components in the areas of 1, 3, 4, 5.3 and 5.7 kHz. This is due to the connection of other massive units of the plant, as well as the presence of a rolling bearing in the rotary plate. As shown by further shock tests, the vibration frequency in the region of 5 kHz (along with other frequencies) is also manifested with greater or lesser intensity when striking other units of the installation, namely, shaft 2, the foundation plate 7, the elastic coupling 4 (Fig. 8.1). Apparently, it can be explained by the fact that the measuring accelerometer is installed directly on the rotary plate and the shock excitations, applied to various units of the installation and spreading over these units, finally arrive on the rotary plate and are perceived by the accelerometer installed on it. In this case, the vibration waves make the rotary plate resonate with varying intensity in the region of its natural frequency. This fact should be taken into account during measurements and further processing of the vibration signal in the process of vibration diagnostics of rolling bearings, carried out using the considered installation.

The experiments carried out in the static mode of the installation made it possible to determine the frequency characteristics of vibrations arising in the bearing mounting unit during impacts at various points of the installation using an IH and IMS. Besides, the experiments clearly showed the character of change of natural frequencies of vibrations of separate units, in particular, of the rotary plate, when they are mounted in the unit. The obtained information on frequencies measured at the bearing mounting unit is taken into account during the experiments on vibration diagnostics of bearings, i.e. during the operation of the unit in dynamic mode. First of all, the data on the frequencies of vibrations of the rotary plate allow us to choose the parameters of band-pass filtering to minimize the influence of vibrations generated by various units of the unit on the vibrations of the bearings tested on the unit.

Let us dwell on the issues of dynamic testing of bearings on a special installation.

Taking into account the diagnostic signs of the technical condition of sub-bearings proposed in the fourth section, the corresponding statistical methods of studying the experimental data using the unit were selected based on analyzing the probabilistic model of bearing vibrations.

The vibration characteristics of the bearings tested at the specified installation were obtained using the IMS described in paragraph 7.1.1. In this case, the shaft of the drive motor of the installation was rotated at a frequency $V = 1000$ rpm, which corresponds to $f_s = 16.6$ Hz. Various samples of conditionally serviceable ball bearings of 309 size with a conventional cage, as well as bearing 309 ESH2 with a plastic cage were tested. LITOL-type grease was used as a lubricant. An accelerometer ABC 017 was used, providing a measurement of vibration accelerations in the frequency range of 20 Hz ... 30 kHz. While measuring the vibrations of the mentioned bearings, a band-pass filter tuned to the frequency of 50 Hz and a low pass filter (LPF) with an upper boundary frequency of 10 kHz were used. Based on the selected frequency range, the investigated vibration signal was converted into a digital code with frequency $f_d = 32$ kHz.

For reliability of information about the character of vibrations of the tested bearing, the measured realization by duration should correspond to at least one revolution of the inner ring of the bearing. Then the sampling volume, by which the investigated realization is represented, should be not less than $N = f_d / f_s = 32{,}000/16.6 = 1928$. Therefore, the vibration characteristics of the tested bearings are determined at a sample volume of $N = 2048$.

A typical vibration spectrogram of a conditionally serviceable bearing of size 309 is presented in Fig. 6a. It follows that the main contribution to the vibrations of the investigated bearing is made by components in the region of 1000 Hz, there are also insignificant frequency components in the band 0...500 Hz and in the area of 1500 Hz.

The results of spectral analysis of vibrations of conditionally serviceable ball bearing 309 ESH2 are shown in Fig. 6b. This spectrogram is close to the vibration spectrogram bearing 309, but unlike the latter, there is no component in the region of 1500 Hz. In addition, the components in the band 0 ... 500 Hz. Finally, the vibration components in the area of 1000 Hz are almost two times less in power in comparison with the bearing 309 having a conventional metal cage. However, some rise in the 1800 Hz area is noted, as well as minor components in the 1800 ... 2100 Hz.

Thus, according to the presented typical spectrograms of bearing vibrations, the following can be noted.

1. The presence of several frequency peaks in the spectrograms of bearing vibrations confirms the results of the fourth chapter, from which it follows that the rolling bearing is a multi-resonant system.

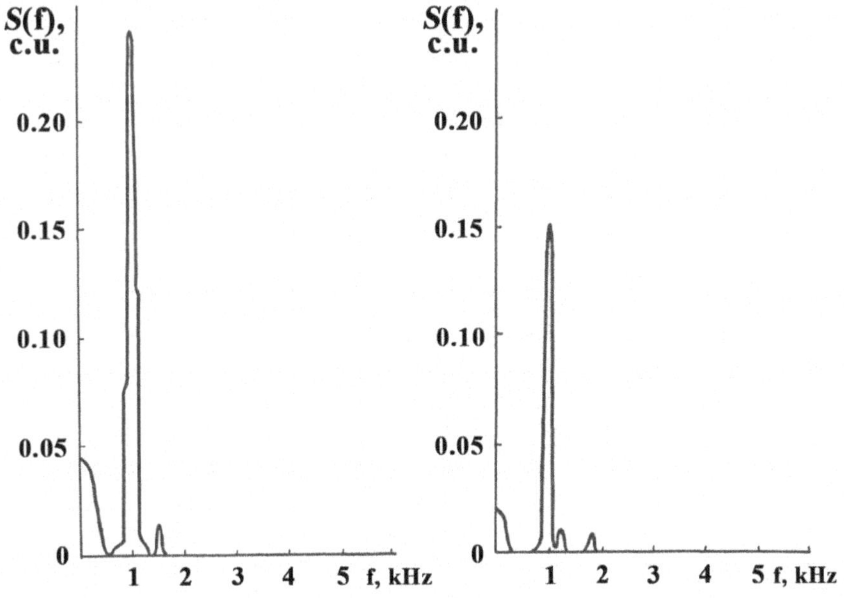

Fig. 8.6 Typical spectrograms of vibrations of conditionally serviceable ball bearings 309 (**a**) and 309 ESH2 (**b**)

2. The character of spectrograms is largely determined by the type and design features of the investigated rolling bearings.
3. When the bearings were tested on the experimental unit, their vibrations did not have components in the 5 kHz region, i.e., the natural frequency of the rotary bearing mounting plate, which was observed during the excitation of vibrations by the impact hammer in the rotary plate itself and during the study of the complete unit with the impact hammer in the static mode, did not manifest itself in the dynamic mode of operation of the unit. This indicates the inexpediency of applying filtering to exclude the influence of frequency in the region of 5 kHz on the vibrations of the bearing itself. The components in the region of 1 kHz, occurring during the tests of the assembly with the impact hammer, are almost an order of magnitude smaller in comparison with the components of natural frequencies of vibrations of the bearing tested on the assembly. Consequently, there is no need to use special methods for separating bearing vibrations and natural vibrations of individual units of the complete unit (shaft, coupling halves, drive motor housing).

The use of spectral analysis results for diagnosing specific defects in bearings will be discussed below [28–32].

Correlation analysis of vibrations does not provide more information than spectral analysis because the correlation function and spectral density are related by the Fourier transform. However, sometimes time domain analysis can be more useful than frequency domain analysis because of the different effects of errors. Figure 8.7 shows the most typical normalized correlograms of vibrations of 309 and 309 ESH2 ball bearings. Thus, the correlograms of vibrations of ball bearing 309 have a form close to exponential-cosine [33–36]. In their character one main frequency is distinguished, i.e. the bearing can be considered as a single-resonance system, which by the expression

$$B(s) = e^{-\beta_1 |s|}\left(\cos \psi_1 s + \frac{\beta_1}{\psi_1} \sin \psi_1 |s|\right) \tag{8.1}$$

is determined by the parameters ψ_1 and β_1. Other frequency components noted in the spectrogram (Fig. 8.6) are not observed.

The correlogram of vibrations of sub-bearing 309 ESH2 is different in character. One main period is also clearly distinguished in it. However, the vibrations decay faster, i.e. if we use expression (8.1) to describe this correlogram, then in the considered case in comparison with the correlogram in Fig. 8.7, we have a larger value of the parameter β_1. Consequently, bearing 309 ESH2 is a less good system in comparison with bearing 309. It can be explained by its design—the presence of a plastic cage.

The most complete picture of the nature of bearing vibrations is given by their histogram analysis [37–41]. As noted above, the histogram analysis program used, in addition to histogram construction, allows obtaining estimates of the initial and central moments of vibrations up to and including the fourth moment. Among them are estimates of dispersion, standard deviation, asymmetry, and excess coefficients. In

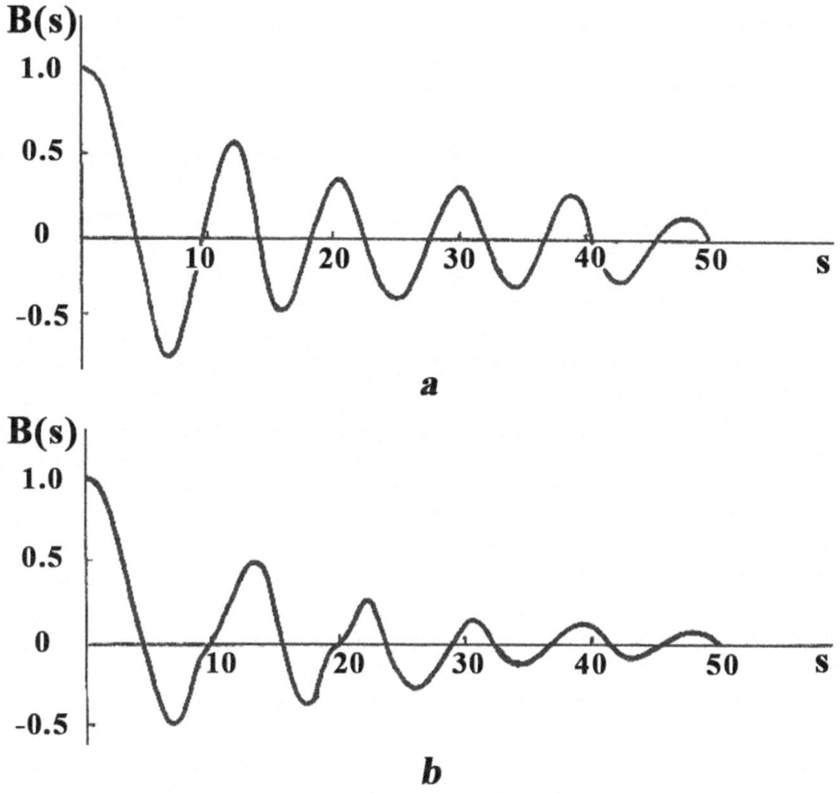

Fig. 8.7 Typical normalized correlograms of vibrations of conditionally serviceable ball bearings 309 (**a**) and 309 ESH2 (**b**)

addition, this program allows smoothing the histograms using the system of Pearson curves, i.e. to find an explicit expression for the smoothing curve with the numerical values of the parameters included in it. Let's consider the most typical characteristics and parameters of vibrations obtained for the investigated bearings with the help of the histogram analysis program. It should be noted at once that histograms of vibrations and curves smoothing them for all bearings 309 and 309 ESH2 turned out to be rather close in character, therefore we shall stop on histograms for ball bearing 309 ESH2.

Figure 8.8 shows a typical histogram and its smoothing curve obtained as a result of the processing of vibrations of the 309 ESH2 bearing. The last one can be referred to as the VII type according to Pearson, which corresponds to the analytical expression

$$p(x) = 0.4439\left[1 + x^2/(2.5168)^2\right]^{-4.6667}.$$

According to [42–46], the most informative diagnostic signs that allow us to determine many defects in bearings are k asymmetry γ and kurtosis coefficients. One

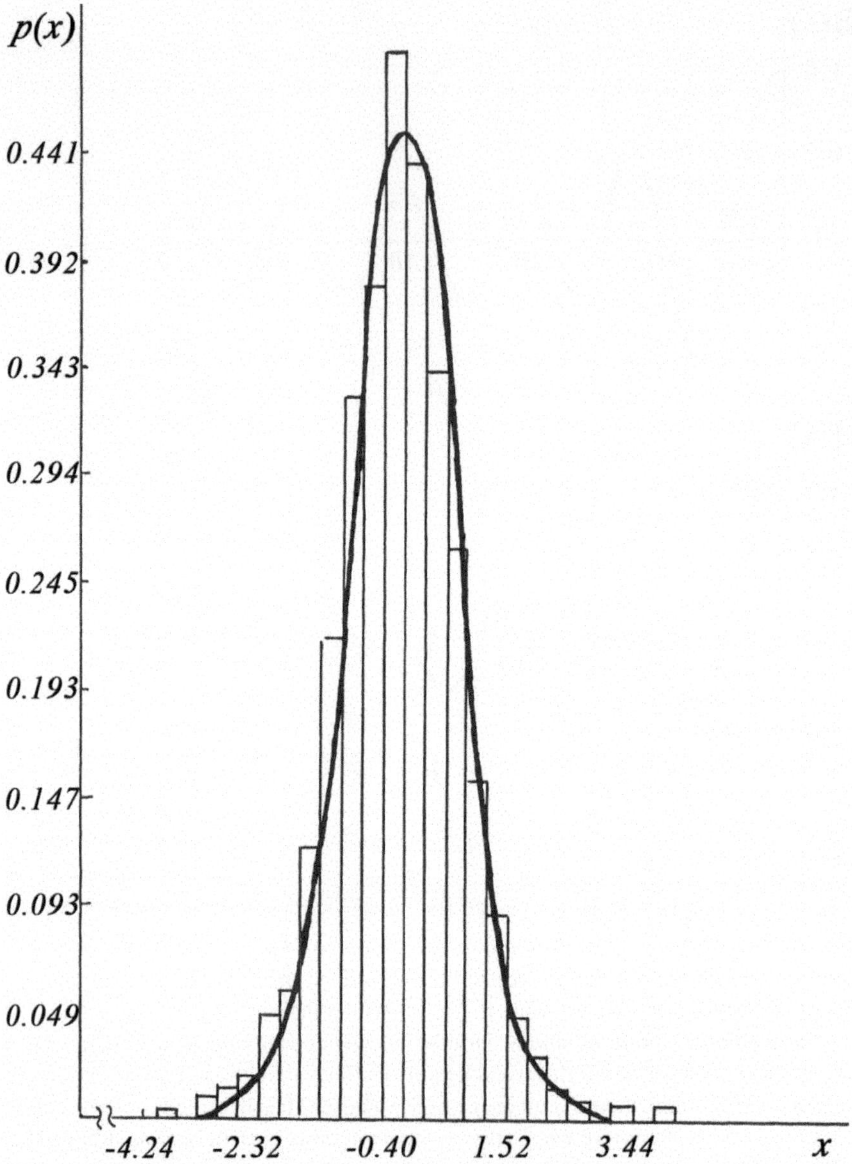

Fig. 8.8 Typical histogram and its smoothing curve, received as a result of the processing of vibrations of bearing 309 ESH2

Table 8.1 Values of coefficients k and γ of bearing 309 ESH2

Bearing test conditions	Number of realizations selected for analysis	Point estimates of the mean value of the coefficients	
		k	γ
Serviceable	60	0.1185 ± 0.0378	0.1730 ± 0.0577
No lubrication	89	0.1142 ± 0.0111	0.6587 ± 0.1196
Skew	75	0.1049 ± 0.0066	1.0711 ± 0.0904
Defective inner ring	85	0.0866 ± 0.0048	1.2227 ± 0.1076

of the main points in the diagnostics of bearings is the establishment of quantitative estimates of the limits of change of selected diagnostic signs corresponding to a certain technical condition of bearings. Thus, for vibrations of the conditionally serviceable ball bearing 309 ESH2, which were observed during 60 experiments, the following point estimates of the averaged value of the asymmetry and excess coefficients are characteristic: $k = 0.1135 \pm 0.0378$; $\gamma = 0.1730 \pm 0.0577$. Here, as well as in Table 8.1 below, the standard deviation of the point estimate is indicated after the $\pm$ signs. To diagnose specific defects in bearings, it is necessary to establish the limits of variation of quantitative estimates of coefficients k and γ at multiple repetitions of experiments for different types of defects.

8.2 Research of Diagnostic Signs Condition of EM Bearing Units

The bearing unit is the main place of vibration transmission from rotating parts of EM to stationary ones, its vibration characteristics are significantly influenced by mechanical unbalance of the rotor, electromagnetic forces, vibrations of the brush-collector unit, as well as aerodynamic noise. As already mentioned, to increase the reliability of diagnostic results, it is necessary to solve the problem of separating the vibrations of bearings from the vibrations generated by the above-mentioned causes in the EM operating. For this purpose, it is possible to apply analog and digital filtering. However, before using the above-mentioned types of filtering, it is necessary to establish the frequency bands in which vibrations are excited due to the above-mentioned units in the operating EM. Moreover, as it was established as a result of numerous experiments, for each type of EM the frequency composition generated by separate nodes is different. Let us dwell on these questions in more detail.

8.2.1 Peculiarities of EM Bearing Units Vibrations

To select filter parameters, it is necessary first of all to determine the frequency bands in which the vibrations generated by various causes in the operating EM are excited. This task is solved in the following way. Having information about the frequency spectrum of vibrations of the bearing itself, we determine the frequency characteristics of bearing vibrations in the bearing unit. Then we conduct experiments on finding frequency characteristics of vibrations excited by the electromagnetic system and brush-collector unit of the machine, for example, on the DC EM P 51, which is used as a drive of the experimental unit. The frequencies of vibrations caused by possible rotor unbalance as well as aerodynamic noise are not determined, since the former are usually strongly manifested in high-speed machines, and the latter—in EM of high power.

In practice, the determination of frequency bands was carried out in two stages: first, the measurement, registration, and subsequent spectral analysis of the investigated vibrations of bearing units on the running EM was carried out; then a similar experiment was carried out on the de-energized EM when its armature was rotating from an external drive. Vibrations were measured both with brushes touching the collector surface and with brushes removed from it.

Figure 8.9 shows typical spectrograms of vibrations of the bearing 309 ESH2, installed in the bearing unit from the collector side in EM P 51, obtained during multiple repetitions of the experiment.

The spectrogram in Fig. 9a corresponds to the vibrations measured at the bearing unit at the independent rotation of the machine armature, in Fig. 8.9b, c—at the rotation of the armature from an external drive in the case when the brushes are in contact with the collector (Fig. 9b) and do not touch the collector (Fig. 9c).

It should be noted that the results of research on bearing vibrations on different types of EM indicate that the parameters of vibration signal filtering during bearing diagnostics should be chosen taking into account the specific type of the machine being diagnosed.

The theoretical basis for obtaining estimates of diagnostic signs characterizing the technical condition of EM bearing units are the results of the study of the probabilistic vibration model of these units, which were described in paragraphs 3.1 and 3.2 of this book. Proceeding from the analysis of the constructed model of vibrations, the most effective signs allowing us to determine the technical condition of rolling bearings of EM are statistical parameters obtained by spectral correlation and histogram analysis of their vibrations. Let us dwell on consideration of these issues.

8.2.2 Correlation and Spectral Diagnostic Features

Correlation and spectral characteristics of vibrations are energetic. Therefore, the diagnostics based on them are carried out within the framework of the energy theory,

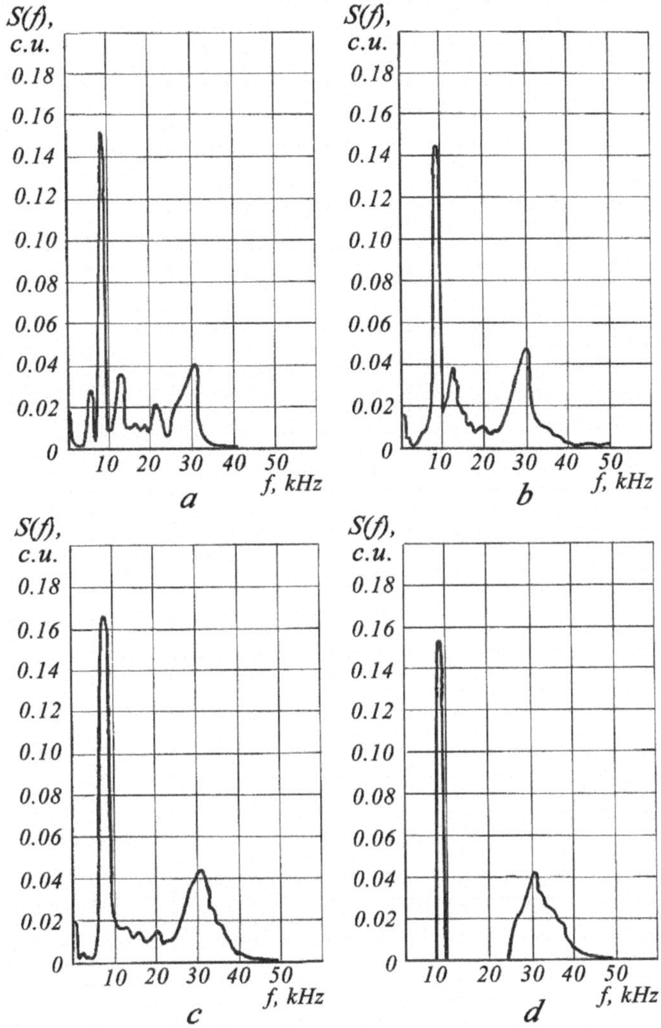

Fig. 8.9 Typical spectrograms of vibrations of bearing 309 ESH2 installed on the side of the collector in EM P-51: **a** at the independent rotation of the machine armature; **b**, **c** at the rotation of the armature from an external drive (**b** brushes are in contact with the collector, **c** brushes do not touch the collector); **d** obtained as a result of vibration filtering

and in this case, in general, the information about the probabilistic structure of the vibration process is not fully used. However, diagnostics within the framework of the energy theory is quite justified, as there is a sufficiently reasoned mathematical apparatus, devices, and software for its realization, and in case of fulfillment of the hypothesis about Gaussianity of vibrations, this approach is not inferior in informativeness to the diagnostic methods based on the use of distribution functions and statistical criteria of hypothesis testing, which were considered earlier.

While using correlation and spectral diagnostic features, the first stage is usually to find estimates of correlation and spectral functions, as described in the previous chapters, and then estimates of diagnostic parameters. For diagnostics using such parameters, it is necessary, first of all, to select diagnostic spaces and form training sets corresponding to certain technical states of the bearing under study, and, finally, to construct solving rules for diagnosing and classifying defects in the research bearings using these sets.

Let us dwell on the issues of estimation of some parameters determined by empirically obtained correlograms and spectrograms of real vibrations of diagnosed bearings. As experimental material, we use the results of correlation and spectral analyses of vibrations of 309 ESH2 type bearings installed on P-51 type DC EM. Vibrations were measured synchronously on both bearing units of the EM using ABC 017 accelerometers. Further processing of these vibrations was carried out with the help of IMS vibrodiagnostics using correlation and spectral analysis programs. Vibrations of both conditionally serviceable bearings and bearings with misalignment introduced into the bearing installed in the machine from the collector side were investigated. The tested EM worked in the X.X mode. The skew was introduced with the help of a special squeezing device, which ensures its installation with an absolute boundary error of no more than $2.5'$. Its value was chosen to take into account that for the majority of EMs of average power the bearing skew of $6' \ldots 10'$ is already inadmissible.

Figure 8.10, shows normalized auto- and mutual correlograms of vibrations of conditionally serviceable bearings, and Fig. 10b—in the presence of skewness $14' \pm 2.5'$ in the bearing. Here $B_1(s)$—the correlogram of vibrations of the bearing installed from the collector side, $B_2(s)$—the same from the drive side, and $B_{12}(s)$—their mutual correlogram. The introduction of misalignment in the bearing entailed a change in the character of the correlogram $B_1(s)$. Thus, if the correlogram $B_1(s)$ of a conditionally serviceable bearing can be analytically described by expression (8.1), i.e. in this case a conditionally serviceable bearing can be considered as a system with one strongly pronounced resonance, then the correlogram of the same bearing with a misalignment of $14' \pm 2.5'$ already characterizes it as a system with several resonances.

To diagnose the technical condition of the research bearings using the obtained correlograms $B_1(s)$, it is necessary to construct quantitative estimates of some parameters describing these correlograms. It was noted that the normalized correlation function can be characterized by a sequence of extrema s_j, $j = 1, 2, \ldots$, and the method of their estimation is described. If the diagnosed bearing unit has at least one clearly expressed resonance, the correlogram has the property of statistical cyclicity with a certain alternation of elements of the sequence of extrema and zeros. In this connection, the visual analysis of such correlograms can be started with the estimation of the period of such cyclicity. Let's construct this estimation for sequences of extremum points in correlograms $B_1(s)$ of vibrations of a conditionally serviceable bearing and

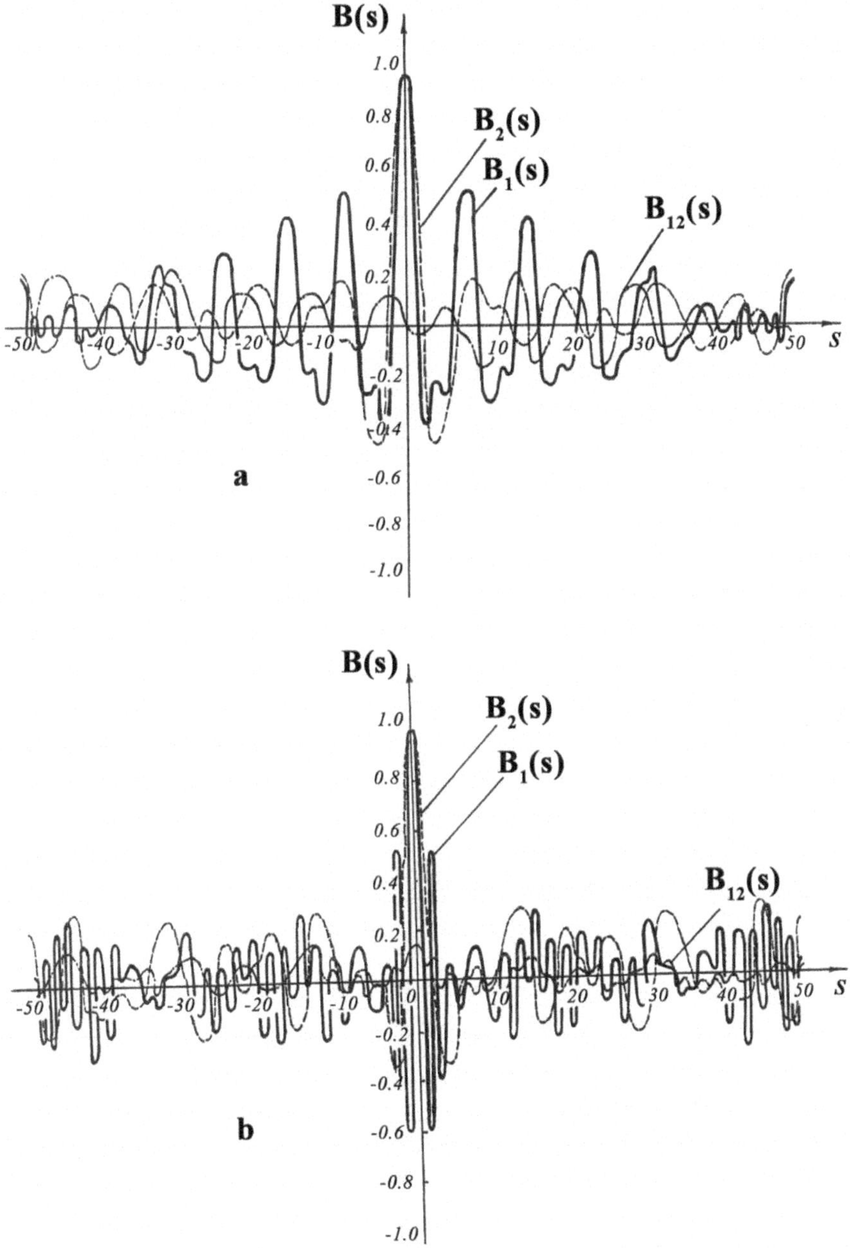

Fig. 8.10 Normalized auto- and mutual correlograms of bearing vibrations: **a** conditionally serviceable, **b** in the presence of misalignment $14' \pm 2.5'$. Indexes 1, 2, and 12 denote respectively correlograms of bearing vibrations from the collector side, from the drive side, and their mutual correlograms

a bearing with misalignment, shown in Fig. 10a, b. In this case, as an estimate of the period of repetition of extrema, we can take $h_s = s/n(s)$, $s \in (0, \infty)$, where $n(s)$ is the number of extrema observed on the interval of shifts s on the correlogram $B_1(s)$.

Each fixed s magnitude h_s is random, and in the experiment, there are observed realizations h_s, which are also denoted by h_s. With increasing s magnitude h_s for resonant systems theoretically tends to some finite limit. However, in practice, one should be limited to some value $s = T_H$, above which the measured values $B_1(s)$ become unreliable due to the limited accuracy of correlometry. Therefore, we can take T_H as the last value s, for which there is a value obtained with a given accuracy $B_1(s)$. In Fig. 8.10 this coincides with the breakpoint of the graph, i.e. $T_H = 50 \cdot 1/5.5 \cdot 10^{-3}$ s.

By direct calculation we find that for $B_1(s)$ Fig. 8.10, the value $n(T_H) = 22$, and in Fig. 8.10b—$n(T_H) = 41$. Therefore, in the first case, the period $h_{T_H} \approx 2.27/5.5 \times 10^{-3}$ s is estimated and in the second case $h_{T_H} \approx 1.22/5.5 \times 10^{-3}$ s, i.e., these periods differ in magnitude by almost a factor of two. The found estimates of the period are point estimates, but the question of their reliability remains open. Nevertheless, in the case under consideration, due to the significant change $n(T_H)$, caused by the skewness of $14' \pm 2.5'$, we can conclude that the period value can be used as a diagnostic sign in practice. Nevertheless, in the case under consideration, due to the significant change h_{T_H}, caused by the skewness of $14' \pm 2.5'$, we can conclude that the period value can be used as a diagnostic sign in practice.

From the preliminary comparison of the vibration $B_1(s)$ correlograms in Fig. 8.10, a and b, it is evident that the introduction of misalignment in the bearing resulted in a change in the damping coefficient β_1. Let us focus on the quantitative confirmation of this preliminary conclusion. As a statistical estimation we can take the expression β_1:

$$\overline{\beta}_1 = -\frac{1}{n}\sum_{j=1}^{n}\frac{\ln\left[(-1)^j B_1(s_j)\right]}{s_j}, \tag{8.2}$$

which was obtained when considering the probabilistic model of rolling bearing vibrations in Chap. 3 of this book.

Let us construct an estimate $\overline{\beta}_1$ for the correlograms $B_1(s)$ in Fig. 8.10a, b, and to simplify the calculation of the estimate, we reduce the duration of the observation interval of these correlograms to the value T_B. Taking into account that in the correlogram $B_1(s)$ of vibrations of a conditionally serviceable bearing the first three negative half-waves have two extrema each, when calculating $\overline{\beta}_1$ we choose only the larger absolute value $B_1(s_j)$. As a result for a conditionally serviceable bearing

$$\overline{\beta}_1 = 0.3321 \cdot 5.5 \cdot 10^{-3}\,\text{s}^{-1},$$

for skewed bearings

$$\overline{\beta}_1 = 0.4483 \cdot 5.5 \cdot 10^{-3}\,\text{s}^{-1}.$$

Note that $\overline{\beta}_1$ it is not quite correct from the mathematical point of view to estimate (2) for the correlogram $B_1(s)$ shown in Fig. 8.10b, since this correlogram corresponds to a system with more than one resonance. The estimation (2) was obtained for the case of one strongly pronounced resonance, i.e., it can be used to estimate β_1 the correlogram of the form $B_1(s)$ shown in Fig. 8.10a.

The obtained exact statistical estimates $\overline{\beta}_1$ for a conditionally corrected bearing and a bearing with skewness differ by almost 25%. This also indicates the possibility of using the parameter β_1 as a diagnostic sign, but to confirm the significance of this difference, it is necessary to use statistical criteria for hypothesis testing.

The mutual correlation analysis of synchronously registered vibrations of bearings both from the collector side and from the drive side in EM P-51 has shown the possibility of using the estimation of the mutual correlation coefficient as a diagnostic sign of bearing misalignment. As a result of averaging over 60 correlograms, the following values of the mutual correlation R coefficient estimation were obtained: $R = 0.0466 \pm 0.02$ for conditionally serviceable bearings and $R = 0.1274 \pm 0.017 \pm 0.017$ in the presence of $14' \pm 2.5'$ misalignment in the bearing on the collector side. As usual, the $\pm$ signs are followed by the RMS value of the point estimate error. Consequently, a misalignment in the bearing on the collector side of the EM P-51 leads, on average, to an almost threefold increase in the coefficient estimate R for the same estimation error. The energy spectrum obtained as a result of spectral analysis of bearing vibrations characterizes the vibration power distribution by frequencies. To establish the diagnostic signs, which can be obtained from the results of this analysis, we will also use the studies of vibrations of bearings 309 ESH2, installed on EM P-51.

Figure 8.11a presents spectrograms of vibrations of conditionally serviceable bearings, and Fig. 8.11b—spectrograms of vibrations when a skew is introduced into the bearing from the collector side.

Here $S_1(f)$—spectrogram of bearing vibrations from the collector side, $S_2(f)$—the same from the drive side, $S_{12}(f)$—mutual spectrogram. According to the spectrogram $S_1(f)$ in Fig. 8.11, it is possible to conclude that the main contribution to the bearing vibrations from the collector side is made by spectral components lying in the area of 750 Hz with amplitudes up to 0.150 units. Small maxima $S_1(f)$ are observed in the regions of 1.35 and 2.8 kHz. The character of the spectrogram $S_2(f)$ indicates a slight shift from $S_1(f)$ the position of the main maximum, which lies in the region of 0.5 kHz. We note some rise $S_2(f)$ in the frequency band 0 .... 0.2 kHz, as well as the presence of a small maximum of 0.032 c.u. at the frequency of 1.6 kHz. At frequencies above 1.6 kHz, there are no significant components in the bearing vibrations from the drive side. Thus, the analysis of spectrograms shows that in vibrations of conditionally serviceable bearing both from the collector side $S_1(f)$ and the drive side it is possible to distinguish three clearly expressed resonances. At the same time, their frequencies differ: $f_{p1}^{(1)} = 750$ Hz; $f_{p2}^{(1)} = 1.35$ kHz; $f_{p3}^{(1)} = 2.8$ kHz—for the bearing on the collector side; $f_{p1}^{(2)} = 100$ Hz; $f_{p2}^{(2)} = 500$ Hz; $f_{p3}^{(2)} = 1.6$ kHz—for the bearing on the drive side.

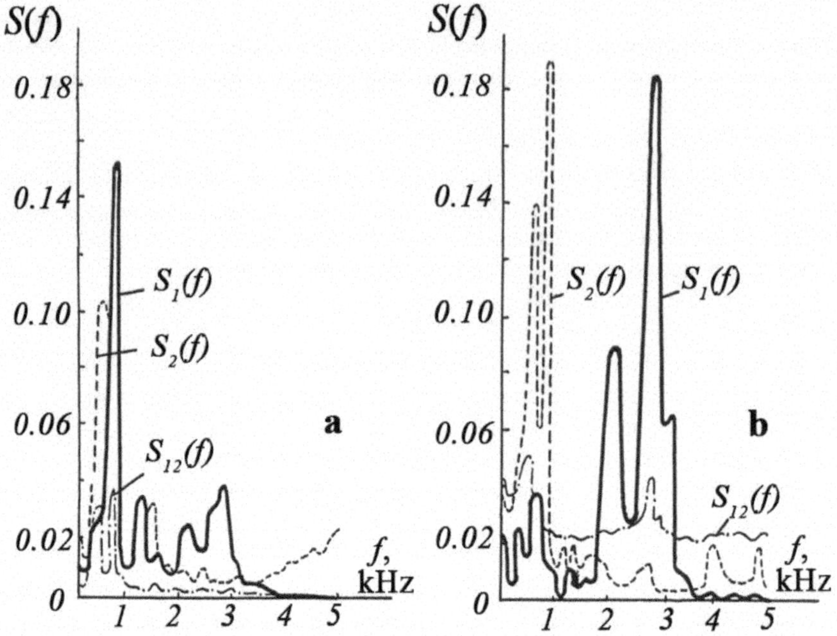

Fig. 8.11 Spectrograms of bearing vibrations: **a** conditionally serviceable, **b** in the presence of misalignment in the bearing from the collector side. Indices 1, 2, and 12 denote respectively correlograms of bearing vibrations from the collector side, from the drive side, and their mutual

The difference in the character of spectrograms $S_1(f)$ and $S_2(f)$, obtained for the same-type bearings 309 ESH2, speaks about the constructive difference of bearing units of the machine P-51, installed from the side of contact rings and the side of the drive.

In Fig. 8.11b the shift of the main maximum of the spectrogram $S_1(f)$ of vibrations of the bearing from the collector side to the area of high (in comparison with $S_1(f)$ the conditionally serviceable bearing) frequencies – 2.7 kHz, that confirms the conclusions made above on estimations h_{T_H}. Two more maxima at frequencies 750 Hz and 2.1 kHz are also observed in the character of the spectrogram $S_1(f)$. For a bearing with misalignment, as well as for a conditionally serviceable bearing, three pronounced resonances with frequencies: $f_{p1}^{(1)}$ = 750 Hz; $f_{p2}^{(1)}$ = 2.1 kHz; $f_{p3}^{(1)}$ = 2.7 kHz can be distinguished in the vibrations.

The skew of the bearing on the collector side was reflected in the character of vibrations of the bearing installed on the drive side, although no changes occurred with it. This is explained by the fact that the investigated bearings operate on the same shaft. Thus, the character of the spectrogram $S_2(f)$ maximum in the area of frequencies of the main maximum has changed—it has become two-vertex with peaks at frequencies of 600 and 800 Hz. It is necessary to note some increase in amplitudes of spectral components in the band 0 ... 100 Hz, as well as the appearance of six maxima not exceeding 0.02 units in the band 1 ... 5 kHz.

From the comparative analysis of the spectrograms $S_{12}(f)$ in Fig. 8.11a, b, it is evident that the introduction of a skew in the bearing from the collector side leads to a change in the frequencies of the maxima in the spectrogram $S_{12}(f)$. Thus, in the spectrogram $S_{12}(f)$ in Fig. 8.11a, two maxima at frequencies 450 and 750 Hz are observed. The introduction of skewness in the bearings leads to a shift of the second (750 Hz) maximum to the frequency area of 2.8 kHz.

Thus, by the example of investigation of vibrations of rolling bearings of EM P-51 the possibilities of correlation and spectral analysis of bearing vibrations for quantitative estimation of the defect of skew type are demonstrated. Based on the conducted experiments diagnostic signs of technical condition of bearings can be used in correlation analysis: the number of points of extrema s_j, $j = 1, 2, \ldots$ on some interval T_H of the correlogram $B(s)$; estimation of the values of attenuation coefficients $j = \overline{1, n}$; estimation of the value of the mutual correlation coefficient of bearing vibrations installed in both bearing units; in spectral analysis: amplitudes and frequencies of the main maxima of the spectrogram; number of the main maxima of the spectrogram in a fixed frequency band.

In conclusion, it should be noted that the idea of using the number of main maxima of the bearing vibration spectrogram as a diagnostic sign was realized, as example. in [47–49].

The most complete information about the nature of the distribution of the investigated bearing vibrations can be obtained by analyzing their higher moments. Let us dwell on this issue in more detail.

8.2.3 Diagnostic Features Based on Estimates of Vibration Moments

Correlation and spectral analyses in diagnostics are based on the study of bearing vibrations within the first two moments. In the third section, it is established that the highest central moments of their vibrations (third, fourth, etc.) and the character of the curve of the probability distribution function can also be used as diagnostic signs of the technical condition of bearings. For this purpose, we will use the results of the histogram analysis of vibrations of bearings 309 ESH2 installed on the DC machine P-51. To select the frequency bands in which the main contribution to vibrations is made by rolling bearings, vibrations are subjected to preliminary band-pass filtering, in which the boundary frequencies are selected by the results of paragraph 2.1 of this chapter.

For the set of statistics in the experiments, vibrations were investigated on 19 samples of 309ESH2-type bearings with different technical states. The measured vibrograms were transformed in the external memory of the computer included in the IMS of vibrodiagnostics in the form of sets of m realizations each:

$$\{\xi_{i,q}, i \in \overline{[1, n]}, \quad q \in \overline{[1, m]}\}, \tag{8.3}$$

with volume $n = 2048$ samples each. Thus $60 \leq m \leq 89$. These sets were obtained both for serviceable bearings and for bearings having misalignments, developed inner ring defects, and operating without lubrication. Pre-processing of the registered sets (3) was carried out on the IMS using histogram analysis. Then, using the registered sets (3), the investigated vibrations were checked for stationarity using t- and F-criteria. Application of these criteria for the registered realizations with normal distribution is equivalent to their stationarity in the strict sense. For realizations with a distribution other than normal, these criteria allow us to check them for stationarity in a broad sense [50, 51].

While checking for stationarity, 60 realizations obtained from the vibrations of a serviceable bearing, 89—in the absence of lubrication in the bearing, 75—in the case of misalignment $14' \pm 2.5'$ (the angle between the plane of the outer ring of the bearing and the shaft axis is specified), and 76—with a developed defect of the inner ring, all with 2048 samples each, were selected for further analysis. As a result of the processing of the selected samples on the IMS of vibrodiagnostics by the program of histogram analysis, in particular, estimates of the initial and central moments (up to the fourth inclusive) were obtained, and also histogram construction and their smoothing by the system of Pearson curves [52–55] were realized.

From the results of experiments, it follows that the most informative diagnostic signs, allowing to establish practically all of the listed defects, are coefficients of asymmetry k and excess γ of investigated vibrations, which are connected with the third and fourth moments by relations (3.49) and (3.50). Let's consider these coefficients as diagnostic signs of the technical condition of bearings.

Table 8.1 shows the observed values of estimates of asymmetry and vibration excess coefficients of bearing 309 ESH2, installed from the collector side in machine P-51 and tested at different technical conditions.

Lack of lubrication in the bearing leads to a slight shift towards an increase in the estimate of the mean value of the excess coefficient compared to the estimate of this parameter for the vibrations of a serviceable bearing. The skew in the bearing with the value of $14' \pm 2.5'$ causes an even greater shift towards an increase in the estimate of the mean value of the coefficient γ and an insignificant decrease in the estimate of the mean value of the coefficient k. For vibrations of the bearing with defects of the inner ring, the largest estimates of the average value of the coefficient γ and the smallest estimates of the average value of the coefficient are characteristic k.

Thus, it is obvious that the coefficients γ can be used as diagnostic signs, allowing us to determine the specified defects of bearings with a certain degree of confidence.

As follows from the results of histogram analysis of bearing vibrations measured at different EM and at the experimental unit, quantitative estimates of asymmetry and excess coefficients depend significantly on the type of bearings, type of defect, and bearing testing conditions, as well as on the object on which the tests were conducted. Proceeding from this, it is necessary to diagnose the technical condition of bearings taking into account a specific type of EM. This task can be solved with a sufficiently high reliability only if the training on determining quantitative estimates of the limits of coefficients k and γ, corresponding to various defects in bearings installed on specific EMs is carried out beforehand.

According to the results of experimental verification and establishment of quantitative estimations of diagnostic signs corresponding to different technical conditions of investigated rolling bearings, it is possible to proceed directly to the formation of training sets and construction of solving rules on diagnostics and classification of specific types of defects in bearings. However, before proceeding to the presentation of this material, let's consider the issues of experimental verification of diagnostic signs used in carrying out shock, acoustic emission diagnostics of electrical equipment, as well as diagnostics of DEG cylinder-piston groups (CPGs).

8.3 Analysis of Shock Vibration Signals, Excited in the Charging Coil

Using the experimental sample of the IMS of shock diagnostics, described in paragrap 1 of the Chap. 7, the experiments on diagnosing the condition of the pressed charge magnetic core were carried out. As an object of research, a part of the pressed magnetic core of a low-power transformer [56–60], was installed on a special test bench (Fig. 8.12).

Pressing of the magnetic core plates was carried out with the help of studs 1, tightening of which was carried out with a torque wrench with a torque of 30 N m. The vibration signal (vibration acceleration) was measured using accelerometer 2, which was installed on the studied magnetic core with the help of special glue "Cyokrin". The accelerometer of ABC 17 type is allowed to measure in the frequency band 20 Hz … 30 kHz vibration wave excited by the IH in the body of the charge magnetic core.

The experiment on diagnostics of the magnetic core plate pressing condition was carried out as follows. With the help of clamping nuts installed on the studs 1, a package of magnetic core plates was compressed with a force of 30 N m. Then the laboratory sample of IMS of shock diagnostics was switched on and adjusted (Fig. 8.4 of the Chap. 7). After tuning the IMS in the direction perpendicular to the plane of the plates, the impact was made, which was recorded by accelerometer A1 installed on the hammer. The signal from accelerometer A1 was fed to the circuit of launching the IMS mockup of shock diagnostics and the vibration signal arising in the charge magnetowire after the hammer impact was measured by accelerometer A2.

After that, the nuts of studs 1, which were used to tighten the plates of the charged magnetic core, were completely released and an experiment similar to the described one was repeated. Figure 8.13 shows the spectrograms measured on the packed magnetic core compressed with a force of 30 N m (Fig. 8.13, a) and in the case when the compressive force is absent (Fig. 8.13, b).

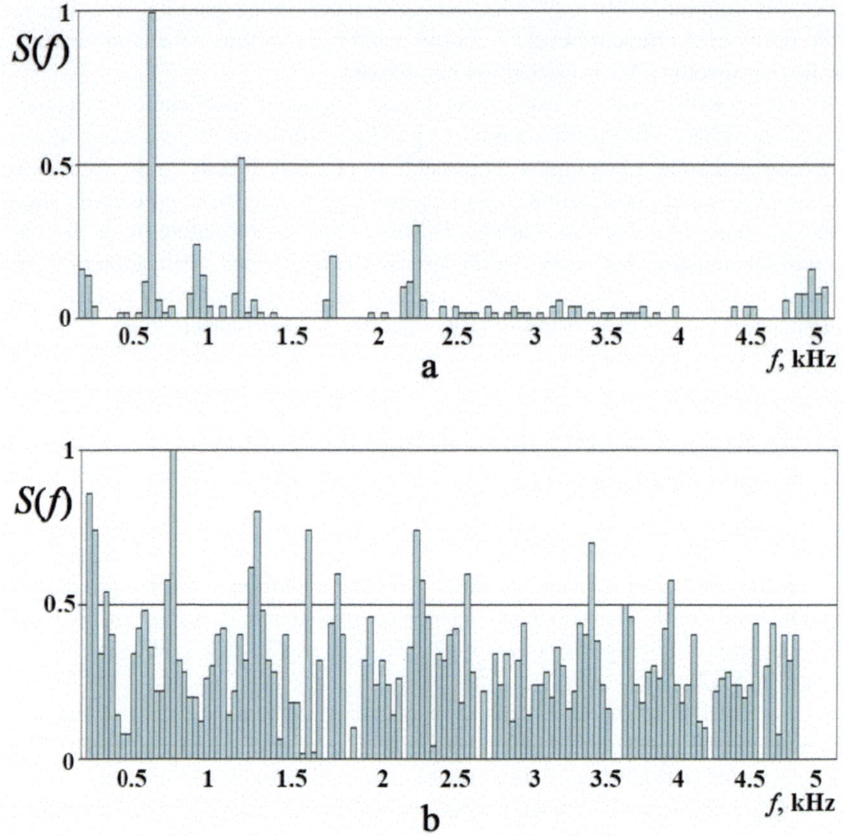

Fig. 8.12 Spectrograms of vibrations measured on the magnetic core: **a** compressive force equal to 30 N m; **b** no compressive force

During the experiments, the analog vibration signal was converted into a digital code with sampling frequency $f_d = 10$ kHz, i.e., the spectrograms presented in Fig. 8.13a, b contain information about the frequencies of the analyzed signal in the band with the upper boundary frequency of 5 kHz.

It can be seen from Fig. 8.13, that for the compressed magnetic core the main vibration frequency is observed in the region of 600 Hz. Less intense components in the area of 1200 and 2300 Hz are also observed.

At removal of the shrinking force in the spectrogram of vibrations of the packed magnetic core (Fig. 8.13b), besides the main frequency in the region of 600 Hz, intensive frequency components are noted practically in the whole analyzed frequency band of 5 kHz. Such multi-resonant structure of vibrations in the charge cores with weakened pressing is explained by vibrations of separate plates forming this magnet conductor.

Fig. 8.13 Layout of a magnetic core

Proceeding from consideration of the spectrograms of shock vibration waves propagating along the body of the charge magnet wire (typical of them are presented in Fig. 8.13a, b), it is possible to note that the most informative diagnostic signs, allowing us to diagnose the degree of pressing of the charge magnet wire, are:

- amplitudes A_f of the main maxima of the spectrogram $S(f)$;
- frequencies f_n of the main spectrogram maxima $S(f)$;
- number n_f of spectrogram main maxima $S(f)$ in a fixed frequency band.

8.4 Analysis of AE Diagnostic Signals in EE's Assemblies

To assess the state of the object with the help of IMS of AE diagnostics, AE signal realizations are measured and according to the results of their statistical processing using the software described in the previous section of this paper, quantitative estimates of some diagnostic features are determined. A characteristic property of AE signals is the change in the process of measurement of quantitative estimates of diagnostic features [61–66]. Let us dwell on the description of the experiments, which were conducted directly on the objects of diagnostics.

Measurements of AE signals were carried out on a wind-electric unit [67–70] type USW 56–100, as well as on standard samples designed for testing on a tensile machine P-20 and made of sheet electrical steel E330. As an example, we will focus on the latter experiment below, because with the help of the P-20 tensile machine, it is possible to create different forces applied to the test sample. This, in turn, allows us to experimentally verify the diagnostic features, quantitative assessments of which should change according to the loads applied to the test piece.

Tests were carried out on standard samples AE-01-P (Fig. 8.14), made of sheet electrical steel of E330 grade (state standard 9925-61). The geometric dimensions of the sample corresponded to the fifth class of accuracy. The surface cleanliness of the material surface in the places of AE primary transducer fastenings corresponded to Rz 40.

The parameters of the AE signals for the studied samples were obtained by testing them on the P-20 tensile machine. This machine is designed to create static loading of

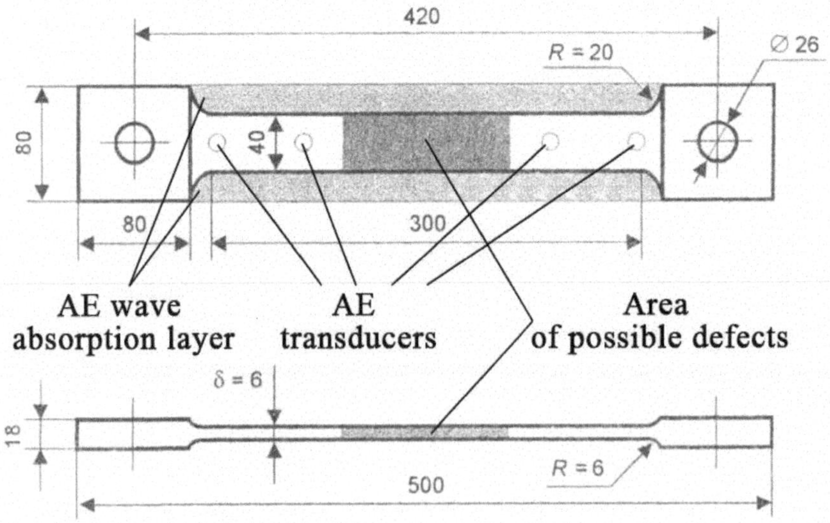

Fig. 8.14 Standard sample AE 01P

samples made of metals and alloys, as well as products made of them during uniaxial tension tests with the ultimate load of 20 tons according to state standard 7855 74. The machine is equipped with a system that allows it to register the load on the sample during its deformation and record the data in the memory of AE diagnostics. The breaking machine operates in the temperature range of 15…35 °C at a relative humidity of 40 … 80%. The speed of the moving gripper varies within the range of 0 … 200 mm/min.

The machine is an installation consisting of a loading device and a control panel. The loading device is designed for the deformation and fracture of the test sample. The control panel is used for controlling the loading process and monitoring. It includes a pumping unit with a control system, dynamometer, and a device for registering the "load-strain" diagram of load and deformation values. The loading device and the control panel are mounted on the foundation and connected by oil pipelines. The loading device has a vertical design with a hydraulic upper and mechanical lower gripper. In the lower part of the console, there is a pumping unit, on which a force-measuring unit is installed. The pumping unit consists of a hydraulic tank and a block of electro-hydraulic equipment. Strain gauges were connected to the sample with an output to the strain gauge station of the control panel. P-20 is powered by a three-phase AC network with a voltage of 380 V and a frequency of 50 Hz.

The laboratory sample of the AE diagnostic IMS was used in the experiments.

AE transducers [71–77] were installed directly on the sample, its loading was carried out, and AE parameters were recorded for a conditionally serviceable sample and with a crack-type defect. In turn, this allowed us to record the intensity λ of AE signals under different loading and at fracture. The structural scheme of the experiment is shown in Fig. 8.15.

Before the tests visual inspection of the sample with fixation of corrosion centres, and mechanical and other damages were carried out, and possible zones of attenuation of AE signals, and places of transducer fastening were determined. Places for the installation of transducers on the surface of the sample were cleaned and thoroughly degreased. The transducers were installed directly on the surface of the sample and acousto-transparent glue was used for their fastening.

After checking the performance of the P 20 machine, the working load was fixed and IMS AE diagnostics was switched on in the measurement mode. The loading of the sample was carried out by the graph shown in Fig. 8.16. Here, the abscissa axis

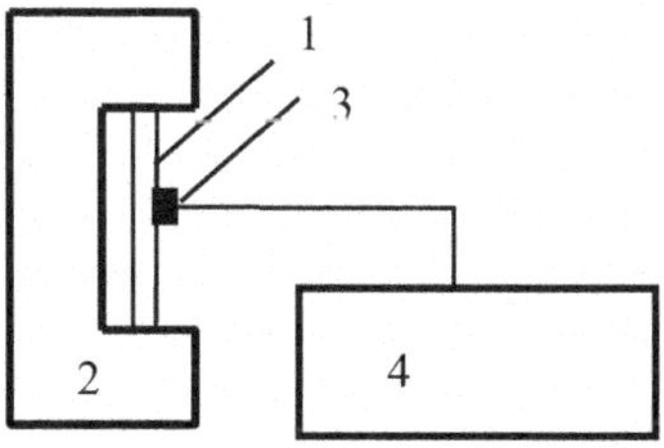

Fig. 8.15 Structural scheme of the experiment on AE diagnostics: 1—test sample AE 01R, 2—tensile machine, 3—AE sensor, 4—AE diagnostics IMS laboratory sample

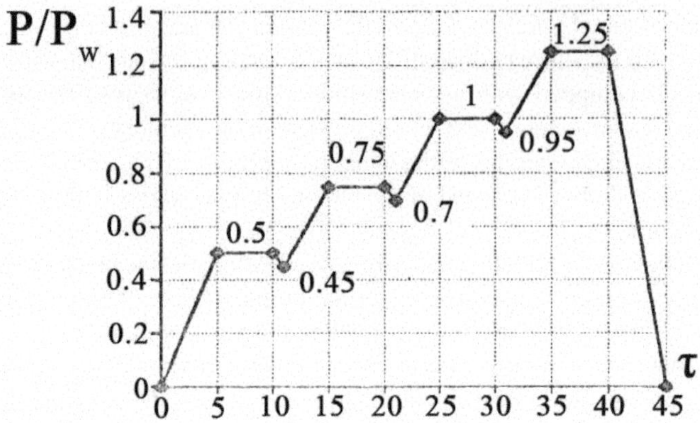

Fig. 8.16 Graph of loading of the sample using the machine P-20

shows the time interval τ during which the loading is carried out, and the ordinate axis shows the ratio of the test pressure to the working pressure P/P_w.

Measurements of AE parameters were made during pressure holding at each stage for five minutes. In this case, the IMS registers the information received from the sample. The tests were short-term. At each stage of loading, IMS AE diagnostics was switched on for 1 s; for 5 min—several dozens of times.

Figure 8.17 shows the graphs of typical realizations of AE signals, which were measured on a new conditionally serviceable test sample made of electrical steel E330 and on a similar sample, which was repeatedly tested on a P-20 tensile machine and which, presumably, had a defect such as an incipient crack. The tests were carried out at four degrees of loading $P/P_w = 0.5; 0.75; 1; 1.25$.

The AE signal realizations shown in Fig. 8.17, represent a sequence of very short in duration but with different pulse amplitudes (some similarity of pulses). As follows from this figure, the intensity of pulses in the new sample is significantly lower compared to the repeatedly tested sample. At the same time, in the realization of the AE signal measured on the new sample (Fig. 8.17a), there are pulses with maximum amplitude, the value of which is almost 5 times higher than the amplitude value of the pulses measured on the repeatedly tested sample (Fig. 8.17b).

Conducting tests on the rupture machine P-20 ultimately pursued the goal of obtaining information about changes in the quantitative estimates of diagnostic features at different degrees of loading of the test specimen. Based on the obtained information the training sets (images) were formed, which are an integral part of the created IMS AE diagnostics and are part of the OS block (Fig. 8.12 of the Chap. 7). These issues are discussed in detail below.

As shown in [42–46], the most informative diagnostic features are the asymmetry k and kurtosis coefficients γ, which are related to the second μ_2, third μ_3, and fourth μ_4 moments of AE processes by relations (3.49) and (3.50). These parameters are used in further AE diagnostics.

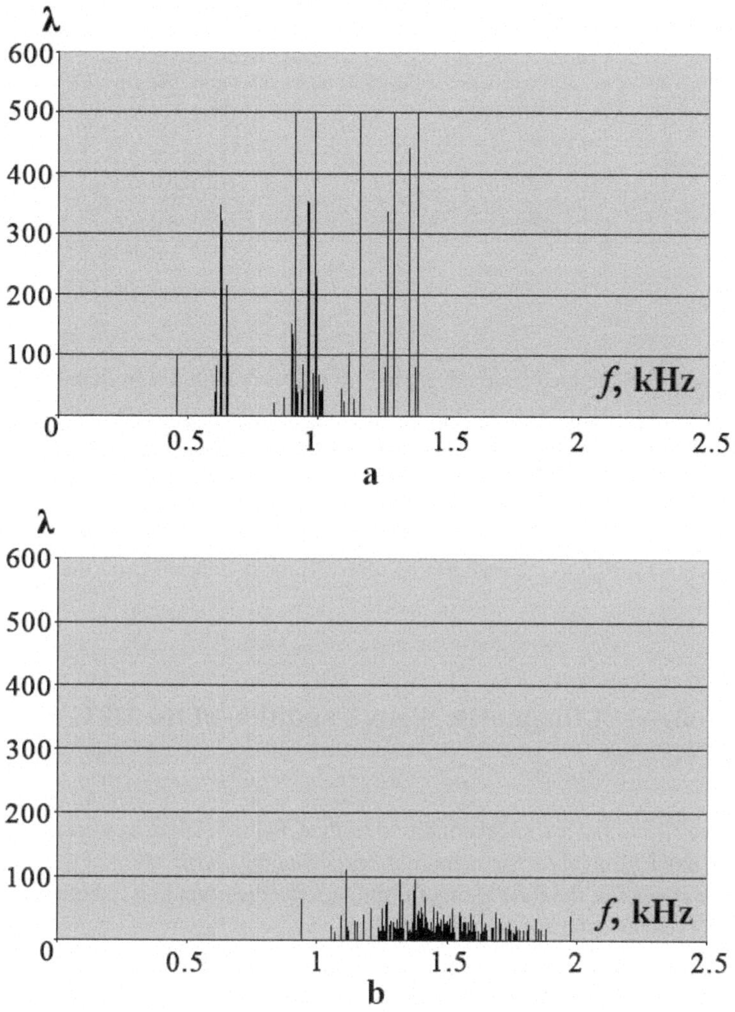

Fig. 8.17 Typical realizations of AE signals were measured on a new conditionally serviceable test sample from electrical steel E330 (**a**) and on a similar sample, which was repeatedly tested on a P-20 tensile testing machine and which, presumably, had a defect of the incipient crack type (**b**)

Table 8.2 shows the observed values of estimates of asymmetry k and excess coefficients γ of AE signals appearing in samples from electrical steel E330 tested on a P-20 tensile machine at different values of tensile loads.

As can be seen from Table 8.2, the increase in the tensile force on the test sample leads to a significant increase (on average) in the estimation of the coefficient of excess $\overline{\gamma}$. More than in 3 times increases the specified estimation at the change of the value of tensile force created by the machine P-20 (from 50 to 125% of the maximum permissible such force). The observed increase $\overline{\gamma}$ in estimates indicates

Table 8.2 Values of coefficients k and γ in samples from electrical steel E330 tested on a P-20 tensile machine

Breaking load on the test piece (in % of max. value)	Number of realizations selected for analysis	Point estimates of the mean value of the coefficients	
		k	γ
50%	60	1.207 ± 0.375	1.430 ± 0.675
75%	80	1.321 ± 0.255	2.457 ± 0.551
100%	90	1.345 ± 0.097	3.621 ± 0.337
125%	70	1.369 ± 0.085	4.715 ± 0.253

the appearance in the measured AE signal of impulses with a significant amplitude with increasing tensile force.

Thus, the results of the experiments confirm the possibility of using the AE signal excess $\overline{\gamma}$ coefficient estimation as a diagnostic sign characterizing the degree of mechanical loading of the test specimen.

Let's pass to consideration of questions connected with experimental verification of diagnostic signs characterizing the technical condition of diesel-electric generators.

8.5 Analysis of Diagnostic Signs Condition of the DEG's CPG

The diagnostics of the technical condition of the CPG of a DEG, as well as the diagnostics of any EE, involves the mandatory experimental verification of the proposed diagnostic attributes, the formation of training sets corresponding to both the conditionally serviceable state of the CPG, and in the case of failure of one or a group of cylinders, and, finally, the construction of decisive rules for the diagnosis and classification of faulty cylinders [78–83]. Let us focus on the experimental verification of the proposed diagnostic features.

As noted above, with the help of the sensor PDF-3 and the data conversion and input device under the control of the program package the samples $\left\{ \theta_j, j = \overline{1, 1200} \right\}$ are entered into the computer memory. Then the transition from the values θ_j, to the values ω_j and $\Delta\omega_j$. With the help of special software, using a given number of cycles of operation of the diesel engine N_D, nested sequences are built estimates $\overline{\Delta\omega_j}$, on which the first 16 Fourier coefficients and their corresponding initial phases [84–88] are determined. The results of this processing are presented in the form of amplitude-frequency (AFC) and phase-frequency (PFC) characteristics.

Figure 8.18a, b shows the AFC and PFC of angular velocity variation on the shaft of DG 5D70, obtained on the test bench of the Production Association "Malyshev Plant". Frequency of rotation of the DEG shaft—1000 rpm, the operating mode of the DEG—idle running (XX).

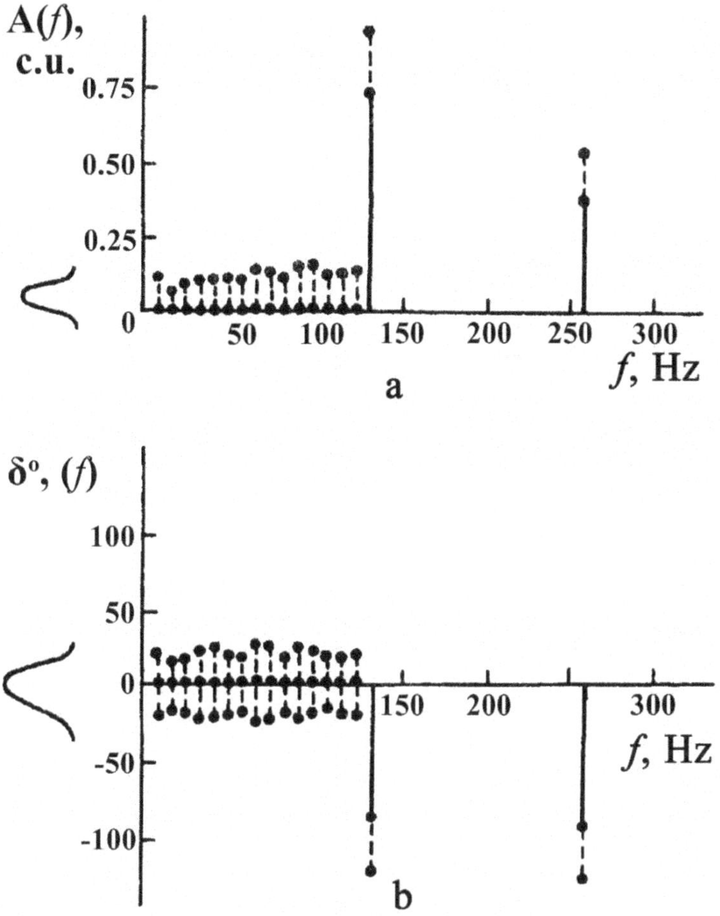

Fig. 8.18 AFC and PFC of angular velocity variation on the DEG shaft, obtained on the test bench of "Malyshev Plant". Frequency of rotation of the DEG shaft—1000 rpm, operating mode of the DEG—idle running

As can be seen from Fig. 8.18 a, for a conditionally serviceable DEG, the main harmonic is manifested at a frequency of 132 Hz. On the discrete corresponding to the main harmonic, squares conventionally show the minimum and maximum values taken by the main harmonic at the observation interval, the duration of which in this case was chosen equal to 500 operating cycles in each cylinder of the diesel engine (1000 revolutions of the crankshaft). In the lower-frequency area, insignificant (compared to the main harmonic) fluctuations of the amplitudes of almost all 15 subharmonics are observed. This peculiar "noise background" of subharmonics is caused by non-uniformity of the processes occurring in the working cylinders of the diesel engine, non-uniformity of the load on the diesel engine shaft, associated with the operation of the electric generator, and other factors. All this leads to the non-ideal periodicity of the shaft rotation process itself. This, in turn, causes fluctuations in the estimates of the phase angles of the corresponding subharmonics. In Fig. 8.18b the range of fluctuations of phase angles of subharmonics is shown by squares in the lower and upper half-planes. The mean value of each of the fluctuating subharmonics shown in Fig. 8.18, is close to zero. Let us briefly dwell on determining the nature of the distribution of fluctuations of subharmonics and main harmonics amplitudes, as well as their corresponding phase angles. This check was carried out as follows. The angular velocity of rotation of the diesel crankshaft during 500 full operating cycles (1000 crankshaft revolutions) in each of the cylinders of the investigated DEG 5D70 were measured and recorded. After checking the recorded data for stationarity using the methods described above, the first 16 Fourier coefficients and their corresponding initial phases were obtained. This operation was performed for each duty cycle in all cylinders of the diesel engine, i.e., in other words, 500 realizations were obtained for analysis. Then, the obtained 500 realizations containing fluctuations in the amplitudes of subharmonics and fundamental harmonics, as well as their corresponding initial phases, were processed using the histogram analysis program described in paragraph 2 of Chap. 7. As a result of this processing, it was found that the constructed histograms of fluctuations of subharmonic and fundamental harmonic amplitudes and their phase angles can be approximated by IV, VII, or XIII types of curves included in the Pearson system, i.e., with a sufficient degree of accuracy their distribution can be considered normal.

Figure 8.18a, b in the left part of the figures schematically show the smoothing distribution curves of the studied fluctuations for the first subharmonic. Figure 8.18, shows that the constructed distribution curve has a mathematical expectation slightly different from zero. At the same time, for a serviceable diesel engine, as shown above, by the results of mathematical modeling, the mathematical expectation should be close to zero.

The distribution curve of fluctuations of phase angles of the first subharmonic (Fig. 8.18b) is close to normal with zero mathematical expectation, which is in full agreement with the previously discussed results of mathematical modeling of DEG operation.

Experiments on the simulation of failure of each of the cylinders, which were carried out by stopping the fuel supply to the corresponding cylinder, showed that the AFC and PFC of the studied angular velocities undergo a significant change (Fig. 8.19). Thus, comparing the AFC of Figs. 8.18a and 8.19a, it can be noted that the failure of one of the cylinders leads to a change in the AFC, which is expressed in the appearance of subharmonics with significant amplitudes and a decrease in the amplitude of the main harmonic. Moreover, at consecutive changes in the number of failed cylinders, the character of the AFC does not change. This fact indicates that the AFC can be used as a diagnostic sign of failure of one of the cylinders, without a specific indication of the number of the failed cylinder.

The nature of changes in the PFC at consecutive failure of each of the diesel cylinders showed that the phase angle of the corresponding subharmonics can be used as a diagnostic sign that allows to diagnose of the number of failed cylinders. In this case, the phase angles of the main harmonic and multiple harmonics

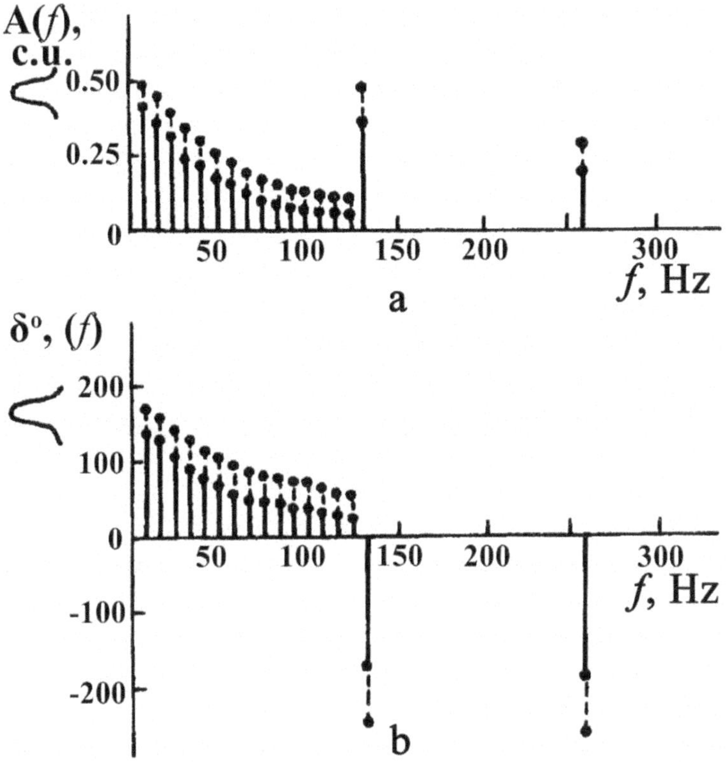

Fig. 8.19 AFC and PFC of angular velocity variation on DEG shaft, obtained at the simulation of failure of cylinder 1L

remain unchanged. Table 8.3 presents the average values of estimates of the initial phase angles (in degrees) of the corresponding subharmonics, obtained during the sequential simulation of failure of each of the cylinders of DEG 5D70.

From the data presented in Table 8.3, we can see that to diagnose the number of the failed cylinder, it is necessary to use the phase angle estimates of odd subharmonics (1, 3, 5, ..., 15), which do not have the same or close values of phase angle estimates for different numbers of failed cylinders.

The choice of even subharmonics (2, 4, ..., 14) for phase angle diagnostics does not provide an unambiguous answer regarding the number of failed cylinders, because the same (or close) average values of phase angle estimates for different numbers of the failed cylinders are observed. For example, for the fourth subharmonic, an average phase angle estimates of – 75° is observed for the 3, 4, 5, and 6 left cylinders. This complicates the task of forming training sets corresponding to different numbers of failed cylinders.

Analysis of the distribution of the phase angle values of the corresponding subharmonics showed that the studied distributions can be approximated by a normal

Table 8.3 Average values of estimates of the initial phase angles

Cylinder	Subharmonic number														
	1	2	3	4	5	6	7	8	9	10	11	12	13	14	15
1L	160	140	120	105	90	80	72	66	62	60	56	52	50	40	35
8R	140	90	50	10	–20	–60	–90	–120	–150	–180	160	140	100	85	60
6L	110	45	–20	–80	–140	170	110	60	10	–40	–90	–130	180	130	80
3R	90	0	–80	–170	110	40	–40	–110	170	95	30	–40	–110	180	100
7L	70	–50	–150	100	0	–95	165	65	–40	–130	140	40	–50	–140	–130
2R	40	–90	140	10	–115	125	5	–115	125	7	–110	130	20	–95	150
4L	20	430	80	–75	140	–10	–155	60	–80	140	5	–135	85	–55	170
5R	0	–180	5	–170	25	–150	45	–115	80	–85	115	–45	150	– 5	–170
8L	–20	140	–80	150	–90	80	–110	65	–120	55	–130	40	–135	40	–140
1R	–40	90	–130	10	155	–60	90	–115	30	–175	–20	135	–75	85	–120
3L	–70	45	160	–75	40	170	–70	60	–170	–40	95	–135	– 5	–125	–100
6R	–90	0	95	–170	–70	35	140	–110	–10	95	–150	–45	65	175	–75
2L	–110	–50	25	100	–180	–100	–20	60	150	–125	–40	40	115	–140	–55
7R	–140	–90	–40	10	65	120	–180	–115	–55	5	70	140	–160	–95	–35
5L	–160	–130	–105	–75	–45	–10	25	65	105	140	–175	–135	–95	–55	– 5
4R	–180	–180	–170	–165	–150	–150	–130	–110	–100	–85	–65	–45	–25	– 5	10

law. Let us briefly consider the changes in the values of phase angles of the first subharmonic at consecutive failure of the cylinders of the investigated DEG.

Figure 8.20 schematically depicts the fluctuations of the phase angle values of the first subharmonic and the distribution curves of these fluctuations (shown to the left of the phase angle graph). In the figures, the maximum and minimum of the observed values are marked with squares, i.e., these values determine the range of variation of the observed samples when each of the cylinders is taken out of service sequentially. The mean value of the distribution of phase angle estimates for each of the experiments is marked by a square, which is located in the middle part of the line of variation of phase angle values of the first subharmonic. As can be seen from the presented figures, the failure of the next in-order cylinder of the investigated DEG causes a shift of the entire distribution curve of phase angle values of the first subharmonic by an average of 20°. Therefore, the average value of the distributions under consideration can be used as a diagnostic sign determining the number of failed cylinders.

Thus, the tests of DG 5D70 carried out with the use of the developed laboratory model of IMS and methods on the stand of "Malyshev Plant" allow to draw the following conclusions:

- the results of experimental studies fully confirmed the data of modeling of DEG operation are given;
- as diagnostic signs of the technical condition of the CPG DEG can be used the results of the analysis of the AFC and PFC of the angular velocities of the shaft of the investigated generator;
- a diagnostic sign of the presence of at least one defective cylinder is the presence of subharmonics in the AFC of the angular velocity of rotation of the DEG shaft. In other words, the diagnostic sign of a technically serviceable diesel engine is the absence of subharmonics in the studied AFC;
- diagnostics of the number of the failed cylinder allows to realize the results of the analysis of the PFC of angular velocities of the DEG shaft.

The results of the conducted experimental studies allow us to proceed directly to one of the key points of diagnostics—selection of diagnostic spaces and formation of training sets in these spaces, corresponding to different numbers of faulty cylinders. These issues are discussed below in paragraph 6.4.

8.6 Construction of Solving Rules for EE Diagnostics

The following methodology for diagnosing various components of EE is based on classical methods of statistical hypothesis testing using the model of linear random processes. The main attention in its development is paid to the study of training populations formed by the most informative diagnostic features in the corresponding spaces, as well as to the decisive rules, which allow to establish the presence of a certain defect in the investigated units of EE with predetermined probabilities of errors of the 1st and 2nd kind.

The fundamental initial data for the selection of diagnostic spaces are diagnostic features, i.e. parameters or characteristics of diagnostic signals, which are the most informative in determining the technical condition of the diagnosed unit.

As it was shown in Chaps. 3 and 4, when analyzing mathematical models of signals accompanying the EE operation, asymmetry k and kurtosis coefficients γ can be considered as diagnostic attributes that determine the technical condition of their nodes.

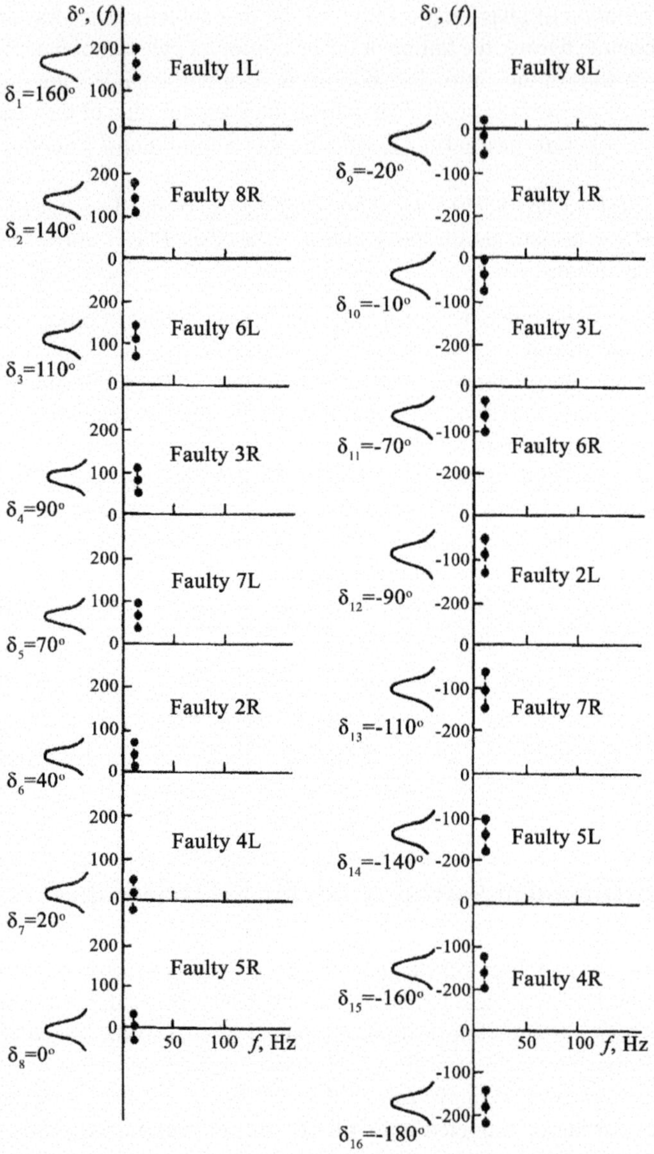

Fig. 8.20 Fluctuations of values of phase angles of the first subharmonics and distribution curves of these fluctuations depending on the number of the faulty cylinder of the DEG

If instead of k and γ we use the associated parameters $\beta_1 = k^2$ and $\beta_2 = \gamma + 3$, then the plane in which the well-known Pearson diagram [89–92] is constructed can be chosen as the diagnostic space (β_1, β_2). For the formation of training sets in the space (β_1, β_2) it is necessary to establish the limits of change of quantitative estimates of $\overline{\beta_1}$ and $\overline{\beta_2}$ depending on the type of defect in the investigated EE node. The designation $\overline{\beta_1}$ means a quantitative estimation of the parameter β_1. The same system of designations remains for the estimates of other parameters.

One of the essential moments of using Pearson diagram with coordinates β_1 and β_2 as a diagnostic space is the possibility of forming in the plane of this diagram some areas obtained by estimations of diagnostic parameters $\overline{\beta_1}$ and $\overline{\beta_2}$. Each of the formed areas can correspond to a certain technical state (conditionally serviceable or having various types of defects) of the object being diagnosed. No less important point of using Pearson diagram as a diagnostic space is the fact that for building training sets on parameters β_1 and β_2 it is necessary to know the first four moments of the investigated diagnostic signals. Obtaining quantitative estimates of these moments is provided by using in the IMS of statistical diagnostics program PIRSON, which was described above. Besides, application of this program allows to receive estimations of dispersion, standard deviation, coefficients of asymmetry and excess, and also to build histograms of the specified signals and to make their smoothing on system of Pearson curves [89–92]. This approach makes it possible to find an explicit analytical expression for the smoothing curve with numerical values of the parameters included in it.

As noted in paragraph 3 of the Chap. 5 of this book, when using the Pearson curve system, the selection of analytical expression is carried out simultaneously for both the evaluation of the distribution of the diagnostic signal under study, and for its classification, because each type of distribution corresponds to a specific number.

A very important point in the practical use of the system of curves and Pearson's diagram (β_1, β_2) is the possibility of relatively quick assessment of the distribution of the diagnostic signal, because the shape of the distribution curve from the Pearson system [5, 6] is determined by two estimates of the parameters $\overline{\beta_1}$ and $\overline{\beta_2}$. Remind that in turn their estimates depend on the values of the first three central moments

$$\beta_1 = \frac{\tilde{\mu}_3^2}{\tilde{\mu}_2^3};\ \beta_2 = \frac{\tilde{\mu}_4}{\tilde{\mu}_2^2},$$

where $\tilde{\mu}_k$ is the k-estimate of the central moment, $k = 2, 3, 4$.

Thus, based on the results of processing the sample under study, the estimates of the parameters $\overline{\beta_1}$ and $\overline{\beta_2}$, a certain point in the diagnostic space (β_1, β_2) is obtained. Then, using the algorithm described in paragraph 3 of Chap. 5 of this book, we select the types of smoothing curves that most closely describe the empirical distribution of the measured diagnostic signal.

While forming training sets in space (β_1, β_2), it is necessary to take into account that the values β_1 and β_2, obtained from the results of processing of diagnostic signal realizations are random variables and, therefore, have some "scatter" relative to their "real" values. Thus, when repeatedly conducting experiments on measurement and subsequent processing of diagnostic signal realizations using the histogram analysis program, the researcher receives some set of points that randomly cover some area in the diagnostic space (β_1, β_2). It should be noted that the specified set of points should be obtained for a quite definite technical state (conditionally serviceable or having a specific type of defect) of the diagnosed unit. The specified set of points is most conveniently represented in the form of scattering ellipses well known in statistics, which provide "coverage" of the specified points with a certain probability level.

The presentation of the basic essence of the diagnostic technique will be carried out on the example of bearing units, the charge magnet core of EM, a sample of electrical steel (AE-diagnostics) and CPG of DEG; diagnostics of other units of EM using this technique is carried out similarly.

8.6.1 *Decision Rules for Vibration Diagnostics and Classification of Defects in EM Rolling Bearings*

Figure 8.21 shows typical training sets in the form of the mentioned scattering ellipses formed in two-dimensional diagnostic spaces (β_1, β_2) as an example.

Rolling bearings type 309 ESH2 [93] with both conditionally serviceable condition and various types of defects (skew, lack of lubrication, inner ring defect) were tested. Each of the scattering ellipses is constructed at the level of 0.05 by averaged estimates $\overline{\beta_1}$ and $\overline{\beta_2}$, corresponding to the specified technical states of the tested bearing.

As can be seen, the lack of lubrication in the bearing 309 ESH2, skewness, as well as the presence of defects in the inner ring lead to a corresponding increase in the estimates $\overline{\beta}_2$ and change in the estimates $\overline{\beta}_1$, which in turn causes a certain regular shift in space (β_1, β_2) of the position of the center of the training set corresponding to specific defects in the bearing. Thus, to a certain position of the training population formed by estimates $\overline{\beta}_1$ and $\overline{\beta}_2$ vibrations of the bearing 309 ESH2, corresponds quite a certain technical condition of the tested bearing.

To build the decisive rules for diagnosing specific defects in bearings, we will use training sets formed by the results of histogram analysis of the investigated vibrations in the diagnostic space (β_I, β_2) (Fig. 8.21).

It is more convenient to diagnose defects using these training sets not in the rectangular coordinate system (β_I, β_2), but in the system (ρ, φ), the transition to which is carried out with the help of relations

$$\rho = \sqrt{\beta_1^2 + (\beta_2 - 1)^2}; \tag{8.4}$$

$$\phi = arctg\,\frac{\beta_2 + 3}{\beta_1 + 4}. \tag{8.5}$$

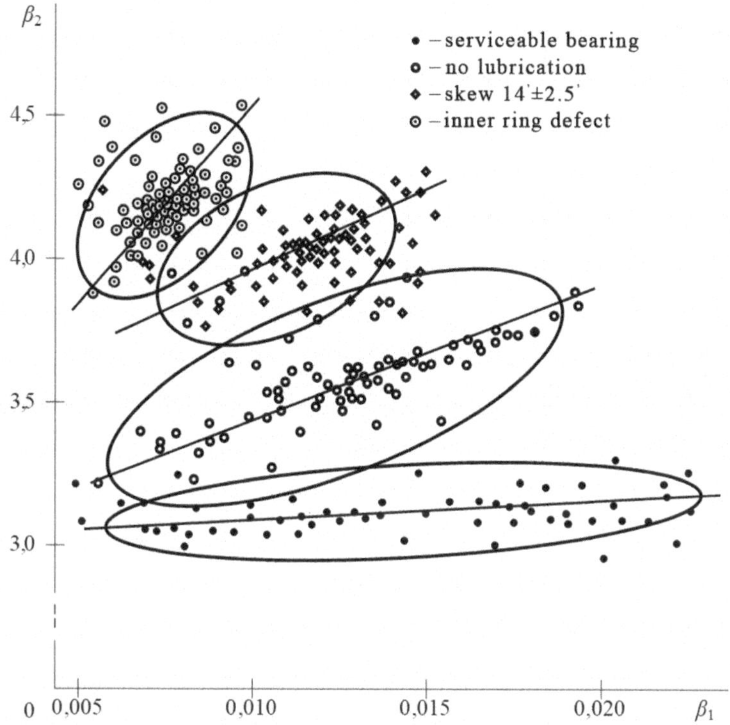

Fig. 8.21 Training sets in diagnostic spaces (β_1, β_2)

From the comparison of Fig. 8.21 with the Pearson diagram, it follows that the areas of the main types of curves (I, IV, VI) have the form of sectors whose vertices begin at the points with coordinates $\beta_1 = 0$, $\beta_2 = 1$ and $\beta_1 = 0$, $\beta_2 = 3$. This is the reason for the transition according to (8.4) and (8.5) to a new coordinate system. In this case, the images corresponding to different technical states of bearings appear stretched along the radius ρ and compressed along the angle φ. Therefore, the cross-section of the distribution density in the plane of the angle at fixed turns out to be more sensitive to changes in the curve type and its parameters than in any other plane.

Let us consider, in particular, the construction of decisive rules for diagnosing misalignments, inner ring defects and lack of lubrication in bearings 309 ESH2 tested at EM P-51.

Figure 8.22 shows typical histograms of the φ vibration parameter of the test bearing installed on the collector side in the machine P-51 operating in the x.x mode. They are plotted using the estimates of the parameter, obtained on the basis of (8.5) by $\overline{\beta}_1$, $\overline{\beta}_2$. However, it is better to construct the solving rules not by histograms, but by their smoothing curves included in the Pearson system. The construction of the latter is provided in the program of histogram analysis considered earlier.

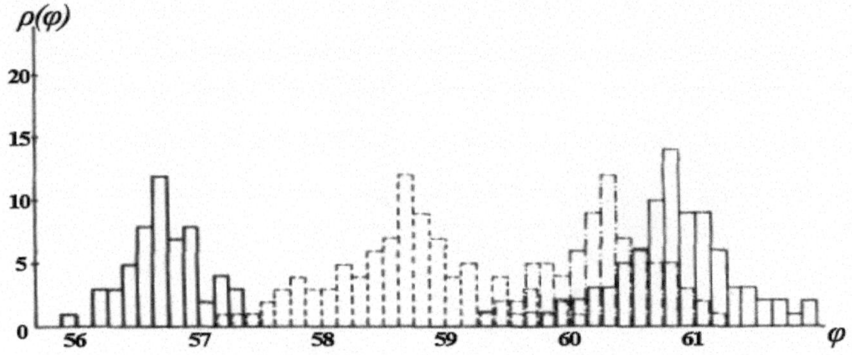

Fig. 8.22 Typical histograms of parameter φ of vibrations of the test bearing installed from the collector side in the P-51 machine operating in the idle mode: thickened line—serviceable bearing; dashed line—lack of lubrication; dashed-dotted line—skew $14' \pm 2.5'$; solid line—inner ring defect

Figure 8.23 shows the curves smoothing the histograms of averaged estimates $\overline{\varphi}$ of vibration of the tested bearing given in Fig. 8.22. These curves are the characteristics of training sets in one-dimensional diagnostic space φ. Distributions of estimation corresponding to different technical states of the tested bearing are shifted relative to each other and have different mathematical expectations: serviceable bearing - $\theta_1 = 56.73°$; lack of lubrication—$\theta_2 = 58.65°$; skew $14' \pm 2.5'$—$\theta_3 = 60.27°$; inner ring defect—$\theta_4 = 60.83°$. The obtained curves belong to XIII, IV or VII types of curves from Pearson's system, i.e. they can be approximated by the normal distribution law.

Consequently, the classical two-alternative Neyman-Pearson procedure of statistical hypothesis testing, which is described by relatively simple mathematical expressions for a normal distribution, can be used to construct decisive rules for diagnosing the above defects in the bearing.

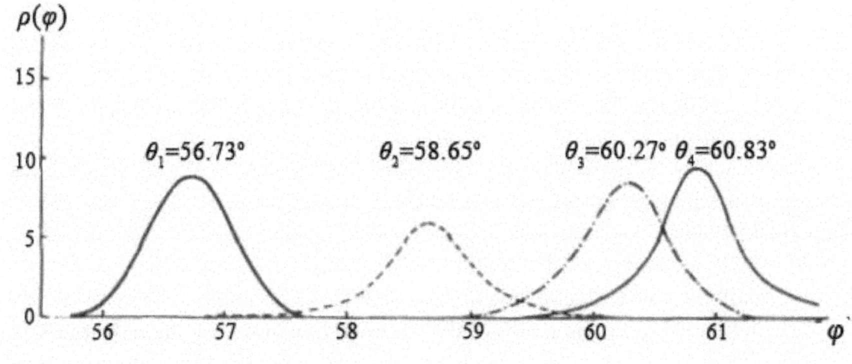

Fig. 8.23 Curves smoothing the histograms presented above

For diagnosing the technical condition of bearings and classifying defects in them, two-alternative solving rules can be constructed, which are reduced to testing the following hypotheses:

H_0: $\theta_1 = 56.73°$ (serviceable) $\leftrightarrow$ H_1: $\theta_2 = 58.65°$ (no lubrication);
H_0: $\theta_1 = 56.73°$ (serviceable) $\leftrightarrow$ H_1: $\theta_3 = 60.27°$ (skew $14' \pm 2.5'$);
H_0: $\theta_1 = 56.73°$ (serviceable) $\leftrightarrow$ H_1: $\theta_4 = 60.83°$ (ring defect);
H_0: $\theta_2 = 58.65$ (no lubrication) $\leftrightarrow$ H_1: $\theta_3 = 60.27°$ (skew $14' \pm 2.5'$);
H_0: $\theta_3 = 60.27°$ (skew) $\leftrightarrow$ H_1: $\theta_4 = 60.83°$ (ring defect).

On the basis of the formulated hypotheses, let's consider a concrete example of diagnosing the technical condition of 309 ESH2 bearings installed on the collector side in the DC EM P-51. Let's build a decisive rule for diagnosing the absence of lubrication in the bearing. Let us start solving the problem by planning the experiment.

Diagnosing the absence of lubrication in the bearing 309 ESH2 is reduced to testing the main hypothesis H_0: $\theta_1 = 56.73°$ (the bearing is serviceable) against the simple alternative H_1: $\theta_2 = 58.65°$ (no lubrication). Having set the probabilities of errors of the first and second kind $\alpha = 0.05$ and $\beta = 0.01$ and having found for them according to the tables of the standard normal distribution the corresponding quantiles $u_{1-\alpha} = -1.645$ and $u_{1-\beta} = -2.326$, we determine the necessary number of observations and the value of the threshold taking into account that the dispersion $\sigma^2 = 0.3933$:

$$n = \frac{\left(u_\alpha + u_\beta\right)^2 \sigma^2}{\left(\theta_2 - \theta_1\right)^2} = \frac{6.202}{3.6864} \approx 2;$$

$$C = \frac{\left(\theta_2 - \theta_1\right)\left(u_\alpha - u_\beta\right)}{2\left(u_\alpha + u_\beta\right)} + \left(\theta_1 + \theta_2\right)/2 = 57.52^0.$$

Having received as a result of the conducted experiment planning all necessary data for construction of the solving rule, let's carry out diagnostics of lack of lubrication in concrete samples of bearings of 309 ESH2 type, installed on DC EM P-51.

To check the technical condition of the bearings, four DC EM P-51 operating in the x.x mode are randomly selected. During the tests 2 samples containing vibrations of tested bearings in each of the selected machines were measured and processed by the program of histogram analysis with the help of IMS vibrodiagnostics. Estimates of vibration parameter values of the bearings installed from the collector side in the tested machines are given in Table 8.4.

The left part of inequality (5.64) after substituting data from the table for the 1st … 4th samples of tested bearings into it takes values equal to 55.91°; 57.41°; 56.53°; 57.02°. 4 samples of tested bearings take values equal to 55.91°; 57.41°; 56.53°; 57.02°. The right part of (5.64) after substitution of the data obtained during planning of the experiment is equal to 57.52°. Thus, on the basis of the obtained results, the

Table 8.4 Estimates of vibration parameter values of the bearings

Measured parameter	Test machine number			
	1	2	3	4
Serviceable bearing				
ϕ_1	55.28°	56.92°	57.10°	56.03°
ϕ_2	56.53°	57.89°	55.95°	58.91°
$\frac{1}{2}\sum_{j=1}^{2}\phi_j$	55.91°	57.41°	56.53°	57.02°
No lubrication				
ϕ_1	55.99°	60.58°	59.70°	60.04°
ϕ_2	59.26°	56.10°	57.35°	60.36°
$\frac{1}{2}\sum_{j=1}^{2}\phi_j$	57.63°	58.34°	58.53°	60.20°

hypothesis H_0 is accepted—bearings 309 ESH2, installed on the collector side, in all four selected EM are serviceable.

After that, grease was removed from all tested bearings. Then similar experiments on measurement and processing of vibrations of bearings working without lubrication in all four tested EMs were carried out. Estimates of vibration parameter values ϕ of tested bearings are given in the lower part of Table 8.4.

After substituting the data from Table 8.4 into (5.64) its left part for the 1st ... 4 of tested bearings takes values equal to 57.63°; 58.34°; 58.53°; 60.20°, i.e. for all four samples of 309 ESH2 bearings the left part of inequality (5.64) is greater than the right one. Consequently, on the basis of the obtained results the hypothesis H_1 is accepted—there is no lubrication in the bearings. Similarly, according to the hypotheses formulated above, the decisive rules for diagnosing other types of defects are constructed.

Thus, it follows from the above examples that the joint use of the third and fourth moments of bearing vibrations allows to choose the Pearson diagram known in statistics as a diagnostic space. The specific transition from the Pearson diagram to the new coordinate system increases reliability and resolving power of diagnostics, and also simplifies the procedure of construction of solving rules due to the Gaussian law of the obtained hypothetical distributions.

A similar approach can be applied to the selection of other diagnostic spaces, which are constructed based on the results of using spectral-correlation methods of processing diagnostic signals. It should be taken into account that these methods are energetic, and therefore the diagnosis based on them is carried out within the framework of the energy theory. Consequently, in the general case the information about the probabilistic structure of the investigated diagnostic signal is not fully utilized. However, diagnostics within the framework of the energy theory is quite justified, as there is a sufficiently reasoned mathematical apparatus, hardware and software for its realization. And in case of fulfillment of the hypothesis about Gaussianity of the investigated signal, this approach is not inferior in informativeness to the diagnostic methods based on the use of distribution functions and statistical criteria for hypothesis testing, which use information about higher moments in the diagnostic signal used.

While using correlation and spectral diagnostic features, usually at the first stage there are estimates of correlation and spectral functions. According to the obtained estimates of correlation and spectral functions, the estimates of diagnostic features are determined. Further, for realization of diagnostic algorithms using such parameters it is necessary, first of all, to choose diagnostic spaces and to form in them training sets corresponding to certain technical states of the investigated EE unit. Below we will dwell on the issues of selecting diagnostic spaces and forming training sets in them based on the results of spectral analysis of diagnostic signals.

As it was shown in Chap. 3, while studying the mathematical model of vibrations of a rolling bearing and a charge magnet wire, it was found that in the spectral analysis of vibrations of the above objects, the frequencies of maximums of the power spectral density of the investigated vibrations can be used as a diagnostic feature. Let us denote by n the number of those maxima that will be taken into account in diagnostics. Only the most pronounced ones will be included in their number. Thus, the training sets will be sets of points corresponding to n-dimensional vectors, the coordinates of which are ordered sequences of resonance frequencies. The coordinate axes of the diagnostic spaces will be denoted by f_j, and the values of the resonances corresponding to them—by f_{pj}.

The method of diagnosing the technical condition of rolling bearings corresponding to two-dimensional diagnostic spaces is proposed. This method can be applied to any section of the main coordinate plane of the above proposed diagnostic space (f_j, f_{j+1}), where $j = \overline{1, n-1}$. Any plane can be chosen as such planes. In this case, only two neighboring maxima are used. Along with the considered one, other diagnostic spaces can be chosen. In particular, the space considering three resonances and the possibility of applying the method in any of the following: $(f_{j+1} - f_j;\ f_{j+2} - f_{j+1}), j = 1, 2$. In this case $n \geq 3$, as coordinate axes are chosen $f_{j+1} - f_j$ and $f_{j+2} - f_{j+1}$.

Figure 8.24, as an example, shows the training sets formed in the proposed diagnostic space at $n = 3$ based on the results of the spectral analysis of vibrations of rolling bearings of the 309 ESH2 type installed on the EM P-51.

The description of experiments on measurement and processing of vibration diagnostic signals on the specified EM is detailed in paragraph 2.2. Both conditionally serviceable bearings were tested, and in case of introduction of misalignment in the bearing installed from the collector side. The most typical spectrograms $S_1(f)$ obtained from the results of the mentioned experiments are shown in Fig. 8.11. Recall that three pronounced maxima can be noted. For example, in Fig. 8.11a, they are located at frequencies 750 Hz, 1.35 kHz and 2.8 kHz, and in Fig. 8.11b—at frequencies 740 Hz, 2.1 kHz and 2.7 kHz. The training sets depicted in Fig. 8.24 are constructed from the mentioned spectrograms $S_1(f)$ in the diagnostic space $(f_3 - f_2; f_2 - f_1)$. The given training sets are presented in the form of scattering ellipses constructed at the level of 0.05 based on the results of processing 60 samples obtained

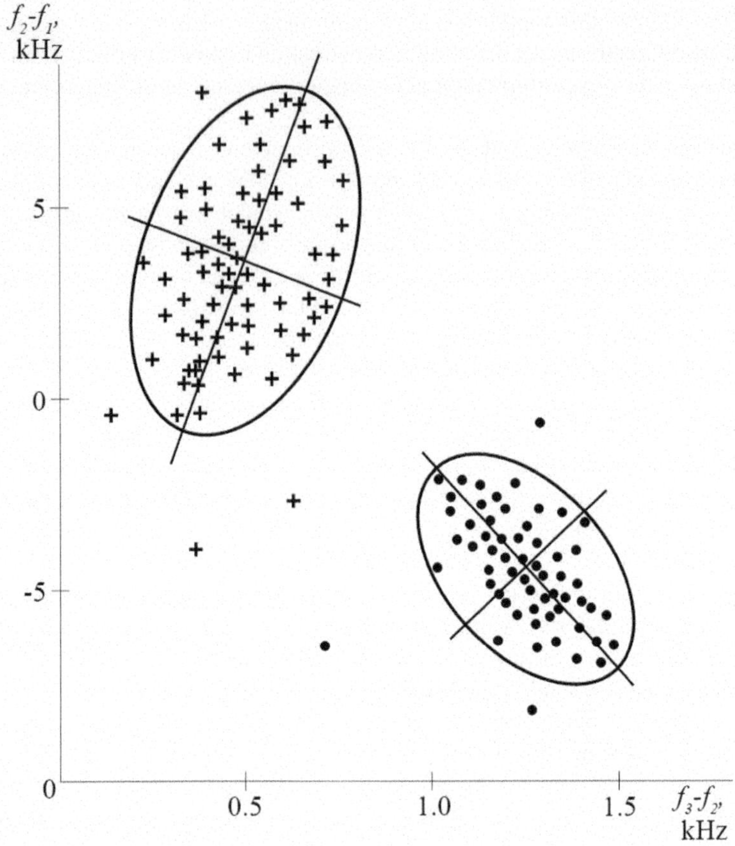

Fig. 8.24 Training sets formed in the diagnostic space ($f_3 - f_2; f_2 - f_1$) according to the results of spectral analysis of vibrations of rolling bearings type 309 ESH2 installed on EM type P-51

for a serviceable bearing and 75 samples for a bearing with a misalignment. The dot "•" denotes the test results corresponding to the serviceable bearing, and the sign "+"—to the bearing with misalignment.

As can be seen from Fig. 8.24, the training sets formed at the level of 0.05 do not overlap, which allows to diagnose the skewness in the bearing with high enough reliability.

Construction of solving rules for diagnosing the technical condition of rolling bearings by training sets shown in Fig. 8.24 is made by analogy with the example considered above, where scattering ellipses were used as training ones (Fig. 8.21). As an example, this issue is considered in [94].

In [95–99], a number of diagnostic attributes derived from the results of correlation analysis of bearing vibrations are proposed, the use of each of which is associated with the choice of appropriate diagnostic spaces. For example, in accordance with the diagnostic attributes based on the results of correlation analysis of rolling bearing vibrations, we can limit ourselves to a one-dimensional diagnostic space for the formation of training sets based on the estimates of the number of repetitions of extrema S in the correlogram under study, estimates of the mutual correlation coefficient $\overline{R}$, estimates of the attenuation coefficient $\beta_j, j = \overline{1, n}$. In this case, the principles of experiment planning, as well as the construction of solving rules are carried out by analogy with the example considered above.

8.6.2 Diagnostics of the Press Condition Packed Magnetic Core with the Help of IMS of Shock Diagnostics

Let us dwell on the issues of selection of diagnostic spaces and construction of training sets corresponding to different degree of pressing of the pressed charge magnetic core of the EM stator core [100–103]. In paragraph 4 of the Chap. 3 of the present work the diagnostic signs, statistical evaluations of which can be obtained by means of statistical spectral analysis, were substantiated. The energy spectrum determined as a result of such analysis characterizes the frequency distribution of the shock vibration wave, which is excited in the studied magnetic core with the help of a special shock hammer.

For experimental verification of the theoretically substantiated diagnostic signs, a series of experiments on a special stand with a charge magnetic core (Fig. 8.12) was carried out. The results of verification, which were described in paragraph 3 of this chapter, allowed us to proceed to the solution of the problem of formation of training sets and construction of decisive rules for diagnosing the degree of pressing of the studied magnetic core using statistical spectral analysis. Let us consider these issues in more detail.

Based on the analysis of the spectrograms $S(f)$ of shock vibration waves propagating along the body of the charged magnetic core (typical ones are presented in Fig. 8.13), it was noted that one of the informative diagnostic features that allow diagnosing the degree of pressing of the charged magnetic core is the number of the main maxima of the spectrogram $S(f)$ in a fixed frequency band, which we denote by υ.

In order to construct training sets in the diagnostic space υ, about 200 experiments were carried out to investigate the vibrations of the charge magnetic core excited by an impact hammer. It should be reminded that these experiments were carried out first on a charged magnetic core compressed with a force of 30 N m, and then—on a magnetic core with the removed tightening force.

Let us explain by the example of spectrograms presented in Fig. 8.13a, b, how the formation of training sets based on the results of spectral analysis takes place. So, at excitation by IH of a vibration wave in a charge magnetic core, pressed with a force of 30 N m, on the spectrogram of its vibrations (Fig. 8.13a) it is possible to distinguish three main frequencies: in the areas of 600, 1200 and 2300 Hz. At the level of 0.5 there are 2 maxima of the spectrogram, thus $\upsilon = 2$. When the compressive

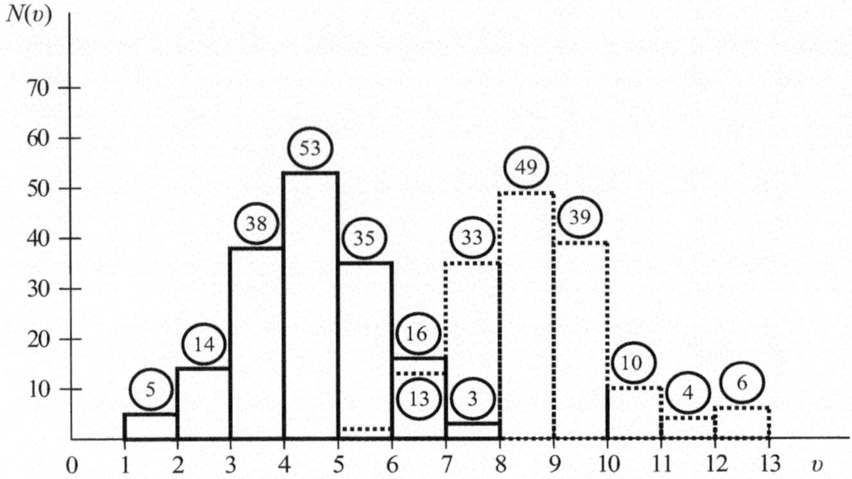

Fig. 8.25 Histograms of the number of maxima υ spectrograms of shock vibration waves of the charged magnetic core, exceeding the level of 0.5

force is removed, besides the main maxima in the spectrogram (Fig. 8.13b), intensive frequency components appear in the whole investigated frequency band up to 5 kHz, and at the level of 0.5 there are 10 maxima, i.e. $\upsilon = 10$.

The level of 0.5, by which the number of spectrogram maxima is fixed, is chosen arbitrarily for the simplicity of counting the maxima. Each of the spectrograms allows us to obtain one point corresponding to different degree of pressing of the charge magnet wire, which is used to form training sets in the selected diagnostic space.

The training sets represent histograms of the number υ (Fig. 8.25). However, it is better to build the solving rules not by histograms, but by their smoothing curves included in the Pearson system. The latter are constructed using the software described in Chap. 7 of this book.

Figure 8.26 shows curves smoothing histograms of averaged υ. These curves are training sets in the one-dimensional diagnostic space υ. Let us dwell on the construction of solving rules for diagnosing the state of pressing of the charge magnetic core.

In order to diagnose the degree of pressing of the charge magnet wire, let us construct a two-alternative decisive rule, which is reduced to checking the hypotheses $H_0 : \theta_1 = 4.47$ (the magnet wire is pressed with a force of 30 N m) against $H_1 : \theta_2 = 8.78$ (there is no tightening force in the magnet wire).

So, let us test the main hypothesis H_0 against the simple alternative H_1 [25, 104, 105]. Having set the probabilities of errors of the first and the second kind $\alpha = 0.05$ and $\beta = 0.01$ having found for them according to the tables of the standard normal distribution the corresponding quantiles $u_{1-\alpha} = -1.645$ and $u_{1-\beta} = -2.326$, we determine the required number of observations N and the value of the threshold C using the expressions given in Chap. 5.

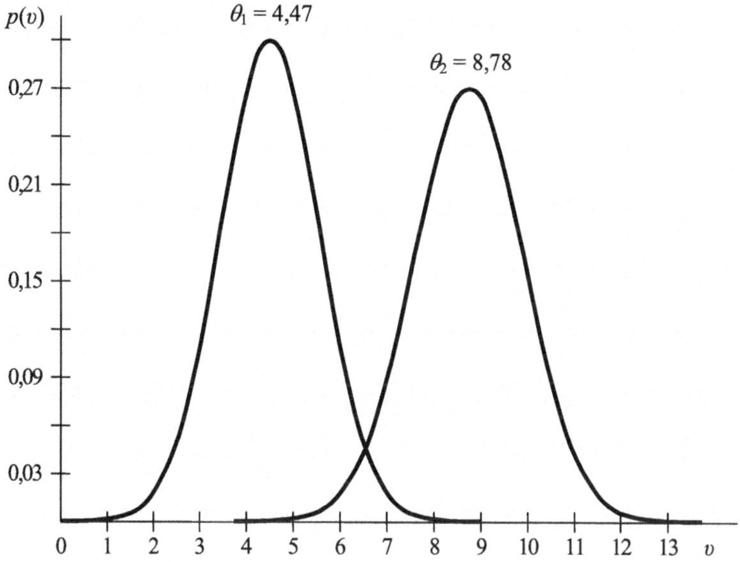

Fig. 8.26 Curves smoothing histograms of the average number υ

The required number of observations taking into account that $u_{1-\alpha} = -u_\alpha, u_{1-\beta} = -u_\beta$:

$$N = \frac{\left(u_\alpha + u_\beta\right)^2 \sigma^2}{(\theta_2 - \theta_1)^2} = \frac{15.769 \cdot 4.36}{(4.31)^2} = 3.70 \approx 4,$$

where σ is the largest dispersion value among the curves shown in Fig. 8.26. For the task under consideration $\sigma = 2.09$.

Threshold value:

$$C = \frac{(\theta_2 - \theta_1)\left(u_\alpha - u_\beta\right)}{2\left(u_\alpha + u_\beta\right)} + \frac{(\theta_1 + \theta_2)}{2} = -\frac{2.935}{7.942} + 6.625 = 6.255.$$

Having received as a result of the conducted planning of experiment all necessary data for construction of the decisive rule, let's carry out diagnostics of degree of pressing of the charged magnetic circuits, similar to the one presented in Fig. 8.12. For check of degree of pressing five such magnetic circuits were randomly chosen, which afterwards were mounted and tested in the stand Fig. 8.12. In the course of tests with the help of the laboratory sample of IMS of shock diagnostics (Fig. 8.4) four samples containing vibrations of the charged magnetic circuits were measured. The number of υ maxima of the spectrogram of these vibrations at the level of 0.5 is given in Table 8.5.

The left part of the inequality (5.64) after substituting in it the data from the table for 1 … 5 samples of tested magnetic circuits takes values equal to 1.75; 4.5; 3.75;

Table 8.5 Estimates of the υ values of vibrations of the tested magnetic circuits

Measured parameter	Number of tested magnetic core				
	1	2	3	4	5
Magnetic conductor pressed with a force of 30 N m					
υ_1	3	5	4	3	5
υ_2	1	3	4	4	6
υ_3	1	5	3	3	4
υ_4	2	5	4	2	4
$0.25\sum_{j=1}^{4}\upsilon_j$	1.75	4.5	3.75	3	4.75
Magnetocoupler without shrinking force					
υ_1	8	11	6	9	7
υ_2	7	8	7	10	9
υ_3	7	9	8	10	8
υ_4	5	8	7	11	9
$0.25\sum_{j=1}^{4}\upsilon_j$	6.75	9	7	10	8.25

3; 4.75. The right part of the inequality (5.64) after substitution of the data obtained during the planning of the experiment is equal to 6.255. Thus, on the basis of the obtained results the hypothesis is accepted H_0—the tested magnetic cores are pressed with the force of 30 N m.

After that, all tested magnetic cores had the shrinking force removed. Then similar experiments on measurement and processing of vibrations caused by shock impact were carried out for all five tested magnetic cores. Estimates of the values of the parameter υ of vibrations of the tested magnetic circuits are given in Table 8.5. After substituting the data from the table into the expression (5.64), its left part for the 1th … 5th tested magnetic circuits. 5 of the tested magnetic circuits takes values equal to 6.75; 9; 7; 10; 8.25, i.e. for all five samples the left part of inequality (5.64) is greater than the right one, therefore the hypothesis is accepted H_1—there is no shrinking force in magnetic circuits.

So, it follows from the considered example that the number of maxima of the vibration spectrogram of the charge magnetic core can be used as a diagnostic sign to determine the degree of its pressing.

Relative simplicity of application of the above considered decisive rules in a complex with the laboratory sample of the IMS of shock diagnostics makes the considered method of shock diagnostics of charge cores suitable for practical use.

8.6.3 Determination of the Degree of Mechanical Loading by Analysis of AE Signal

Earlier it was shown that the most informative diagnostic features in AE diagnostics are asymmetry k and kurtosis γ coefficients, which are associated with the second μ_2, third μ_3 and fourth μ_4 moments of AE processes. It is these parameters that are used in the further AE-diagnostics of EE nodes. Application of these parameters as diagnostic signs, as in the previous cases, allows us to use as a diagnostic space the Pearson diagram with coordinates. It is the space of this diagram that is used to form training sets corresponding to different degrees of loading of the diagnosed object.

Let us briefly discuss the formation of training sets in the diagnostic space (β_1, β_2) corresponding to different values of the load applied to the test specimen. For this purpose, let us set the limits of change of quantitative estimates $\overline{\beta_1}$ and $\overline{\beta_2}$ depending on the degree of loading of the test specimen made of electrical steel. The line above the symbol $\overline{\beta}$ means the estimation of the parameter β. Obtaining quantitative estimates was carried out according to the results of experiments, using the laboratory sample of AE IMS.

Figure 8.27a, b show histograms and smoothing curves, which were constructed by realizations presented in Fig. 8.17a, b.

The histogram of the AE signal measured on a new sample (Fig. 8.27a) can be approximated by the Pearson type I [106, 107], the curve of which is analytically described by the expression

$$p(x) = \left|1 + \frac{x}{0.6525}\right|^{-0.8675} \times \left|1 - \frac{x}{2.5193}\right|^{-0.4885}.$$

The histogram shown in Fig. 8.27b, corresponds to the sample subjected to multiple tests. It is approximated by a Pearson curve of type I or III. This curve corresponds to the expression

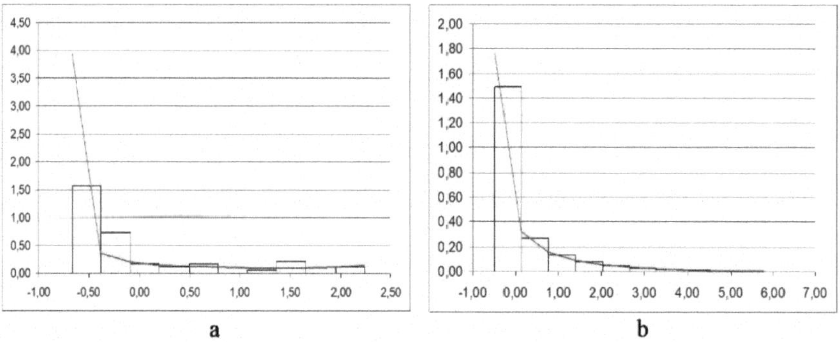

Fig. 8.27 Histograms and curves smoothing them, constructed on realizations of AE signals for new sample (**a**) and sample with a defect (**b**)

$$p(x) = \left|1 + \frac{x}{1.4587}\right|^{-0.5301} \cdot \exp\left(-0.5301\frac{x}{1.4587}\right).$$

It should be noted that the overwhelming majority of histograms based on the results of processing the realizations of the AE process were approximated by I, III or IX types of Pearson curves.

Let us consider the issues of forming training sets based on the results of the experiments described in paragraph 4 of this book.

Figure 8.28 shows the training sets in the diagnostic space (β_1, β_2) formed by the results of the study of the AE process realizations for two samples: conditionally serviceable (Fig. 8.28a) and with an incipient crack (Fig. 8.28b).

Each training population is a scatter ellipse, constructed at the level of 0.05 by the points, the coordinates of which are the averaged estimates $\overline{\beta_1}$ and $\overline{\beta_2}$. Numbers 1 and 5 denote the training population corresponding to the degree of loading P/P_w = 0.5; 2 and 6—$P/P_w = 0.75$; 3 and 7—$P/P_w = 1$; 4 and 8—$P/P_w = 1.25$.

No change in the type of approximating curve was observed for stepwise increase in load $P/P_w = 0.5; 0.75; 1; 1.25$. However, an increase (on average) in the coefficient estimate was recorded $\overline{\beta_2}$.

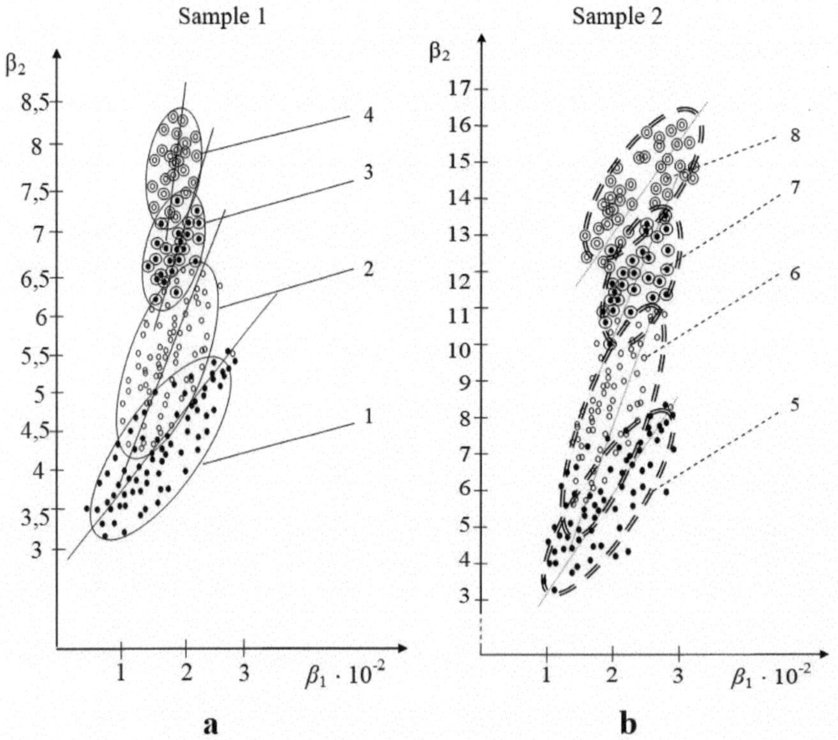

Fig. 8.28 Training sets in two-dimensional diagnostic spaces (β_1, β_2)

The averaged center of gravity for ellipse 1 (Fig. 8.28a) by parameter estimation $\overline{\beta_2} \approx 4.43$; for ellipse 2: $\overline{\beta_2} \approx 5.45$; for ellipse 3: $\overline{\beta_2} \approx 6.62$; for ellipse 4: $\overline{\beta_2} \approx 7.71$. A similar increase on the averaged value of the parameter estimate $\overline{\beta_2}$ was observed for sample 2 (Fig. 8.28b). In addition, as can be seen from the comparison of Fig. 8.28a, b, the increase in load leads to a corresponding increase in estimates $\overline{\beta_2}$ and $\overline{\beta_1}$ in the case of loading to the level of $P/P_w = 1.25$. This causes some displacement in space (β_1, β_2) of the position of the center of the ellipse 8 (Fig. 8.28b). This fact is explained by the fact that with increasing load in the registered realizations of the AE process, a significant increase in the intensity λ of AE pulses is noted, as well as the appearance of AE pulses with significant amplitude.

It should also be noted that the formation of training sets requires multiple experiments under identical conditions. Thus, ellipse 1 was constructed based on the results of processing 60 samples at the value of the applied load $P/P_w = 0.5$, which corresponds to the unloaded state of the material from which the EO unit is made; ellipse 2–80 samples at $P/P_w = 0.75$, i.e., when the load is increased to the level of 0.75 of the nominal value; ellipse 3–90 samples at $P/P_w = 1$ (nominal load value), ellipse 4–70 samples at $P/P_w = 1.25$ (critical load). For sample № 2 for all cases the sample volume was 70.

The final stage of EE technical condition diagnostics is reduced to the construction of diagnostic solving rules (DSR) for classification of defects or loading degree in EE units [108–111].

To build the DSR to determine the degree of loading of the EE node, we use the training sets (TS) (Fig. 8.28a), which were formed in the diagnostic space (β_1, β_2) based on the results of the histogram analysis of the AE process observed in the studied sample made of electrical steel E330. Due to the fact that the scattering ellipses have the maximum displacement along the β_2 axis, we will plot the DSR along the OS displaced along the β_2 axis.

It is better to construct DSR not by histograms, but by Pearson curves smoothing them. Figure 8.29 shows smoothing curves of histograms of averaged estimates of the parameter β_2 for different degrees of loading of the experimental sample. These curves are the characteristics of training sets in the one-dimensional diagnostic space formed by the set of values of the parameter β_2. The distribution of estimates, which correspond to different degrees of loading of the test specimen, are shifted relative to each other and have mathematical expectations: $M_1\beta_2 = 4.43$; $M_2\beta_2 = 5.45$; $M_3\beta_2 = 6.62$; $M_4\beta_2 = 7.71$.

The obtained curves (Fig. 8.29) belong to the Pearson's XIII, IV or VIII types, i.e. they can be approximated by the normal distribution law with a sufficient degree of accuracy. In this case, the Neyman-Pearson procedure of statistical hypothesis testing can be used to construct DSR. As an example, for the construction of the DSR to determine the degree of loading in the test specimen, let's choose two OS corresponding to the degree of loading of the specimen 100% and 125% (Fig. 8.30).

In order to diagnose and classify the degree of loading of EO nodes in them, two-alternative DSR can be constructed, which are summarized by testing the following hypotheses:

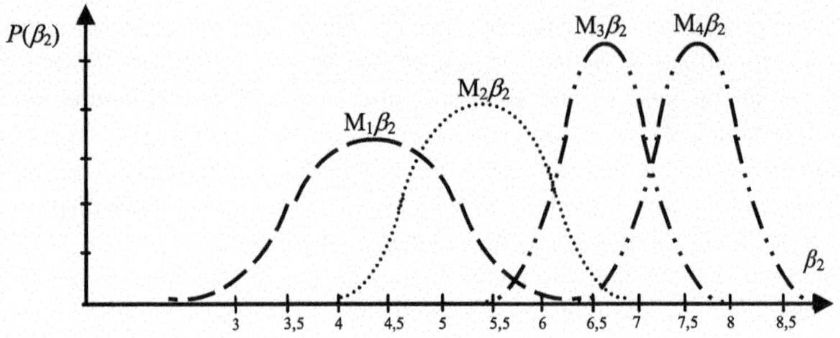

Fig. 8.29 Smoothing curves of histograms of averaged estimates of parameter β_2 for different degrees of loading of the experimental sample

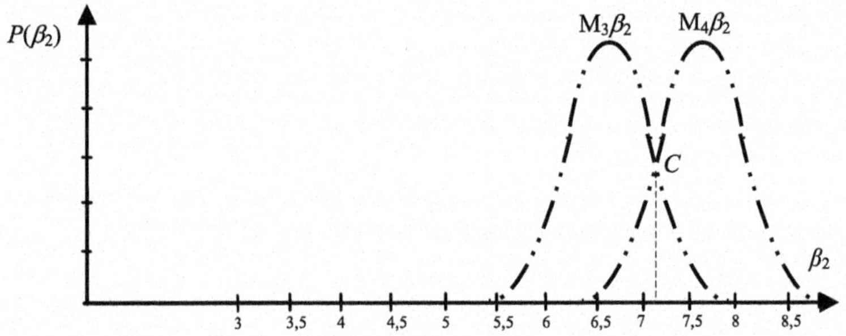

Fig. 8.30 TS corresponding to the degree of loading of the sample—100 and 125%

(1) hypothesis H_0: $\Theta_3 = 6.62$–100% sample load;
(2) hypothesis H_1: $\Theta_4 = 7.71$–125% sample load.

Setting the probabilities of errors of the first and second kind $\alpha = 0.05$ and $\beta = 0.01$, and finding for them according to the tables of the standard normal distribution of the corresponding quantiles $u_{1-\alpha} = -1.645$ and $u_{1-\beta} = -2.326$, we determine the required number of observations, taking into account that the dispersion $\sigma^2 = 0.4755$:

$$n \approx \frac{(U_\alpha + U_\beta)^2 \sigma^2}{(\Theta_4 - \Theta_3)^2} \approx 3.$$

In this case, the value of the threshold

$$C = \frac{(\Theta_4 - \Theta_3)(U_\alpha - U_\beta)}{2(U_\alpha + U_\beta)} + \frac{(\Theta_4 + \Theta_3)}{2} \approx 7.25.$$

In the process of research were measured and processed with the help of IMS AE diagnostics on 3 samples, which contain the signal coming from the AE sensor for each of the samples.

5 conditionally serviceable samples were tested and 3 realizations of the AE signal were measured on the samples that were loaded to a load of 100%. After processing of realizations by the program of histogram analysis were obtained estimates Θ, the values of which are given in the upper part of Table 8.6.

The left part of the inequality (5.64) after substituting the data from Table 8.6 into it takes average values, which are respectively equal to 6.62; 6.63; 6.61; 6.60; 6.62. The right part of inequality (5.64) after substituting in it the data, which were obtained during the planning of the experiment, is equal to 7.25. Thus, based on the results obtained, the hypothesis $H_0 - (\Theta_3 = 6.62$—100% loading on the sample) is accepted.

After that, a tensile loading force of 125% of the breaking force load was set on P-20. Then, similar experiments were performed to process the AE process, at 125% loading in all 5 tested samples. Estimates of the values of the parameter Θ of the AE process are given at the bottom of Table 8.6.

After substituting the data from the right part of Table 8.6 into (5.64), its left part for 1…5 tested samples takes average values of 7.71. For all 5 tested samples, the left part of inequality (5.64) is greater than the right part. Consequently, on the basis of the obtained results, hypothesis H_1 is accepted.

Similarly, using the Neyman-Pearson solving procedure, the solving rules can be constructed, allowing to classify the degree of applied mechanical load to the diagnosed object and, as a consequence, to prevent the development of microcrack type defect in it.

Let's move on to the issue of the formation of TS and construction of DSR for diagnostics of DEG's CPG.

Table 8.6 Θ estimates, obtained in 5 conditionally serviceable samples

Measured parameter	Test sample's number				
	1	2	3	4	5
1 experiment. 100% loading					
Θ_1	6.59	6.48	6.87	6.78	6.94
Θ_2	6.73	6.65	6.53	6.62	6.40
Θ_3	6.54	6.76	6.43	6.40	6.52
$\frac{1}{3}\sum_{j=1}^{3}\Theta_j$	6.62	6.63	6.61	6.60	6.62
2 experiment. 125% loading					
Θ_1	7.74	7.76	7.75	7.72	7.72
Θ_2	7.50	7.68	7.69	7.63	7.61
Θ_3	7.86	7.81	7.72	7.78	7.68
$\frac{1}{3}\sum_{j=1}^{3}\Theta_j$	7.70	7.75	7.72	7.71	7.67

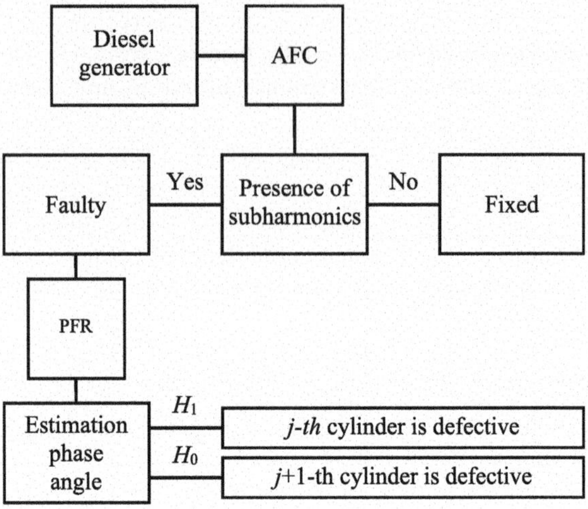

Fig. 8.31 Schematic diagram of the process of DEG's CPG diagnostics

8.6.4 Diagnostics of DEG's CPG

Let us dwell on the issues of DSR constructing for determining the number of the failed cylinder. It should be noted the following. The answer to the question, whether there is at least one faulty cylinder in the DEG, we get immediately by analyzing the AFC, where the absence or insignificant fluctuation (with a serviceable) or the presence (with a faulty cylinder) of subharmonics is used as a diagnostic sign. Construction of the decisive rule on diagnostics of a technical condition of a diesel-generator can be schematically represented in the form of Fig. 8.31. According to this figure the diagnostics of a diesel-generator CPG is realized in the following way:

- the presence of subharmonics is determined by the AFC of angular velocities of the crankshaft;
- a decision is made about the presence of faulty cylinders;
- if there is a faulty cylinder in the DEG, the PFC of angular velocities of the DEG's crankshaft is plotted;
- on estimations of phase angles of subharmonics with the use of two-alternative procedure of Neumann-Pearson the decisive rule on diagnostics of the concrete failed cylinder is constructed.

Let us consider the peculiarities of using the Neyman-Pearson procedure for diagnostics of the cylinder-piston group of a diesel generator. For this purpose, we will use the training sets formed in the diagnostic space δ based on the results of observation of fluctuations of phase angles of the first subharmonic (Fig. 8.32).

As noted above, the constructed curves belong to the IV, VII or XIII types of curves included in the Pearson system, i.e., they can be approximated by the normal distribution law. Consequently, as in the previous cases, the classical two-alternative procedure of statistical hypothesis testing by the Neyman-Pearson criterion can be

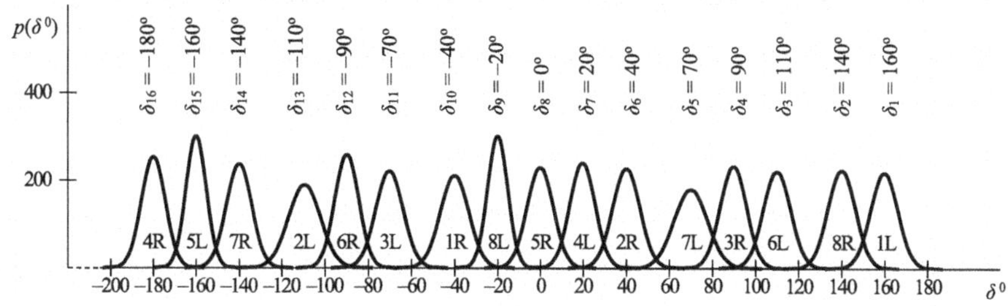

Fig. 8.32 Fluctuations of phase angles of the first subharmonic at failure of the cylinder with a certain number

used to construct decisive rules for the diagnosis of a particular cylinder. As in the above cases, for the diagnosis of failed cylinders with the indication of their number, we construct two-alternative decisive rules, which are reduced to the subsequent testing of the following hypotheses:

H_0: $\delta_1 = 160°$ (defective 1L) $\leftrightarrow$ H_1: $\delta_1 = 140°$ (defective 8R);
H_0: $\delta_2 = 140°$ (defective 8R) $\leftrightarrow$ H_1: $\delta_3 = 110°$ (defective 6L);

....

H_0: $\delta_{15} = -160°$ (defective 5L) $\leftrightarrow$ H_1: $\delta_{16} = -180°$ (defective 4R).

Let's consider a concrete example of diagnostics of the technical condition of the cylinder-piston group of DEG 5D70, tested on the stand of "Malyshev Plant".

Let's build a decisive rule for diagnosing the failure of the first left cylinder. In this case, let's stipulate two points at once.

The first one is connected with the definition of observation during the experiment, namely, each observation is a realization of angular velocities of the diesel crankshaft, registered for two revolutions of the specified shaft. This corresponds to a full operating cycle in all cylinders of the diagnosed diesel engine. When conducting experiments, each such realization is separated from each other by means of the mark of the top dead center of the diesel crankshaft position. This mark is formed by a special sensor. Thus, one observed realization is enclosed between two top dead center marks.

The second point is related to the peculiarity of the experiments on the simulation of cylinder faults on the running diesel generator. When planning the experiments on diagnostics of faulty cylinders using the training sets shown in Fig. 8.32, the previously mentioned circumstance that there should be at least one faulty cylinder in the diagnosed diesel engine was taken into account. Proceeding from this, in the course of experiments the sequential failure of all 16 cylinders of the diesel generator 5D70 was carried out.

In accordance with the order of operation of the cylinders of the DEG 5D70 and taking into account the training sets presented in Fig. 8.32, the diagnostics of the inoperative 1st left cylinder is reduced to the sequential testing of the following hypotheses:

H_0: $\delta_1 = 160°$ (defective 1L) $\leftrightarrow$ H_1: $\delta_2 = 140°$ (defective 8R);
H_0: $\delta_1 = 160°$ (defective 1L) $\leftrightarrow$ H_1: $\delta_3 = 110°$ (defective 6L);
H_0: $\delta_1 = 160°$ (defective 1L) $\leftrightarrow$ H_1: $\delta_4 = 90°$ (defective 3R);
H_0: $\delta_1 = 160°$ (defective 1L) $\leftrightarrow$ H_1: $\delta_5 = 70°$ (defective 7L);
H_0: $\delta_1 = 160°$ (defective 1L) $\leftrightarrow$ H_1: $\delta_6 = 40°$ (defective 2R);
H_0: $\delta_1 = 160°$ (defective 1L) $\leftrightarrow$ H_1: $\delta_7 = 20°$ (defective 4L);
H_0: $\delta_1 = 160°$ (defective 1L) $\leftrightarrow$ H_1: $\delta_8 = 0°$ (defective 5R);
H_0: $\delta_1 = 160°$ (defective 1L) $\leftrightarrow$ H_1: $\delta_9 = -20°$ (defective 8L);
H_0: $\delta_1 = 160°$ (defective 1L) $\leftrightarrow$ H_1: $\delta_{10} = -40°$ (defective 1R);
H_0: $\delta_1 = 160°$ (defective 1L) $\leftrightarrow$ H_1: $\delta_{11} = -70°$ (defective 3L);
H_0: $\delta_1 = 160°$ (defective 1L) $\leftrightarrow$ H_1: $\delta_{12} = -90°$(defective 6R);
H_0: $\delta_1 = 160°$ (defective 1L) $\leftrightarrow$ H_1: $\delta_{13} = -110°$ (defective 2L);
H_0: $\delta_1 = 160°$ (defective 1L) $\leftrightarrow$ H_1: $\delta_{14} = -140°$(defective 7R);
H_0: $\delta_1 = 160°$ (defective 1L) $\leftrightarrow$ H_1: $\delta_{15} = -160°$ (defective 5L);
H_0: $\delta_1 = 160°$ (defective 1L) $\leftrightarrow$ H_1: $\delta_{16} = -180°$ (defective 4R).

As an example, consider the problem of statistical hypothesis testing for the first case, i.e., for hypotheses H_0: $\delta_1 = 160°$ (faulty 1L) vs. H_1: $\delta_2 = 140°$ (faulty 8P). For the other 15 cases, the tests are constructed in a similar way. We start solving the problem by planning the experiment. Having set the probabilities of errors of the first and second kind $\alpha = 0.05$ and $\beta = 0.01$ and having found for them according to the tables of the standard normal distribution [112] the corresponding quantiles $u_{1-\alpha} = -1.645$ and $u_{1-\beta} - 2.326$, let us determine by (6.62) the necessary number of observations taking into account that the dispersion $\sigma^2 = 45.326$:

$$n \approx \frac{\left(u_\alpha + u_\beta\right)^2 \sigma^2}{(\delta_2 - \delta_1)^2} = \frac{714.75}{400} = 2.$$

Based on expression (6.65), the value of threshold C is equal to

$$C = \frac{(\delta_2 - \delta_1)\left(u_\alpha - u_\beta\right)}{2\left(u_\alpha + u_\beta\right)} + \frac{(\delta_1 + \delta_2)}{2} = 1.72° + 150° = 151.72°.$$

As a result of the experiment planning, we have obtained all the necessary data for the construction of the solving rule for diagnostics of the first left cylinder.

As it was mentioned above, the experimental verification of the proposed solving rules was carried out using a DEG 5D70, installed on the stand of "Malyshev Plant". The tested DEG worked in the idling mode and in the process of its operation all 16 cylinders were put out of operation sequentially (one by one) by stopping the fuel supply to the corresponding cylinder. At the same time the angular velocity of the diesel crankshaft was measured and registered. Then the registered data were processed with the help of IMS and determined the AFC and PFC of the investigated angular velocities, which were used to find estimates of phase angles of the first subharmonics corresponding to different inoperative cylinders of the diesel engine.

Table 8.7 δ estimates of the first subharmonic angle

Observed parameter	Cylinder condition	
	Defective 1st left	Defective 8th right
$\delta^{(1)}$	151.81°	142.34°
$\delta^{(2)}$	152.44°	139.56°
$\frac{1}{2}\sum_{j=1}^{2}\delta^{(j)}$	152.125°	140.95°

Table 8.7 shows the δ estimates of the first subharmonic angle corresponding to the inoperative 1st left and then 8th right cylinders.

This experiment was carried out as follows. The diesel generator operated at idle speed with shaft speed of 1000 rpm. In this mode, the fuel supply to the 1st left cylinder was stopped and at the same time the initial data were measured and recorded, which are summarized in the middle part of Table 8.7.

The left part of inequality (5.64) after substituting the data of Table 8.7 into it takes the value of 152.125°. The right part of this inequality, as it was established when planning the experiment, is equal to 151.72°. Consequently, according to the obtained results, the hypothesis is accepted—the 1st left cylinder in the tested DEG 5D70 is defective.

Then in the diesel generator the fuel supply to the 1st left cylinder was resumed with simultaneous cessation of fuel supply to the 8th right cylinder. The results of the experimental data obtained for this case are summarized in the right part of Table 8.7.

After substituting the data of Table 8.7 into the inequality (5.64), its right part takes the value of 140.95°, which is less than the previously calculated threshold C equal to 151.72°. Thus, on the basis of the obtained results the hypothesis is accepted H_1—in the tested diesel generator the 8th right cylinder is defective. Similarly, according to the hypotheses formulated above, the decisive rules for diagnosing all other cylinders of the diesel generator are constructed.

In conclusion, it should be noted that at connected to the tested DEG and configured IMS the whole operation of diagnosing the diesel CPG, from the moment of the beginning of measurement of initial data and up to the issuance of the result of diagnosis on the computer display, takes a few seconds.

8.7 Prediction of EE Failures

To build a forecast of the possible number of failures we will use elements of the theory of statistical spline functions, the basic analytical calculations of which were described in paragraph 7.2 of Chap. 1. The results of practical application of statistical spline functions to the solution of the problem of forecasting the reliability of electrical equipment are considered in [113–117]. As an object of study, we choose a powerful power autotransformer of 330 kV class.

The initial data for determining the possible moment of failure of autotransformers (AT) or their units using the forecasting methods described in paragraph 7.2 of Chap. 1 is the number of failures for the observation period of at least 8 consecutive years. The forecast of the moment of occurrence and total number of failures is calculated 1...14 years ahead relative to the last year of observation. For this purpose, computer software developed using analytical expressions given in the mentioned in paragraph 7.2 of the Chap. 1 is used. The results of the calculation are given in the form of a table and a graph. On the basis of the analysis of the obtained results, conclusions are made about the residual resource of AT. Let us consider the application of the proposed approach by example.

As a result of collection and generalization of data on reliability of 330 kV class ATs, a polygon of the number of failures for 14 years of observation (from 1979 to 1992) was obtained. This polygon is given in Table 8.8.

We are interested in what number of failures can be expected during 1993...1997 with a forecast confidence probability equal to $P = 0.80$. The program input data for making this forecast are:

- number of observations (years of failures) $n = 14$;
- forecast interval for 5 years ahead, $n_{for} = 5$;
- number of intervals to be interpolated by spline $r = 6$;
- number of intervals of division of the time interval, on which the data on failures was collected (division is performed in order to find the optimal interpolation spline), $n_{div} = 13$;
- confidence probability of estimation of the predicted number of failures $P = 0.8$.

As a result of calculation by the program printouts in the form of a table and a graph (Fig. 8.33) are received. The table shows both initial data and results of calculations:

- points of spline junction (years);
- points of lower and upper boundaries of the spline approximation channel (number of failures);
- function at the points of spline junction (number of failures);
- dispersion at the points of spline junction.

The spline approximation (Fig. 8.33) represents some channel consisting of six sections. The spline junction points on the upper boundary of the channel are denoted by 1H ... 7H, on the lower boundary—1L ... 7L. The width of the channel is determined by the accepted confidence probability of the forecast: for comparison, we present Fig. 8.34, which shows the results of the forecast at $P = 0.95$ and the same other parameters. At $P = 0.95$ the channel is wider than at $P = 0.80$. Plots 1 ... 6 of

Table 8.8 Number of AT failures for 14 years of observation

Year	79	80	81	82	83	84	85	86	87	88	89	90	91	92
Number failures	4	4	4	10	5	8	5	7	5	5	2	4	8	9

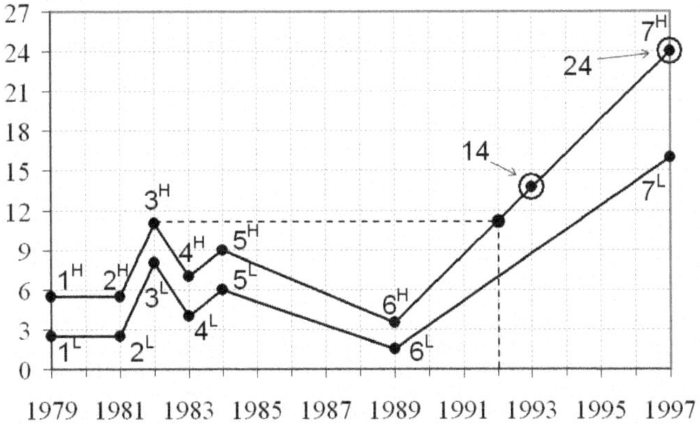

Fig. 8.33 Spline approximation of the number of failures of AT 330 kV at $P = 0.8$

the graphs characterize the observation interval, and plot 6 … 7—forecast interval. As follows from the graphs, the forecast of failures is made until 1997, i.e., 5 years ahead of 1992—the last year of observation. As the forecast interval increases, the channel into which the possible results of observations can fall widens. Thus, in 1995 one can expect 12 … 19 failures at $P = 0.8$ (Fig. 8.33); 10 … 21 failures at $P = 0.95$ (Fig. 8.34); in 1997 one can expect from 16 to 24 failures at $P = 0.8$ (Fig. 8.33); from 13 to 27 failures at $P = 0.95$ (Fig. 8.34).

Residual resource depends on timely repairs of AT. The solution of the problem of determining the time of AT repair can be divided into two stages. The first stage is a determination of the acceptable level of failures for the predicted period, the second stage is selection of the period between repairs.

Three cases are considered when determining the acceptable level of AT failures.

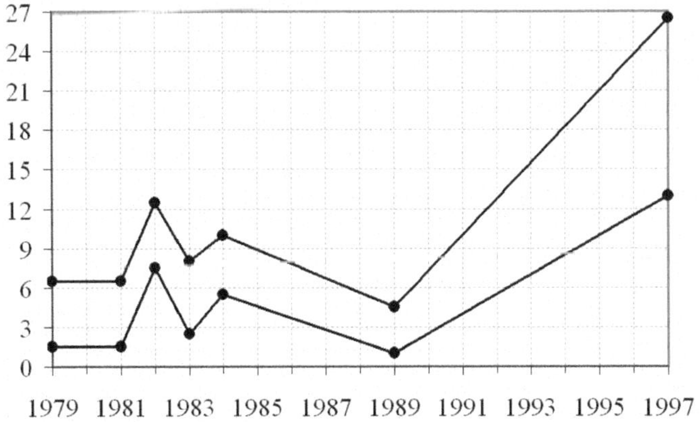

Fig. 8.34 Spline approximation of the number of failures of AT 330 kV at $P = 0.95$

In the first case, this level is selected as the largest value of the upper limit of the confidence interval. This means that in the process of operation it is allowed to work practically up to wear, i.e. the maximum possible at the given probability level of failures is chosen. In other words, it corresponds to the lowest level of AT reliability.

In the second case, the minimum value of the upper boundary of the confidence interval, except for the last point on the observation site, is selected as the acceptable failure level. Such a choice can provide high operational reliability of AT, but under the condition of frequent maintenance or repair works.

Based on these two extreme cases, it is possible to make a compromise decision taking into account economic considerations. For example, it is possible to choose the repair period based on the average number of failures obtained in the first and second cases.

The second stage consists in determining the period between repairs of AT. It is most simply done by a graphical printout of results of processing of data on failures. Let's briefly consider this question. From the point corresponding to the maximum value of the upper boundary of the observation interval, a horizontal line is drawn to the right up to the intersection with the upper boundary of the approximating channel. The distance between abscissa of the obtained point and the maximum point of the forecast is the period between repairs. This period corresponds to low reliability of AT and small value of its residual resource. Then the point corresponding to the average number of failures is marked on the upper boundary of the forecast interval. The distance between the abscissa of this point and the year of the last observation will correspond to the interrepair period, which provides high reliability and increase of the residual life.

Let's concretize the proposed rules for determining the AT repair terms by the example of Fig. 8.33.

According to the first stage we determine the permissible failure levels. The maximum level is at point 7H (24 failures). The minimum is at point 6L (4 failures). Average level = (24 + 4)/2 = 14 failures. The point 14 is marked at the upper boundary of the forecast interval.

In accordance with the second stage, based on the acceptable levels of failures, we obtain two values of the period between repairs. To ensure the level equal to 14 failures, the duration of the period between repairs will be 1993–1992 = 1 year. The period between repairs for a level corresponding to 24 failures will be 1997–1992 = 5 years. Consequently, to ensure higher reliability and prolong the residual life of ATs, they should be repaired once a year.

After considering the above examples, we can conclude that the use of statistical splines in solving the problem of forecasting failures of electrical equipment makes it possible to obtain the time interval in which the failure of the equipment under study will occur with a given level of confidence probability. As a result, it provides possibility of more accurate planning of terms of maintenance and repair of EE. This approach allows to more fully utilize the technical resource of such equipment at its reliable operation.

References

1. Immovilli, F., Bellini, A., Rubini, R., Tassoni, C.: Diagnosis of bearing faults in induction machines by vibration or current signals: a critical comparison. IEEE Trans. Ind. Appl. **46**(4), 1350–1359 (2010)
2. Melnyk, O., Onyshchenko, S., Onishchenko, O., Lohinov, O., Ocheretna, V.: Integral approach to vulnerability assessment of ship's critical equipment and systems. Trans. Maritime Sci. **12**(01), 3–3 (2023)
3. Burda, E.A., Zusman, G.V., Kudryavtseva, I.S., Naumenko, A.P.: An overview of vibration analysis techniques for the fault diagnostics of rolling bearings in machinery. In: Shock and Vibration (2022)
4. Abboud, D., Elbadaoui, M., Smith, W.A., Randall, R.B.: Advanced bearing diagnostics: a comparative study of two powerful approaches. Mech. Syst. Signal Process. **114**, 604–627 (2019)
5. Babak, V.P., Babak, S.V., Eremenko, V.S., Kuts, Y.V., Myslovych, M.V., Scherbak, L.M., Zaporozhets, A.O.: Models and measures for the diagnosis of electric power equipment. In: Models and Measures in Measurements and Monitoring, pp. 99–126 (2021)
6. Babak, V., Zaporozhets, A., Kovtun, S., Myslovych, M., Kuts, Y., Scherbak, L.: Information support for identification of the technical state of electric power facilities. In: Systems, Decision and Control in Energy V, pp. 129–153. Springer Nature Switzerland, Cham (2023)
7. Wrzochal, M., Adamczak, S., Piotrowicz, G., Wnuk, S.: Industrial experimental research as a contribution to the development of an experimental model of rolling bearing vibrations. Strojniški vestnik-J. Mechan. Eng. **68**(9), 552–559 (2022)
8. Adamczak, S., Stępień, K., Wrzochal, M.: Comparative study of measurement systems used to evaluate vibrations of rolling bearings. Procedia Eng. **192**, 971–975 (2017)
9. Rauber, T.W., da Silva Loca, A.L., de Assis Boldt, F., Rodrigues, A.L., Varejão, F.M.: An experimental methodology to evaluate machine learning methods for fault diagnosis based on vibration signals. Expert Syst. Appl. **167**, 114022 (2021)
10. Song, L., Wang, H., Chen, P.: Vibration-based intelligent fault diagnosis for roller bearings in low-speed rotating machinery. IEEE Trans. Instrum. Meas. **67**(8), 1887–1899 (2018)
11. Perales, N., Galar, D.: Inspection and Analysis of the Functioning of the Bearings Used on Railways: A Study of the Life of a Bearing Under Real Operating Conditions. Luleå Tekniska Universitet (2014)
12. Dhanola, A., Garg, H.C.: Tribological challenges and advancements in wind turbine bearings: a review. Eng. Fail. Anal. **118**, 104885 (2020)
13. Chen, L., Bu, D.T., Feng, Z.B., Liu, H.B.: Fatigue life analysis of remanufactured radial rolling bearing with the replaced loading zone. J. Eng. (2023)
14. Liu, G., Cong, J., Wang, P., Du, S., Wang, L., Chen, R.: Study on vertical vibration and transmission characteristics of railway ballast using impact hammer test. Constr. Build. Mater. **316**, 125898 (2022)
15. Jain, A.K., Rastogi, V., Agrawal, A.K.: Experimental investigation of vibration analysis of multi-crack rotor shaft. Procedia Eng. **144**, 1451–1458 (2016)
16. Wang, P., Zhang, Y., Feng, L., Hou, W., Wang, J., Li, W., Feng, K.: Study on the pneumatic hammer phenomenon of aerostatic bearings based on the empirical mode method: Numerical and experimental analysis. Tribol. Int. **181**, 108305 (2023)
17. Kim, M., Park, H.W., Lee, S.K.: Investigation of machining stability considering thermal and rotation effect: Effectiveness of impact excitation for a rotating spindle. Int. J. Precis. Eng. Manuf. **23**(10), 1143–1162 (2022)
18. Melnyk, O., Onishchenko, O., Onyshchenko, S., Shumylo, O., Volyanskyy, S., Bondar, A., Cheredarchuk, N.: Application of fuzzy controllers in automatic ship motion control systems. Int. J. Electr. Comput. Eng. **13**(4), 2088–8708 (2023)
19. Kovtun, S., Ponomarenko, O., Nazarenko, O.: Quality of the information flow management at stochastic energy consumption conditions. Syst. Res. Energ. **3**(74), 78–84 (2023)

20. Babak, V.P., Babak, S.V., Myslovych, M.V., Zaporozhets, A.O., Zvaritch, V.M.: Technical provision of diagnostic systems. In: Diagnostic Systems for Energy Equipments, pp. 91–133 (2020)
21. Wang, T., Celik, O., Catbas, F. N., Zhang, L.: Practical calibration techniques for the modal impact hammer. In: Sensors and Instrumentation, Volume 5: Proceedings of the 33rd IMAC, A Conference and Exposition on Structural Dynamics, pp. 23–29. Springer International Publishing (2015)
22. Gharaibeh, M.: Identification of printed circuit boards mechanical properties using response surface methods. Microelectron. Int. **39**(1), 38–47 (2022)
23. Will, D.T., Zhu, W.D.: Experimental modal analysis and operational deflection shape analysis of a cantilever plate in a wind tunnel with finite element model verification. In: Experimental Techniques, pp. 1–20 (2023)
24. Onishchenko, O.A., Melnyk, O.M., Yarovenko, V.A., Aleksandrovska, N.I., Kurdiuk, S.V., Parmenova, D.G., Storchak, O.O.: Study of efficiency and advancement of marine engine oil purification and filtration technologies. J. Chem. Technol. **31**(4), 762–774 (2023)
25. Babak, V., Eremenko, V., Zaporozhets, A.: Research of diagnostic parameters of composite materials using Johnson distribution. Int. J. Comput. **18**(4), 483–494 (2019)
26. Eremenko, V., Babak, V., Zaporozhets, A.: Method of reference signals creating in non-destructive testing based on low-speed impact. Tekhnichna Elektrodynamika **4**, 070 (2021)
27. Babak, V.P., Babak, S.V., Myslovych, M.V., Zaporozhets, A.O., Zvaritch, V.M.: Principles of construction of systems for diagnosing the energy equipment. In: Diagnostic Systems for Energy Equipments, pp. 1–22 (2020)
28. Wang, Y., Xiang, J., Markert, R., Liang, M.: Spectral kurtosis for fault detection, diagnosis and prognostics of rotating machines: a review with applications. Mech. Syst. Signal Process. **66**, 679–698 (2016)
29. Zaporozhets, A., Eremenko, V., Babak, V., Isaienko, V., Babikova, K.: Using Hilbert transform in diagnostic of composite materials by impedance method. Periodica polytechnica Electr. Eng. Comput. Sci. **64**(4), 334–342 (2020)
30. Zaporozhets, A., Eremenko, V., Isaenko, V., Babikova, K.: Approach for creating reference signals for detecting defects in diagnosing of composite materials. In: Conference on Computer Science and Information Technologies. pp. 154–172. Springer International Publishing, Cham (2019, September)
31. Babak, V., Zaporozhets, A., Kulyk, M., Kuts, Y., Scherbak, L.: Application of discrete Hilbert transform to estimate the characteristics of cyclic signals: information provision. In: Systems, Decision and Control in Energy IV: Volume I. Modern Power Systems and Clean Energy, pp. 93–115. Springer Nature Switzerland, Cham (2023)
32. Kedadouche, M., Thomas, M., Tahan, A.J.M.S.: A comparative study between empirical wavelet transforms and empirical mode decomposition methods: application to bearing defect diagnosis. Mech. Syst. Signal Process. **81**, 88–107 (2016)
33. Roy, S.S., Dey, S., Chatterjee, S.: Autocorrelation aided random forest classifier-based bearing fault detection framework. IEEE Sens. J. **20**(18), 10792–10800 (2020)
34. Javorskyj, I., Kravets, I., Matsko, I., Yuzefovych, R.: Periodically correlated random processes: application in early diagnostics of mechanical systems. Mech. Syst. Signal Process. **83**, 406–438 (2017)
35. Rai, A., Upadhyay, S.H.: A review on signal processing techniques utilized in the fault diagnosis of rolling element bearings. Tribol. Int. **96**, 289–306 (2016)
36. Peng, D., Teng, W., Liu, Y.: Improved energy spectrum via spectral correntropy-based coherence-gram for bearing fault diagnosis. IEEE Trans. Instrum. Measur. (2023)
37. Natarajan, S.: Vibration signal analysis using histogram features and support vector machine for gear box fault diagnosis. Int. J. Syst. Control Commun. **8**(1), 57–71 (2017)
38. Khan, S.A., Kim, J.M.: Automated bearing fault diagnosis using 2D analysis of vibration acceleration signals under variable speed conditions. Shock Vibr. (2016)
39. Unal, M., Sahin, Y., Onat, M., Demetgul, M., Kucuk, H.: Fault diagnosis of rolling bearings using data mining techniques and boosting. J. Dyn. Syst. Meas. Contr. **139**(2), 021003 (2017)

40. Dzyuba, V., Zaporozhets, A.: Mathematical approaches to forecasting and researching the technical state of cylindrical shells of energy objects' elements based on vibration monitoring systems. In: Systems, Decision and Control in Energy III, pp. 107–119. Springer International Publishing, Cham (2021)
41. Saidi, L., Ali, J.B., Fnaiech, F.: Application of higher order spectral features and support vector machines for bearing faults classification. ISA Trans. **54**, 193–206 (2015)
42. Ming, A., Zhang, W., Fu, C., Yang, Y., Chu, F., Liu, Y.: L-kurtosis-based optimal wavelet filtering and its application to fault diagnosis of rolling element bearings. J. Vib. Control **30**(7–8), 1594–1603 (2024)
43. Liang, K., Zhao, M., Lin, J., Ding, C., Jiao, J., Zhang, Z.: A novel indicator to improve fast kurtogram for the health monitoring of rolling bearing. IEEE Sens. J. **20**(20), 12252–12261 (2020)
44. Gao, Z., Lin, J., Wang, X., Xu, X.: Bearing fault detection based on empirical wavelet transform and correlated kurtosis by acoustic emission. Materials **10**(6), 571 (2017)
45. Hemmati, F., Orfali, W., Gadala, M.S.: Roller bearing acoustic signature extraction by wavelet packet transform, applications in fault detection and size estimation. Appl. Acoust. **104**, 101–118 (2016)
46. Ren, B., Yang, M., Chai, N., Li, Y., Xu, D.: Fault diagnosis of motor bearing based on speed signal kurtosis spectrum analysis. In: 2019 22nd International Conference on Electrical Machines and Systems (ICEMS), pp. 1–6. IEEE (2019, August)
47. Wang, Y., Peter, W.T., Tang, B., Qin, Y., Deng, L., Huang, T., Xu, G.: Order spectrogram visualization for rolling bearing fault detection under speed variation conditions. Mech. Syst. Signal Process. **122**, 580–596 (2019)
48. Babak, V., Zaporozhets, A., Zvaritch, V., Scherbak, L., Myslovych, M., Kuts, Y.: Models and measures in theory and practice of manufacturing processes. IFAC-PapersOnLine **55**(10), 1956–1961 (2022)
49. Obuchowski, J., Wyłomańska, A., Zimroz, R.: The local maxima method for enhancement of time–frequency map and its application to local damage detection in rotating machines. Mech. Syst. Signal Process. **46**(2), 389–405 (2014)
50. Babak, V.P., Babak, S.V., Eremenko, V.S., Kuts, Y.V., Myslovych, M.V., Scherbak, L.M., Zaporozhets, A.O.: Problems and features of measurements. In: Models and Measures in Measurements and Monitoring, pp. 1–31 (2021)
51. Babak, V., Zaporozhets, A., Kovtun, S., Kuts, Y., Fryz, M., Scherbak, L.: Information provision for monitoring the current state of electric power facilities. In Advanced System Development Technologies I, pp. 283–309. Springer Nature Switzerland, Cham (2023)
52. Babak, V., Zaporozhets, A., Kuts, Y., Myslovych, M., Fryz, M., Scherbak, L.: Models and characteristics of identification of noise stochastic signals of research objects. In: CEUR Workshop Proceedings, vol. 3309, pp. 349–362 (2022, December)
53. Zvaritch, V., Myslovych, M., Gyzhko, Y.: Application of linear random processes to construction of diagnostic system for power engineering equipment. In: IFIP International Conference on Advances in Production Management Systems, pp. 617–622. Springer International Publishing, Cham (2021, August)
54. Myslovych, M., Zvarych, V., Ostapchuk, L., Hyzhko, Y., Hutorova, M.: Some practical issues of creating teaching complexes providing informational support for multilevel diagnostic systems for electrotechnical equipment. Comput. Probl. Electr. Eng. **2**(10), 21–27 (2020)
55. Zvaritch, V.M., Zaitsev, I.O., Myslovych, M.V., Levytskyi, A.S., Zakusilo, S.A.: Some features of the systems for monitoring and diagnostic hydro units technical condition with considering smart grid technology. In: IOP Conference Series: Earth and Environmental Science, vol. 1254, No. 1, p. 012026. IOP Publishing (2023, October)
56. Babak, V., Bohachev, I., Zaporozhets, A., Khaidurov, V., Havrysh, V., Kalinichenko, A.: Some features of modeling ultrasound propagation in non-destructive control of metal structures based on the magnetostrictive effect. Electronics **12**(3), 477 (2023)
57. Bohachev, I., Babak, V., Zaporozhets, A.: Novel small-aperture transducers based on magnetostrictive effect for diagnostic systems. Tech. Electrodyn. **3**, 69–78 (2022)

58. Borucki, S., Frącz, P., Boczar, T., Zmarzły, D.: Diagnostics of power transformers cores using a modified vibroacoustic method. In: 2012 IEEE International Symposium on Electrical Insulation, pp. 179–183. IEEE (2012, June)
59. Shengchang, J., Yongfen, L., Yanming, L.: Research on extraction technique of transformer core fundamental frequency vibration based on OLCM. IEEE Trans. Power Delivery **21**(4), 1981–1988 (2006)
60. Rivera, H.L., García-Souto, J.A., Sanz, J.: Measurements of mechanical vibrations at magnetic cores of power transformers with fiber-optic interferometric intrinsic sensor. IEEE J. Sel. Top. Quantum Electron. **6**(5), 788–797 (2000)
61. Bohachev, I., Kovtun, S., Kuts, Y., Sozonov, S., Khaidurov, V.: Enhanced phase method of signal detection for ultrasonic magnetostriction defectoscopy of power equipment. Syst. Res. Energy **2**, 72–82 (2023)
62. Babak, V., Zaporozhets, A., Kuts, Y., & Scherbak, L.: Some features of Hilbert transform and their use in energy informatics. Probl. General Energy 1–2 (2022)
63. Wu, J.D., Chuang, C.Q.: Fault diagnosis of internal combustion engines using visual dot patterns of acoustic and vibration signals. NDT and E Int. **38**(8), 605–614 (2005)
64. Ibarra-Zarate, D., Tamayo-Pazos, O., Vallejo-Guevara, A.: Bearing fault diagnosis in rotating machinery based on cepstrum pre-whitening of vibration and acoustic emission. Int. J. Adv. Manuf. Technol. **104**, 4155–4168 (2019)
65. Glowacz, A., Tadeusiewicz, R., Legutko, S., Caesarendra, W., Irfan, M., Liu, H., Xiang, J., et al.: Fault diagnosis of angle grinders and electric impact drills using acoustic signals. Appl. Acoust. **179**, 108070 (2021)
66. Stief, A., Ottewill, J.R., Orkisz, M., Baranowski, J.: Two stage data fusion of acoustic, electric and vibration signals for diagnosing faults in induction motors. Elektronika ir Elektrotechnika **23**(6), 19–24 (2017)
67. Hameed, Z., Hong, Y.S., Cho, Y.M., Ahn, S.H., Song, C.K.: Condition monitoring and fault detection of wind turbines and related algorithms: a review. Renew. Sustain. Energy Rev. **13**(1), 1–39 (2009)
68. Elforjani, M.: Diagnosis and prognosis of real world wind turbine gears. Renewable Energy **147**, 1676–1693 (2020)
69. Lin, Y., Zhuang, L., Xuan, L., Hong, W., Sun, T., Yang, P., Yan, B.: The research on condition monitoring and fault diagnosis method of wind power spindle bearing. J. Phys. Conf. Ser. **2254**(1), 012048. IOP Publishing (2022, April)
70. Lang, W., Hu, Y., Li, Q., Wireko-Brobby, A., Alkahtani, M.: ASSGCN model based vibration signal reconstruction in motor vibration testing. IEEE Trans. Ind. Electron. (2023)
71. Ozevin, D., Greve, D.W., Oppenheim, I.J., Pessiki, S.P.: Resonant capacitive MEMS acoustic emission transducers. Smart Mater. Struct. **15**(6), 1863 (2006)
72. Ono, K.: Calibration methods of acoustic emission sensors. Materials **9**(7), 508 (2016)
73. Eaton, M., Pullin, R., Holford, K., Evans, S., Featherston, C., Rose, A.: Use of macro fibre composite transducers as acoustic emission sensors. Rem. Sens. **1**(2), 68–79 (2009)
74. McLaskey, G.C., Glaser, S.D.: Acoustic emission sensor calibration for absolute source measurements. J. Nondestr. Eval. **31**, 157–168 (2012)
75. Grosse, C.U., Ohtsu, M., Aggelis, D.G., Shiotani, T. (eds.): Acoustic emission testing: basics for research–applications in engineering. Springer Nature (2021)
76. Sikorski, W.: Development of acoustic emission sensor optimized for partial discharge monitoring in power transformers. Sensors **19**(8), 1865 (2019)
77. Zaporozhets, A., Eremenko, V., Serhiienko, R., Ivanov, S.: Methods and hardware for diagnosing thermal power equipment based on smart grid technology. In: Advances in Intelligent Systems and Computing III: Selected Papers from the International Conference on Computer Science and Information Technologies, CSIT 2018, September 11–14, Lviv, Ukraine, pp. 476–489. Springer International Publishing (2019)
78. Zaporozhets, A.: Analysis of control system of fuel combustion in boilers with oxygen sensor. Periodica Polytech. Mechan. Eng. **63**(4), 241–248 (2018)

79. Zaporozhets, A., Babak, V., Sverdlova, A., Isaienko, V., Babikova, K.: Development of a system for diagnosing heat power equipment based on IEEE 802.11 s. In: Systems, Decision and Control in Energy II, pp. 141–151. Springer International Publishing Cham (2021)
80. Shyshko, Y., Cherniavskyi, A., Shyshko, D., Borychenko, O., Zaporozhets, A., Zaitsev, I.: Increasing the efficiency of loading devices for biomass boilers. Energies **17**(3), 590 (2024)
81. Zaporozhets, A.O.: Experimental research of a computer system for the control of the fuel combustion process. In: Control of Fuel Combustion in Boilers, pp. 89–123 (2020)
82. Hvozdeva, I., Myrhorod, V., Budashko, V., Shevchenko, V.: Problems of improving the diagnostic systems of marine diesel generator sets. In: 2020 IEEE 15th International Conference on Advanced Trends in Radioelectronics, Telecommunications and Computer Engineering (TCSET), pp. 350–354. IEEE (2020, February)
83. Zacharewicz, M., Kniaziewicz, T.: Method of evaluation of the technical condition of the Diesel-electric unit. Diagnostyka **20**(2), 113–119 (2019)
84. Babak, V.P., Babak, S.V., Eremenko, V.S., Kuts, Y.V., Myslovych, M.V., Scherbak, L.M., Zaporozhets, A.O.: Models and Measures for Standardless Measurements of the Composite Materials Characteristics. Models and Measures in Measurements and Monitoring, pp. 157–190 (2021)
85. Casado, A.J., Nieto, F.J., Blázquez, F., Platero, C.A.: A monitoring system for diesel engine driven generators based on electric power output oscillation assessment. IEEE Trans. Ind. Appl. **53**(3), 3182–3188 (2016)
86. Liu, Z., Zhang, K., Yao, X., Wei, H., Fang, S., Wang, T.: Monitoring system for diesel generator operation condition via sound signal. In: 2022 2nd International Conference on Consumer Electronics and Computer Engineering (ICCECE), pp. 298–302. IEEE (2022, January)
87. Drewing, S., Witkowski, K.: Spectral analysis of torsional vibrations measured by optical sensors, as a method for diagnosing injector nozzle coking in marine diesel engines. Sensors **21**(3), 775 (2021)
88. Mahtani, K., Guerrero, J.M., Beites, L.F., Platero, C.A.: Auto-adaptive stator ground fault protection for synchronous generators in diesel-electric locomotives. In: 2023 IEEE 14th International Symposium on Diagnostics for Electrical Machines, Power Electronics and Drives (SDEMPED), pp. 116–122. IEEE (2023, August)
89. Ahmad, Z., Prosvirin, A.E., Kim, J., Kim, J.M.: Multistage centrifugal pump fault diagnosis by selecting fault characteristic modes of vibration and using Pearson linear discriminant analysis. IEEE Access **8**, 223030–223040 (2020)
90. Zaporozhets, A., Kataiev, D.: Method of compensating for instrumental uncertainty in measurements using a coordinate measuring arm. Syst. Res. Energy **1**(76), 45–53 (2024)
91. Zhiyong, L., Hongdong, Z., Ruili, Z., Kewen, X., Qiang, G., Yuhai, L.: Fault identification method of diesel engine in light of pearson correlation coefficient diagram and orthogonal vibration signals. Math. Probl. Eng. (2019)
92. Nowicki, J., Hebda-Sobkowicz, J., Zimroz, R., Wyłomańska, A.: Dependency measures for the diagnosis of local faults in application to the heavy-tailed vibration signal. Appl. Acoust. **178**, 107974 (2021)
93. Martchenko, B.G., Myslovitch, M.V., Zvaritch, V.N.: Vibration signal expert system for fault detection of power equipment rolling bearings. IFAC Proc. Vol. **32**(2), 7861–7866 (1999)
94. Hertsyk, S., Gyzhko, Y., Zvaritch, V., Myslovych, M., Ostapchuk, L.: Features of the construction of autonomous diagnostic systems of electrical equipment, taking into account the modes of its operation. In: 2019 IEEE 20th International Conference on Computational Problems of Electrical Engineering (CPEE), pp. 1–3. IEEE (2019, September)
95. Janjarasjitt, S., Ocak, H.A.S.A.N., Loparo, K.A.: Bearing condition diagnosis and prognosis using applied nonlinear dynamical analysis of machine vibration signal. J. Sound Vib. **317**(1–2), 112–126 (2008)
96. Klausen, A., Robbersmyr, K.G.: Cross-correlation of whitened vibration signals for low-speed bearing diagnostics. Mech. Syst. Signal Process. **118**, 226–244 (2019)
97. Randall, R.B., Antoni, J., Chobsaard, S.: The relationship between spectral correlation and envelope analysis in the diagnostics of bearing faults and other cyclostationary machine signals. Mech. Syst. Signal Process. **15**(5), 945–962 (2001)

98. Laissaoui, A., Bouzouane, B., Miloudi, A., Hamzaoui, N.: Perceptive analysis of bearing defects (contribution to vibration monitoring). Appl. Acoust. **140**, 248–255 (2018)
99. Ming, A.B., Qin, Z.Y., Zhang, W., Chu, F.L.: Spectrum auto-correlation analysis and its application to fault diagnosis of rolling element bearings. Mech. Syst. Signal Process. **41**(1–2), 141–154 (2013)
100. Capolino, G.A., Antonino-Daviu, J.A., Riera-Guasp, M.: Modern diagnostics techniques for electrical machines, power electronics, and drives. IEEE Trans. Industr. Electron. **62**(3), 1738–1745 (2015)
101. Bertenshaw, D.R., Smith, A.C., Ho, C.W., Chan, T., Sasic, M.: Detection of stator core faults in large electrical machines. IET Electr. Power Appl. **6**(6), 295–301 (2012)
102. Frosini, L.: Novel diagnostic techniques for rotating electrical machines—a review. Energies **13**(19), 5066 (2020)
103. Choi, S., Haque, M.S., Tarek, M.T.B., Mulpuri, V., Duan, Y., Das, S., Toliyat, H.A., et al.: Fault diagnosis techniques for permanent magnet AC machine and drives—a review of current state of the art. IEEE Trans. Transp. Electrif. **4**(2), 444–463 (2018)
104. Zaporozhets, A.O.: Correlation analysis between the components of energy balance and pollutant emissions. Water Air Soil Pollut. **232**, 1–22 (2021)
105. Mallat, J.: Understanding the null hypothesis (H0) in non-inferiority trials. Crit. Care **21**(1), 101 (2017)
106. Solomon, H., Stephens, M.A.: Approximations to density functions using Pearson curves. J. Am. Stat. Assoc. **73**(361), 153–160 (1978)
107. Fox, J.: Regression Diagnostics: An Introduction. Sage Publications (2019)
108. Henao, H., Capolino, G.A., Fernandez-Cabanas, M., Filippetti, F., Bruzzese, C., Strangas, E., Hedayati-Kia, S., et al.: Trends in fault diagnosis for electrical machines: a review of diagnostic techniques. IEEE Ind. Electron. Mag. **8**(2), 31–42 (2014)
109. Kim, Y., Park, T., Kim, S., Kwak, N., Kweon, D.: Artificial intelligent fault diagnostic method for power transformers using a new classification system of faults. J. Electr. Eng. Technol. **14**, 825–831 (2019)
110. Li, W., Monti, A., Ponci, F.: Fault detection and classification in medium voltage DC shipboard power systems with wavelets and artificial neural networks. IEEE Trans. Instrum. Meas. **63**(11), 2651–2665 (2014)
111. Mellit, A., Tina, G.M., Kalogirou, S.A.: Fault detection and diagnosis methods for photovoltaic systems: a review. Renew. Sustain. Energy Rev. **91**, 1–17 (2018)
112. Beyer, W.H.: Handbook of Tables for Probability and Statistics. CRC Press (2019)
113. Klyuev, R.V., Morgoev, I.D., Morgoeva, A.D., Gavrina, O.A., Martyushev, N.V., Efremenkov, E.A., Mengxu, Q.: Methods of forecasting electric energy consumption: a literature review. Energies **15**(23), 8919 (2022)
114. Hong, T., Fan, S.: Probabilistic electric load forecasting: a tutorial review. Int. J. Forecast. **32**(3), 914–938 (2016)
115. Daliento, S., Chouder, A., Guerriero, P., Pavan, A.M., Mellit, A., Moeini, R., Tricoli, P.: Monitoring, diagnosis, and power forecasting for photovoltaic fields: a review. Int. J. Photoenergy (2017)
116. Ahmed, A., Khalid, M.: A review on the selected applications of forecasting models in renewable power systems. Renew. Sustain. Energy Rev. **100**, 9–21 (2019)
117. Ye, L., Xia, X., Chang, Z.: Dynamic prediction for accuracy maintaining reliability of superprecision rolling bearing in service. Shock Vibr. (2018)

Chapter 9
Tasks for Creating the Environmental Monitoring Systems for Energy Objects

Abstract This chapter delves into the innovative development and assessment of environmental monitoring systems and atmospheric modeling concerning power engineering facilities. Emphasizing the pivotal role of unmanned aerial vehicles (UAVs) and unmanned aerial systems (UAS), alongside advanced mathematical and statistical models, this research aims to enhance the accuracy, efficiency, and reliability of environmental impact assessments on air quality, particularly concerning radioactive aerosol propagation and deposition. Authors contribute to a multidimensional analysis that focuses on: (1) developing a new class of environmental monitoring systems leveraging UAV technology for dynamic air quality data collection around power engineering facilities; (2) evaluating mathematical models for atmospheric transport of radionuclides, including vector random field models, Gaussian models, and integral models, to accurately predict the environmental and public health impacts of radioactive releases; (3) innovating "in-situ" experimentation methodology using UAS equipped with filter injection devices for direct and reverse problem-solving in radioactive aerosol concentration determination; (4) enhancing information support within environmental monitoring systems through comprehensive analysis of their structure, function, and life cycle stages, aimed at improving system reliability and effectiveness. The research spans theoretical, simulation, and experimental studies, intending to mitigate the environmental ramifications of power engineering facilities over an expected timeframe. Key environmental parameters monitored include radionuclide distribution in the atmosphere and their deposition, encompassing essential environmental dynamics such as wind speed variations, turbulent diffusion, and gravitational deposition. This chapter underscores the necessity for accurate models and systems in environmental monitoring, providing crucial insights for designing, implementing, and modernizing such systems to ensure minimal human health and environmental impact while addressing the challenges of traditional monitoring methods.

Keywords Environmental monitoring · Atmospheric modeling · Power engineering facilities · UAV · UAS · Radioactive aerosols · Mathematical models · Information support

V. Babak et al., *Statistical Diagnostics of Electric Power Equipment*, Studies in Systems, Decision and Control 573, https://doi.org/10.1007/978-3-031-76253-6_9

9.1 Scientific and Technical Problems and Tasks of Environmental Monitoring of Power Engineering Facilities

Relevance of research. The modern world and the development of society create a number of problems related to the environment, the state and change of which become one of the main objects of research. If we summarize in an integral form the results of society development, the tendency of environmental change in general for the worse (pollution, including radiation, changes in the natural balance, etc.) does not change and we can say that it only progresses in time, especially in the areas of industrial facilities, primarily power facilities, including nuclear power plants, thermal power plants and others. It should also be noted that society is making more and more efforts to reduce the negative consequences of environmental changes, as well as efforts to restore the disturbed natural state of the environment [1–5].

It is worth noting one characteristic feature of the modern approach to environmental research. Today, the importance of the results of modern environmental research has sharply increased, since traditional sciences such as meteorology and others have been joined by areas of research in energy, medicine, industrial sectors and agriculture. This is due to the fact that now the characteristics of meteorological atmospheric fields, along with natural processes, also reflect the activities of artificial objects, including energy facilities [5–10].

The accumulated experience of research in various branches of science and technology makes it possible to justify the directions of research on the state and main characteristics of the environment. The main directions of such research include:

- theoretical studies;
- simulation (modelling), including computational (virtual) experiments;
- experimental, including a wide range of full-scale measurement experiments.

If we speak from the point of view of metrology as a science of measurements, then in most cases the results of field measurements are the basis for the final decision on the conducted research. The performance of a full-scale measurement experiment itself takes place in necessary, sometimes, in extreme cases, and the results of such an experiment acquire special weight and significance. However, the role of the results of theoretical and modelling studies should not be diminished, because such results determine the objectives, place, scope, development and methodology of using hardware and software tools in conducting in-situ measurement experiments on the state and basic characteristics of the environment [11–13].

The development of science, technology, and the use of modern information technologies create significant potential opportunities for environmental research.

At present, international and state programs on environmental protection are being implemented, first of all, the consequences of emissions from industrial enterprises, energy facilities are assessed, both during their normal operating modes and during various types of abnormal situations (disasters, accidents, etc.) and their impact on the environment and human health.

Monitoring as a research direction. Scientific and technical problems of environmental monitoring are very diverse, multi-planned and in their statements display a global international character. Of particular relevance and importance for Ukraine has a scientific and technical problem of monitoring the state and characteristics of the environment in the areas of operation of energy facilities [6, 7, 14].

This is evidenced by the works on harmonization of domestic regulatory and legislative support of environmental research with international standards, increase in the number of scientific publications for the last period. In fact, monitoring is considered as a further promising direction of using a set of different technical systems, modern information technologies, the potential application of which has increased significantly today.

Monitoring, in general, is a direction of research of various kinds, primarily technical objects in time and space, their relationship and interaction with the environment, to determine the current and predictable characteristics, parameters of the states of objects and the environment according to the data and results of processing observations, measurements, control, diagnostics and forecasting in order to use the results obtained for reliable and accurate solutions to the tasks at hand.

Carrying out the monitoring process requires the appropriate use of various resources, including information, technical, economic and a number of others, the realization of appropriate conditions and the application of means [15].

In each specific case, the general formulation of the monitoring process acquires a clear content, physical meaning and displays the following:

- the subject area of the research process and its specifics, characteristic features;
- purpose, object, subject and tasks of research;
- physical and mathematical models of the investigated objects;
- methods, ways, actions and operations of solving the set monitoring tasks;
- a priori and a posteriori information support for solving the tasks;
- databases, results of data processing, tests, operation of hardware and software of the monitoring system;
- evaluation of research results, decision-making on the achievement of the research goal and formation of output information.

The monitoring process can be illustrated by the structure shown in Fig. 9.1.

The results of solving the problem of monitoring the state and characteristics of the environment in the areas of operation of power facilities are based on the results of a significant number of tasks. As noted above, the solutions to these problems include the creation and use of a variety of information resources, hardware-software information measuring systems, special vehicles to ensure the monitoring process and other equipment.

Presented material base on the analysis of materials [16–22]:

- international, state programs on environmental protection;
- scientific publications, reports on environmental studies in Japan, USA, Canada, Ukraine, first of all the Institute of NPP Safety Problems of the National Academy

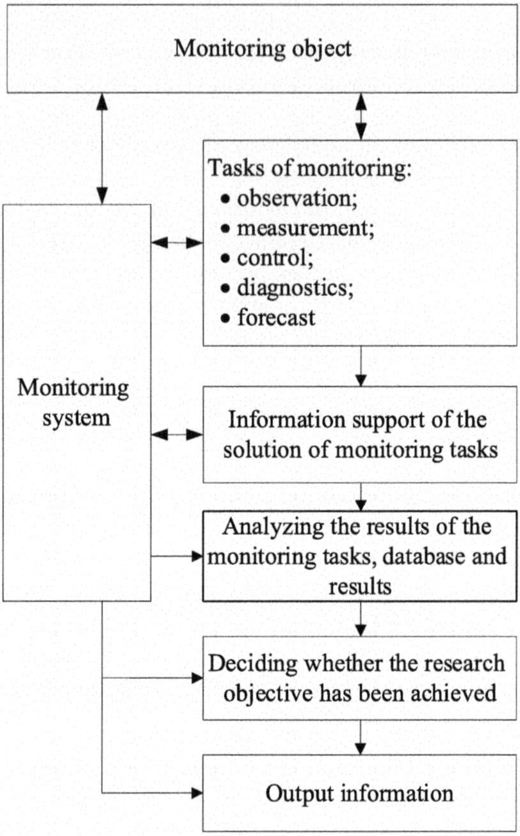

Fig. 9.1 Structure of the monitoring process

of Sciences of Ukraine; complex analysis and its results of radiological assessments of the consequences of the Chernobyl accident to identify the main pathways of radionuclide spread and migration in environmental objects, taking into account the characteristics of the terrain;

- on the use of developed models of multicomponent radionuclide propagation in non-stationary and horizontally heterogeneous meteorological fields and other models, including the proposed IAEA Gaussian model, in modeling the dynamics of air and soil contamination formation;
- on application of means, systems of environment monitoring in Ukraine, other countries, first of all, means and systems of monitoring of radiation contamination of Chernobyl accident, "Fukushima" accident and other nuclear disasters, including the use of new means of transportation—unmanned aerial vehicles;
- on the statement of perspective scientific and technical problems of environmental monitoring using the results of the latest achievements of science and technology, proposals and substantiated recommendations on the prospects for the creation of information and hardware-software means of environmental monitoring systems, we make the following conclusion.

Taken together, the results of the performed analysis determine the relevance and importance of creating a system for monitoring the characteristics of the environment using an unmanned aircraft complex. Let us formulate this scientific and technical problem in the following wording: *development of the theory and principles of construction of computerized information and measurement systems of environmental monitoring of power engineering facilities on the basis of unmanned aircraft systems* [23–28].

Creation of such a mobile monitoring system gives an opportunity to:

- to use in the creation of the system the latest achievements of science and technology in various subject areas of research by developing the theory and principles of construction of a new class of environmental monitoring systems;
- to develop modern hardware and software tools for measurements, control, diagnostics and forecasting of characteristics and parameters of the environment;
- to ensure the efficiency and mobility of using the system based on the use of unmanned aerial systems in various areas of operation of industrial enterprises and energy facilities, including hard-to-reach, mountainous and forest areas;
- to use the remote research method, while ensuring the safety of researchers and the natural experiment as a whole in emergency situations of the functioning of energy facilities (accidents, catastrophes, etc.);
- to provide an assessment and forecast of environmental pollution in the studied areas based on measurement data.

Let us dwell on the solution to such a complex, essentially multifaceted, problem in more detail.

Object of study. The environment is an integral concept that unites a significant number of components. Each specific area, locality, region is characterized by geographical features, terrain, topology of population distribution, industrial and other objects (agricultural land, urban and rural buildings), natural objects (forests and reservoirs).

The IAEA has proposed the following classification of the spatial scale of problems of atmospheric transfer of radioactive substances in the event of a radiation accident (accident class):

- local (up to 20 km);
- mesoscale (from 20 to 200 km);
- regional scale (from 200 to 2000 km);
- transboundary scale (2000 km or more).

The airspace of the environment (the atmosphere) plays an increasingly important role in the life of earthly civilization. The atmosphere is a spatiotemporal channel for the transmission of optical, interacting meteorological fields of pressure (geopotential), temperature and wind, electromagnetic and other types of fields. Air meteorological fields and their dynamics in time and space determine weather conditions on Earth. Energy facilities, primarily nuclear power plants (NPPs), thermal power plants (TPPs), and high-voltage power transmission lines are sources of various fields, primarily fields of radionuclide emissions, aerosols and other components of air pollution, powerful electromagnetic fields. Such fields, the sources of which are energy facilities, are informational in the scientific and technical problems of environmental protection, and determinations of the state and characteristics of such fields are the main objects of environmental research, i.e. the result of solving monitoring problems [6, 15, 17, 24].

An important point in determining the object of air environmental research is its stochastic nature, as a result of the action of a significant number of random physical factors.

Thus, the object of research of air environment monitoring systems of power engineering facilities is the formation and propagation of vector physical field, the mathematical model of which in general can be described by a vector random field of the following form [29, 30]

$$\begin{aligned} \Xi_n(\boldsymbol{\omega}; x, y, z; t) &= (\xi_1(\omega_1; x, y, z; t); \\ &\quad \xi_2(\omega_2; x, y, z; t), ..., \xi_n(\omega_n; x, y, z; t)); \\ \boldsymbol{\omega} &= (\omega_1, \omega_2, ..., \omega_n), \boldsymbol{\omega} \in \boldsymbol{\omega}, x, y, z \in Q, t \in T, \end{aligned} \tag{9.1}$$

where each component $\{\xi_j(\omega_j; x, y.z; t), j = \overline{1, n}\}$ of the vector field $\Xi_n(\boldsymbol{\omega}; x, y, z; t)$ is a random field describing a particular physical field in a limited three-coordinate volume of airspace Q on the observation time interval T. The randomness or stochasticity of the field formation (9.1) is due to the vector space of random events $\boldsymbol{\Omega}$, and each particular random field is $\xi_j(\omega_j, x, y, z, t)$, respectively $\boldsymbol{\omega}_j \in \boldsymbol{\Omega}_j$.

Subject of research. The solution of the formulated scientific and technical problem of air environment monitoring is based on the results of solving a significant number of problems, which in fact constitute the subject of research.

Let's group these tasks by research directions.

(A) *Theoretical Research*

Development of mathematical models of investigated physical fields of air pollution of power engineering facilities.

Determination of spatial and temporal characteristics of the investigated fields to solve monitoring tasks.

Justification of the method of measurements of the given spatial and temporal characteristics of the investigated fields.

Development of information support of measurements, control, diagnostics and prognosis, including a complex of algorithms and software for measurement data, statistical evaluation of the obtained results and characteristics of accuracy (uncertainty) of measurements.

Development of the structure and characterization of the information-measuring system of measurement, selection of the type and characteristics of modern sensor subsystems—primary measuring transducers of air environmental pollution.

Justification of the principles of building a new class of monitoring systems based on the use of unmanned aerial systems, the realization of which provides the solution of various tasks of monitoring in time and space of the air environment of energy facilities.

(B) *Simulation (Modeling) Studies*

Development of information support for simulation computational (computer) experiments based on the results of theoretical studies to substantiate:

- the optimal variant of the information-measuring system structure according to the given criterion in the class of possible structures of systems and sensor subsystems;
- type and characteristics of the unmanned aviation complex as a vehicle of the system of mobile monitoring of the air environment;
- efficiency of algorithmic-software support of statistical processing of measurement data, test results, determination of calculated technical characteristics, reliability characteristics of monitoring systems.

(C) *Experimental Research*

Realization of the results of theoretical and simulation studies, creation of the monitoring system and use of the developed information support to obtain:

- databases of real measurements;
- results of statistical evaluation of characteristics and parameters of the environment;
- results of comparative analysis of theoretical, simulation and experimental studies;
- decision on the achievement of the goal and objectives of the research as a solution to the scientific and technical problem of environmental monitoring.

Let us dwell on the description of information support of the research of the scientific and technical problem formed above.

9.2 Information Support of Environmental Monitoring System

The term *information support* needs some interpretation and concretization, as information support is an integral product of a significant number of components, constituents of the so-called *soft equipment* of the studied monitoring systems. For example, information support in information systems is a set of forms of documents, databases and programmatically realized decisions regarding the volume, placement and forms of information existence. Requirements of integrity, completeness, reliability, protection from unauthorized access, unification, minimization of volume during transmission, and preservation are put forward to this kind of information support [15, 31–35].

While analyzing the work of modern technical systems, hardware is referred to the so-called *aging means*, due to the fact that aging processes, changes in characteristics, parameters of elements, etc. significantly limit the time of their operation. Information support is referred to *the non-aging ("youthful") means*, because in the process of operation such support, primarily software, reduces the number of errors and has a significant time period of application.

It is also known that the potential capabilities of modern computing facilities can be realized by different versions of software, which can largely provide the solution of research problems without changing the hardware of the systems and thus provide cost and time savings for the development of new and modernization of existing systems.

The composition of information support in the creation of monitoring systems includes various components, including:

- physical and mathematical models, which homomorphically display the functioning of systems when performing a sequence of n necessary functions under specified modes of operation in time and space take into account the conditions of application of monitoring systems;
- databases of calculation and operational parameters, characteristics, including reliability characteristics of technical devices, modules, subsystems, which are used in the creation of the monitoring system and others;
- databases of test, measurement, control and diagnostics data during operation of various environmental monitoring systems, results of their processing;
- algorithmic support of computer modeling tasks, including computational tasks of computer modeling of different variants of the structure of the created monitoring system and others;
- algorithmic support for the operation of different subsystems of the monitoring system being created: automatic design (CAD), statistical processing of measurement data, control and diagnostics, automatic control of operating modes, system-operator dialog control system and others;
- software for realization of mathematical and algorithmic support.

Lets analyze the structure of information support of monitoring systems, which is presented in Fig. 9.2.

This structure of information support of the monitoring system is a typical one, which is created and applied at the stages of the life cycle of each such system.

Along with the maintenance of existing monitoring systems, information support should serve as a basis for the creation of scientific and technical basis for the development of prospective systems. If we try to introduce a certain classification of information support of monitoring systems, different classification features can be taken as a basis. So the following classification signs characterize both current and prospective variants of performance of the given functions by monitoring systems, for example:

- *functional-current information support, which remains largely unchanged during the development, implementation and operation of the current model of the system*;
- *innovative information support, which reflects both the results of operation of existing monitoring systems and new developments in different fields of science and technology.*

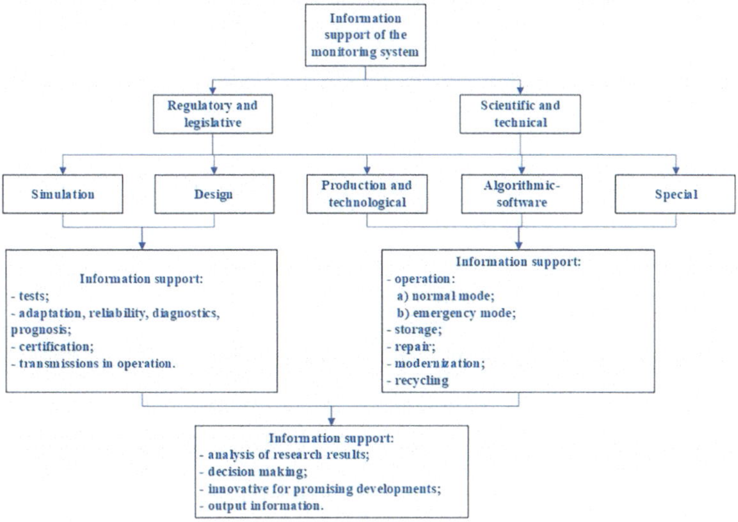

Fig. 9.2 Structure of monitoring system information support

Along with this, the following classification looks fully justified:

- *a priori support* at the first stages of the life cycle of systems, namely at the stages of design, manufacturing, testing, when mainly calculated characteristics and parameters of the system are used;
- *a posteriori support* is mainly used at the stages of operation and partially tests of the system and in terms of meaning is adjusted support, the use of which makes it possible to adjust the calculated characteristics and obtain the real characteristics of the system, including the characteristics of reliability, residual life by the results of statistical evaluation of real data of measurements, control, diagnostics and prognosis.

Other signs of classification of information support reflect the monitoring process in more detail:

- *theoretical, algorithmic and software, or scientific and technical*, at all stages, especially the appropriate role of its use at the initial stages, when the physical and mathematical models, systems are developed, the justification of the structure of systems based on the results of computer modeling options is carried out, the design characteristics and parameters of the system performance of specified functions are determined;
- *production-technological support* at the stages of production preparation, manufacturing of experimental and serial samples of the system;
- *normative-legislative support* is used at all stages, but the main role is played during testing, certification, transfer into operation and at the stage of operation, repair and modernization of the system.

It is known that any classification of objects is based on the classification tasks, some certain properties, characteristics of objects and therefore cannot reflect their entire diversity, i.e. it is in some sense conditional. The considered classification of information support for monitoring systems displays some types of such support on the time interval of the life cycle of the monitoring system under study.

9.3 Realization of Information Support at the Life Cycle Stages of the Monitoring System

While creating a monitoring system, a variety of scientific, technical, industrial, economic, social and other tasks are solved. Let us consider the tasks of development and use of information support, based on the sequence in time of the stages of the life cycle of the monitoring system during its creation and operation [15, 36–40].

At the initial stage of creation of the monitoring system a significant amount of work is done to substantiate, develop and coordinate the terms of reference for the system between the customer (department, firm, organization) and the executor—the developer and manufacturer of the system (firm, design bureau, manufacturer). The terms of reference for a similar system, which has passed the stage of testing and operation, is taken as a basis. The technical specification shows all stages of the life cycle of the system, the main characteristics and parameters of the system when performing the specified functions, hourly intervals of design, testing, operation and other characteristics, including reliability characteristics, warranty period of operation and others.

At the preparatory stage, the time schedule for the creation of the monitoring system is determined, in fact, with the involvement of the number of developers, the level of their qualification, the volume of special means and equipment, the allocation of financial resources, etc.

Using the methods of *systems theory*, the potentially achievable characteristics of the created system are determined, taking into account the characteristic features of its operation when performing the specified functions.

In other words, at this stage the plan of creation and content of information support is formed.

At the design stage, a significant amount of theoretical, simulation (modeling) and experimental (laboratory) research is performed to substantiate the system structure and to develop design, technological and technical documentation for the implementation of the monitoring system into a modern hardware and software system of measurement, control, diagnostics and forecasting of environmental characteristics in the investigated areas of power facilities.

In this case, the mathematical apparatus of creating technical systems, on the basis of which the information support is created, is used in full. The basic provisions and methods of practically all natural sciences, first of all physics and mathematics, as well as the latest achievements of technical sciences are used.

Thus, physical, mathematical, simulation models of the monitoring system to be created are developed, and the problems of the system operation are solved in order to fulfill the specified functions. Based on the results of simulation, mainly computer modeling, the optimal variant of the system structure is justified in some sense.

At this stage mainly a priori *information support* is used, calculation characteristics are determined, including reliability characteristics. It should be noted the level of information support of automation, design, which provides:

- improvement of the quality of design of technical and design documentation due to a significant reduction of design errors;
- significantly reduce the time of documentation development on the basis of using modern automated information design technologies;
- development of documentation for manufacturing of parts, modules of the system on modern equipment with numerical program control.

Information support for testing of prototypes of means, modules, subsystems and serial sample of the system, statistical processing of test data, production preparation is being developed.

Stage of manufacturing and testing of research samples, modules, subsystems. In the presence of design and engineering documentation for the monitoring system to be created on the basis of information support, the following works are performed:

- preparation of production for manufacturing of research samples of modules, subsystems;
- during manufacturing of research samples the operations of incoming control of component parts, elements of electrical circuits of modules and constructions are performed;
- the methodology and corresponding test plans developed at the design stage are implemented using information-measuring systems to obtain measurement data, characteristics and parameters of information signals of the investigated research samples, modules, subsystems;
- the methodology of statistical processing of test data and determination of experimental characteristics of the modules and subsystems under study, developed at the design stage, is implemented;
- based on the results of comparative analysis of calculated and experimental characteristics, necessary corrections are made to the project technical and design documentation.

On the basis of the results of the performed researches the adjustment of the information support, first of all, of the design and construction documentation for the development of the serial sample of the system is carried out.

The stage of production sample manufacturing, certification and transfer to operation of the system. At this stage of the system life cycle, the following works are performed on the basis of information support:

- preparation of production for manufacturing of serial samples of the system with specified system characteristics;
- at manufacturing of serial samples of the system the methodology, which was developed at the design stage, adjustment of different modules of subsystems, checking of their functioning for realization of fulfillment of set functions, is carried out;
- testing of the manufactured serial sample of the operating monitoring system to perform the specified functions (normal operation mode), certification of the system using metrologically certified measuring instruments.

Operation stage. This stage of the life cycle of the monitoring system is the most time-consuming and, accordingly, information support takes part in the following operations:

- transportation, adjustment and commissioning of the system;
- current monitoring of the system functioning in specified modes using subsystems of measurements, control, diagnostics and control;
- formation of the database of measurements of the system functioning with further statistical processing of the measurement data of characteristics, parameters of information signals;
- realization of the methodology, which was developed at the stage of design, construction of long-term monitoring and determination of the predicted characteristics of the system reliability;
- implementation of the system repair methodology;
- adjustment of mathematical and software of computer modeling of options for modernization of the current system, conducting a simulation experiment and formation of recommendations for modernization of the current system in order to improve the efficiency of its use;
- implementation of recommendations on modernization of further operation of the monitoring system.

The final stage of the system life cycle. This stage is the most science-intensive in the life cycle of the monitoring system and consists in the formation of research data of all stages, their systematization and generalization of research results of a particular type of monitoring system. Such generalization of research results serves as a foundation for the development of all types of information support in the creation of new, promising, highly reliable monitoring systems.

At this stage the work on utilization of the monitoring system, which has exhausted its technical resource, is also carried out.

9.4 Generalized Models of Information Fields of Environmental Air

Creation of information support for monitoring systems is usually started with the development of mathematical models of environmental information fields. It should be noted that such models are initially the main object of environmental research and their development began intensively in the 30 s of the twentieth century. The initial models were published in scientific articles by famous Soviet mathematicians A. N. Kolmogorov, A. M. Obukhov and others [41–45]. The first mathematical models were stochastic, since their physical nature was formed under the action of a significant number of random factors.

The decisive factor for a different approach to models was a number of accidents associated with nuclear facilities in the USA, the Soviet Union, England, and the Chernobyl accident. Other models, primarily models of airborne radionuclide transport in the presence of a strong radionuclide source arising from nuclear accidents, have begun to be developed [46–50].

Let us first consider *the theoretical model* of the initial information field.

In the process of creating *a physical model*, first of all, physical properties, characteristics of the object of environmental measurements, its elements, modules, their actions and interactions with each other are displayed taking into account physical laws. In this case the model is written using mathematical formulas with dimensions of units of physical quantities.

Mathematical model formalizes to a greater extent the process of mapping the properties of the measurement object into the model. Mathematical methods are effective methods of research and, depending on the formulation of the measurement problem for the same measurement object, can use different mathematical models on the one hand, and on the other hand—use the same model for the study of objects of different physical nature.

Each of the models, physical and mathematical, complement each other and make it possible to conduct the measurement process more efficiently, for example, to justify ways to improve the accuracy of measurement results.

Let's formulate the definition of theoretical model of initial information of measurement object in the following wording.

Theoretical model of initial information of measurement object is a set of knowledge, assumptions, hypotheses, initial and boundary conditions, which homomorphically display the main properties, characteristics in space and time of the initial information, built in the form of an integral, consistent, logical structure, formulated with the use of necessary objects, terms and symbols of this science for solving a given class of problems.

On the example of mathematical model of initial information of the object of environmental measurements we will consider the main stages of its use in the research of the measurement process.

At *the first stage,* on the basis of comprehensive analysis of a priori information about the measurement object, its main properties and characteristics to be measured, *the model is created* using mathematical objects, terms and symbols, and proposed for further use.

At *the second stage* the proposed model is used for solving various measurement problems in order to determine the result and uncertainty of measurements.

Measurements with application of the proposed model, including justification of the answer to the question whether the results of measurements necessary for theory and practice meet the requirements for accuracy characteristics. According to the results of the analysis one of the possible options is accepted:

(a) the proposed model is recommended for further research;
(b) the proposed model should be improved, changes obtained during its practical use should be introduced into it and the improved model should be recommended for further use;
(c) to develop a new model based on the results of research obtained in practice for further use.

When using the scientific and technical problem of creating environmental monitoring systems for power engineering facilities, such a generalized model is a vector random field of the form (9.1), i.e., a vector random field of the form (9.1).

At *the third stage*, the obtained measurement results are analyzed using the proposed model, including justification of the answer to the question—whether the results of measurements necessary for theory and practice meet the requirements for accuracy characteristics. According to the results of the analysis one of the possible options is accepted:

(a) *the proposed model* is recommended for further research;
(b) the proposed model should be improved, changes obtained during its practical use should be introduced into it and *the improved model* should be recommended for further use;
(c) to develop *a new model* based on the results of research obtained in practice for further use.

When using the scientific and technical problem of creating environmental monitoring systems for power engineering facilities, such a generalized model is a vector random field of the form (9.1), i.e., a vector random field of the form (9.1):

$$\begin{aligned}\Xi_n(\boldsymbol{\omega}; x, y, z; t) = (&\xi_1(\omega_1; x, y, z; t),\\ &\xi_2(\omega_2; x, y, z; t), \dots, \xi_n(\omega_n; x, y, z; t));\\ \boldsymbol{\omega} = (&\omega_1, \omega_2, \dots, \omega_n); \boldsymbol{\omega} \in \boldsymbol{\Omega}, x, y, z \in Q, t \in T,\end{aligned}$$

where, as noted above, each component $\left\{\xi_j\left(\omega_j; x, y.z; t\right), j = \overline{1, n}\right\}$ of the vector field $\Xi_n(\boldsymbol{\omega}; x, y, z; t)$ is a random field describing a particular physical field in a limited three-coordinate volume of airspace Q at the observation time interval T [29, 30].

One of the important points of using the model (9.1) is the choice of the coordinate system. Let us briefly dwell on it.

Coordinate systems. First, let us concretize the object of measurement—the area of the environment in the locations of power engineering facilities. One of the typical geometrical figures of such area is a parallelepiped with uneven continuous bottom edge, which reflects the inhomogeneity of the underlying surface relief. Such a surface is formed on a particular terrain by land and water surfaces. In each particular case of environmental studies, the linear dimensions of the faces of the research environmental parallelepiped vary in the dimensions recommended by the IAEA:

- for local surveys up to 30 km;
- at mesoscale surveys up to 200 km;
- in regional studies from 200 to 2000 km;
- in transboundary studies from 2000 km and more.

Since environmental studies are conducted in time and space, it is customary to use two Cartesian rectangular coordinate systems:

A. the fixed coordinate system OXYZ, in which the origin of coordinates O is coincided with the intersection of the lower faces of the parallelepiped ABCDA′B′C′D′, the OZ axis is directed upward in the direction of the local vertical, the OX axis—to the east, and the OY axis—to the north;
B. mobile coordinate system OX′Y′Z′, which is associated with movement through space within the studied parallelepiped ABCDA′B′C′D′, the origin of which O is coincided with the origin of coordinate system OXYZ, the axis OZ′ is directed upwards in the direction of the local vertical, the axis OX′ is directed along the wind direction during the measurement period, and the axis OY′ is directed so that the mobile coordinate system OX′Y′Z′ is right-handed, at that

$$0 \le x \le L_x,\ 0 \le y \le L_y,\ \sigma \le z \le z_m,$$

where L_x and L_y—linear dimensions of the parallelepiped's faces on the horizontal of the (x, y) plane, and the function $z = \sigma(x, y)$ describes the terrain relief (inhomogeneities) of the underlying surface, a—defines the height of the studied parallelepiped ABCDA′B′C′D′ as the upper boundary of the considered area.

Consideration of two systems—stationary A) and mobile B) Cartesian rectangular coordinates will make it possible to use simultaneously both stationary (fixed in the parallelepiped under study) and mobile (car, aircraft—airplanes, helicopters, unmanned aerial complexes) measuring instruments with subsequent joint processing of measurement data.

9.5 Stochastic Model of Formation of Meteorological Fields

General characteristics, regularities. A characteristic feature of the atmosphere is the disordered turbulent nature of the motions occurring in it. The fields of meteorological elements are extremely variable. The dependence of the instantaneous values of the fields on the spatial ordinates, as well as the time course of these values, becomes very complex and confusing, and when these values are registered under identical conditions, they appear different each time. Due to this fact, it is not possible to describe the fields in such a way that one can specify their instantaneous values in each point of space at each moment of time [6, 51–53].

A natural approach to the study of the structure of meteorological element fields is the probabilistic or statistical approach, in which each field is treated as a random field and the methods of random function theory are applied to describe it, as was mentioned in paragraph 5 of this Chapter.

The basis of this approach is the refusal to consider the features of individual instantaneous field values and the transition to the consideration of some averaged properties of a statistical set of field realizations corresponding to some set of fixed external conditions.

In the experimental determination of statistical characteristics of a random field, it is assumed that there is some set of its realizations corresponding to the same conditions of the experiment, or in the case of a homogeneous field possessing ergodic property, the presence of one realization of the field in a sufficiently large spatial and temporal area.

Statistical set of realizations. Formation of such a set of homogeneous realizations of the studied meteorological field is a necessary statistical material in the process of conducting a full-scale measurement experiment to assess the state and characteristics of the field.

Meteorological fields physically do not allow mass repetition under identical external conditions. The researcher does not have at his disposal a statistical ensemble of realizations of the studied fields due to the variability of these conditions.

In meteorology, a single process is usually divided into parts, which are conventionally taken as different realizations, i.e., observations conducted in different spatial areas or at different points in time are used as realizations of the random field. At the same time, as realizations corresponding to the same external conditions, observations conducted in similar in some sense spatial areas or time intervals are taken, which can be used for joint statistical processing.

In the theory of random functions, the situations corresponding to the same external conditions are usually called those in which the laws of distribution of the random field are preserved. In practice, these distribution laws are usually not known, so the selection of similar situations is made on the basis of everyday meteorological experience and the results of previous studies.

The way in which the choice of similar situations to be averaged over is solved in each particular case determines the information obtained about the structure of the field under consideration. Another requirement for a set of realizations is the independence of the individual realizations. If the realizations are closely related to each other, they will all contain very little new information compared to one of them, and, therefore, an increase in the number of realizations in this case will not lead to a noticeable refinement of the obtained statistical characteristics.

Based on the above requirements and the physical nature of meteorological processes, we can note some basic provisions that should be taken into account when combining experimental data into one statistical population.

The selection of time points corresponding to similar situations should be based on the daily and annual variations of the meteorological elements. The presence of the diurnal variation means that moments belonging to one specific time of day can be considered analogous. Due to the annual course, moments belonging to different seasons of the year cannot be considered analogous. Strictly speaking, only realizations obtained on the same day and hour of each year should be considered analogous. However, this turns out to be practically disadvantageous, since in this case we will be able to operate only with a very small set of realizations, averaging over which will be insufficient to reliably obtain statistical characteristics. Therefore, in practice, it is common to combine into one population the realizations related not to one day but to some interval of the year, such as a month or a season, i.e., all available realizations related to a certain time of day and to the season under consideration, obtained from observations conducted over a number of years, are combined into one population. In order for the realizations to be independent, a sufficient time interval between observations should be chosen. For example, it is known that air pressure varies little during a day, hence the dependence between its values at different moments of the day is large. This dependence remains noticeable during the next two days, therefore, when selecting a set of realizations of the pressure field, observations with an interval of at least three days are usually used.

In addition to taking into account the daily and annual variations, when combining different realizations into one statistical set, it is possible to further classify the empirical material by some special features. For example, when studying the wind field, one separates the realizations corresponding to different circulation conditions, singling out, for example, jet currents separately or classifying the realizations by wind speed value, etc. Separation by circulation types is also sometimes made when studying the pressure field (geopotential).

While combining similar spatial volumes, i.e., realizations obtained at different geographic locations, it is assumed that these locations must belong to the same climatic regions.

Field homogeneity and isotropy. When studying the spatial structure of meteorological fields, it is very important to observe the conditions of field homogeneity and isotropy, which imposes certain restrictions on the spatial extent of the field under study [24, 54–56]. It is assumed that in a real turbulent flow, which, generally speaking, is not homogeneous and isotropic, it is possible to identify a scale within which the conditions of homogeneity and isotropy should be approximated for the fields of meteorological elements. *Such fields are called locally homogeneous and isotropic.*

Depending on the scale of the studied fields in meteorology, *microstructure*, *mesostructure* and *macrostructure* are divided.

The first one describes the features of the fields in the intervals from fractions of millimeters to hundreds of meters. In this region there is local homogeneity and isotropy in three dimensions.

The statistical *mesostructure* describes the features of the fields in the interval from a kilometer to tens of kilometers. In this region the distinction between vertical and horizontal directions is clearly manifested. Homogeneity and isotropy are approximated only in the horizontal direction.

Variability and mutual relations at spatial scales of the order of hundreds of kilometers or more are described by the statistical *macrostructure*. Macro-processes are related to the atmospheric formations of synoptic and even global character, the physical nature of which is fundamentally different from the nature of disordered small-scale turbulent pulsations.

In many cases it is still convenient to consider macro-processes as random and to describe them by analogy with small-scale processes as a kind of macroturbulent exchange. However, this analogy has a formal character.

In this area, the conditions of homogeneity and isotropy are roughly approximated only in the horizontal plane. In the region of meso- and macroturbulence, one can speak about homogeneity and isotropy only for deviations of meteorological elements from the climatic norm, since the climatic norms themselves can undergo significant differences at these scales. In this case, it is practically impossible to use the ergodic property, which greatly facilitates the study of a homogeneous field and allows us to determine statistical characteristics from a single realization of sufficient duration. Indeed, the mathematical expectation of a meteorological field depends on the coordinates, so one cannot use a single realization to calculate the mathematical expectations, but a large number of realizations. In addition, empirical studies of the structure of large-scale meteorological fields use data from observations at weather stations, which are usually few in the spatial area, combined in one statistical population, i.e., we have values of realization only in a small number of discrete points, and, therefore, averaging over a single realization will not be effective.

When studying the field structure, such statistical characteristics as mathematical expectation, correlation or structural functions are estimated, which are necessary for solving a wide range of problems.

Let us give examples of the obtained results of experimental studies of specific meteorological fields.

Pressure field (geopotential). When studying the pressure field, the points corresponding to similar situations are considered to be the points of the globe having the same latitude and differing only in longitude [18, 57].

The studies have shown that in the middle latitudes the conditions of homogeneity and isotropy with respect to the structural functions of the geopotential field are fulfilled quite well. However, the field dispersions still undergo some changes depending on longitude. Usually, the dependence of statistical characteristics on longitude is not taken into account, i.e. the field is considered homogeneous in longitude. At the same time, the assumption of longitude homogeneity makes statistical processing very easy, since it makes it possible to consider as corresponding to similar situations all observation points located near the same parallel and, consequently, makes it possible to significantly increase the number of realizations over which averaging is performed.

It was found that the mean values (norms) of the height H turn out to be significantly different for different circulation types. The differences between the structural functions for different circulation forms turned out to be small and can be neglected, i.e., the structural functions obtained for different circulation forms can be averaged and a single structural function for all circulation forms can be used.

Example. As an example, the statistical structural function of the form, which is most consistent with the results of experimental studies of the plane geopotential field at altitude (in most cases, the height H is chosen discretely: 1000, 850, 700, 500 and 300 mb (millibars)), is a statistical structural function of the form.

$$B_H(l) = 400\big(1 - \exp\big(-0.188 l^{1.3}\big)\cos 0.54l\big),$$

where l expressed in km.

By approximating this structure function, the corresponding autocorrelation function of the plane pressure field was determined.

$$R_H(l) = 200\exp\big(-0.188 l^{1.3}\big)\cos 0.54l.$$

The results of studies of meteorological pressure fields (geopotentials) are given in a number of publications, including [58–62].

Air temperature field. The field of air temperature deviations from normal, like the geopotential field, is considered to be homogeneous and isotropic in the horizontal plane or on a given isobaric surface. Consequently, correlation and structure functions on a given surface are considered as functions of a single argument—the horizontal distance between observation points. The spatial structure is characterized, in addition to the autocorrelation and structure functions for each standard isobaric surface, by the correlation and structure functions for each pair of surfaces [63–67].

A comparison of the normalized correlation functions obtained in the experimental studies shows that they closely coincide for the same levels, especially at distances up to 1000–1500 km.

At the same time, the dispersions in the cases under consideration differ significantly. For example, the temperature dispersion at the 700 mb level was 24 deg for Europe and 32 deg for America.

As an example of an approximating analytical relationship for the statistical autocorrelation functions of temperature for winter are obtained:

- at the 700 mb isobaric surface
  $R_T(l) = D\exp\big(-0.747 l^{0.96}\big)J_0(0.96l)$;
- at the isobaric surface 850 mb
  $R_T(l) = D\exp\big(-0.553 l^{0.97}\big)J_0(0.83l)$;
- above ground

$R_T(l) = D\exp\left(-0.825l^{0.92}\right).$
where $J_0(\gamma l)$—Bessel function of zero order, l expressed in 10^3 km.

Wind field structure. The regularities of the wind field structure are given in a number of publications on meteorology. The fundamental theoretical works in this direction are those of A. N. Kolmogorov and A. M. Obukhov [41–45].

In these works, it was established theoretically for a locally homogeneous and isotropic field that the structural function of wind speed pulsations is described by the formula

$$B_u(l) = Al^{2/3},$$

where A—some proportionality coefficient.

This relation is called the "*2/3 law*". The limited spatial scales within which the "*2/3 law*" is satisfactorily fulfilled are natural, since the real turbulent wind field can be considered homogeneous and isotropic only for sufficiently small spatial volumes [68–72]. At increasing scales, anisotropy begins to affect, manifested in the inequality of horizontal and vertical directions for real atmospheric motions of large scales. In [68], an analysis of the conditions of applicability of the "*2/3 law*" is given and it is shown that, outside the domain of this law, the structural function of wind pulsations is described by the relation.

$$B_u(l) = Cl,$$

where C—proportionality coefficient, i.e., the structure function of wind pulsations is proportional to the distance.

This relation is called the "law of the first degree".

The results of experimental processing confirmed the satisfactory feasibility in the real atmosphere of the "law of the first degree" in the distance range $l = 500$–1400 km. As for the conditions of macroturbulence, the complexity of the processes occurring here complicates the theoretical study of the structure of macroscale meteorological fields.

In the following years, the results of studies of meteorological fields—geopotential, air temperature and wind fields—were used in the development of information support for weather forecasting.

9.6 Models for Assessing the Consequences of Atmospheric Releases NPPs Under Conditions of Different Types of Radionuclide Fields in the Air

To analyze and forecast the impact of radioactive releases from nuclear power facilities both on the environment and on public health, one of the decisive elements is the assessment of radionuclide distribution in the atmosphere and their deposition on the underlying surface. This is due to the fact that the conditions of air transport play a determining role in the formation of radioactive contamination fields of air and soil in case of radioactive releases from NPPs and, as a consequence, the values of dose loads on the population. Mathematical models of radioactivity propagation in the atmosphere are an important tool for assessment of radioactive contamination of the atmosphere and underlying surface in case of accidental releases of NPPs during the early phase of a radiation accident. The choice of an adequate model of atmospheric transport, taking into account the peculiarities of both the release source and the radioactivity spreading area, is of decisive importance in solving this problem.

In the near zone of the release source (local scale), assessments of contamination of surface air and underlying surface by releases from nuclear power facilities are predominantly carried out using a Gaussian model [73–76].

To describe the long-range transport of impurity (at distances of the order of a thousand and more kilometers), simplified models are predominantly used, with the help of which it is possible to obtain spatially averaged characteristics of atmospheric pollution.

Intermediate in spatial scale and the most complicated for modeling are the processes of impurity diffusion at distances of the order of tens and hundreds of kilometers. This is due to the fact that the mathematical model should take into account the daily variability of turbulence variability in the atmospheric boundary layer, orographic and thermal heterogeneity of the underlying surface, etc.

According to the normative document of Ukraine NRSU-97 [77], the period of the early phase of a radiation accident includes the following events:

- gas-aerosol releases of radioactive material from the emergency source;
- processes of airborne transport and intensive terrestrial migration of radionuclides;
- radioactive fallout and the formation of a radioactive trace.

Under these conditions, there will certainly be a shortage of time and instrumental resources to carry out full-scale measurements of the consequences of radioactive contamination of the environment in the zone of influence of the release source. Therefore, mathematical models of impurity propagation in the atmosphere and its deposition on the underlying surface are an important tool to obtain a first assessment of the radiation situation in the vicinity of the release source and to give a forecast of its development for the nearest period. They can and should be used to solve the following main practical problems:

(1) operational forecast of the release propagation and assessment of the amount of radioactive contamination of the environment for the organization of radiological reconnaissance during the early phase of the accident;
(2) reconstruction of release parameters using data from measurements on the ground by solving the inverse problem of radioactivity propagation in the atmosphere;
(3) reconstruction of deposition fields and dynamics of volumetric activity of short-lived nuclides in the initial period of the accident (which is especially important for retrospective dosimetry of the population of the affected regions).

Underlying surface data used for real time assessments of the consequences of a radiation accident should first of all answer two basic questions:

1. Which areas (in which direction and at what distance) are likely to be in the contamination zone and when will the radioactive release reach them?
2. What doses can be expected there, based on available estimates of the activity and nuclide composition of the release?

To answer the questions posed, the description of the main physical mechanisms determining the distribution of impurity in the atmosphere and its deposition on the underlying surface is detailed, as well as the choice of appropriate mathematical algorithms for this purpose. Therefore, to forecast the consequences of activity releases to the atmosphere in NPP emergencies, it is necessary to have a set of several computational models of radionuclide transport for different types of accidents and given spatial and temporal scales, which also take into account possible features of the release source, meteorological conditions of the release propagation and physical and geographical conditions of the NPP location territory.

At present, the regulatory documents developed by the IAEA and in Ukraine [78–81] are used to address these environmental problems.

We note the characteristic features of the use of the so-called IAEA Gaussian jet model (hereinafter referred to simply as the Gaussian model) recommended as a calculation tool for assessing and predicting the consequences of releases from NPPs [82–84].

The Gaussian model is an effective tool for estimating the radionuclide concentrations in the surface air and their deposition densities on the underlying surface of releases from high-altitude point sources (e.g. nuclear power plant vents). The wide applicability of this model is explained by the following factors:

- simplicity of the mathematical formulation of the model and, consequently, ease of its realization in the form of software;
- relatively small number of input parameters of the model. In particular, all input meteorological parameters can be obtained from the measurement data of a standard meteorological station of the Hydrometeorological Service network or estimated from these data;
- reliability of calculation results, confirmed by the data of many verifications of the model on the basis of special in-situ diffusion experiments or measurements of emissions from industrial enterprises.

However, such simplicity and availability of application of the Gaussian model has a downside: its use beyond the limits of applicability of the model, leading to a significant increase in the errors of calculated values up to incorrect determination of the location and size of the area of increased pollution. Therefore, it is necessary to have a clear understanding of the limits of applicability of the Gaussian model and the errors of estimates made with its help, both within and outside these limits.

The main limitations of the Gaussian model are:

(1) the model can be used at distances up to 10 km from the source (depending on the complexity of the relief). To assess the possibility of using the model at greater distances, a special study should be carried out in each specific case, taking into account the influence of physiographic features of the site on the dispersion of the impurity in the atmosphere;
(2) stationarity of meteorological conditions during the emission propagation period. At an average transport velocity of 3 m/s, a short-term emission will leave the limits of the calculated area with a radius of 10 km in about 1 h, and with a radius of 30 km—in 3 h. This condition is well enough fulfilled for most weather conditions, except for calculations in transitional periods (morning and evening hours; periods of sharp weather change associated with the passage of atmospheric fronts, etc.);
(3) horizontal homogeneity of the underlying surface. Within the NPP observation area, this condition can be considered to be fulfilled with an acceptable degree of accuracy for terrain without significant relief features and in the absence of large bodies of water. Their presence can significantly change the turbulent and wind regime of the lower layer of the atmosphere and, as a consequence, affect the distribution of the release in the air;
(4) horizontal homogeneity of meteorological conditions within the calculation area. To a large extent, this condition is related to the previous two;
(5) stationarity of the emission source. The Gaussian model is a stationary jet model and is applicable for modeling a long (continuous) emission of constant power, an emission of finite duration or a short-term ("instantaneous") emission. In the latter case, the model does not give the radionuclide concentration in air, but the concentration integrated over the time of the release.

The Gaussian model requires the following input information:

- measurement data from the nearest meteorological station to the emission source or values of quantities that can be derived from these data: surface wind speed and direction, stability category, mixing layer height, precipitation intensity;
- roughness parameter of the underlying surface; quantity or intensity of the release;
- emission characteristics (isotopic and physicochemical composition); dry deposition rate;
- and source height.

In addition to these "explicit" sources of uncertainty in the results of Gaussian model calculations, there are a number of "implicit" sources that may be much more important in their contribution to the errors in the Gaussian model estimates. The "implicit" sources of estimation errors in this case mean either the use of the Gaussian model actually outside its domain of applicability, or incorrect estimation of values of some input parameters of the model.

The main "implicit" sources of errors in the Gaussian model are: wind direction rotation with height in the boundary layer; wind direction fluctuations;

- horizontal inhomogeneity of the underlying surface;
- relief;
- non-stationarity of meteorological parameters;
- influence of the sampling period on the results of concentration measurements in the jet.

In the following subsections, models for the assessment of the consequences of atmospheric releases and NPPs under inhomogeneous and nonstationary radionuclide activity fields in air will be briefly discussed as alternatives when the Gaussian model is not applicable.

The conditions for the application of the Gaussian model were defined earlier. However, any significant deviation of real conditions from these hypotheses (rapidly changing weather conditions, unsteady release, the influence of terrain, etc.) leads to the need to develop other models that would allow taking into account their influence while maintaining the relative simplicity of the model.

Let's consider other models for estimation of consequences of atmospheric emissions of NPPs.

Semi-infinite source model. For estimates of the effective equivalent external dose rate from radionuclides contained in atmospheric air, the so-called semi-infinite source model is usually used, relating the activity of the radionuclide in surface air and the dose rate through the value of the external dose coefficient from the radioactive cloud. However, this approach is justified only at such distances from the source when the bulk activity field of the radionuclide near a given point can be considered homogeneous. At small distances from the source of a release to the atmosphere under conditions of high spatial gradients of radionuclide activity, the model of a homogeneous semi-infinite source may lead to large errors in the estimation of the external dose rate from a radioactive cloud.

General model of an arbitrarily shaped source. In general, the effective equivalent external dose rate H from an arbitrarily shaped radioactive cloud at a point with coordinates (x, y), near the ground surface can be calculated by the formula

$$\dot{H}(x, y) = \frac{K\mu_a \sum_i R_i(E_i) n_i E_i}{4\pi\rho w} \iiint_V \frac{A_V(x, y, z) B_D}{r^2} \exp(-\mu r) dV,$$

where E_i—energy of i-th photon; n_i—absolute yield in the decay scheme, photon/decay; $R_i(E_i)$—transition factor from absorbed dose in air to effective equivalent dose; $K = 1.602 \cdot 10^{-13}$—energy equivalent; μ_a and μ—linear coefficients of absorption and attenuation of energy in air, respectively; ρ—1.293 kg/m^3, air density at normal conditions; $w = 1$—energy equivalent of gray referred to the mass in 1 kg of air; B_D—dose accumulation factor in air, r—distance from the point (x, y) to the elementary volume dV.

Integral model. The leading scientists of Ukraine in the field of radiation protection and NPP safety, whose research results are published in numerous publications, including, have developed models for calculating the consequences of possible accidental releases at distances up to 30 km from the source, including non-uniform and rapidly changing activity fields of radionuclides in the air.

Let us describe the developed models in more detail.

The unsteady integral model of atmospheric transport is used in situations characterized by rapidly changing emission dynamics, as well as in conditions of meteorological fields changing during the period of impurity transport. A prolonged (continuous) unsteady release of radionuclides from a source is represented as a sequence of discrete releases of puffs. The total activity of each puff is calculated as the product of the source intensity at that moment by the duration of the time interval between two consecutive puffs.

The cloud propagation is considered in the Cartesian coordinate system with the origin at the location of the emission source. The horizontal trajectory of the cloud motion is represented by a sequence of points whose values are calculated from

$$x_{i+1} = x_i + U_x(t_i)\Delta t,\ y_{i+1} = y_i + U_y(t_i)\Delta t,$$

where U_x, U_y—components of wind speed at the moment of time t_i; Δt—time step of trajectory calculation. The total distance l traveled by the cloud during its propagation time t is equal to $l(t) = \sum_{i=1}^{N} \left|\vec{U}(t_i)\right| \Delta t$, $N = t/\Delta t$—the number of time steps in the calculation of the cloud trajectory.

The radial symmetry of each cloud in the horizontal plane is assumed, i.e. that the values of the mean-square deviation of the coordinates of impurity particles in the transport direction and in the perpendicular direction are the same and equal to σ_y. The values σ_y and σ_z (deviation of coordinates of impurity particles in the vertical direction) are calculated as functions of the path l, travelled by the cloud, according to the dependences, $\sigma_y = p_y \cdot l^{q^y} \sigma_z = p_z \cdot l^{q_z}$, where constants p_y, q_y, p_z and q_z are determined by the values of atmospheric stability category according to Pasquill, wind speed and effective emission height.

Then the value of the volume specific activity of radionuclide C_V in air determined by the spread of a single cloud is determined by the formula

$$C_V(x, y, z, t) = \frac{QF(l)}{(2\pi)^{3/2}\sigma_y^2(l)\sigma_z(l)} \exp\left[-\frac{(y - y_c(l))^2 + (x - x_c(l))^2}{2\sigma_y^2(l)}\right] \times \left\{\exp\left[-\frac{\left(z - h_{eff}\right)^2}{2\sigma_z^2(l)}\right] + \exp\left[-\frac{\left(z + h_{eff}\right)^2}{2\sigma_z^2(l)}\right]\right\}, \quad (9.2)$$

where Q—radionuclide activity in the cloud; x_c, y_c—coordinates of the cloud centre; h_{eff}—effective height of the release; $F(l)$—source depletion function, which takes into account the influence of the processes of dry deposition of radionuclides on the underlying surface, radionuclide leaching by atmospheric precipitation and radioactive decay.

Thus, the values of the volumetric (as a function of time) nuclide activity, as well as the sediment density to the earth's surface at a given point, can be calculated by summing the contributions of each cloud.

In the integral model, the external dose rate $\dot{H}$ from a radioactive cloud is calculated as the sum of the contributions from photon $\underset{ph}{\dot{H}}$ and from beta $\underset{r}{\dot{H}}$ radiations

$$\dot{H} = \underset{ph}{\dot{H}} + \underset{r}{\dot{H}}.$$

The effective equivalent external dose rate from photon radiation $\underset{ph}{\dot{H}}$ near the earth's surface from a cloud propagating in the atmosphere, in which the activity is distributed according to (9.2), is determined from the formula

$$\underset{ph}{\dot{H}}(x, y) = \eta \frac{K_{\mu_a} \sum\limits_i R_i(E_i) n_i E_i}{4\pi \rho w} F(l)$$
$$\cdot Q \int\limits_0^\infty \left[erfc\left(\frac{\mu}{2\sqrt{s}} \right) + \frac{k\mu}{\sqrt{\pi s}} \exp\left(-\frac{\mu^2}{4s} \right) \right]$$
$$\times \exp(-\phi(x, y, s)) ds.$$

In it the function $\phi(x, y, s)$ is equal to

$$\phi(x, y, s) = \frac{(x - x_c)^2 s}{1 + 2s\sigma_y^2(l)} + \frac{(y - y_c)^2 s}{1 + 2s\sigma_y^2(l)}$$
$$+ \ln\left[1 + 2s\sigma_y^2(y)\right] + \psi(s),$$

and the function $\psi(s)$ is defined by the expression

$$\psi(s) = \begin{cases} 1 + \frac{h_{eff}^2 s}{2s\sigma_z^2(l)} + \frac{1}{2}\ln\left[1 + 2s\sigma_z^2(l)\right], \text{ at } l \le l_{crit}; \\ \ln\left(2H_m\sqrt{\frac{s}{\pi}}\right), \text{ at } l \ge 2l_{crit}. \end{cases}$$

In the above formulas η is a coefficient determining the relative contribution of photon radiation to the external dose factor from a radioactive cloud B_a; $erfc(x)$ is an additional integral of probabilities; $k = (\mu - \mu_a)/\mu$—coefficient in the dose factor of accumulation in air, which is given as $B_D(E, r) = 1 + k(E)\mu(E)r$.

In the intermediate interval $l_{crit} < l \leq 2l_{crit}$ the value of $\psi(s)$ is determined by linear interpolation from l.

Here are the results of test calculations obtained for the above models:

- of an arbitrarily shaped source;
- of a semi-infinite source;
- of the integral model, which are of the greatest practical interest for environmental monitoring tasks.

Test calculations of the external dose rate using the three models mentioned above were carried out. The purpose of such calculations was to compare the results obtained by different models and to determine the areas of applicability of each of them. It is known that the arbitrarily shaped source model gives the most accurate value of external dose, so the results obtained using this model were used to assess the accuracy of the results of calculations using the semi-infinite and integral models.

To test dose rate models, the propagation of a single cloud with activity reserve $Q = 10^{10}$ Bq under steady-state meteorological conditions was calculated. Four sets of atmospheric stability meteorological conditions were considered: (1) wind speed $U = 1\ \mathrm{m\,s^{-1}}$, stability category A; (2) wind speed $U = 3\ \mathrm{m\,s^{-1}}$, stability category C; (3) wind speed, stability category $U = 5\ \mathrm{m\,s^{-1}}D$; 4) wind speed, stability category $U = 2\ \mathrm{m\,s^{-1}}F$. Calculations were performed for source heights of $h_{eff} = 10$ and 150 m, and transport distances equal to $x_0 = 100$, 500, 1000 and 3000 m. At the same time, for each x_0 the dose rate dependence on the distance to the cloud centre was calculated in the interval $0 \leq y_0 \leq 10\sigma_y(x_0)$.

Test calculations showed that for all used parameter sets the dose rate values obtained by the integral model and the arbitrarily shaped source model do not differ by more than 5%. At the same time, the calculation time of the integral model is several orders of magnitude shorter than that of the arbitrarily shaped source model. Thus, testing has shown that the integral model provides an estimate of external dose with high accuracy in a wide range of problem parameters and is suitable for use in operational calculations of the consequences of radioactive releases into the atmosphere.

In some results of comparison of calculations by the integral model and the semi-infinite source model are presented. At a distance of 100 m from the emission source, the semi-infinite source model overestimates the dose rate value by 2–4 times in the centre of the cloud for ^{133}Xe and by 6 to 13 times for ^{132}I at an effective source height of 10 m. With increasing distance from the cloud centre, the dose rate values decrease according to the Gaussian distribution of activity in the cloud: at the distance $y_0 = 3\sigma_y$ the dose rate decreases by 2 times, and for $y_0 = 6\sigma_y$—by 8 times. Note that in this case the σ_y value is 36 m for category A, 22 m for category D and 78 m for category F. According to the integral model, the dose rate decreases by only one order of magnitude at the distance $y_0 = 3\sigma_y$ and by two orders of magnitude at the distance $y_0 = 6\sigma_y$ from the centre of the cloud. Similar results are obtained for other nuclides. Consequently, the traditionally used semi-infinite source model underestimates the dose rate at distances $6\sigma_y$ from the cloud centre by two to five orders of magnitude. Thus, in the case of a high activity release, the use of the semi-infinite model leads to a significant reduction in the value of the horizontal dimensions of the zone of exposure to radiation from a radioactive cloud propagating in the atmosphere.

The estimation of the dependence of the dose rate on the absolute distance to the cloud centre for the same calculations is considered. For category D, the semi-infinite source model underestimates the dose rate by more than two orders of magnitude at a distance of 100 m from the cloud centre. For category A, a discrepancy of more than one order of magnitude is found at distances greater than 130 m. Since for category F the σ_y value has the highest value, the discrepancy between the results of both models starts at distances greater than 300 m, the absolute values of the dose rate are significantly smaller compared to the centre of the cloud.

Even greater differences between the results are observed with increasing height of the emission source h_{eff}. The results of the calculations for $h_{eff} = 150$ m are shown. Even in the centre of the cloud, the semi-infinite source model underestimates the dose rate by eight orders of magnitude for stability category A. For the other categories, the dose rates from the semi-infinite model are even smaller. Thus, the semi-infinite source model is fundamentally unsuitable for estimating dose rates from a high-altitude source at transport distances comparable to the height of the source.

According to the results of test calculations obtained by computer modelling of the external dose rate from a radioactive cloud in the air using the three proposed models, the following conclusions were made.

The proposed set of models is suitable for estimations of radioactive release propagation in the atmosphere at distances up to 30 km from the source for unsteady meteorological conditions of transport and arbitrary release dynamics. It allows to calculate the volumetric and surface activity of radionuclides, as well as the external dose rate from radioactive releases into the atmosphere.

It is shown that the proposed integral model for calculating the external dose rate from a radioactive cloud propagating in the atmosphere allows to significantly refine the dose estimate, especially at distances comparable to the height of the release source. The error of the model results does not exceed 5% compared to the general expression for a source of arbitrary shape.

9.7 Mathematical Model of the Atmospheric Transport of Radionuclides for the Measuring Their Volumetric Activity by a Monitoring System on the Basis of an Unmanned Aerial System

Let us focus on more specific models of environmental monitoring tasks of power engineering facilities [2, 6, 7, 15, 85].

Basic positions and coordinate systems. A point source of radionuclide release that is located at some height above the Earth's surface is considered. The coordinate systems, as noted earlier in paragraph 5 of this chapter, are chosen as follows.

Fixed Cartesian rectangular coordinate system (*OXYZ*), in which the *OZ* axis is directed upwards (in the direction of the local vertical), the *OX* axis—to the east, and the *OY* axis—to the north, the origin of the coordinate system t. *O* corresponds to the location of the emission source on the *XOY* plane.

Simultaneously with the coordinate system *OXYZ*, an additional "mobile" coordinate system *Oxyz is* introduced, the origin of which coincides with the above-mentioned point *O*, the *Ox* axis is directed along the wind direction, the *Oz axis is* directed upwards (along the local vertical direction), and the *Oy* axis is directed so that the coordinate system is right-handed.

$$0 \leq x \leq L_x,\ 0 \leq y \leq L_y,\ \delta \leq z \leq z_m,$$

where L_x, L_y—horizontal dimensions of the area, function $z = \delta(z, y)$ describes terrain relief (roughness), z_m—position of the upper boundary of the considered area.

The study area is defined as a parallelepiped with an irregular lower boundary that reflects the inhomogeneity of the underlying surface topography.

The transport of radioactive impurities and aerosols, further radionuclides, in the atmosphere is carried out by wind-driven air flows taking into account its small-scale fluctuations. The averaged flux has advective and convective components, and the averaging fluctuation motion can be interpreted as diffusion against the background of the associated main averaged motion.

Let us formulate the problem of radionuclide transport in the atmosphere in a general form.

Let $C(x, y, z, t)$ be the concentration in the atmosphere of a radioactive aerosol that moves in the air stream with some velocity $\vec{U} = u\vec{i} + v\vec{j} + w\vec{k}$. This velocity is some function of coordinates x, y, z and time t; $\vec{i}, \vec{j}, \vec{k}$ are unit orthogonal to the axes of the coordinate system OXYZ, respectively; u, v, w are projections of the velocity vector $\vec{U}(x, y, z, t)$ on the axes of the coordinate system.

Models of radionuclide propagation in the atmosphere. The process of transport of radionuclides together with the air atmospheric flux can be described in the form of an equation representing the equality to zero of the full derivative of the concentration $C(x, y, z, t)$ of the impurity:

$$\frac{dC(x, y, z, t)}{dt} = 0,$$

taking into account that $\frac{\partial x}{\partial t} = u$, $\frac{\partial y}{\partial t} = v$, $\frac{\partial z}{\partial t} = w$, we get:

$$\frac{\partial C(x, y, z, t)}{\partial t} + \frac{\partial C(x, y, z, t)}{\partial x} + v\frac{\partial C(x, y, z, t)}{\partial y} + w\frac{\partial C(x, y, z, t)}{\partial z} = 0.$$

Since the law of conservation of mass defined by the continuity equation is well fulfilled for the lower atmosphere, we obtain:

$$\frac{\partial u}{\partial x} + \frac{\partial v}{\partial y} + \frac{\partial w}{\partial z} = 0, div\left(\vec{U}\right) = 0.$$

Thus:

$$\frac{\partial C(x, y, z, t)}{\partial t} + div\left(\vec{U} C(x, y, z, t)\right) = 0. \quad (9.3)$$

Equation (9.3) for the case of radionuclide propagation can be made more specific by considering that part of the aerosol reacts with the external environment and decays with a time constant $\tau = \frac{1}{\alpha}$, and by considering the source of contamination described by some function $f(x, y, z, t)$. In this case (9.4) can be transformed to the form:

$$\frac{\partial C(x, y, z, t)}{\partial t} + div\left(\vec{U} C(x, y, z, t)\right) + \alpha C(x, y, z, t) = f(x, y, z, t). \quad (9.4)$$

The exact solution of Eq. (9.4), which is a nonlinear partial differential equation, can be obtained by numerical methods for the case when the values of the function $\vec{U} = u\vec{i} + v\vec{j} + w\vec{k}$ at given points of space and at specific moments of time are known. If the information about the components of the velocity vector is insufficient, it is reasonable to use various approximations and assumptions [6, 17].

To describe the process of radionuclide diffusion in the atmosphere—their transport by the wind flow taking into account gravitational deposition of such impurities, the semiempirical equations of turbulent diffusion for the surface air layer are widely used as one of the concretised representations of expression (9.3). Let us give the form of recording of such equation at a choice of the mobile coordinate system *Oxyz*. The direction of the *Ox* axis is chosen according to the direction of the velocity vector of the acting wind: $\vec{U} = u\vec{i} + 0\vec{j} + w\vec{k}$. For such conditions we obtain:

$$\begin{aligned} &\frac{\partial C(x, y, z, t)}{\partial t} + u\frac{\partial C(x, y, z, t)}{\partial x} + w\frac{\partial C(x, y, z, t)}{\partial z} \\ &= \frac{\partial}{\partial x}\left(k_x \frac{\partial C(x, y, z, t)}{\partial x}\right) + \frac{\partial}{\partial y}\left(k_y \frac{\partial C(x, y, z, t)}{\partial y}\right) \\ &+ \frac{\partial}{\partial z}\left(k_z \frac{\partial C(x, y, z, t)}{\partial z}\right) + \alpha C(x, y, z, t), \end{aligned} \quad (9.5)$$

where $k_x(t), k_y(t), k_z(t)$—turbulent diffusion coefficients along the x, y, z axes, respectively; u—wind speed projection on the *Ox axis*; $w = w_w + w_g$—wind speed projection on the *Oz axis*, w_w—vertical component of wind speed (e.g. due to convection flows), w_g—radionuclide deposition rate; $\alpha C(x, y, z, t)$—component characterised by half-life.

Equation (9.5) is formulated for a part of the space $0 \le x \le L_x$, $0 \le y \le L_y$, $\delta \le z \le z_m$ and on the boundary $z = \delta$ the boundary conditions $C(x, y, z, t)$ are set. Such boundary conditions taking into account gravitational deposition of radionuclides can be concretised in the following form:

$$\left(k_z\frac{\partial C}{\partial z} + w_g C = \beta C\right)\Big|_{z=\delta}. \quad (9.6)$$

Here β is some constant that has the dimension of velocity, and in (9.6) there is a vertical flux component due to turbulent diffusion $k_z\frac{\partial C}{\partial z}$, as well as a component due to gravitational deposition $w_g C$.

Analytical solution of Eq. (9.5) is possible for the case of certain assumptions concerning its coefficients. Thus, for the case of constant coefficients and under the condition of the emission source located at a certain height, as well as under the boundary condition $\beta = 0$ at the level of $z = \delta(z = 0)$ (complete reflection of particles from the surface), the analytical solution of Eq. (9.5) has the following form:

$$C(x, y, z, t) = \frac{c(t)}{4\pi x\sqrt{k_y k_z}}\exp\left(\frac{uy^2}{4k_y x}\right)\left(e^{\left[-\frac{u(z-h)^2}{4k_z x}\right]} + e^{\left[-\frac{u(z+h)^2}{4k_z x}\right]}\right), \quad (9.7)$$

where $ct = C(0, 0, h, t)$—radionuclide concentration at the outlet of the contamination source with coordinates $(x = 0,\ y = 0,\ z = h)$. We neglect the component of the equation $\frac{\partial}{\partial x}\left(k_x\frac{\partial C(x,y,z,t)}{\partial x}\right)$, describing the variability of diffusion.

Of greatest practical interest is the case where it is necessary to analyse the radionuclide concentration distribution from the source of the release, in the direction of the wind operating in the atmosphere. Then the solution of (9.7), the concentration distribution along the axis of the mobile coordinate system *Oh,* will take the following form:

$$C(x, 0, 0, t) = \frac{c}{2\pi x\sqrt{k_y k_z}}e^{-\frac{uh^2}{4k_z x}}. \quad (9.8)$$

In this case, the distribution of aerosols in transverse directions to the direction of the acting wind (along the coordinate y) in expression (9.7) has a normal distribution with standard deviation $\left(u/2k_y x\right)^{-1/2}$. Using expression (9.8) it is possible to estimate the value of the largest concentration of radioactive aerosol $C_{\max}$, which is located at the distance $x_{C_{\max}} = uh^2/4k_z$:

$$C_{\max} = \frac{2c}{\pi euh^2}\sqrt{\frac{k_z}{k_y}}.$$

In the case of complete radionuclide absorption by the Earth's surface (which corresponds to the boundary condition (9.6) at $\beta = \infty$), the expression (9.7) will take the following form:

$$C(x, y, z, t)|_{\beta=\infty} = \frac{c(t)}{4\pi x\sqrt{k_y k_z}}\exp\left(-\frac{uy^2}{4k_y x}\right)\left(e^{\left[-\frac{u(z-h)^2}{4k_z x}\right]} - e^{\left[-\frac{u(z+h)^2}{4k_z x}\right]}\right).$$

Using the expression (9.7) it is also possible to determine the deposition rate of radionuclides $\sigma(x, y)$, which is contained in the air stream:

$$\sigma(x, y) = \left(k_z \frac{\partial C(x, y, z)}{\partial z}\right)_{z=\delta=0} = \frac{cuh}{4\pi x^2 \sqrt{k_y k_z}} e^{-\frac{uy^2}{4k_y x} - \frac{uh^2}{4k_z x}}. \tag{9.9}$$

Function (9.9) characterises the pollution trace on the Earth surface. At that, the maximum deposition rate is reached along the sym axis of the trace at the distance $x_m = \frac{uh^2}{8k_z}$ from the emission source and is as follows

$$\sigma_m = \frac{4k_z}{\pi e^2 uh^3} \sqrt{\frac{k_z}{k_y}}.$$

Expressions (9.7–9.9) are usually supplemented by a number of practically confirmed relations for estimation and determination of parameters of turbulent diffusion and deposition in the atmosphere of radionuclides. For example, it was proposed: at, $u = const k_y = const$ to define the turbulent diffusion coefficient in the form $k_z = k(z/z_1)^{1-1/p}$, where p is a parameter that characterises the thermal stability of the atmosphere (for example, for the case of convection $p < 0$), and at the height value $z = 1$ to take $k_1 = \frac{u}{40}$.

In practice, it is not always convenient to use the obtained expressions, which are derived with reference to rectangular Cartesian coordinate systems. This is due to the fact that it is necessary to accurately determine the relative coordinates of the emission source and measurement location, as well as the subsequent transfer of the obtained results to absolute geographic coordinates. Therefore, it seems reasonable to supplement expressions (9.6–9.8) with similar expressions in the curvilinear cylindrical coordinate system. In this case, we will deal only with the direction to and distance from the source of the release, which greatly facilitates the procedure of remote analysis of the distribution of radioactive aerosols.

For this purpose, we introduce a curvilinear cylindrical coordinate system (ρ, ϕ, z), the origin of which is, as before, point O—the point of location of the emission source.

For the case of application of the cylindrical coordinate system, the basic Eq. (9.5) will take the following form:

$$\frac{\partial C(\rho, \phi, z, t)}{\partial t} + V_\rho \frac{\partial C(\rho, \phi, z, t)}{\partial \rho} + V_\phi \frac{\partial C(\rho, \phi, z, t)}{\partial \phi}$$
$$+ V_z \frac{\partial C(\rho, \phi, z, t)}{\partial z} = \frac{\partial}{\partial \rho}\left(k_\rho \frac{\partial C(\rho, \phi, z, t)}{\partial \rho}\right) + \frac{1}{\rho^2}\frac{\partial}{\partial \phi}\left(k_\phi \frac{\partial C(\rho, \phi, z, t)}{\partial \phi}\right)$$
$$+ \frac{\partial}{\partial z}\left(k_z \frac{\partial C(\rho, \phi, z, t)}{\partial z}\right) + \alpha C(\rho, \phi, z, t), \tag{9.10}$$

where $V_\rho = u\cos\varphi + v\sin\varphi$, $V_\varphi = u\sin\varphi + v\cos\varphi$, $V_z = w$—projections of the velocity vector in the cylindrical coordinate system; k_ρ, k_ϕ, k_z are the corresponding turbulent diffusion coefficients.

The solution of the last equation allows to obtain new analytical expressions for determining the radionuclide concentration in the zone of the release centre using known radionuclide concentrations, wind speed and direction at some distance from the release site.

The construction of radiation safety monitoring systems is based on the measurement of radioactive contamination parameters by means of a predetermined number of measurements or observation time. Such systems have proved to be good for analysing contamination levels during normal operation of sources, but they have a significant disadvantage, since all information on the radiation situation refers to the elapsed time, which is inadmissible in emergency situations.

The determination of activity during a radiation accident can be solved using modern technologies and techniques, among which the use of unmanned aerial systems (UAS) is proposed [85–87].

The sampling of radioactive aerosols above or near the source of a release can be achieved by installing a filter ejection device (FED) on board an unmanned aerial vehicle (UAV). The concentration of harmful contaminants contained at a particular point in the Earth's atmosphere can be measured using the FED.

The performance of the FED we created as a function of its airspeed [88, 89] is shown in Fig. 9.3.

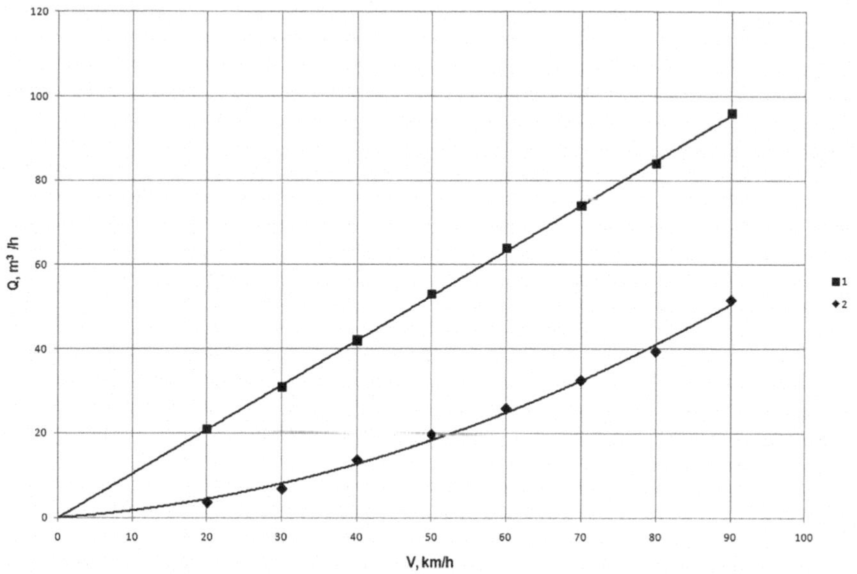

Fig. 9.3 FED productivity Q as a function of air velocity V

In Fig. 9.2, the squares indicate volumes for given velocities through the cross-section of the FED, and the dependence passing through the rhombics (number 2 on the graph) is its calibration characteristic. It follows from the graph that, for example, at the UAV air speed 70 km/h, 32 m^3 of air will pass through the filter fabric of the FED during one hour of flight. In two minutes of flight we have about 1 m^3, which allows to measure the bulk activity ^{137}Cs with a value of 1 Bq/m^3. It was shown that during a forest fire in the Chernobyl zone, the average volume activity of ^{137}Cs during the fire period reached 18.5 Bq/m^3.

The UAV is also equipped with a navigation system and a flight control system, which provide the flight control centre with all the necessary flight and navigation information. Thus, considering the UAV as a tool for pollution monitoring it is possible to determine the concentration of harmful impurities $C(x, y, z)$, and using its flight data to determine the values of speed and wind direction at the UAV flight altitude relative to the reference coordinate system.

Thus, by the example of flying around a possible source of harmful aerosols, it is possible to determine their concentration directly at the exit from the source of pollution and to establish a picture of their distribution in the surface layers of the atmosphere. In this case, the wind direction and its speed are determined on the basis of flight and navigation parameters of the UAV flight from the following relations:

$$K_w = arc\sin\left(\frac{V}{U}\sin(K_\varphi)\right), \vec{U} = \sqrt{V^2 + W^2 - 2VW\cos\left(K_\varphi\right)}, \tag{9.11}$$

where K_w—wind angle relative to the north direction, V—UAV airspeed, U—wind speed value; W—UAV speed relative to the ground surface (V and W are measured by the aircraft's onboard navigation system), K_φ—UAV drift angle, determined as the difference between the heading angle and the UAV heading angle (based on data from the aircraft's onboard navigation system).

From the above relations we solve the "direct" problem of determining the distribution of harmful impurities in the atmosphere—the case when it is possible to determine the concentration of harmful impurities directly at the source of emission.

For the case when there is no possibility of direct measurement near the emission source, the "reverse" problem is solved—when concentration measurements are made at some distance from the location of the emission source. In this case, on the basis of UAV flight data and relations (9.11) the wind direction and speed are determined, and the concentration of harmful impurity in this point of space is measured with the help of FED. At the same time it is possible to determine the concentration of impurities at the outlet of the remote source.

$$c = C(\rho, \varphi, 0)2\pi\rho\cos\varphi\sqrt{k_\varphi k_z}\left[e^{\frac{V_\rho h^2}{4k_z\rho\cos\varphi}}\right]. \tag{9.12}$$

Here, for a known distance to the source, the concentration of the harmful impurity at the source outlet is determined along the wind direction.

For the case when the location of the emission source does not lie on the wind line, by solving Eq. (9.10) we obtain the following relation:

$$c = C(\rho, \varphi, z)\, 4\pi\rho\cos\varphi\sqrt{k_\varphi k_z}$$
$$\times\left[\exp\left(-\frac{V_\rho(\rho\sin\varphi)^2}{4k_\varphi\rho\cos\varphi}\right)\left(e^{\left[-\frac{V_\rho(H-h)^2}{4k_z\rho\cos\varphi}\right]}+e^{\left[-\frac{V_\rho(H+h)^2}{4k_z\cos\varphi}\right]}\right)\right]^{-1}. \quad (9.13)$$

Here ρ, φ, z—coordinates of the UAV measurement location relative to the position of the emission source (distance to the object angle, flight altitude).

Expressions (9.12) and (9.13), which are derived from the solution of Eq. (9.10), allow estimating the concentration of radioactive aerosols in the area of the release source. The initial data are: the value of the local measured aerosol concentration and the coordinates of the sampling location relative to the aerosol release centre.

Results of mathematical modelling. Mathematical modelling of solutions of direct and inverse problems of determining the parameters of radioactive aerosols propagation carried by the air stream has been carried out using the obtained relations.

Thus, to solve the direct problem of determining the concentration of radioactive aerosols, provided that the source of the release, for example, is at a height of 130 m with an intensity of 1000 Bq/m^3 at a wind speed of 5 m/s; on the basis of (9.6), we obtain the following distribution diagram as one moves away from the source of the release at 70 m height. Figure 9.4 shows the distribution of radioactive aerosol along the wind direction for 10 km from the source of the release.

By carrying out mathematical modelling of the expressions defining the concentration of radioactive aerosols, its three-dimensional distribution can be obtained, as illustrated by Fig. 9.5.

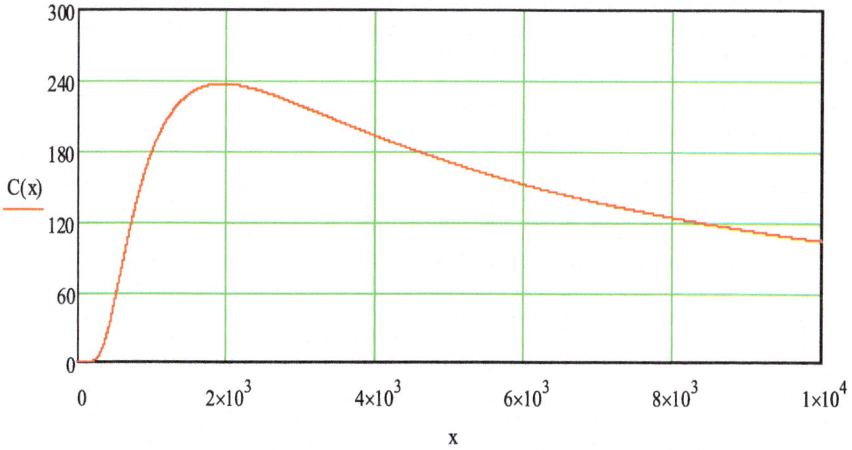

Fig. 9.4 Concentration distribution along the wind direction

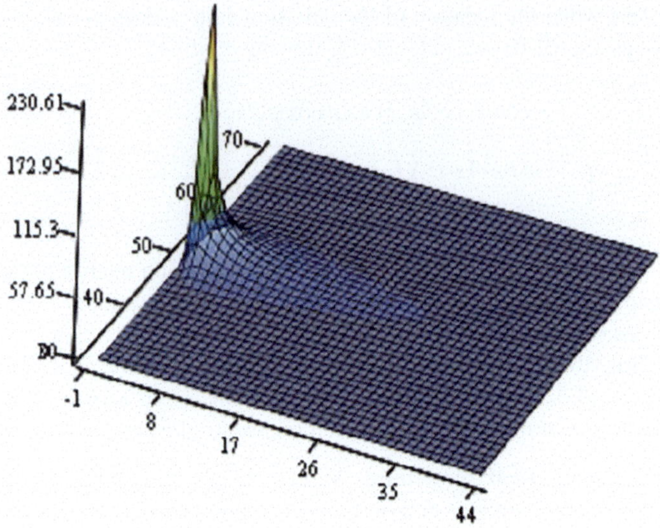

Fig. 9.5 Spatial distribution of the concentration of radioactive aerosol concentration in the release area

Of greatest practical interest are the diagrams of the level lines corresponding to certain concentrations of radioactive aerosol (Fig. 9.6), which are also the result of solving the mathematical model (9.10).

Thus, the given mathematical model, its solutions, the methodology of in-situ experimentation, and the results of mathematical modelling are a rather convenient approach to the analysis of radioactive aerosol propagation.

On the basis of the analysis of mathematical models of radioactive aerosol propagation in the surface layer of the atmosphere, a number of their solutions for particular cases of aerosol propagation have been established.

The established relations make it possible to determine the degree of pollution of the airspace area near the emission sources.

As a result of the conducted research, new relations have been derived and a new methodology has been developed, which can be used for remote analysis of emission sources using the obtained relations (9.12) and (9.13) for solving such a problem, as well as a modern means of diagnostics and environmental monitoring—UAV, which is equipped with a filter-ejection device and an appropriate navigation system (see Chap. 10).

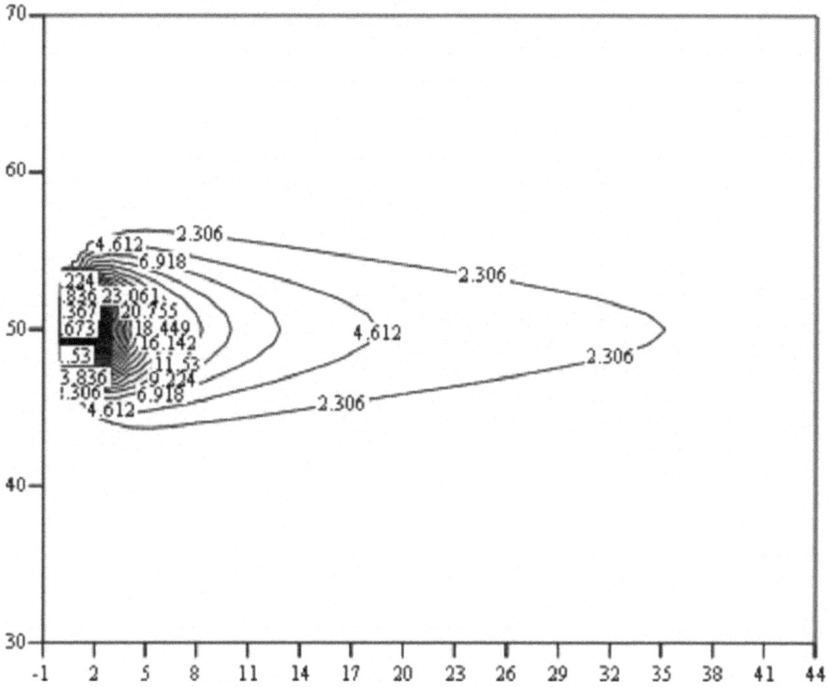

Fig. 9.6 Lines of radioactive aerosol concentration levels

Acknowledgment These researches have been performed within the scientific grant "Information Technology for Energy Auditof Buildings as a Component of the Energy Security of the Country" (0123U103703), provided by the National Research Foundation of Ukraine (program "Science for the Recovery of Ukraine in the War and Post-War Periods"). The authors thank the National Research Fund for the opportunity to conduct research that will contribute to improving the state's energy security.

References

1. Sulich, A., Sołoducho-Pelc, L.: Changes in energy sector strategies: a literature review. Energies **15**(19), 7068 (2022)
2. Babak, V., Zaporozhets, A., Zvaritch, V., Scherbak, L., Myslovych, M., Kuts, Y.: Models and measures in theory and practice of manufacturing processes. IFAC-PapersOnLine **55**(10), 1956–1961 (2022)
3. Zaporozhets, A.O.: Correlation analysis between the components of energy balance and pollutant emissions. Water Air Soil Pollut. **232**, 1–22 (2021)
4. Lah, O.: Factors of Change: The influence of policy environment factors on climate change mitigation strategies in the transport sector. Transp. Res. Procedia **25**, 3495–3510 (2017)

5. Babak, V., Mokiychuk, V., Zaporozhets, A., Redko, O.: Improving the efficiency of fuel combustion with regard to the uncertainty of measuring oxygen concentration. Eastern Eur. J. Enterp. Technol. **6**(8), 84 (2016)
6. Babak, V.P., Babak, S.V., Eremenko, V.S., Kuts, Y.V., Myslovych, M.V., Scherbak, L.M., Zaporozhets, A.O.: Models and measures for atmospheric pollution monitoring. In: Models and Measures in Measurements and Monitoring, pp. 227–266 (2021)
7. Babak, V.P., Babak, S.V., Eremenko, V.S., Kuts, Y.V., Myslovych, M.V., Scherbak, L.M., Zaporozhets, A.O.: Models of measuring signals and fields. In: Models and Measures in Measurements and Monitoring, pp. 33–59 (2021)
8. Ionut, N.: Electromagnetic pollution in urban areas. In: 2014 International Conference and Exposition on Electrical and Power Engineering (EPE), pp. 565–570. IEEE (2014, October)
9. Liu, T., Xie, Y., Su, S., Wang, L., Song, Y., Chen, Y., Xiang, J., et al.: Synergistic removal effects of ultralow emission air pollution control devices on trace elements in a coal-fired power plant. Energy Fuels **36**(5), 2474–2487 (2022)
10. Maliarenko, O., Ivanenko, N., Maistrenko, N., Teslenko, O., Zaporozhets, A.: Forecasted petroleum products consumption in ukraine and estimated greenhouse gas emissions from their use. In: Modern Technologies in Energy and Transport, pp. 113–135. Springer Nature Switzerland, Cham (2023)
11. Babak, V.P., Babak, S.V., Myslovych, M.V., Zaporozhets, A.O., Zvaritch, V.M.: Methods and models for information data analysis. In: Diagnostic Systems for Energy Equipments, pp. 23–70 (2020)
12. Babak, V.P., Babak, S.V., Eremenko, V.S., Kuts, Y.V., Myslovych, M.V., Scherbak, L.M., Zaporozhets, A.O.: Problems and features of measurements. In: Models and Measures in Measurements and Monitoring, pp. 1–31 (2021)
13. Dunn, P.F., Davis, M.P.: Measurement and Data Analysis for Engineering and Science. CRC Press (2017)
14. Stogniy, B.S., Kurylenko, O.O., Pavlovs'kyj, V.V., Sopel, M.F., Steljuk, A.O., Luk'janenko, L.M.: Development of emergency automatics for power system with significant renewable generation. Sci. Innov. **12**(№ 4), 24–28 (2016)
15. Zaporozhets, A., Babak, V., Popov, O., Scherbak, L., Kuts, Y.: Methodological support of air pollution monitoring system. In: Systems, Decision and Control in Energy V, pp. 713–737. Springer Nature Switzerland, Cham (2023)
16. Blaabjerg, F., Yang, Y., Yang, D., Wang, X.: Distributed power-generation systems and protection. Proc. IEEE **105**(7), 1311–1331 (2017)
17. Zaporozhets, A.O., Khaidurov, V.V.: Mathematical models of inverse problems for finding the main characteristics of air pollution sources. Water Air Soil Pollut. **231**(12), 563 (2020)
18. Fang, S., Da Xu, L., Zhu, Y., Ahati, J., Pei, H., Yan, J., Liu, Z.: An integrated system for regional environmental monitoring and management based on internet of things. IEEE Trans. Industr. Inf. **10**(2), 1596–1605 (2014)
19. Sanada, Y., Torii, T.: Aerial radiation monitoring around the Fukushima Dai-ichi nuclear power plant using an unmanned helicopter. J. Environ. Radioact. **139**, 294–299 (2015)
20. Kaneko, M., Iwata, H., Shiotsu, H., Masaki, S., Kawamoto, Y., Yamasaki, S., Utsunomiya, S., et al.: Radioactive Cs in the severely contaminated soils near the Fukushima Daiichi nuclear power plant. Front. Energy Res. **3**, 37 (2015)
21. Qin, L., Yu, Q., Huang, Y., Zhang, L., Yan, X., Wu, W., Fan, H., et al.: A novel fluorescent sensor with an overtone peak reference for highly sensitive detection of mercury (II) ions and hydrogen sulfide: mechanisms and applications in environmental monitoring and bioanalysis. Anal. Chim. Acta **1287**, 342086 (2024)
22. Raj, D., Maiti, S.K.: Risk assessment of potentially toxic elements in soils and vegetables around coal-fired thermal power plant: a case study of Dhanbad, India. Environ. Monitor. Assess. **192**(11), 699 (2020)
23. Zaporozhets, A.O., Sverdlova, A.D., Ivaschenko, T.G., Kovach, V.O., Artemchuk, V.O.: Electronic circuits of measuring modules of air pollution monitoring system based on low-cost sensors. In: IOP Conference Series: Earth and Environmental Science, vol. 1049, no. 1, p. 012016. IOP Publishing (2022, June)

24. Babak, V.P., Babak, S.V., Eremenko, V.S., Kuts, Y.V., Myslovych, M.V., Scherbak, L.M., Zaporozhets, A.O.: Monitoring the air pollution with UAVs. In: Models and Measures in Measurements and Monitoring, pp. 191–225 (2021)
25. Babak, S., Zaporozhets, A., Gryb, O., Karpaliuk, I.: Analysis of UAVs and their technical parameters for overhead power lines monitoring. In: Systems, Decision and Control in Energy IV: Volume I. Modern Power Systems and Clean Energy, pp. 181–193. Springer Nature Switzerland, Cham (2023)
26. Asadzadeh, S., de Oliveira, W.J., de Souza Filho, C.R.: UAV-based remote sensing for the petroleum industry and environmental monitoring: State-of-the-art and perspectives. J. Petrol. Sci. Eng. **208**, 109633 (2022)
27. Manfreda, S., McCabe, M.F., Miller, P.E., Lucas, R., Pajuelo Madrigal, V., Mallinis, G., Toth, B., et al.: On the use of unmanned aerial systems for environmental monitoring. Rem. Sens. **10**(4), 641 (2018)
28. Green, D.R., Hagon, J.J., Gómez, C., Gregory, B.J.: Using low-cost UAVs for environmental monitoring, mapping, and modelling: examples from the coastal zone. In: Coastal Management, pp. 465–501. Academic Press (2019)
29. Babak, V.P., Babak, S.V., Eremenko, V.S., Kuts, Y.V., Myslovych, M.V., Scherbak, L.M., Zaporozhets, A.O.: Examples of using models and measures on the circle. In: Models and Measures in Measurements and Monitoring, pp. 127–156 (2021)
30. Babak, V.P., Babak, S.V., Eremenko, V.S., Kuts, Y.V., Myslovych, M.V., Scherbak, L.M., Zaporozhets, A.O.: Models and measures for measuring random angular quantities. In: Models and Measures in Measurements and Monitoring, pp. 61–97 (2021)
31. Babak, V., Zaporozhets, A., Kovtun, S., Myslovych, M., Kuts, Y., Scherbak, L.: Information support for identification of the technical state of electric power facilities. In: Systems, Decision and Control in Energy V, pp. 129–153. Springer Nature Switzerland, Cham (2023)
32. Graafland, J., Smid, H.: Environmental impacts of SMEs and the effects of formal management tools: evidence from EU's largest survey. Corp. Soc. Responsib. Environ. Manag. **23**(5), 297–307 (2016)
33. Schaltegger, S., Burritt, R.: Contemporary Environmental Accounting: Issues, Concepts and Practice. Routledge (2017)
34. McKinley, D.C., Miller-Rushing, A.J., Ballard, H.L., Bonney, R., Brown, H., Cook-Patton, S.C., Soukup, M.A., et al.: Citizen science can improve conservation science, natural resource management, and environmental protection. Biol. Cons. **208**, 15–28 (2017)
35. Babak, V., Zaporozhets, A., Kovtun, S., Kuts, Y., Fryz, M., Scherbak, L.: Information provision for monitoring the current state of electric power facilities. In: Advanced System Development Technologies I, pp. 283–309. Springer Nature Switzerland, Cham (2023)
36. Lee, J., Kao, H.A., Yang, S.: Service innovation and smart analytics for industry 4.0 and big data environment. Procedia Cirp. **16**, 3–8 (2014)
37. Zaporozhets, A., Babak, V., Sverdlova, A., Isaienko, V., Babikova, K.: Development of a system for diagnosing heat power equipment based on IEEE 802.11 s. In: Systems, Decision and Control in Energy II, pp. 141–151. Springer International Publishing, Cham (2021)
38. Rainer, R.K., Prince, B., Sánchez-Rodríguez, C., Splettstoesser-Hogeterp, I., Ebrahimi, S.: Introduction to Information Systems. Wiley (2020)
39. Zheng, P., Wang, H., Sang, Z., Zhong, R.Y., Liu, Y., Liu, C., Xu, X., et al.: Smart manufacturing systems for Industry 4.0: conceptual framework, scenarios, and future perspectives. Front. Mech. Eng. **13**, 137–150 (2018)
40. Ismagilova, E., Hughes, L., Dwivedi, Y.K., Raman, K.R.: Smart cities: advances in research—an information systems perspective. Int. J. Inf. Manage. **47**, 88–100 (2019)
41. Ren, Y., Zhang, H., Wei, W., Cai, X., Song, Y.: Determining the fluctuation of PM2. 5 mass concentration and its applicability to Monin–Obukhov similarity. Sci. Total Environ. **710**, 136398 (2020)
42. Velasteguí, A.H., Sinaluisa, J.L., Cahueñas, N.P., Nogales, J.M.: Small-scale variation of atmospheric dynamics applying chaos theory, case study. Atmósfera **38**, 473–484 (2024)

43. Ortiz-Suslow, D.G., Wang, Q.: An evaluation of Kolmogorov's—5/3 power law observed within the turbulent airflow above the ocean. Geophys. Res. Lett. **46**(24), 14901–14911 (2019)
44. Malik, N.A., Hussain, F.: New scaling laws predicting turbulent particle pair diffusion, overcoming the limitations of the prevalent Richardson–Obukhov theory. Phys. Fluids **33**(3) (2021)
45. Katul, G.G., Porporato, A., Shah, S., Bou-Zeid, E.: Two phenomenological constants explain similarity laws in stably stratified turbulence. Phys. Rev. E **89**(2), 023007 (2014)
46. Simsek, V., Pozzoli, L., Unal, A., Kindap, T., Karaca, M.: Simulation of 137Cs transport and deposition after the Chernobyl nuclear power plant accident and radiological doses over the Anatolian Peninsula. Sci. Total. Environ. **499**, 74–88 (2014)
47. Talerko, M., Kovalets, I., Lev, T, Igarashi, Y., Romanenko, O.: Simulation study of radionuclide atmospheric transport after wildland fires in the Chernobyl Exclusion Zone in April 2020. Atmos. Pollut. Res. **12**(3), 193–204 (2021)
48. Talerko, M., Garger, E., Lev, T., Nosovskyi, A.: Atmospheric transport of radionuclides initially released as a result of the Chernobyl accident. In: Behavior of Radionuclides in the Environment II: Chernobyl, pp. 3–74 (2020)
49. Zhuang, S., Fang, S., Dong, X., Xu, Y., Sheng, L., Li, X.: Local atmospheric transport behaviors of representative radionuclides during the Fukushima accident: a 200-m-resolution cross-scale study from site to 20 km. J. Environ. Radioact. **265**, 107212 (2023)
50. Shestopalov, V.M., Bublias, V.N., Bohuslavsky, A.S., Bixio, A.C., Putti, M.: Modeling of radionuclide fast migration paths at a typical depression in the Chernobyl exclusion zone. In: Groundwater 2000, pp. 129–130. CRC Press (2020)
51. Arockia Bazil Raj, A., Selvi, J.A.V., Raghavan, S.: Real-time measurement of meteorological parameters for estimating low-altitude atmospheric turbulence strength (Cn2). IET Sci. Measur. Technol. **8**(6), 459–469 (2014)
52. Mirza, A.K., Dacre, H.F., Lo, C.H.B.: A case study analysis of the impact of a new free tropospheric turbulence scheme on the dispersion of an atmospheric tracer. Quarter. J. Roy. Meteorol. Soc. (2024)
53. Sutherland, P., Melville, W.K.: Field measurements of surface and near-surface turbulence in the presence of breaking waves. J. Phys. Oceanogr. **45**(4), 943–965 (2015)
54. Gomes-Fernandes, R., Ganapathisubramani, B., Vassilicos, J.C.: The energy cascade in near-field non-homogeneous non-isotropic turbulence. J. Fluid Mech. **771**, 676–705 (2015)
55. Sirangelo, B., Ferrari, E.: Analysis of the spatial correlation structure exhibited by daily rainfall in Southern Italy. Theoret. Appl. Climatol. **118**, 203–209 (2014)
56. Friedrich, J., Moreno, D., Sinhuber, M., Wächter, M., Peinke, J.: Superstatistical wind fields from pointwise atmospheric turbulence measurements. PRX Energy **1**(2), 023006 (2022)
57. Xu, G., Shen, W., Wang, X.: Applications of wireless sensor networks in marine environment monitoring: A survey. Sensors **14**(9), 16932–16954 (2014)
58. Bellinguer, K., Girard, R., Bontron, G., Kariniotakis, G.: Short-term forecasting of photovoltaic generation based on conditioned learning of geopotential fields. In: 2020 55th International Universities Power Engineering Conference (UPEC), pp. 1–6. IEEE (2020, September)
59. Zaporozhets, A., Redko, O., Babak, V., Eremenko, V., Mokiychuk, V.: Method of indirect measurement of oxygen concentration in the air. Naukovyi Visnyk Natsionalnoho Hirnychoho Universytetu **5**, 105–114 (2018)
60. Dobslaw, H.: Homogenizing surface pressure time-series from operational numerical weather prediction models for geodetic applications. J. Geodetic Sci. **6**(1) (2016)
61. Weyn, J.A., Durran, D.R., Caruana, R.: Can machines learn to predict weather? Using deep learning to predict gridded 500-hPa geopotential height from historical weather data. J. Adv. Model. Earth Syst. **11**(8), 2680–2693 (2019)
62. Browell, J., Drew, D.R., Philippopoulos, K.: Improved very short-term spatio-temporal wind forecasting using atmospheric regimes. Wind Energy **21**(11), 968–979 (2018)
63. Smoliak, B.V., Wallace, J.M., Lin, P., Fu, Q.: Dynamical adjustment of the Northern Hemisphere surface air temperature field: methodology and application to observations. J. Clim. **28**(4), 1613–1629 (2015)

64. Williamson, S.N., Hik, D.S., Gamon, J.A., Kavanaugh, J.L., Flowers, G.E.: Estimating temperature fields from MODIS land surface temperature and air temperature observations in a sub-arctic alpine environment. Remote Sensing **6**(2), 946–963 (2014)
65. Cifuentes, J., Marulanda, G., Bello, A., Reneses, J.: Air temperature forecasting using machine learning techniques: a review. Energies **13**(16), 4215 (2020)
66. Ehsan, M.A., Nicolì, D., Kucharski, F., Almazroui, M., Tippett, M.K., Bellucci, A., Kang, I.S., et al.: Atlantic Ocean influence on Middle East summer surface air temperature. NPJ. Clim. Atmosph. Sci. **3**(1), 5 (2020)
67. Ingleby, B.: Global assimilation of air temperature, humidity, wind and pressure from surface stations. Q. J. R. Meteorol. Soc. **141**(687), 504–517 (2015)
68. Yaacob, M.R., Schlander, R.K., Buchhave, P., Velte, C.M.: Experimental evaluation of Kolmogorov's-5/3 and 2/3 power laws in the developing turbulent round jet. J. Adv. Res. Fluid Mechan. Therm. Sci. **45**(1), 14–21 (2018)
69. Nguyen, T.A., Nguyen, A.V.: Transient volume of evaporating sessile droplets: 2/3, 1/1, or another power law? Langmuir **30**(22), 6544–6547 (2014)
70. Pandey, S., Quetz, A., Aryal, A., Dubenko, I., Blinov, M., Rodionov, I., Ali, N.: Giant field-induced adiabatic temperature changes in In-based off-stoichiometric Heusler alloys. J. Appl. Phys. **121**(13) (2017)
71. Hu, D., Wu, H.: Volume evolution of small sessile droplets evaporating in stick-slip mode. Phys. Rev. E **93**(4), 042805 (2016)
72. Granero-Belinchon, C.: Neural network based generation of a 1-dimensional stochastic field with turbulent velocity statistics. Physica D **458**, 133997 (2024)
73. Aghaie, M., Minuchehr, A., Alahyarizadeh, G.: Evaluation of atmospheric dispersion of radioactive materials in a severe accident of the BNPP based on Gaussian model. Prog. Nucl. Energy **113**, 114–127 (2019)
74. Shamsuddin, S.D., Omar, N., Koh, M.H.: Development of radionuclide dispersion modeling software based on Gaussian plume model. Matematika 149–157 (2017)
75. Balanya, S.A., Ramos, D., Ramirez-Hereza, P., Toledano, D.T., Gonzalez-Rodriguez, J., Ariza-Velazquez, A., Gutiérrez, N.D., et al.: Gaussian processes for radiation dose prediction in nuclear power plant reactors. Chemom. Intell. Lab. Syst. **230**, 104652 (2022)
76. Aliyu, A.S., Ramli, A.T., Saleh, M.A.: Assessment of potential human health and environmental impacts of a nuclear power plant (NPP) based on atmospheric dispersion modeling. Atmósfera **28**(1), 13–26 (2015)
77. Bondarenko, O., Movchan, Y., Gulevets, D., Goutsal, O., Drapaliuk, A.: Establishment of the Chornobyl radioecological biosphere reserve as recovery biotechnology implementation of the exclusion zone. Proc. Nat. Aviat. Univ. **4**, 89–95 (2016)
78. Krynychnay, I.P., Hozhilo, I.I., Djachenko, O.I.: Ukraine International cooperation in nuclear and radiation safety: public-administrative aspect. Publ. Admin. Aspect. **37–38**, 32–39 (2016)
79. Zabulonov, Y., Popov, O., Burtniak, V., Iatsyshyn, A., Kovach, V., Iatsyshyn, A.: Innovative developments to solve major aspects of environmental and radiation safety of Ukraine. In: Systems, Decision and Control in Energy II, pp. 273–292. Springer International Publishing, Cham (2021)
80. Duliba, Y., Chudyk, N.:Good lessons based on bad experience: confronting risks and governing nuclear safety in Ukraine. Polityka Energetyczna **26**(3) (2023)
81. Balashevska, Y., Gumenyuk, D., Ovdiienko, I., Pecherytsia, O., Shevchenko, I., Yesypenko, Y., Zhabin, O.: Strengthening the SSTC NRS scientific and technical potential through participation in the IAEA coordinated research projects. Nucl. Radiat. Saf. **1**(89), 5–13 (2021)
82. de Vicente, S.G., Mazon, D., Xu, M., Pinches, S., Churchill, M., Dinklage, A., Verdoolaege, G., et al.: Summary report of the 4th IAEA technical meeting on fusion data processing, validation and analysis (FDPVA). Nucl. Fusion **63**(4), 047001 (2023)
83. Liu, Y., Mui, T., Xie, Z., Hu, R.: Benchmarking FFTF LOFWOS Test# 13 using SAM code: baseline model development and uncertainty quantification. Ann. Nucl. Energy **192**, 110010 (2023)

84. Ninokata, H., Pellegrini, M., Kamide, H., Ricotti, M.E.: Modelling of multi-physics phenomena in fast reactor design: safety and experimental validation. In: Fast Reactors and Related Fuel Cycles: Safe Technologies and Sustainable Scenarios (FR13), pp. 151–167. IAEA (2015)
85. Zaporozhets, A.: Overview of quadrocopters for energy and ecological monitoring. In: Systems, Decision and Control in Energy I, pp. 15–36. Springer International Publishing, Cham (2020)
86. Čerba, Š, Lüley, J., Vrban, B., Osuský, F., Nečas, V.: Unmanned radiation-monitoring system. IEEE Trans. Nucl. Sci. **67**(4), 636–643 (2020)
87. Popov, O., Bondar, O., Ivaschenko, T., Puhach, O., Iatsyshyn, A., Skurativskyi, S.: Features of the modern UAV-based complexes use to solve radiation control problems. In: Systems, Decision and Control in Energy IV: Volume II. Nuclear and Environmental Safety, pp. 35–57. Springer Nature Switzerland, Cham (2023)
88. Babak, S.: Radiation monitoring of environment using unmanned aerial complex. Adv. Sci. J. **12**, 41–44 (2014)
89. Chen, Z., Li, H., Yu, H., Zhao, Y., Zhang, C., Zhang, H.: Research on UAV flight parameter identification method based on launch force and airspeed. Sensors **24**(5), 1597 (2024)

Chapter 10
Unmanned Aerial Measurement Systems for Monitoring the Environmental Area of NPP and TPP

Abstract This chapter investigates the utility of Unmanned Aerial Measurement Systems (UAMS) for environmental monitoring around nuclear and thermal power plants. UAMS offer improved safety, flexibility, and cost-effectiveness over traditional methods. The systems can be categorized by various factors, including the type of measurement signals, data collection methods, and the number of UAVs deployed. UAMS are crucial in analyzing harmful emissions and radiation, ensuring environmental and personnel safety. The study discusses UAMS's role and future development, emphasizing its potential to enhance international and state environmental protection programs. The research includes case studies, such as the Chornobyl and Fukushima Daiichi NPP incidents, demonstrating UAVs' pivotal role in remote monitoring and emergency response.

Keywords Unmanned aerial measurement systems · Environmental monitoring · Radiation monitoring · Emergency response · Environmental safety · Chornobyl · Fukushima Daiichi · UAV capabilities

10.1 Development of Information and Measurement Systems Based on UAV Complexes

The development of modern information and measurement systems (IMS) is taking place in various directions. On one hand, this process is driven by the continuous improvement of hardware and software infrastructure, the emergence of new sensors with enhanced metrological characteristics that use new principles and materials based on nanotechnology achievements, increased performance, speed, and reliability of computer technology, enabling the processing of large volumes of information and the use of complex processing algorithms [1–5]. Additionally, it involves the development of communication tools for transmitting and protecting measurement information over significant distances and the creation and improvement of software tools oriented towards the use of both standard and functionally oriented software environments (such as LabVIEW) [6–10].

V. Babak et al., *Statistical Diagnostics of Electric Power Equipment*, Studies in Systems, Decision and Control 573, https://doi.org/10.1007/978-3-031-76253-6_10

On the other hand, the development of hardware and software for IMS encourages the expansion of measurement issues and their application to objects with special operating conditions. Examples of such objects are energy objects. These objects use various IMS at the stages of design and commissioning. These systems are intended to provide control over the condition of technological equipment, continuous monitoring of equipment parameters and operating mode characteristics, the quality of generated electrical energy, and so on.

In recent times, the issue of monitoring (control, diagnostics, forecasting) the state of the environment around spatially distributed and technologically hazardous objects, including NPPs and TPPs, has become increasingly relevant and important [2, 11–15]. This environmental problem, both pre-existing and emerging, has become a well-known issue of national security, especially after accidents at the Chornobyl NPP (Ukraine) in 1986 and the Fukushima Daiichi NPP (Japan) in 2011 [16–19].

The information obtained through monitoring can be extremely valuable not only for issues related to the safety of the population and personnel at such facilities but also indirectly by providing data for analyzing the development of various technological processes for monitoring potential harmful emissions and radiation in both normal and emergency (disaster, accident, etc.) operating modes.

Currently, there are various international and state environmental protection programs. Examples of accidents at the Chornobyl NPP and the Fukushima Daiichi NPP only emphasize the importance of implementing these programs, expanding them, and developing specialized technical means for radiation monitoring and monitoring of harmful emissions into the environment. The experience of dealing with the consequences of these accidents has shown the potential for using unmanned aerial complexes (UAC) as part of such technical means [20–22]. However, due to the lack of specifically developed IMS based on UAC, their usage remains sporadic. Nevertheless, due to their potential capabilities, such systems have prospects for further development. These systems can provide monitoring and data collection automatically over large areas around monitoring objects, in conditions hazardous to humans and in remote or hard-to-reach areas. Moreover, many of the functions mentioned can be implemented through remote operator involvement in the measurement process.

Let’s summarize the experience of using modern IMS in conjunction with UAC and consider the possibilities of creating a new class of unmanned aerial measurement systems (UAMS).

We will analyze the most common functional properties of such a class of systems to address various measurement problems, including remote monitoring of parameters and characteristics of the environment around spatially located hazardous facilities, including NPPs and TPPs. Based on the results of this analysis, we will establish a classification of the systems under consideration.

The work [23] focuses on systems built around stationary ground control points. The placement of these points takes into account the environmental characteristics of the area, landscape type, meteorological conditions, characteristics of potential threat sources, as well as economic and physical-technical factors.

The development and operation of such systems are not always justified for various reasons. First and foremost, there are limited access possibilities to certain control points (for example, in swampy areas, river flooding, presence of rocky terrain, etc.) [24–28]. There's also a need for quick reconfiguration for different types of emissions or radiations and the necessity to protect personnel from the harmful effects of controlled objects. After all, there's always a nonzero probability of controlled objects transitioning into abnormal operating conditions with difficult-to-predict consequences of accidents and incidents.

Another approach to building such systems is based on using UACs as part of them. As mentioned earlier, the first broad experience of using UACs in environmental monitoring systems occurred during the study of the consequences of accidents at Chornobyl and Fukushima-1 NPPs [22, 29–33]. In these cases, UAVs were primarily used for video surveillance in the vicinity of NPPs. Thus, UAVs in the system are capable of moving the sensors of the IMS in space along any closed trajectory. Although in such systems UAVs are used only as a means of transporting sensors in space, their inclusion in the system, together with new scientific and technological advancements, provides these systems with new features and significantly expands their functional capabilities [34].

This is facilitated by:

1. Innovative information support based on known and new models of studied information fields, methods and algorithms for processing measurement data, software, determining the dynamics of field characteristics in space and time with the necessary reliability and accuracy.
2. Widespread use of computer technology methods and tools for data formation, transmission, protection, processing, and storage.
3. Development of methods and means for interference-resistant and secure communication.
4. Utilization of new structural materials, especially composite nanocoatings, in creating the hardware components of systems, particularly UACs.
5. Miniaturization, increased reliability, reduced energy consumption, and cost reduction of system modules.
6. Advanced energy technologies to ensure reliable power supply for system modules.
7. Development of navigation tools and systems, air traffic organization using GPS technology.
8. Advancement in specialized geographic information technologies and systems.
9. Advancements in image processing and transmission, pattern recognition technologies.

Here are the integrated functional capabilities that can be considered as classification features of the new class of IMS – UAMS:

1. UAMS serves as a new measuring tool and, consequently, a measuring resource for conducting various natural experiments in space and time for the study of processes and phenomena occurring on Earth and in the atmosphere.
2. UAMS ensures the safety of conducting natural experiments by using remote measurement methods in emergency, disaster, and other non-standard situations.

3. A wide range of subject areas for using UAMS, including the provision of monitoring processes according to a specified program for the dynamics of parameters and characteristics of the environment of spatially located technologically hazardous objects. This includes NPPs, TPPs, HPPs, and various hard-to-reach locations (mountainous, swampy, aquatic, and forested areas).
4. UAMS allows reducing the volume of financial and human resources compared to the use of other systems when solving measurement problems.

To conduct a more detailed analysis of the capabilities of UAMS and their development prospects, let's consider their classification presented in Fig. 10.1.

UAMS are classified into passive and active systems based on the method of generating measurement signals. In passive UAMS, the sensor generates a measurement signal proportional to a parameter or characteristic of the environment, such as aerosol concentration or radiation level, and so on [35–39].

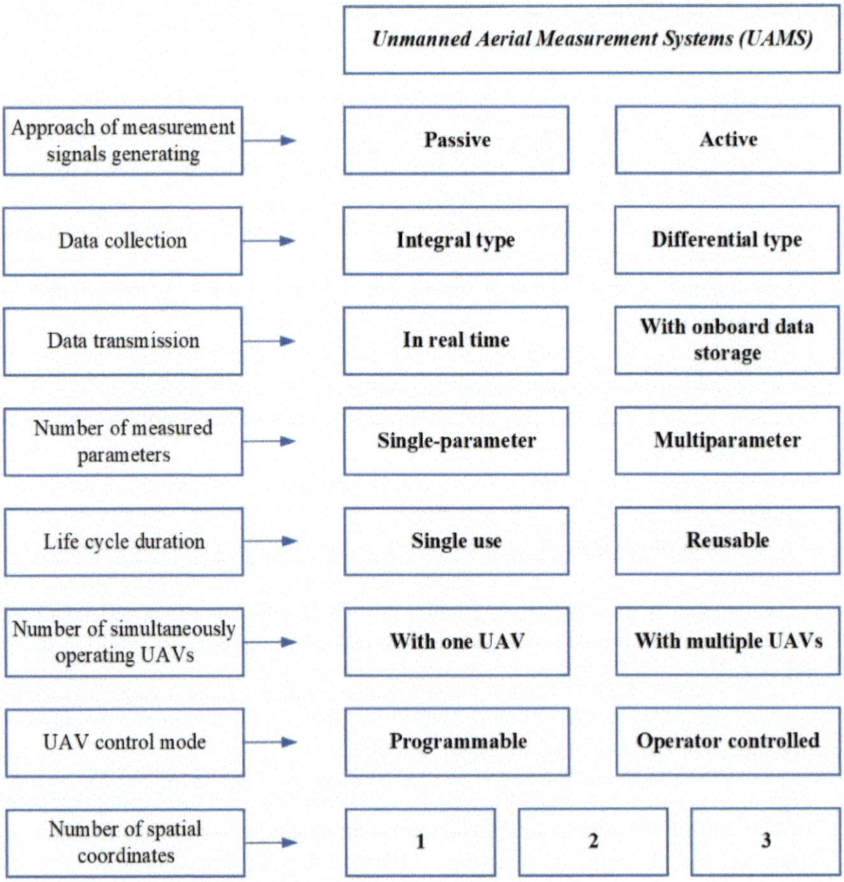

Fig. 10.1 Classification of the main types of UAC-based IMS

In passive systems, the source of the information signal is the environment being studied. In active systems, it is necessary to create an artificial test signal (e.g., optical, ultrasonic, or electromagnetic in the radio frequency range) whose parameters depend on the characteristics of the surrounding environment in a certain way. This can include optical transparency of the atmospheric environment, the speed of aerosol dispersion, and so on. Various configurations of active systems are possible, with full or partial air-based elements of the UAMS (for example, ground-based stationary or mobile deployment of the signal generator and sensor on a UAC or vice versa, as well as the placement of both the signal generator and sensor on two separate UAC).

Based on the data collection method, UAMS are classified into integral and differential types. This distinction is determined by both the measurement tasks and the purpose of the sensor, as well as the data collection modes. Integral UAMS provide a single integrated value of the studied parameter as the sensor moves along the entire trajectory of the UAC. In contrast, differential UAMS provide a specific set or series of numerical values for the parameter or characteristic being studied. These values can be obtained through their uniform or non-uniform discretization in time or space. The discretization of measurements should provide the necessary probability of detecting unacceptable concentrations of harmful emissions or radiation in narrow segments of the UAV flight path (up to 5 to 10 degrees).

Based on the data transmission method, systems can be classified into real-time systems and systems with data storage on board. Real-time systems must have data transmission channels from "onboard" to "ground". These are advisable to use in cases where transport delays in transmitting measurement information are unacceptable, for example, when there is a real threat of severe accidents with environmental safety implications. Systems with data storage on board and subsequent processing after the entire flight of the UAC should be recommended in cases where, due to information security considerations, transmitting them over the radio channel is undesirable, and the time resource and inertia of the studied processes allow for the analysis of measurement information with significant delays. The absence of a radio channel provides these systems with greater energy efficiency.

By the number of measured parameters, UAMS are divided into single-parameter and multi-parameter ones. From the names of these types of UAMS, it is evident that the first transport only one sensor, while the second can transport several sensors. This possibility is limited only by the size of the sensors, the possible energy consumption for their maintenance, and the ability of the UAC to carry a certain payload.

By the duration of the life cycle, UAMS can be divided into single-use and reusable ones. When considering the economic efficiency of different systems, it is necessary to take into account not only the costs of designing, manufacturing, and operating the systems but also their disposal costs. This requirement becomes particularly relevant, for example, during accidents at NPPs when there is a question of burying used and radiation-contaminated mechanisms and equipment. It is understandable that for single-use UAVs, it is advisable to choose the cheapest models with low reliability. It should be noted that UACs are capable of not only transporting sensors to the areas of research objects but also automatically delivering contaminated UAVs with all equipment to places for disposal.

Based on the number of UAVs simultaneously in the air and collecting primary measurement data, UAMS is classified as having one or several UAVs. Increasing the number of simultaneously operating drones allows for significantly reducing observation time, which leads to a more accurate fulfillment of the condition of the stationarity of experimental conditions and, ultimately, to an increase in the reliability of the obtained information.

Based on the number of spatial coordinates, UAMS can be classified as one-dimensional, two-dimensional, or three-dimensional systems. This division is determined by the characteristics of the objects and the tasks of their control. For example, research on the support structures of power transmission lines needs to be performed by their height, which is one-dimensional. Transporting the sensor during the solution of such a task can be accomplished using a multicopter. Monitoring the concentration of aerosols around a NPP is an example of a two-dimensional task. This can be addressed using a single UAV. Determining the characteristics of the transport of hazardous substances in the airspace requires a three-dimensional measurement approach and the simultaneous use of multiple UAVs.

Based on the mode of UAV control, they are classified as programmable and operator-controlled systems.

Therefore, the creation of a new class of IMS based on UACs significantly expands the possibilities of obtaining important measurement information about processes and phenomena occurring on the ground and in the atmosphere. In general, UAMS significantly complement ground-based monitoring and control systems and provide the opportunity to measure the parameters and characteristics of the environment in emergency situations of both natural and man-made nature. The importance and necessity of developing such systems are associated with the fact that the processes of radionuclide transport in the air play a crucial role in shaping the fields of radioactive contamination of the air, land surfaces, and water bodies in the presence of harmful emissions from NPPs [40–42].

10.2 History and Current State of Unmanned Aviation

Undoubtedly, the basic element in any UAS is the UAV, because it solves the main task of transporting multifunctional equipment to solve the assigned tasks.

The first UAVs appeared due to the need to effectively solve military tasks—delivery of weapons (bombs, torpedoes, etc.), tactical reconnaissance and battlefield management [43–50]. The first documented use of UAVs was the delivery of balloon bombs by Austrian troops to besieged Venice in 1849. These were vehicles that utilised the aerostatic principle of generating lift. The invention and spread of the radio telegraph allowed Nikola Tesla to create the first miniature radio-controlled craft in 1899, and the development of aviation allowed Charles Kettering to propose the use of unmanned flying machines in 1910. According to his idea, a flying machine with a clockwork mechanism was to drop its wings at a given place and fall like a bomb. Several copies were produced under the order of the US Department of Defence, but their use in combat was not recorded. The first reusable UAV, in the modern sense, was created in 1933 in Great Britain on the basis of a biplane (Fig. 10.2), controlled remotely from a ship and served as a target aircraft for training of the British Air Force until 1943. The US Navy also used UAVs of its own design as target aircraft from 1940. In the USSR in 1930–1940, aircraft designer Nikitin developed a flying-wing type rocket-propelled glider in two versions: manned training and target practice and unmanned with full automation.

During the Second World War, German scientists made a breakthrough in UAV development, giving the world the jet engine and the Fau-1 rocket as the first UAV to be used en masse during combat operations. "V-1" (Fig. 10.3) was the first aerial vehicle and became the prototype for cruise missiles. The control system of the projectile aircraft was an autopilot, which kept the vehicle on the course and altitude set at launch throughout the flight. The launch system had two options: from a ground-based catapult or from a carrier aircraft. Although the first launches were made from a carrier aircraft, the majority of launches were made from a ground-based catapult, which ensured fewer abnormal situations at the time of launch.

After the end of the Second World War, interest in UAVs increased significantly. There was a rapid development of technologies for the creation of cruise and ballistic missiles, guided missiles, "smart bombs", and so on. At the same time, the development of UAVs in the usual sense, in the form of unmanned aviation, was also developing.

Fig. 10.2 The first reusable Queen Bee UAV

Fig. 10.3 V-1 aeroplane launcher

Between 1950 and 1980, UAVs were built to perform such tasks in addition to strike combat missions:

- acting as target training aircraft for the Air Force;
- photographic reconnaissance, radio-technical reconnaissance;
- signal retransmission, electronic warfare;
- performing decoy functions for enemy air defence systems, etc.

At that time, the USSR created several dozen types of UAVs, the most successful and widespread being the TU-123 "Yastreb", TU-143 "Flight", and TU-141 "Strizh" (Fig. 10.4), which were in service with the USSR Army from 1964 to 1979. The TU-143 "Flight" was delivered to African and Middle Eastern countries in the 1970s, in the 1980s it was operated in Czechoslovakia, Romania, Iraq and Syria, and was used during the Lebanon War. The TU-141 "Strizh" was in service with the Ukrainian Air Force.

It should be noted that in 1970–1980, the USSR was the leader in UAV production—only TU-143 produced about 950 units. The USSR was the leader in UAV production—about 950 TU-143 alone were produced.

The U.S., in the early 1960s, used UAVs to track the deployment of Soviet missiles in Cuba; in the 1970s, during the Vietnam War, UAVs were used for photographic reconnaissance and electronic warfare, relaying signals and as decoys to complicate the air situation.

Israel used UAVs already during the Arab–Israeli conflict in 1973 and in 1982 during the fighting in Lebanon for surveillance and reconnaissance purposes, as well as for decoys and to identify the positions of enemy surface-to-air missile systems.

Fig. 10.4 TU-141 "Strizh" UAV

In 1991, during the Persian Gulf War, UAVs were massively used by both sides, many of them were shot down, which once again demonstrated their main advantage over conventional aircraft—the loss of UAVs does not result in human casualties.

Since the 1960s, the first attempts to use UAVs for civilian tasks have been noted.

One of the first of these machines was the Soviet YAK-9V drone; aircraft of this type have long carried out dosimetric control in the atmosphere in nuclear test areas.

In general, the history of UAV development can be divided into four time phases:

1. 1869—the beginning of the twentieth century: the first attempts and experimental experiments on the creation of UAVs, the beginning of the creation of theoretical foundations of aerodynamics, flight theory and aircraft design in the works of scientists (K.E. Tsiolkovsky, I.V. Meshchersky). During this period, the aerostatic principle of flight prevailed in UAVs.
2. The beginning of the twentieth century—1945: almost all UAVs were developed for military purposes and were either projectile or target planes for air force training.
3. 1945–1960s: development and creation of reconnaissance UAVs. Period of expansion of UAV nomenclature by purpose.
4. 1960s—present day: expansion of the range and improvements of UAVs, beginning of the use of UAVs for civilian tasks.

There is currently no universally accepted classification of UAVs due to the sheer number of different systems in terms of design, size and operational characteristics. Most UAVs are categorised by weight, duration, altitude and range, and intended use.

In general, the following definitions of UAVs can be given.

A UAV is an aircraft with associated equipment that does not carry an operator (pilot) and is capable of flying autonomously or being piloted remotely.

It should be noted that this scientific and technical direction has developed rapidly in the last 20–30 years. This is due, first of all, to the creation of new technologies used in the production of UASs, namely:

- development and production of modern structural materials, primarily composites, with the use of nanocoatings;
- modern computer technologies, including multiprocessor systems for data collection, processing and storage;
- theory of automatic control systems as a branch of cybernetics, associated with the theory of information transmission, encryption, data compression;
- communications facilities and systems, including space communications;
- Earth remote sensing technologies (radar, optoelectronic systems, multispectral sensors);
- energy technologies, use of alternative energy sources: high-capacity batteries, solar energy, fuel cells;
- means and systems of navigation, air traffic management through the introduction of automatic dependent surveillance (ADS);
- Geographic Information Systems (GIS);
- image processing, pattern recognition technologies;
- human–machine interface development;
- artificial intelligence development.

Although attempts at civilian applications of UASs have been noted since the 1960s, until 2000 almost all UASs were developed for military missions. The U.S. Department of Defence leads the global military market in the development, ownership, and use of UASs. As of early 2012, the US Department of Defence used more than 7,000 UASs. The range of UASs in use is very broad: from the smallest two-kilogram Raven, which flies for about an hour at distances up to 50 km and at altitudes typically no higher than 300 m, to the largest 12-tonne Global Hawk, which flies for about 30 h at altitudes up to 20 km.

The market of civil UASs began to emerge at the end of the twentieth century, mainly through the efforts of enthusiastic aeromodellers. At the beginning of the twenty-first century, commercial companies began to enter the civil UASs market, and as of the beginning of 2015, there are hundreds and thousands of them all over the world.

The development of civilian UAS production based on modern ultralight UAVs was facilitated by the following determining reasons:

1. Computer Science. Creation of cheap, mass, fast single-chip microcomputers.
2. Navigation.
    - 2.1. Full-scale reliable deployment of the GPS (Global Positioning System) satellite constellation of the global 24-h navigation satellite system.
    - 2.2. Worldwide distribution of cheap, mass-produced, miniaturised GPS receivers.

   2.3. Appearance of cheap, mass and sufficiently accurate solid-state barovisotometers.

3. Orientation devices. Appearance of cheap, mass-produced, miniaturised micromechanics products: accelerometers and angular velocity sensors.
4. Optoelectronics. Widespread use of miniature matrixes to create television and infrared cameras.
5. Chemical power sources. Distribution of cheap and reliable nickel–cadmium, nickel-metal hydride and lithium batteries with high specific capacity.
6. Aeromodelling.

   6.1. Microminiaturisation, increased reliability and comparative cheapening of transmitters, receivers, steering machines, engines and other aircraft equipment.
   6.2. Spread among aeromodellers the vacuum technology of manufacturing large aircraft airframe parts from carbon and glass fabric, epoxy resin and other available materials.

The field of civilian applications of drone technology has enormous potential.

Nevertheless, the civil UAS market continues to develop in anticipation of the adoption of uniform rules and standards for the use of civil UASs in national airspaces.

According to Teal Group estimates, the volume of the global UAS market in 2011 was 3.4 billion US dollars, with the share of civil UASs amounting to 293 million US dollars or about 9%. In 2014, the volume of the global UAS market was already about 6 billion US dollars, and the share of civil UASs was more than 10% (Fig. 10.5).

Teal Group forecasts that from 2015 to 2025, the civilian unmanned systems market will grow at an average rate of 19% and the military market at a rate of 5%, the total global production volume of UAS by 2024 will be 13 billion dollars, including the production volume of civilian unmanned systems will be more than 3 billion dollars, accounting for more than 25% of the global market.

The emergence of these technologies led to the miniaturisation of UAS elements, cheaper production and created the prerequisites for the emergence of a class of ultra-light UAVs, and, consequently, the effective and large-scale application of unmanned technologies for civilian purposes.

It should be noted that the United States accounts for about 60 per cent of the UACs produced, and in the near term, the share of US UACs in the global market is unlikely to change dramatically, despite the rapid growth of unmanned systems production in the Asian region (Fig. 10.6).

Based on adjusted data, the drone market size is valued at 17.31 billion dollars in 2024 and is expected to reach 32.95 billion dollars by 2029, growing at a CAGR of 13.74% during the forecast period (2024–2029) [51].

As we have already noted, the applications of civilian UASs have virtually unlimited potential.

Currently, the market for civilian UAVs in Europe is estimated at about 1.5 billion dollars, mainly due to the use of UAVs with high flight duration at high and medium

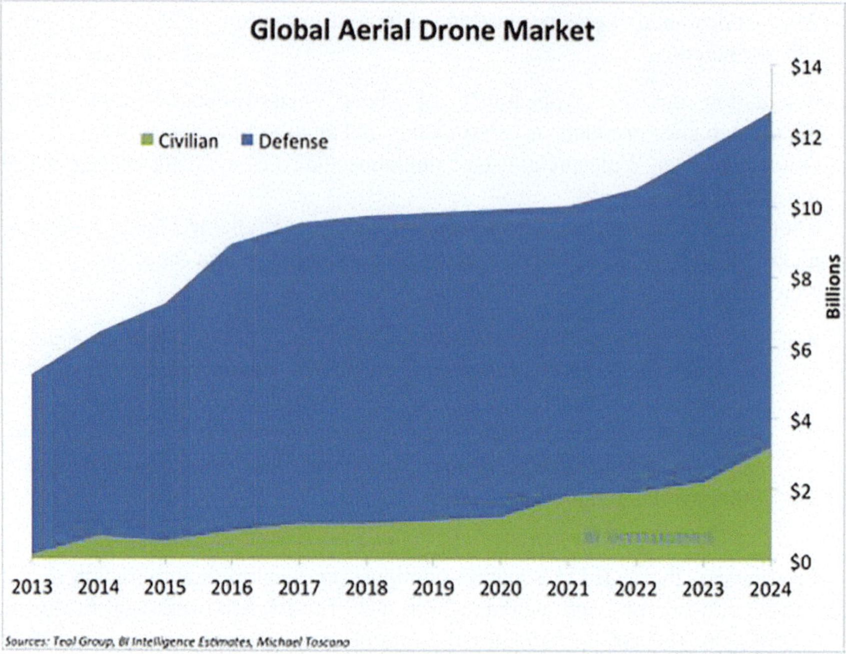

Fig. 10.5 Global UAS production volume forecast for 2015–2025

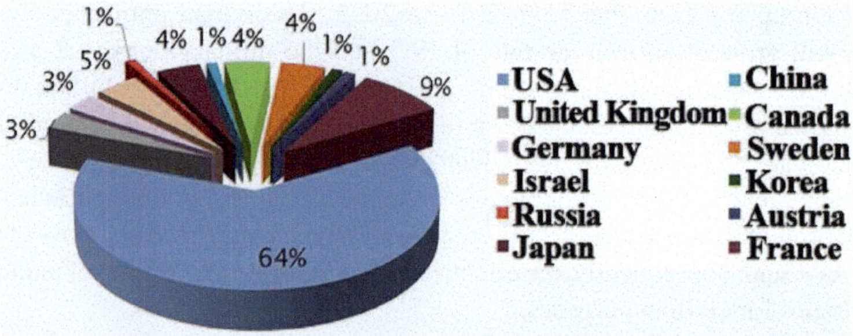

Fig. 10.6 World production volume of UAS by countries

altitudes, as well as mini-UAVs [52]. And they are already being actively used for civilian missions (Fig. 10.7).

The growth of the civilian segment directly depends on the speed of adoption of unified standards for the use of UAS in civil aviation space and certification of both equipment and operators of unmanned aircraft systems.

In the civilian segment of the market, UASs can be used to solve such problems:

- forestry: forest fire monitoring, deforestation control, aerial photogrammetry, etc.;
- water management: monitoring of water body spills, anti-poaching;

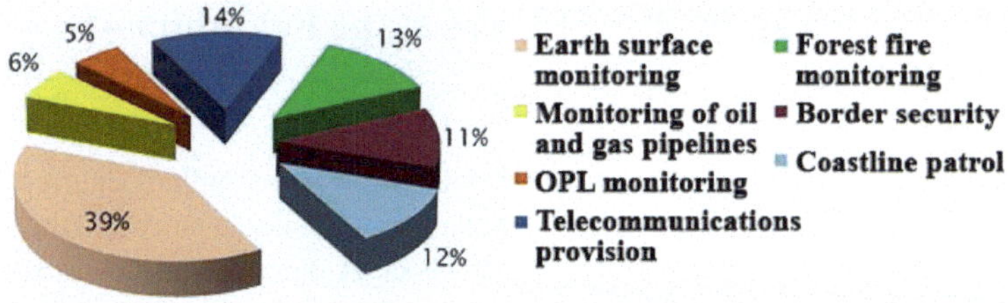

Fig. 10.7 Civilian applications of UASs

- agriculture: fertiliser application, monitoring of crops, sowing and harvesting quality, remote analysis of vegetative indices, monitoring of agricultural machinery efficiency, precision farming;
- pipeline transport: pipeline condition monitoring, identification of pipeline ruptures, infrastructure monitoring and prevention of unauthorised access to facilities;
- emergencies: search for victims of natural disasters, assessment of infrastructure damage, etc.;
- transport and public order: monitoring the condition of transport infrastructure (roads, bridges, railway tracks), monitoring traffic flow and congestion, filming traffic accidents and accidents, monitoring crowded places and observing public order;
- land management: three-dimensional mapping, cadastral surveys, remote land measurement;
- ecology: air quality analyses, monitoring of radiation and other types of air pollution;
- critical infrastructure: inspection and video surveillance of critical infrastructure facilities, control of the territory of such facilities and prevention of unauthorised access to them;
- energy sector: monitoring and inspection of energy facilities, analysing emissions from energy facilities, monitoring and inspection of power transmission lines;
- scientific research; etc.

For monitoring of power facilities, UASs have been used relatively recently, and one of the most powerful impulses for the development of this area of research, as previously noted, was one of the largest accidents at power facilities—the radiation accident at the Fukushima-1 NPP (Japan).

10.3 Experience in the Use of UAVs at the Fukushima-1 NPP

The Fukushima Daiichi accident was a major radiation accident that occurred on 11 March 2011 as a result of the strongest earthquake in Japanese history and the subsequent tsunami. The earthquake and tsunami knocked out external power supplies and backup diesel generators. This caused the failure of all normal and emergency cooling systems and led to core meltdowns in Units 1, 2 and 3 in the first days of the accident. Without touching upon the chronology of events at this power plant, let us consider some methods of remote monitoring of emergency units and NPPs in general, using UAS.

The first series of images of the Fukushima Daiichi NPP were taken on 20 and 24 March 2011 using the UAS, which included a small, remotely piloted, fixed-wing UAV (Fig. 10.8) from Air Photo Service of Niigata Prefecture [53]. The images obtained showed the extent of the destruction at the NPP (Figs. 10.9 and 10.10). No further use of this vehicle was reported.

Further, helicopters of the Japan Self-Defence Forces and the US RQ-4 Global Hawk UAS, which has a Global Hawk UAV [54], were used for radiation measurements, photography and video surveillance. Its appearance is shown in Fig. 10.11.

This UAV is built on a normal aerodynamic scheme and is an autonomous jet UAV with a sensor package that includes synthetic aperture, electro-optical, infrared sensors with telescopic capabilities for high resolution imaging and others.

Global Hawk helped Japan collect real-time images that allowed authorities to better prioritise and direct their resources to the disaster.

Global Hawk has already been used for similar tasks in the aftermath of a major earthquake in Haiti.

Fig. 10.8 Air photo service UAV

Fig. 10.9 Photo of destruction at Fukushima-1 NPP from 20.03.2011

Fig. 10.10 Photo of destruction at Fukushima-1 NPP from 24.03.2011

Fig. 10.11 Global Hawk UAV

The use of UAS with small UAVs made it possible to determine the real situation at various points at the Fukushima Daiichi NPP and in the surrounding area. Ultimately, this ensured the safety of the remediation work.

In addition to the Global Hawk considered, Japan has purchased three Helicopter UAV UASs from the French company Helipse. There is no information about their use. Additionally, the USA supplied Japan with four Honeywell's RQ-16 T-Hawk UASs of the RQ-16 T-Hawk type. A UAV of this type is shown in Fig. 10.12 [55].

To date, hundreds of T-Hawk UAVs have completed more than 12,000 test and operational flights and have proven themselves in a variety of situations.

These UAVs, remotely piloted by company employees, have made dozens of sorties, providing rescuers with photos and video of the nuclear reactor. However, one of them crashed and exploded on the roof of the second reactor in June, but this fact does not detract from the effectiveness of their use in this situation.

All four vehicles operating at Fukushima were equipped with radiation measurement equipment. Data on the radiation background at different altitudes were obtained: on 17.03.2011 08:38 the EDR at a height of 97 m above Fukushima-1 reached 87.7 mSv/h, and at a height of 305 m—4.13 mSv/h (normal radiation background is 0.1…0.2 μSv/hour).

Thus, the example of Fukushima has shown the feasibility and efficiency of using small UASs for radiation monitoring of the environment and video observations.

The accumulated experience of practical application of UASs in the zones of emergency NPPs allows formulating and realising design requirements for such UASs.

Fig. 10.12 T-Hawk UAV

10.4 UASs in the Tasks of Control of Power Engineering Facilities

The use of UAVs during emergency situations at power facilities allows to obtain real-time photographs and video images to assess the real situation at the facilities and to make decisions on the choice of effective means to combat the consequences of accidents. UAVs in such UASs can be equipped with conventional photo and video cameras, infrared cameras for operation in night conditions, thermal imaging cameras for obtaining information on temperatures in various parts of facilities, devices for measuring radiation background and air analysers for obtaining data on the state of the atmosphere around power facilities, etc. [21, 56].

At the same time, it makes sense to use UAS not only during emergency situations, but also to monitor the environmental parameters of power facilities that operate in normal mode in order to obtain information about the state of the facilities and the environment in real time, inspect the infrastructure, control unauthorised access and prevent emergency situations.

In order to select the most appropriate UAS for these purposes, they first need to be classified [21, 26, 57, 58].

To date, there is no single generally accepted classification of UASs, but taking into account their differences and features that can be used for environmental monitoring of power facilities, the following classification is proposed (Fig. 10.13).

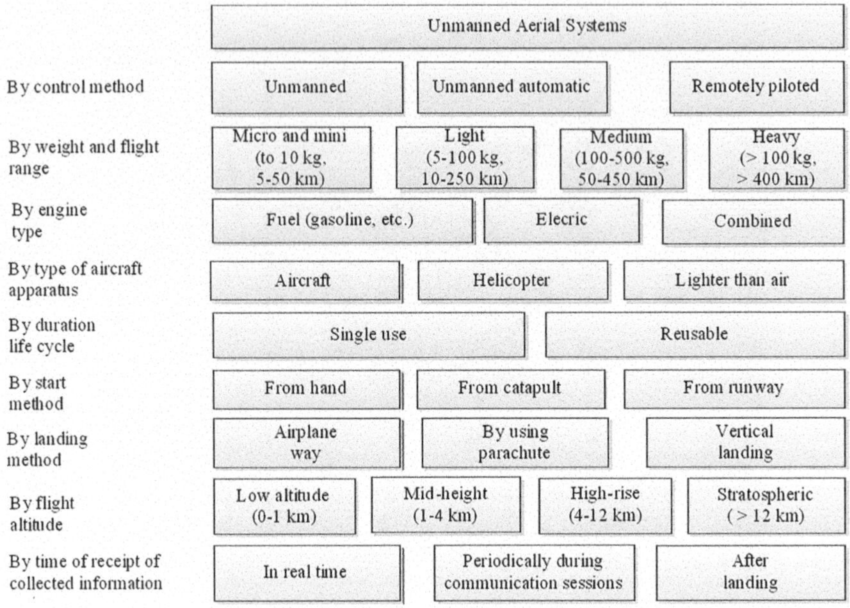

Fig. 10.13 Classification of UASs

According to the control method, UAVs in the UAS can be divided into unmanned unguided (missiles, probes, balloons), unmanned automatic and remotely piloted. The first ones can be used for a single launch to the epicentre of a man-made accident at energy facilities to obtain real-time data on the state of the environment at the accident site. Unmanned automatic systems can be used for periodic monitoring at long distances from the operator or in locations where communication between the UAV and the ground control station/operator may be interfered with.

Remotely piloted systems are best used in changing conditions and at a distance from the operator that guarantees constant communication with the UAV, as well as in cases where the flight programme is unknown and there is a need to change the task for the UAV quickly.

It is worth noting that some automatic systems have the function of operative intervention and change of the flight programme in real time mode. Most modern autopilots used in UAVs have both manual remote control and automatic flight according to a pre-recorded programme. This combination of automatic and remote control of UAVs is most suitable for environmental monitoring of energy facilities.

In terms of mass and range, a distinction is made between mini and micro, light, medium and heavy UASs. For the purpose of our work, the optimal UASs will be those with UAVs of the mini class, weighing up to 10 kg and with a range of up to 50 km, which can carry up to 4 kg of payload.

UAVs can have fuel (petrol), electric and combined engines. Petrol engines are more powerful and are usually used on all classes of UAVs larger than the mini class.

They allow a significant increase in range and flight time, but are expensive and labour intensive to maintain, as well as having a thermal footprint and exhaust emissions, so are unacceptable for environmental monitoring. Electric motors, mainly used in mini UAVs, are quiet, easy to use, relatively inexpensive and lightweight (up to 500g). Combined engines can use, in addition to traditional fuel and electric thrust, super flywheel elements and others. For the purposes of remote environmental monitoring of power engineering facilities, the use of UAS with an electric motor is optimal, as they have sufficient power to lift a UAV weighing up to 10 kg and do not emit harmful emissions into the atmosphere.

UASs can have aircraft-type UAVs, helicopter-type UAVs, and vehicles that are lighter than air. The latter include probes, balloons and airships. Aircraft-type UAVs (Fig. 10.14) have greater range and speed, but cannot stay in one point of three-dimensional space for long. Helicopter-type UAVs (multi-rotor platforms), on the contrary, are not characterised by high speed or range of flight, but can stay in one point of space for a long time (Fig. 10.15).

UAS with aircraft-type aircraft makes sense to use for monitoring large areas, at long distances from the operator and for periodic overflights along the same route, while multi-rotor platforms are better to use at small distances from the operator and for monitoring of environmental parameters at a given point of three-dimensional space, including in enclosed spaces.

Fig. 10.14 Aircraft-type UAV

Fig. 10.15 Helicopter-type UAV (multi-rotor platform)

To fulfil the tasks of monitoring the environmental parameters of TPPs and NPPs, the most effective will be the choice of UAS with two aircrafts simultaneously: aircraft and helicopter types, which will ensure the fulfilment of the full range of tasks of monitoring the environmental parameters of power facilities.

According to the possibility of use, or life cycle, one can distinguish single-use and reusable UASs. When considering the issue of economic efficiency of various systems, it is necessary to take into account the costs not only for design, production and operation of the systems, but also for their utilisation. This requirement becomes relevant, for example, during accidents at nuclear power plants, when the issue of disposal of spent and radiation contaminated mechanisms and technical equipment arises. Obviously, as a disposable UAV it makes sense to choose the cheapest models, albeit with low reliability. It should be noted that UASs can provide not only the transportation of sensors in the areas of the objects under study, but also the automatic delivery of contaminated UAVs with all the equipment to the places of their disposal. To monitor environmental parameters of power facilities operating in normal mode, reusable, high quality and reliable UAVs should be used.

In terms of the launch method, UASs are differentiated as those that launch from the hand, from a catapult, from a runway. It should be noted that these launch methods apply only to aircraft-type UAVs, as all multi-rotor platforms have a vertical take-off and landing function. The difference in launch methods mainly depends on the takeoff weight of the aircraft. Mini class (up to 5kg) can be easily launched by hand, while heavier vehicles require a catapult or a runway.

According to the landing method, UASs are divided into those that land by aeroplane, parachute or vertically.

Multirotor platforms use a vertical landing method, while aircraft-type UAVs can land by aircraft or parachute. Parachute landings are more acceptable from the point of view of preserving the UAV's hull and are necessary when sensors are located on the lower or front of the aircraft fuselage and when there is not enough space for an aircraft-type landing.

In terms of practical flight ceiling, UASs can be low-altitude (up to 1 km), medium-altitude (1…4 km), high-altitude (4…12 km) and stratospheric (>12 km). For monitoring of environmental parameters of power facilities, a practical flight ceiling of up to 1 km is sufficient, but modern small UASs can climb up to 3 km, which may be necessary during accidents at power facilities to obtain images of the environment around the epicentre of the accident or to analyse the air at certain altitudes.

According to the data transmission method, we can distinguish between real-time systems and systems with data accumulation on board. The former should have onboard-to-ground data transmission channels. They should be used in cases when transport delays in the transmission of measurement information are unacceptable, for example, when there is a real threat of an accident with severe consequences and a threat to the safety of people and the environment. Systems with data accumulation on board and their subsequent processing after the UAS flight completion should be recommended in cases when for safety reasons their transmission via radio channel is undesirable, and the time resource and inertia of the processes under study allow to analyse the measurement information with significant delays. The absence of radio diocanal provides such systems with greater energy efficiency.

Thus, for monitoring of environmental parameters of energy facilities it is recommended to use UAS with two UAVs of aeroplane and helicopter types, mini class, weight up to 10 kg, flight range up to 50 km, low-altitude (practically ceiling up to 2 km), with electric engine, remotely piloted, but with functions of automatic flight along a given route, vertical take-off and landing (for multirotor platform) and hand launch and aircraft landing (for aircraft-type UAVs), with real-time data transmission (for photo and video images) and on-board data accumulation (for other types of sensors).

The monitoring complex is initially developed as a holistic system consisting of a number of subsystems (Fig. 10.16) and includes:

- control station, which houses the operator's workstation and software applications that provide operator control of the entire complex;
- UAV carrying payload equipment of various types;
- a communication system that ensures the transmission of control commands from the control station to the UAV, as well as the transmission of useful information from the UAV to the ground control station in real time;
- additional equipment designed to provide technical support for ongoing research.

UAVs have a high degree of artificial intelligence automation. They have the ability to "communicate" with an operator at a ground control station and can transmit data

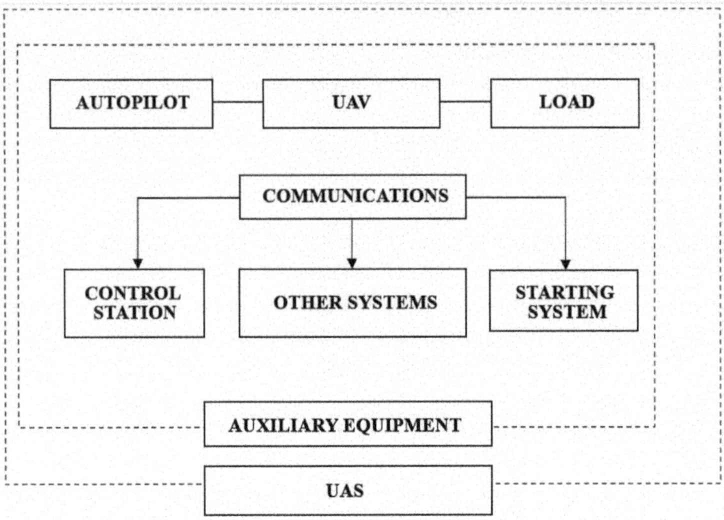

Fig. 10.16 Structure of the UAV-based monitoring complex

such as optical or thermal images of the terrain to that station, along with primary information about the UAV's position—altitude, heading, speed, roll, etc.

UAVs also can transmit a service data package to the ground control station covering aspects such as fuel quantity, and temperature of components such as engine or electronics. If a fault occurs in any of the subsystems or components, the UAV can develop automatic corrective action and/or operator warning. For example, if radio communication between the UAV and the ground control station is disrupted, the UAV may be programmed to search for a radio receiver and re-establish contact or switch to a different frequency band.

The methodology of monitoring using UAVs is developed depending on the object of research, its area, extent, level and nomenclature of polluting components of the environment.

Atmospheric air monitoring is carried out in two stages. This methodology can be rationally applied when monitoring the state of atmospheric air in areas where point and area sources of atmospheric pollution are located [12, 56, 59].

The monitoring programme includes control of the concentration of pollutants in the upper atmosphere and the total contribution of the pollution source in the lower atmosphere, the trajectory of the overflight is determined by the area of the pollution source as well as by the technical capabilities of the unmanned aerial vehicles, determining the maximum length of the possible flight of the UAV.

For monitoring of long objects (electric transmission lines, gas and oil pipelines) it is proposed to use the overflight of an unmanned aerial vehicle of the aircraft type along the monitoring object. The software of the ground control station allows using any topographic base as a map. The georeferencing can be done by two or several points. It is also possible to use electronic maps as a topographic base. The programme provides introduction, automatic control and editing of the overflight route. Altitude can be set for each point of the route. There is a possibility of landing point task as well as an algorithm of the drone's behaviour in abnormal situations. The ground control station also programs the sampling frequency along the flight route, the frame overlap factor (when conducting aerial photography of the area). When recording points and any actions related to the change of the drone control mode, an electronic log (log) of the events that occurred with is kept. The log also records incoming telemetry data.

The operational display panel allows to monitor all important parameters to analyse the operation and serviceability of onboard systems. Built-in software is used to view flight logs (including real-time), to analyse the behaviour of the drone and onboard equipment in flight, and to quickly adjust parameters.

UAV avionics is a complex of hardware and software on board the UAV [60]. That is, the onboard control equipment (OCE), which provides all modes of flight and functional task fulfilment. Avionics has a radio channel of communication with the ground control equipment (GCE). The UAV as a control object, steering actuators of the controls, OCE form the automatic flight control system (AFCS)—Fig. 10.17.

OCE and GCE shall provide the following UAV flight modes: takeoff and landing in automatic mode (manual takeoff and landing mode with radio control by the operator is also possible); flight in semi-automatic mode with radio control with correction of operator's actions by onboard control equipment; flight in automatic mode by control points with simultaneous sending of telemetry to ground control equipment.

In manual mode, the operator, visually assessing the UAV's behaviour (state), uses the UAS to deflect the controls (rudders, engine controls) driven by rudder machines

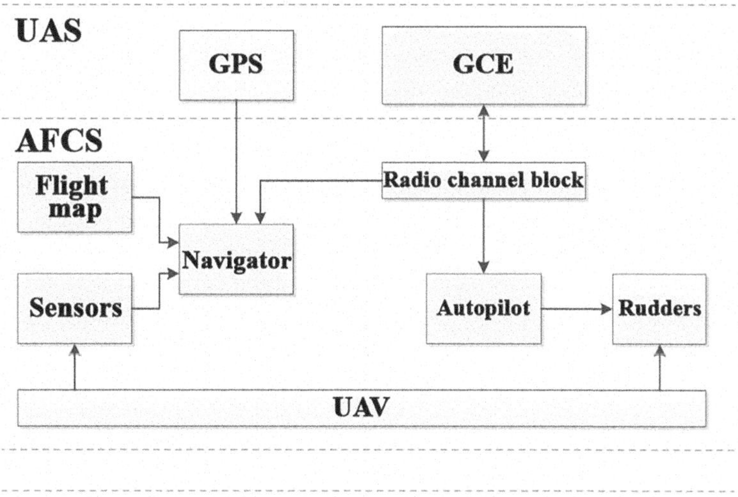

Fig. 10.17 Schematic of the flight control system as part of the UAS

(RMs). Semi-automatic mode is possible within the radio channel range, which for small-sized UAVs without special radio antenna equipment is within 2500 m.

Semi-automatic control (piloting) mode in this case is carried out with the help of information about the UAV's spatial position, which comes via radio channel and is displayed on the GCE virtual dashboard. The operator's actions in this control mode are corrected by the AFCS, which fulfils the functions of an autopilot and prevents potentially dangerous UAV movement parameters.

In semi-automatic mode, the UAV's OCE provides two information streams via radio channel:

- from the manual control unit to the autopilot control unit;
- from sensors and orientation system through the navigator module, telemetry interface to the flight parameter visualisation device.
- In the semi-automatic flight mode of the UAV, the autopilot module controls the OGE commands:
- receives a command from the OGE over the radio channel;
- receives current information about the OGE's angular orientation and angular velocities in two axes (roll and pitch) from the attitude control system.

In this case, the orientation systems are used. If the UAV is oriented in the horizontal plane within the specified values of angles and angular velocities in all three axes, the command is transmitted to the RM, otherwise the command to stabilise the horizontal flight mode, which is formed by the autopilot algorithm, is transmitted to the RM.

The automatic mode is to ensure flight along the route predetermined with the help of control points. In this mode it is possible that the UAS is not in radio contact with the control and communication equipment. It is switched on automatically when the UAV leaves the radio channel visibility zone.

The commands from the control unit are ignored. In this flight mode, automatic control of flight altitude, speed, heading, and deviation from the specified trajectory is performed based on the indications of the orientation, navigation, and sensor systems.

In the automatic flight mode, control is based on the guidance-stabilisation principle. The navigator module solves the task of guidance, i.e. produces a guidance command (including the required flight direction and the current flight direction calculated from the signals of the orientation, navigation and sensor systems), which is transmitted to the "Autopilot". The "Autopilot" module solves the problem of stabilisation, i.e. processing the guidance command and ensuring the stability of the movement by generating control commands for the autopilot algorithm. In case of exceeding the set thresholds (in terms of angles and angular velocities), a stabilisation command of the horizontal flight mode is given, which is performed by the autopilot algorithm as in the case of semi-automatic control. In automatic mode, the navigator module has the task of periodically checking for communication via radio channel. In case of its presence, the navigator module sends telemetry data to the ground. The main functions of the UAV's AFCS in semi-automatic and automatic modes are performed by the autopilot, which implements the laws of pitch, yaw and roll channel control.

Fully automatic control of UAVs is possible if there is information about the current state of the UAV in space, as well as information about the specified motion of the UAV, corresponding to the conditions of piloting accuracy. The trajectory control of UAVs for various purposes can be command (by commands that come from outside), software (the trajectory is formed and set on board in the form of time dependencies), adaptive terminal, in which the control is carried out to achieve the final result (a number of restrictions can be fulfilled).

In addition to trajectory control, as a rule, angular stabilisation and angular position control of the UAV are performed. The most important tasks in the creation of such UAV control are: ensuring stability of motion in all flight modes, taking into account possible disturbances, deviations of initial data; achieving the accuracy of realisation of the target purpose of the UAV; ensuring survivability of control at specified failures caused by external influences in the control system.

While flying in automatic mode, the UAV must fly at low altitudes with terrain envelopment. The accuracy of altitude maintenance within three metres must be ensured. Creation of an intelligent system of automatic control of UAV, the core of which is the onboard control equipment (avionics), is possible only if the integrated system of orientation and adjustment of autopilot coefficients for specific UAV flight modes are used in the control system (as a system for assessing the state of the UAV). For example, there must be options for getting out of critical flight modes, such as critical roll, which may occur as a result of a wind gust during a turn manoeuvre.

The main purpose of completing (combining) the orientation and navigation systems is to improve the accuracy of determining the navigational and angular parameters of UAV orientation. Not only systems, but also separate primary information sensors (pressure sensors, magnetoresistors, accelerometers, etc.) that produce the same parameters can be combined. When combining several navigation sensors, two comp lexing schemes, known as the compensation and filtering method, are the most widely used.

To improve the accuracy of UAV angular orientation determination, it is possible to combine different types of orientation systems [61, 62]. Figure 10.18 shows the block diagram of combining the triaxial magnetometer module, pyrohorizon and GPS/GLONASS satellite radio navigation system (SRNS) module. The triaxial magnetometer produces normalised X, Y, Z readings on each channel. The pyrohorizon determines the pitch angles ϑ_n and roll angles γ_n. The GPS/GLONASS module receives signals from the satellite navigation system and transmits the value of geographic coordinates φ, λ, track angle ψ and magnetic slope angles D and inclination I to the orientation angle calculation unit. Based on the pitch and roll angles determined by the pyrometric sensors, the calculation block calculates the heading angle. Thus, there are course angle values calculated from the readings of the pyrometers and magnetic sensors and the course that is given by the GPS module. This allows the implementation of a Kalman filter to estimate the systematic errors of the magnetic sensors. Thus, bundling improves the accuracy of pitch and roll angles and eliminates errors due to the presence of residual uncompensated magnetic interference and error due to the glide angle of the UAV.

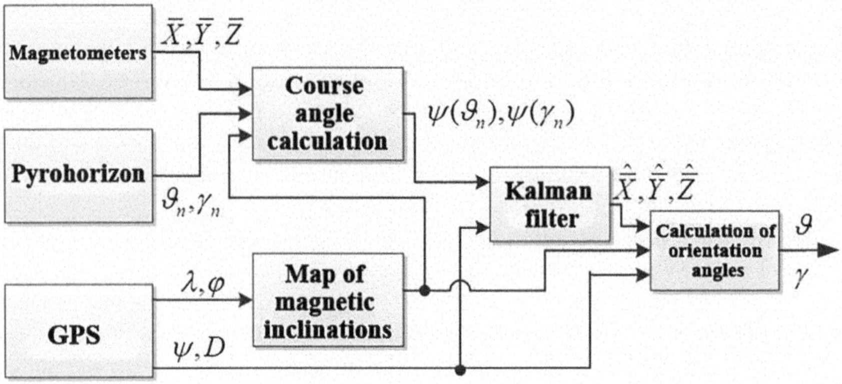

Fig. 10.18 Block diagram of completing the magnetometer module, pyrogorizon and GPS module

The use of optimal Kalman filtering methods assumes that the parameters of the stochastic description of perturbations and measurement errors are known exactly [63–65]. In practice, in case of parameter uncertainty, the filter tuning is performed on an a priori model of perturbations and measurement errors, which leads to additional losses in the estimation accuracy. In this case, the covariance matrix is not an estimate of the state vector accuracy, i.e. the filter incorrectly forms the accuracy characteristics (covariance of the estimation error) together with the state vector estimate.

To improve the accuracy of the state vector, a neural network algorithm for approximating arbitrary probability distribution density function is developed along with Kalman filtering.

Neural networks can learn from real measurements rather than an error model, thus circumventing model uncertainties.

Fulfilment of the above functions is ensured by the use of variable payloads of modular construction, which are installed on the internal and/or external suspensions of the UAV. In essence, a UAV is an aerial platform for transporting payloads, which may include: sensors (sensors) for collecting information, including: photo, television, thermal and multispectral cameras (EO/IR); laser scanners and range finders (LRF/LD); synthetic aperture radar (SAR); equipment for relaying communications; a multifunctional computerised information and measurement system, and the like.

In determining (selecting) options for the layout of payload modules, their technical characteristics, advantages and disadvantages, as well as the conditions for performing flight, instrumentation or other tasks as intended (weather, terrain, time of day and time of year, and other factors) are taken into account.

As a summary of the above, we note that AFCS and UAV payloads play an extremely important role in ensuring autonomous (controlled) UAV flight along a certain trajectory at a given altitude at a given speed, withdrawal to a certain area and fulfilment of tasks in accordance with the pre-flight plan and changes to it in the process of its implementation.

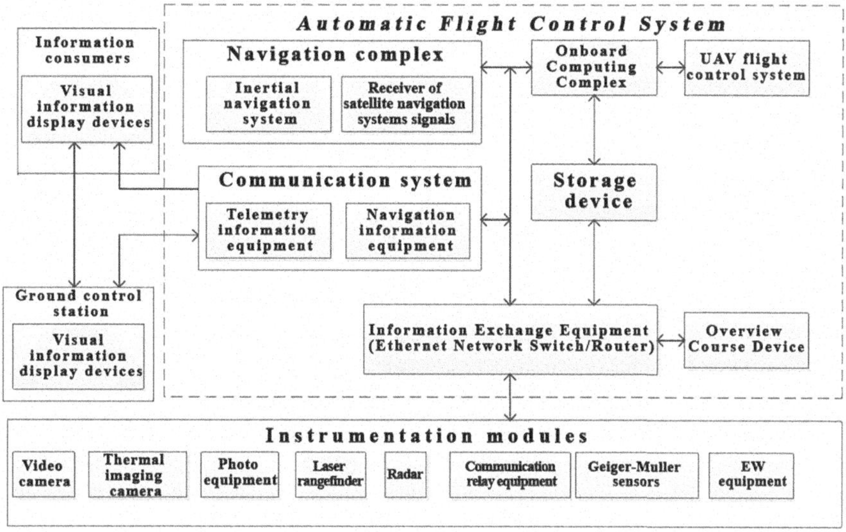

Fig. 10.19 Structure of the UAV flight control system and multifunctional control equipment

Also, one of the main functions of the AFCS is to control the payload operation and the transmission (dissemination) of relevant information to the ground control centre and its consumers. The above-mentioned system enables the operator to control the UAV flight and its payload in manual mode from a standard remote control panel, in automatic mode—by signals from the AFCS subsystems, and in semi-automatic mode—by operator commands.

The UAV's AFCS is a complex, multi-level integrated system, which includes (Fig. 10.19): on-board computing complex; navigation complex; UAV flight control system; communication system; information exchange equipment and other components necessary for UAV operation [66–68].

Onboard computing complex (OCC) is a set of information interconnected hardware and software means of information transmission, storage and processing, which are designed to convert input data into output data, collection and processing of information of the measuring subsystems of the complex in accordance with certain tasks of functioning.

The architecture of modern OCCs is a multi-level, hierarchical and heterogeneous computing system.

At the lower level of the hierarchy specialised calculators are used, which provide primary processing of information from one or a group of homogeneous sensors.

At the middle level of the hierarchy, more powerful universal digital systems are used, which solve the main functional tasks based on complex processing of sensor information.

At the upper level of the hierarchy, as a rule, universal digital systems are used, which are designed to solve the tasks of control, monitoring, indication and communication.

OCC should solve the tasks of navigation, orientation and vector gravimetry, optimal estimation of various parameters and their correction, as well as the tasks of interrogation of sensors and systems for recording arrays of initial information into memory, ensuring the functioning of the indication system and control of the state of peripheral devices, self-monitoring, etc. Today the approach to construction of small-sized onboard computing complexes based on the use of microcontrollers functioning under the control of real-time operating systems is widespread.

The UAV navigation system is an integrated system that processes navigation information coming from the inertial system (INS, inertial component) and the satellite radionavigation system (SRNS, satellite component) [69–71].

The key element of the AFCS to ensure the autonomy of its operation is the inertial component. This is usually a platformless inertial navigation system (PINS), which fulfils the function of determining the UAV's position in space and has inertial sensors, a barometric altimeter and a triaxial magnetometer, and after comparing the data from the above-mentioned sensors with the data from the SRNS receiver, the system produces a complete navigation solution in terms of coordinates and orientation angles. The satellite component provides the determination of UAV and reconnaissance object coordinates using global satellite navigation system (GPS/GLONASS) signals, as well as the determination of UAV heading angle.

The communication system is designed to ensure stable communication between ground control centre personnel and UAV information consumers.

The main elements of the UAV communication system are:

- telemetry information equipment (transmitter and antenna-feeder device), which is designed to transmit telemetry information in real time to the ground control centre and distribute it within radio visibility;
- command and navigation information equipment (receiver and antenna feeder)—designed to receive UAV flight control commands and payloads. To communicate over long distances and improve noise immunity, both the UAV and the ground control centre use acutely directional antenna systems.

The information exchange equipment provides switching, passing and routing of information flows between components and elements of on-board and target equipment of the UAS, including distribution of information between on-board sensors of information collection, video information transmitter and its storage device.

Flight task input and pre-launch control of the functioning of the main components and systems of the UAV are performed before take-off through the external port of this equipment.

The data storage device is designed to store information from the time the UAV takes off until it lands. This device can be removable or built-in. The information that is read from it allows for a more detailed analysis of the information received by the UAV's sensors during flight missions.

The overview course device is designed for visual control of the UAV flight control process and provides the necessary coverage of the area over which the UAV is travelling. The information received from the overview course device is transmitted to the UAV operator (payload operator) and is used to control the UAV flight and the operation of multifunctional instrumentation. The overview course device includes a television camera with a wide-angle lens, which can be replaced or supplemented with a thermal imaging camera, digital camera or radar, depending on the mission.

The main requirements that are imposed on the UAV's AFCS are:

- low cost;
- miniaturisation (minimisation of weight and dimensional characteristics); reduction of power consumption;
- ensuring automated UAV flight execution, stabilisation of orientation angles and following the specified trajectory in all control modes at all stages of flight, in particular during take-off, climb, descent and landing; ensuring the possibility for operators to remotely switch from manual to automatic UAV (payload) control modes and vice versa;
- the UAV automated control programme can be changed during the flight by the ground control station personnel; use of inexpensive, commercially available technical means and equipment, as well as own innovative software;
- increase in the memory capacity of the central computer system required for accumulation of measurement information; options of payload module layout should guarantee the fulfilment of tasks as intended under difficult operating conditions, including abrupt changes in temperature conditions;
- improving noise immunity, etc.

10.5 Examples of Construction of Measuring Channels Systems of UAV-Based NPP and TPP Environmental Characteristics Monitoring

Consider the concept of building, creating and researching information channels for a small UAS that enable the following functions:

- measurement of external radiation dose rate (exposure dose rate) in airspace;
- sampling of radioactive aerosols in different layers of the plume and at different distances from the source of the release [72].

Typically, these samplers have an inlet tube, filter, air duct and an electrically driven pump and wind wheel connected to it. In these devices, under the influence of the vacuum created by the pump, part of the aerosol stream is sucked into the inlet tube, passes through the filter where aerosol separation takes place, the duct and is discharged into the atmosphere through the hollow blades of the wind wheel,

The sampler is orientated in the horizontal plane towards the wind on a vertical stand due to the selforientation effect of the wind wheel, as well as by means of a weather vane. The isokinetic condition is maintained by a special system of automatic regulation based on signals from gas pressure sensors in the inlet tube and in the undisturbed flow. Their main disadvantage is impossibility to install them in a UAV.

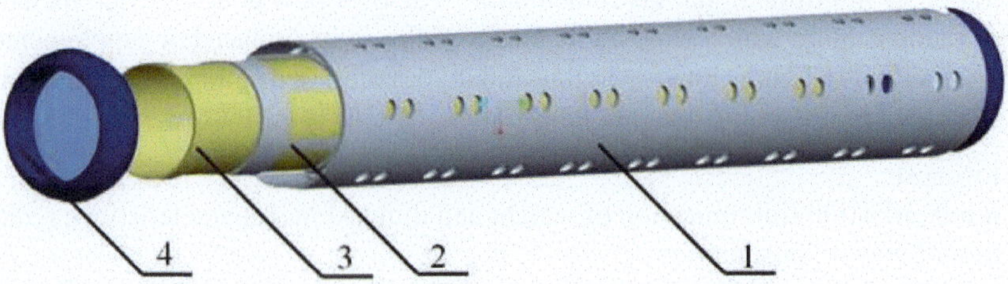

Fig. 10.20 Scheme of the FES: 1—outer casing (ejector), 2—inner mesh casing, 3—filter, 4—inlet flange

The proposed isokinetic sampler lacks this disadvantage by changing the sampling principle. For a better understanding of the principle of operation of such an isokinetic sampler, which is designed as filter ejection device (FED) [73], let's consider its schematic diagram (Fig. 10.20).

The FED is mounted on the UAV parallel to the longitudinal axis of the aircraft so that the input flange 4 with a hole D_{in} is directed in the direction of flight. While the UAV is flying, the air flow impinging on the FED is divided into two parts, one of them is directed to the middle of the FED through the hole in the inlet flange. For low flight altitudes, we assume the static pressure to be atmospheric P_a. Then, taking into account the Bernoulli equation, the dynamic pressure (air pressure) arising inside the FED is as follows:

$$\Delta P_d = P_a + (\rho V^2 / 2 \times C_{aero}), \tag{10.1}$$

where ρ—air flow density 1.225 kg/m^3 $\rho V^2/2$—velocity head at a given speed (V) of the UAV flight, C_{aero}—coefficient determined by aerodynamic resistance of the FED. From inside the FED due to ΔP_d air passes through the filter 3, internal mesh housing 2 and goes outside the FED through the ejector 1.

The second part of the air flow, flowing around the outer surface of the FED, leads to a decrease in static pressure to the value of

$$\Delta P_e = P_a - \rho V^2 / 2. \tag{10.2}$$

Due to ejection, the air from inside the FED will also pass through the filter fabric 3 to the outside. Both of these mechanisms determine the capacity Q (m^3/h) of the FED for a given flight speed V. Let us turn to the PEF manufactured and investigated by the authors. Its appearance is shown in Fig. 10.21.

The characteristics of this FED are as follows: total length $L = 300$ mm; effective length (excluding flanges and glue joints) $L_{eff} = 250$ mm; ejector diameter $D_e =$ 53 mm; effective area of the ejector (excluding the area of the solder) $S_e = 370$ cm^2; the diameter of the FED inlet—$D_{in} = 37$ mm ($S_{in} = 10.7$ cm^2), while the effective area of the filter cloth is $S_{eff} = 160$ cm^2; curb weight—310 g.

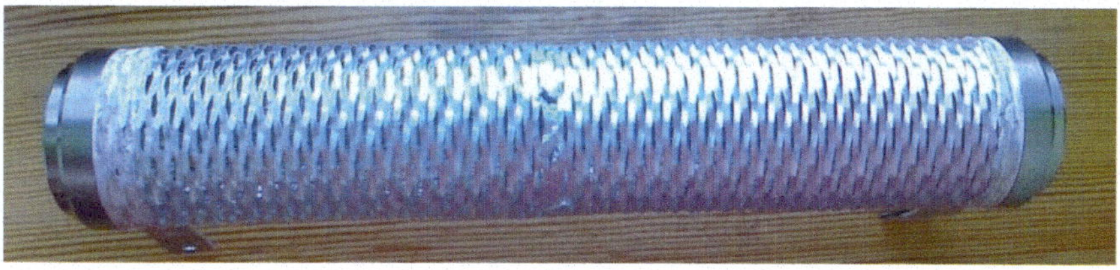

Fig. 10.21 Manufactured FED

The Petryanov fabric of the FPP-15–1.5 type used in such cases was used as a filter fabric. The determining value is the resistance equal to 15 Pa at air flow velocity through the fabric of 1 cm/s.

The first part of the experimental studies consisted in determining the efficiency of the FED Q (m^3/h) under the action of air pressure without taking into account ejection. For this purpose, the FED was placed inside a cylinder of larger diameter, which from the side of the inlet was hermetically connected to the front flange face, and the outlet of this cylinder was connected to a Turnigy Mini (Sirius35) anemometer through a branch pipe with cross-sectional area $S_{out} = 7.54\ \text{cm}^2$.

The anemometer measured the velocity of the outlet air flow on the Beaufort scale in m/s. Anemometer measurement range $V_a = 0.3$–30 m/s, measurement error—± 5%. The FED was installed on the vehicle. In this case, the output $Q = S_{out} \times V_a$.

The second part of the experimental studies consisted in determining the efficiency of the FED Q (m^3/h) due to ejection under the action of the air flow streamlined by the FED. For this purpose, the flanges of the FED were swapped, a conical nozzle was installed on the rear flange, and the opening of the front flange (as in the first experiment) was connected to a Sirius35 anemometer through a spigot.

In this case we have P_a inside the FED and $\Delta P_e = P_a - \rho V^2/2$. outside. As in the previous case, the flow goes from the inside to the outside. In the first case due to increase of pressure inside, now due to decrease of pressure outside.

Assuming the performance of this FES as the sum of the two experimentally obtained, we determine the calibration characteristic, which is shown in Fig. 10.22.

Here, the squares (number 1 on the graph) denote the volumes for these velocities through a cross section of area $S_{in} = 10.7\ \text{cm}^2$, and the dependence passing through the rhombuses (number 2 on the graph) is the calibration characteristic of the FED. For example, during an hour of flight at a speed of 80 km/h with our FED it is possible to analyse 40 m^3.

Thus, having installed such a FED on the UAV, at the flight speed of 70–80 km/h, radioactive aerosols from 1 m^3 of atmospheric air will be sampled through the specified type of filter in 1.5–2 min.

To measure the radiation background (exposure dose rate—EDR), a scheme was developed and a computer system was manufactured that allows to determine the EDR in the specified spatial coordinates in automatic mode with simultaneous registration of longitude, latitude and altitude of the measurement site. The system is essentially an autonomous device that can be installed on any mobile object. Its size

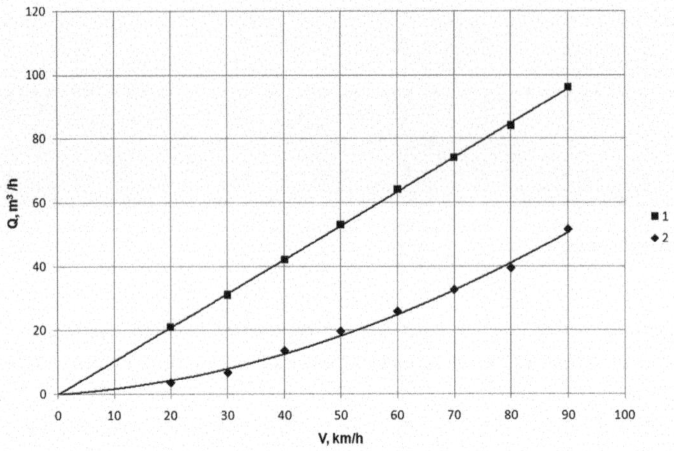

Fig. 10.22 FED's calibration characteristic

and weight characteristics allow its use on UAVs as well. In this case, the mode of switching on and off the recording of information can be carried out both by commands from the remote control and by commands from the autopilot. Functional scheme of the system is shown in Fig. 10.23.

Let us turn to the description of the components of the above system diagram.

The basis of the system is a single-chip microcontroller with RISC-architecture of "ATXmega" family, produced by Atmel company, hereinafter referred to as "microcontroller". The microcontroller is a coordinating link between separate blocks of the system and controls its functions as a whole, according to the programme written in ROM.

The basic function of the system is measurement of EDR gamma radiation, for which the following blocks are provided: high voltage source, Geiger-Muller counter and pulse shaper.

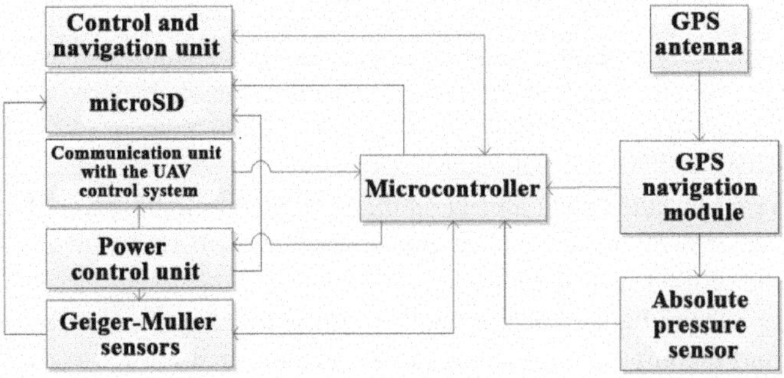

Fig. 10.23 Computerised system of remote EDR control

Geiger-Muller sensors are proportional counters of ionising radiation particles. The applied counter type "SI22B" is able to register mainly gamma-quanta, to a much lesser extent—beta-particles and is practically insensitive to alpha-particles. The number of ionising radiation registration events is directly proportional to the EDR. Thus, the EDR can be calculated as the number of current pulses from the counter per unit time. However, taking into account that the registration events have a random time character, especially at low EDR, mathematical filtering by the linear-weighted moving average method was used to smooth the measurement results. In addition, in order to reduce the scatter at low filtering degree, the parallel inclusion of two Geiger-Muller counters SI22G in one external sealed unit was chosen.

The pulse shaper is a converter of current pulses from the Geiger-Muller counter into logic level pulses that are electrically compatible with the microcontroller ports.

The memory card of "micro SD" standard is a universal carrier of digital information, is a non-volatile electrically reprogrammable ROM and is intended for periodic saving of the report on EDR levels and relative altitude with reference to the geographical location of the device. The device supports various types of memory cards of "micro SD" standard, including "SD HC", as well as memory cards of different capacities.

A photo of the system assembled and ready for installation on a UAV or other mobile object is shown in Fig. 10.24.

Controls and indications include LED status indicators, electromagnetic or piezo-electric capsules, as well as a clock microswitch for selecting the degree of filtering of measurement results.

The absolute pressure sensor measures the elevation of the device relative to the calibration point and is an integral barometer with a digital interface, made by "MEMS" micromechanical technology. At initial initialisation of the device by means of the integral barometer sensor measures atmospheric pressure, taking it as a reference point (zero altitude). Later, during the system operation, the difference between the pressure measured at initialisation and periodically measured pressure is mathematically converted into height in metres.

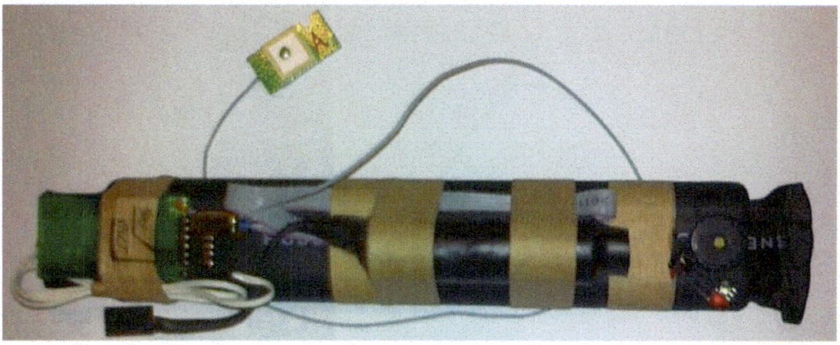

Fig. 10.24 Photo of the computerized EDR remote control system mounted on the UAV

GPS/GLONASS navigation module determines geographical coordinates by means of one of the GPS or GLONASS satellite navigation systems or both systems simultaneously.

The specified module is a miniature printed circuit board, enclosed in a shielding case, with a circuit of a radio receiver of satellite signals, as well as a microprocessor, which calculates the geographical location according to these signals and transmits the data via NMEA standard to the DHA microcontroller, where they are further processed. The measurement time in the system can be set to 3, 5, 10, 15 or 30 s. At the same time the data acquisition clock remains unchanged and is 1 s. Since we use the linear-weighted moving average method in the measurement process, the result of each data acquisition cycle is the average EDR value on the path segment $S = t_{mes} \times V_0$ taking into account the previous t results (t_{mes}—measurement time in the system, V_0 –speed of the object on which the system is installed).

The computerised system of remote EDR control was calibrated at the UPD-inter facility at the Institute of NPP Safety Problems of the National Academy of Sciences of Ukraine. Calibration was carried out in the EDR range from 20 μR/h to 50 mR/h by ^{137}Cs.

An experimental test of the system was carried out. It was installed in a car, which was travelling to Kyiv along one of the sections of the Zhytomyr-Kyiv highway. The file was formed in the memory card in the format ".KML" After its closure the card was extracted from the system and connected to a PC. The results of the file visualisation using "GoogleEarth" software are presented in Fig. 10.25.

The marked points are every second regardless of the speed of movement. To the left of each of them is the EDR value in μSv/h (1 Sv≈100 R). However, multiplying

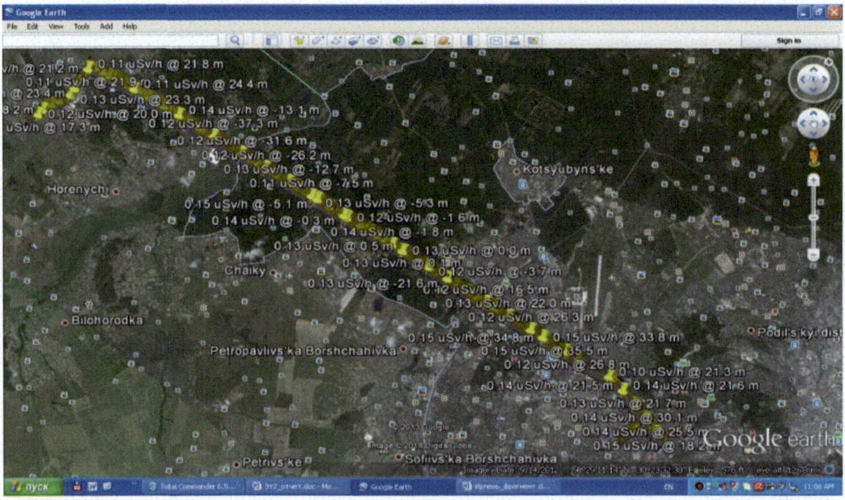

Fig. 10.25 Result of data visualisation during experimental verification of EDR remote monitoring systems outside the city

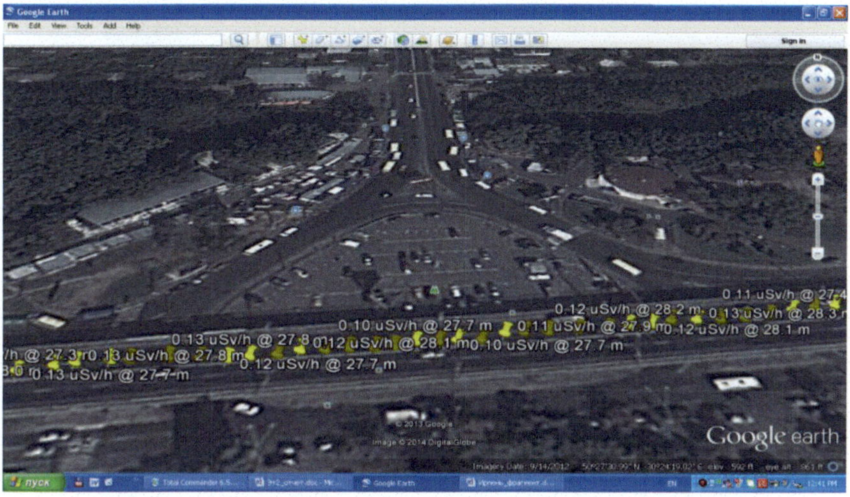

Fig. 10.26 Result of data visualization during experimental validation of EDR remote control systems within the city

μSv/h by 100 we get the usual μR/h. The right side shows the height in metres relative to the value obtained during the initial calibration of the system.

From the presented experimental results, first of all, the EDR values are of interest. Within the norm they should not exceed 30 μR/h (0.3 μSv/h). The programme used allows scaling of the obtained image.

A fragment of such scaling when driving in Kyiv, along Peremohy Avenue in the area of Nyvky metro station is shown in Fig. 10.26.

It can be seen that the EDR value here is 0.10...0.13 μSv/h. These values are more than twice less than the maximum permissible values.

10.6 Experimental Results of Radiation Monitoring at the Border of the Chornobyl Exclusion Zone

Based on modern information technologies, a domestically produced unmanned information and diagnostic complex and a system of environmental radiation monitoring and video reconnaissance were created. The system includes onboard equipment for measuring EDR at specified geographic coordinates and altitudes in real time, an isokinetic aerosol sampler, and equipment for video surveillance in radiation-hazardous and hard-to-reach areas.

The experiments were conducted using the UAV shown in Fig. 10.27.

The FED is mounted on the fuselage of the UAV. Aerosol sampling was performed near the village of Potoki, which is located near the Stari Sokoly checkpoint and is 31 km from the Chornobyl NPP. The results of the visualisation of the experimental

Fig. 10.27 Photo of the experimental UAV

data are shown in Figs. 10.28 and 10.29. The flights were performed in May–June 2013 at maximum upwelling flows. The wind was northeasterly (from the side of the Chornobyl NPP) with a speed of 3…5 m/sec. At altitudes of 90 and 180 m, 30 m^3 of air each was sequentially filtered, and at altitude of 70 m to 60 m^3. All three filters were submitted for analysis to the Nuclear and Radiation Safety Department (Chornobyl) of the Institute of NPP Safety Problems of the NAS of Ukraine. The analyses were carried out by γ -spectroscopy. For 90 m and 180 m the sensitivity was  0.02 Bq/m^3, and for 70 m–0.007 Bq/m^3. For all cases no activity was detected.

In the same place, at the border of the Chornobyl NPP Exclusion Zone, in March 2014, experiments on EDR level measurement at heights of 100 … 200 m were carried out (Figs. 10.28 and 10.29). A computerised system of remote EDR control (Fig. 10.24) was installed inside the fuselage of the UAV (Fig. 10.27). The EDR values obtained were 0.05…0.11 μSv/h, which is much less than the maximum permissible values.

The presented UAV weighing about 2.5 kg, with flight duration up to 1 h at the speed of 70 km/h, not requiring prepared sites for take-off and landing and flight preparation time not more than 10 min, appears to be a convenient means of remote control of radiation situation in the vicinity of NPPs and TPPs.

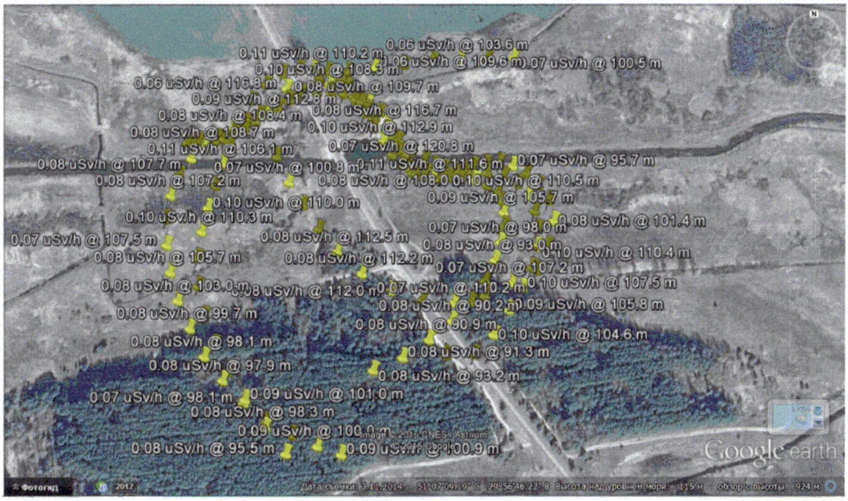

Fig. 10.28 Result of visualisation of experimental data of EDR values at a height of 100 m

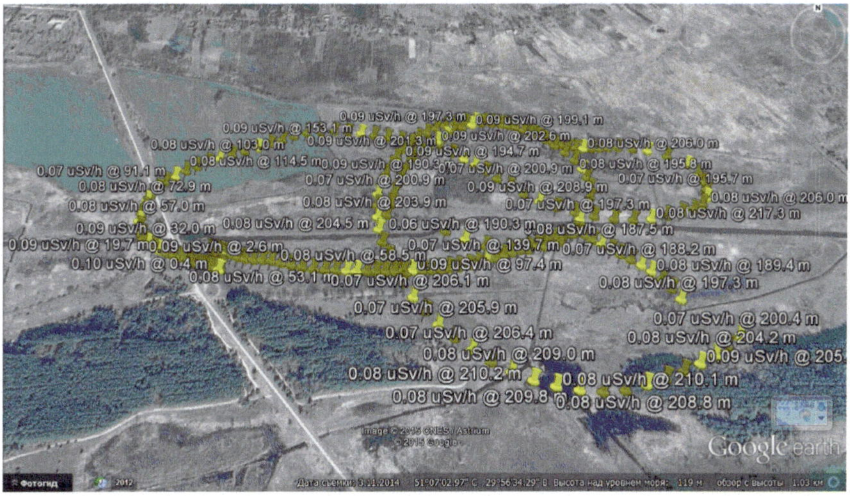

Fig. 10.29 Result of visualisation of experimental data of EDR values at a height of 200 m

References

1. Oleg, S., Oleksii, K., Serhii, P.: Informational and analytical system for diagnostics of the electric power equipment condition. In: 2020 IEEE 7th International Conference on Energy Smart Systems (ESS), pp. 105–110. IEEE (2020)
2. Zaporozhets, A., Eremenko, V., Serhiienko, R., Ivanov, S.: Methods and hardware for diagnosing thermal power equipment based on smart grid technology. In: Advances in Intelligent Systems and Computing III: Selected Papers from the International Conference on

Computer Science and Information Technologies, CSIT 2018, September 11–14, Lviv, Ukraine, pp. 476–489. Springer International Publishing (2019)
3. Aboulian, A., Green, D.H., Switzer, J.F., Kane, T.J., Bredariol, G.V., Lindahl, P., Leeb, S.B.: NILM dashboard: A power system monitor for electromechanical equipment diagnostics. IEEE Trans. Industr. Inf. **15**(3), 1405–1414 (2018)
4. Babak, V.P., Babak, S.V., Eremenko, V.S., Kuts, Y.V., Myslovych, M.V., Scherbak, L.M., Zaporozhets, A.O.: Problems and features of measurements. Models and Measures in Measurements and Monitoring, pp. 1–31 (2021)
5. Henao, H., Capolino, G.A., Fernandez-Cabanas, M., Filippetti, F., Bruzzese, C., Strangas, E., Hedayati-Kia, S.: Trends in fault diagnosis for electrical machines: a review of diagnostic techniques. IEEE Ind. Electron. Mag. **8**(2), 31–42 (2014)
6. Babak, V.P., Babak, S.V., Myslovych, M.V., Zaporozhets, A.O., Zvaritch, V.M.: Simulation and software for diagnostic systems. Diagnost. Syst. Energy Equipm. 71–90 (2020)
7. Esmalifalak, M., Liu, L., Nguyen, N., Zheng, R., Han, Z.: Detecting stealthy false data injection using machine learning in smart grid. IEEE Syst. J. **11**(3), 1644–1652 (2014)
8. Kovtun, S., Kuts, Y., Malko, V., Fryz, M., Shcherbak, L., Kuts, V.: Application of Hilbert transform for power quality indicators monitoring in general purpose grids. Syst. Res. Energy, 2(77):71–83 (2024)
9. Hotra, O., Kulyk, M., Babak, V., Kovtun, S., Zgurovets, O., Mroczka, J., Kisała, P.: Organisation of the structure and functioning of self-sufficient distributed power generation. Energies **17**(1), 27 (2023)
10. Zagorodny, A., Bogdanov, V., Ermoliev, Y., Kulyk, M.: Development of top-down and bottom-up methodology using risk functions for systems with multiplicity of solutions. In: Systems, Decision and Control in Energy IV: Volume I. Modern Power Systems and Clean Energy, pp. 3–23. Springer Nature Switzerland, Cham (2023)
11. Zaporozhets, A., Kovtun, S., Dekusha, O.: System for monitoring the technical state of heating networks based on UAVs. In: Conference on Computer Science and Information Technologies, pp. 935–950. Springer International Publishing, Cham (2019)
12. Babak, V.P., Babak, S.V., Eremenko, V.S., Kuts, Y.V., Myslovych, M.V., Scherbak, L.M., Zaporozhets, A.O.: Monitoring the air pollution with UAVs. Models Measures Measurem. Monitor. 191–225 (2021)
13. Popov, O., Kupriianchuk, S., Sizov, A., Matvieieva, I., Kovach, V.: Modeling of radiation characteristics of fuel-containing materials of the shelter object for calculation dose rate. In: Systems, Decision and Control in Energy IV: Volume II. Nuclear and Environmental Safety, pp. 23–34. Springer Nature Switzerland, Cham (2023)
14. Diviziniuk, M., Popov, O., Telelym, V., Kovach, V., Artemchuk, V.: General characteristics of radar stations for physical protection of nuclear objects. In: Systems, Decision and Control in Energy IV: Volume II. Nuclear and Environmental Safety, pp. 113–124. Springer Nature Switzerland, Cham (2023)
15. Babak, V.P., Babak, S.V., Myslovych, M.V., Zaporozhets, A.O., Zvaritch, V.M.: Principles of construction of systems for diagnosing the energy equipment. Diagnost. Syst. Energy Equipm. 1–22 (2020)
16. Godun, R., Sushchenko, K., Kupriianchuk, S., Iatsyshyn, A., Lahoiko, A.: Neutron Incident at the Chornobyl NPP Shelter Object: Study of the Causes and Impacts of Safety in the New Confinement. In Systems, Decision and Control in Energy IV: Volume II. Nuclear and Environmental Safety, pp. 91–111. Springer Nature Switzerland, Cham (2023)
17. Onda, Y., Taniguchi, K., Yoshimura, K., Kato, H., Takahashi, J., Wakiyama, Y., Smith, H.: Radionuclides from the Fukushima Daiichi nuclear power plant in terrestrial systems. Nat. Rev. Earth Environ. **1**(12), 644–660 (2020)
18. Tominaga, T., Hachiya, M., Tatsuzaki, H., Akashi, M.: The accident at the Fukushima Daiichi nuclear power plant in 2011. Health Phys. **106**(6), 630–637 (2014)
19. Steinhauser, G., Brandl, A., Johnson, T.E.: Comparison of the Chernobyl and Fukushima nuclear accidents: a review of the environmental impacts. Sci. Total. Environ. **470**, 800–817 (2014)

20. Rezinkina, M.M., Sokol, Y.I., Zaporozhets, A.O., Gryb, O.G., Karpaliuk, I.T., Shvets, S.V.: Monitoring of energy objects parameters with using UAVs. In: Control of Overhead Power Lines with Unmanned Aerial Vehicles (UAVs), pp. 1–8. Springer International Publishing, Cham (2021)
21. Zaporozhets, A.: Overview of quadrocopters for energy and ecological monitoring. In: Systems, Decision and Control in Energy I, pp. 15–36. Springer International Publishing, Cham (2020)
22. Popov, O., Bondar, O., Ivaschenko, T., Puhach, O., Iatsyshyn, A., Skurativskyi, S.: Features of the modern UAV-based complexes use to solve radiation control problems. In: Systems, Decision and Control in Energy IV: Volume II. Nuclear and Environmental Safety, pp. 35–57. Springer Nature Switzerland, Cham (2023)
23. Zaporozhets, A., Babak, V., Isaienko, V., Babikova, K.: Analysis of the air pollution monitoring system in Ukraine. In: Systems, Decision and Control in Energy I, pp. 85–110. Springer International Publishing, Cham (2020)
24. Rezinkina, M., Rezinkin, O., Zaporozhets, A.: UAVs Application in power engineering. In: 2021 IEEE 6th International Conference on Actual Problems of Unmanned Aerial Vehicles Development (APUAVD), pp. 161–164. IEEE (2021)
25. Babak, S., Zaporozhets, A., Gryb, O., Karpaliuk, I.: Analysis of UAVs and their technical parameters for overhead power lines monitoring. In: Systems, Decision and Control in Energy IV: Volume I. Modern Power Systems and Clean Energy, pp. 181–193. Springer Nature Switzerland, Cham (2023)
26. Mohsan, S.A.H., Khan, M.A., Noor, F., Ullah, I., Alsharif, M.H.: Towards the unmanned aerial vehicles (UAVs): a comprehensive review. Drones **6**(6), 147 (2022)
27. Alioua, A., Djeghri, H.E., Cherif, M.E.T., Senouci, S.M., Sedjelmaci, H.: UAVs for traffic monitoring: a sequential game-based computation offloading/sharing approach. Comput. Netw. **177**, 107273 (2020)
28. Shakhatreh, H., Sawalmeh, A.H., Al-Fuqaha, A., Dou, Z., Almaita, E., Khalil, I., Guizani, M.: Unmanned aerial vehicles (UAVs): a survey on civil applications and key research challenges. IEEE Access **7**, 48572–48634 (2019)
29. Martin, P.G., Martin, P.G.: Response, Contamination and Release Estimates. The 2011 Fukushima Daiichi Nuclear Power Plant Accident: An Analysis from the Metre to the Nanometre Scale, 23–61 (2019).
30. Briechle, S., Molitor, N., Krzystek, P., Vosselman, G.: Detection of radioactive waste sites in the Chornobyl exclusion zone using UAV-based lidar data and multispectral imagery. ISPRS J. Photogramm. Remote Sens. **167**, 345–362 (2020)
31. Burtniak, V., Zabulonov, Y., Stokolos, M., Bulavin, L., Krasnoholovets, V.: Application of a territorial remote radiation monitoring system at the Chornobyl nuclear accident site. J. Appl. Remote. Sens. **12**(4), 046007–046007 (2018)
32. Sanada, Y.: Experience for fukushima environmental radiation monitoring and application for preparedness of post-accident. In: 2021 IEEE Nuclear Science Symposium and Medical Imaging Conference (NSS/MIC), pp. 1–5. IEEE (2021)
33. Sato, Y., Ozawa, S., Terasaka, Y., Minemoto, K., Tamura, S., Shingu, K., Torii, T.: Remote detection of radioactive hotspot using a Compton camera mounted on a moving multi-copter drone above a contaminated area in Fukushima. J. Nucl. Sci. Technol. **57**(6), 734–744 (2020)
34. Farinha, A., Zufferey, R., Zheng, P., Armanini, S.F., Kovac, M.: Unmanned aerial sensor placement for cluttered environments. IEEE Robot and Automat Lett **5**(4), 6623–6630 (2020)
35. Koparan, C., Koc, A.B., Privette, C.V., Sawyer, C.B.: In situ water quality measurements using an unmanned aerial vehicle (UAV) system. Water **10**(3), 264 (2018)
36. Yoon, H., Shin, J., Spencer, B.F., Jr.: Structural displacement measurement using an unmanned aerial system. Comput-Aided Civ Infrastruct Eng **33**(3), 183–192 (2018)
37. Balestrieri, E., Daponte, P., De Vito, L., Lamonaca, F.: Sensors and measurements for unmanned systems: an overview. Sensors **21**(4), 1518 (2021)
38. Li, B., Li, K., Chen, J.: A UAV migration-based decision-making scheme for on-demand service in 6G network. Alex. Eng. J. **69**, 25–33 (2023)

39. Olanipekun, O.A., Montalvo, C.J., Salau, T.A., Simolowo, E.O.: Unmanned aerial vehicles: a 21st century review of advanced air mobility platforms. In: AIAA Aviation 2023 Forum, p. 4044, (2023)
40. Babak, S.: Radiation monitoring of environment using unmanned aerial complex. Adv. Sci. J. **12**, 41–44 (2014)
41. Čerba, Š, Lüley, J., Vrban, B., Osuský, F., Nečas, V.: Unmanned radiation-monitoring system. IEEE Trans. Nucl. Sci. **67**(4), 636–643 (2020)
42. Xiang, W., Xinxin, W., Jianhua, Z., Jianchao, F., Xiu, S., Dejun, Z.: Monitoring the thermal discharge of hongyanhe nuclear power plant with aerial remote sensing technology using a UAV platform. In: 2017 IEEE International Geoscience and Remote Sensing Symposium (IGARSS), pp. 2958–2961. IEEE (2017)
43. Prisacariu, V.: The history and the evolution of UAVs from the beginning till the 70s. J. Defense Res. Manage. (JoDRM) **8**(1), 181–189 (2017)
44. Prisacariu, V.: A brief history of UAVs in the 1970s. Sci. Res. Educat. the Air Force-AFASES, 2022 (2022)
45. Kahvecioglu, S., Oktal, H.: Historical development of UAV technologies in the world: the case of Turkey. Sustainable Aviat.: Energy Environ. Issues, pp. 323–331 (2016)
46. Gibradze, G., Guchua, A., Gogua, G.: The role of modern technologies in military conflicts of the 21st century. Ukrainian Policymaker **11**, 26–34 (2022)
47. Grant, Jr, R.L.: Military innovation in the third age of US unmanned aviation, 1991–2015 (Doctoral dissertation, Monterey, California. Naval Postgraduate School) (2020)
48. McKinnon, G.: The 1960 U-2 crisis reconsidered: technology, masculinity, and US airpower's 'Unmanning'. Diplomatic History, dhae029 (2024)
49. Hall, A.R., Coyne, C.J.: The political economy of drones. Defence and Peace Econom **25**(5), 445–460 (2014)
50. Kolejka, J., Plánka, L.: Technical report: the development and experience with UAV research applications in former Czechoslovakia (1960s–1990s). Pure Appl. Geophys. **175**(9), 3375–3390 (2018)
51. UAV Market Size. Report on Industry Size & Market Share Analysis - Growth Trends & Forecasts (2024–2029) Source: https://www.mordorintelligence.com/industry-reports/uav-market
52. Kunertova, D.: Military drones in Europe: the European defense market and the spread of military UAV technology (2019)
53. Brumfiel, G.: Japan's long road ahead. Nature **472**(7341), 14–14 (2011)
54. Brown, S.T., Lambrigtsen, B., Denning, R.F., Gaier, T., Kangaslahti, P., Lim, B.H., Tanner, A.B.: The high-altitude MMIC sounding radiometer for the global hawk unmanned aerial vehicle: instrument description and performance. IEEE Trans. Geosci. Remote Sens. **49**(9), 3291–3301 (2011)
55. Fleming, J., Langford, M., Tweedie, S., Ng, W., & Goossen, E. (2012) Development of an air data system for ducted fan unmanned aircraft. In: 50th AIAA Aerospace Sciences Meeting including the New Horizons Forum and Aerospace Exposition (p. 668)
56. Babak, V.P., Babak, S.V., Eremenko, V.S., Kuts, Y.V., Myslovych, M.V., Scherbak, L.M., Zaporozhets, A.O.: Models and measures for atmospheric pollution monitoring. Models and measures in measurements and monitoring, pp 227–266 (2021)
57. Jawhar, I., Mohamed, N., Al-Jaroodi, J., Agrawal, D.P., Zhang, S.: Communication and networking of UAV-based systems: classification and associated architectures. J. Netw. Comput. Appl. **84**, 93–108 (2017)
58. Korchenko, A.G., Illyash, O.S.: The generalized classification of unmanned air vehicles. In: 2013 IEEE 2nd International Conference Actual Problems of Unmanned Air Vehicles Developments Proceedings (APUAVD), pp. 28–34. IEEE (2013)
59. Zaporozhets, A.O., Khaidurov, V.V.: Mathematical models of inverse problems for finding the main characteristics of air pollution sources. Water Air Soil Pollut. **231**(12), 563 (2020)
60. Babak, S., Myslovych, M., Sysak, R.: Module structure of UAV-based computerized systems for remote environment monitoring of energy facilities. In: 2016 17th International Conference Computational Problems of Electrical Engineering (CPEE), pp. 1–3. IEEE (2016)

61. Babak, V.P., Babak, S.V., Eremenko, V.S., Kuts, Y.V., Myslovych, M.V., Scherbak, L.M., Zaporozhets, A.O. (2021) Models and Measures for Measuring Random Angular Quantities. Models Measures Measurem. Monitor. 61–97
62. Babak, V. P., Babak, S.V., Eremenko, V.S., Kuts, Y.V., Myslovych, M.V., Scherbak, L.M., Zaporozhets, A.O.: Examples of using models and measures on the circle. Models and measures in measurements and monitoring, 127–156, (2021)
63. Luo, C., McClean, S.I., Parr, G., Teacy, L., De Nardi, R.: UAV position estimation and collision avoidance using the extended Kalman filter. IEEE Trans. Veh. Technol. **62**(6), 2749–2762 (2013)
64. Huang, T., Jiang, H., Zou, Z., Ye, L., Song, K.: An integrated adaptive Kalman filter for high-speed UAVs. Appl. Sci. **9**(9), 1916 (2019)
65. Xiong, J.J., Zheng, E.H.: Optimal kalman filter for state estimation of a quadrotor UAV. Optik **126**(21), 2862–2868 (2015)
66. Kwak, J., Sung, Y.: Autonomous UAV flight control for GPS-based navigation. IEEE Access **6**, 37947–37955 (2018)
67. Lee, C., Kim, H.R.: Optimizing UAV-based radiation sensor systems for aerial surveys. J. Environ. Radioact. **204**, 76–85 (2019)
68. Han, J., Xu, Y., Di, L., Chen, Y.: Low-cost multi-UAV technologies for contour mapping of nuclear radiation field. J. Intell. Rob. Syst. **70**, 401–410 (2013)
69. Balamurugan, G., Valarmathi, J., Naidu, V.P.S.: Survey on UAV navigation in GPS denied environments. In: 2016 International Conference on Signal Processing, Communication, Power and Embedded System (SCOPES), pp. 198–204. IEEE (2016)
70. Kuroki, Y., Young, G.S., Haupt, S.E.: UAV navigation by an expert system for contaminant mapping with a genetic algorithm. Expert Syst. Appl. **37**(6), 4687–4697 (2010)
71. Zhang, G., Hsu, L.T.: Intelligent GNSS/INS integrated navigation system for a commercial UAV flight control system. Aerosp. Sci. Technol. **80**, 368–380 (2018)
72. Zaporozhets, A.O.: Correlation analysis between the components of energy balance and pollutant emissions. Water Air Soil Pollut. **232**, 1–22 (2021)
73. Chepur, M.L., Babak, S.V., Marinoshenko, O.P., Morozov, A.V., Kanchenko, V.Y., Karnaushenko, R.V.: Method for determining the activity of volume aerosols during communal radiation accident using unmanned aircraft. Nucl. Phys. Atomic Energy **14**(4) (2013)

SPRINGER NATURE

GPSR Compliance

The European Union's (EU) General Product Safety Regulation (GPSR) is a set of rules that requires consumer products to be safe and our obligations to ensure this.

If you have any concerns about our products, you can contact us on ProductSafety@springernature.com

In case Publisher is established outside the EU, the EU authorized representative is:

Springer Nature Customer Service Center GmbH
Europaplatz 3
69115 Heidelberg, Germany

The manufacturer's authorised representative in the EU is Springer Nature Customer Service Centre GmbH, Europaplatz 3, 69115 Heidelberg, Germany. If you have any concerns regarding our products, please contact ProductSafety@springernature.com

Printed and bound by CPI Group (UK) Ltd, Croydon, CR0 4YY
01/10/2026
02225958-0012